Gmelin Handbook of Inorganic Chemistry

8th Edition

Gmelin Handbook of Inorganic Chemistry

8th Edition

Gmelin Handbuch der Anorganischen Chemie

Achte, völlig neu bearbeitete Auflage

Prepared and issued by — Gmelin-Institut für Anorganische Chemie der Max-Planck-Gesellschaft zur Förderung der Wissenschaften

Director: Ekkehard Fluck

Founded by — Leopold Gmelin

8th Edition — 8th Edition begun under the auspices of the Deutsche Chemische Gesellschaft by R. J. Meyer

Continued by — E. H. E. Pietsch and A. Kotowski, and by Margot Becke-Goehring

Springer-Verlag
Berlin · Heidelberg · New York · London · Paris · Tokyo 1990

Gmelin-Institut für Anorganische Chemie
der Max-Planck-Gesellschaft zur Förderung der Wissenschaften

Volumes published on "Sulfur" (Syst. No. 9)

Sulfur A 1
History (in German) – 1942

Sulfur A 2
Occurrence. Technology of Sulfur and Its Compounds. Colloidal Sulfur. Toxicity (in German) – 1953

Sulfur A 3
The Element. Preparation in Pure State. Properties (in German) – 1953

Sulfur B 1
Hydrides and Oxides of Sulfur (in German) – 1953

Sulfur B 2
Sulfur-Oxygen Acids (in German) – 1960

Sulfur B 3
Compounds (concluded) (in German) – 1963

Sulfur-Nitrogen Compounds 1
Compounds with Sulfur of Oxidation Number VI (in German) – 1977

Sulfur Suppl. Vol. 1
Thionyl Halides (in German) – 1978

Sulfur Suppl. Vol. 2
Sulfur Halides (in German) – 1978

Sulfur Suppl. Vol. 3
Sulfur Oxides (in German) – 1980

Sulfur Suppl. Vol. 4 a/b
Sulfanes – 1983

Sulfur-Nitrogen Compounds 2
Compounds with Sulfur of Oxidation Number IV – 1985

Sulfur-Nitrogen Compounds 3
Compounds with Sulfur of Oxidation Number IV – 1987

Sulfur-Nitrogen Compounds 4
Compounds with Sulfur of Oxidation Number IV – 1987

Sulfur-Nitrogen Compounds 5
Compounds with Sulfur of Oxidation Number IV – 1990

Sulfur-Nitrogen Compounds 6
Compounds with Sulfur of Oxidation Number IV – 1990 (present volume)

Gmelin Handbook of Inorganic Chemistry

8th Edition

S

Sulfur-Nitrogen Compounds

Part 6

Compounds with Sulfur of Oxidation Number IV

With 35 illustrations

AUTHORS Norbert Baumann, Hans-Jürgen Fachmann, Brigitte Heibel, Reimund Jotter, Alfons Kubny

EDITORS Norbert Baumann, Hans-Jürgen Fachmann, Brigitte Heibel, Reimund Jotter, Joachim Wagner

CHIEF EDITOR Alfons Kubny

System Number 9

Springer-Verlag
Berlin · Heidelberg · New York · London · Paris · Tokyo 1990

LITERATURE CLOSING DATE: 1988
IN MANY CASES MORE RECENT DATA HAVE BEEN CONSIDERED

Library of Congress Catalog Card Number: Agr 25-1383

ISBN 3-540-93609-2 Springer-Verlag, Berlin · Heidelberg · New York · London · Paris · Tokyo
ISBN 0-387-93609-2 Springer-Verlag, New York · Heidelberg · Berlin · London · Paris · Tokyo

Printed in Germany

Typesetting, printing, and bookbinding: LN-Druck Lübeck

Preface

The present volume is the second of the series describing acyclic sulfur-nitrogen compounds with sulfur of oxidation number IV. It continues the description of sulfur-nitrogen compounds with 2-coordinate sulfur that started with poly(sulfur nitride), sulfur nitride compounds and their metal complexes in the volume "Sulfur-Nitrogen Compounds" Part 5, 1990.

In this volume, sulfinyl imides and the sulfinyl imide anion, their transition metal complexes, amino-sulfinyl radicals, salts of N-sulfinyl-aminium cations, and thiosulfinyl imides are dealt with.

The description of the sulfinyl imides (N-sulfinylamines) begins with the parent compound O=S=NH and the sulfinyl imide ion (1−), $[O{=}S{=}N]^-$, which was studied in the gas phase, in solutions, and in the solid state where it forms salts. The next sections deal with sulfinyl imides carrying substituents at the nitrogen atom, which involve the elements of the main groups VII to III (e.g., sulfinyl halogenoimides, sulfinyl organyloxyimides, N-sulfinyl-sulfenamides, -sulfinamides, -sulfonamides, -hydrazines, -silanamines, -germanamines, -stannanamines, and boron-substituted sulfinyl imides). The description of N-sulfinyl-organylamines fills a large part of the volume. N-Sulfinylamines carrying substituents with S^{VI} bound to the nitrogen atom are not described in this volume but were dealt with in "Sulfur-Nitrogen Compounds" Part 1, 1977 (e.g., N-sulfinyl-sulfonamides, $O{=}S{=}N{-}SO_2X$ (X = Hal, CF_3)). Exceptions are N-sulfinylamines carrying S^{VI} substituents at the nitrogen atom with $S^{VI}{-}C$ bonds (e.g., $O{=}S{=}N{-}SO_2R$ (R = alkyl, aryl)). These latter compounds are included in this volume.

The next sections give a detailed survey of the reactions of N-sulfinylamines with organic compounds, with radicals, carbenes, ylides, and carbanions, and with organometallic compounds and metal complexes. The numerous reactions of N-sulfinylamines with organic compounds have considerable importance in preparative chemistry. Usually, the S=N bond is involved while the S=O bond remains unaffected in most cases.

The transition metal complexes of sulfinyl imides and the sulfinyl imide anion in which the ligands coordinate via N, S, O, or via N=S double bond are described in the following chapter. Amino-sulfinyl radicals and salts of the N-sulfinyl-aminium cations complete the description of the class of sulfinyl imides.

In the last chapter of this volume thiosulfinyl imides and derivatives (e.g., salts of the trisulfur nitride anion and the tetrasulfur nitride anion), which are well-characterized, are described. Only a relatively small number of thiosulfinyl imides are known.

The names of the compounds have been selected from those given in the literature and in most cases synonyms are given.

The literature up to 1988 has been fully evaluated and most of the references up to end-1989 are included.

The series will be continued by describing other acyclic $S^{IV}{-}N$ compounds with 2-coordinate sulfur, such as sulfur diimides and with 3- and 4-coordinate sulfur.

Frankfurt am Main
April 1990

Alfons Kubny

Remarks on Abbreviations and Standards

Physical data are cited in short form using abbreviations:

D_m	measured density
D_x	density calculated from X-ray data
m.p.	melting point; dec.: melting with decomposition
b.p.	boiling point, often measured at low pressure (given in Torr behind the b.p.)
IR	(infrared) spectrum. The medium of measurement (the physical state or a solvent) is given in parentheses. For the absorption maxima (vibration frequencies ν in cm^{-1}), intensity and shape are abbreviated as usual: s (strong), vs (very strong), w (weak), vw (very weak), vvw (very very weak), m (medium), br (broad), sh (shoulder). The assigned bands are usually labeled with the symbols ν for stretching, δ for in-plane bending, γ for out-of-plane bending, ϱ for rocking, and τ for torsion vibration; the symbols $_s$ and $_{as}$, e.g., in ν_s and ν_{as}, mean symmetric and antisymmetric.
Ra	(Raman) spectrum. For abbreviations, see IR.
UV	(ultraviolet) spectrum includes the visible spectrum. The medium of measurement is set in parentheses. The wavelength of the absorption maximum λ_{max} is given in nm followed by the extinction coefficient ε or log ε, ε in $L \cdot cm^{-1} \cdot mol^{-1}$, in parentheses; sh means shoulder, br means broad.
PES	photoelectron spectroscopy
X-PES	X-ray photoelectron spectroscopy
UPS	ultraviolet photoelectron spectroscopy
NMR	(nuclear magnetic resonance) spectrum, noise decoupling is indicated by braces { }. Solvent and standard are given in parentheses. For δ (chemical shift in ppm) downfield shift is indicated by a positive sign as recommended by IUPAC. Standard substances (if not otherwise cited) are TMS (= $Si(CH_3)_4$) as internal standard for ^1H and ^{13}C NMR, neat CH_3NO_2 as external standard for ^{14}N NMR and ^{15}N NMR, $CFCl_3$ as internal standard for ^{19}F NMR, and 85% H_3PO_4 as external standard for ^{31}P NMR. The multiplicity of the signal is given in parentheses: s (singlet), d (doublet), t (triplet), q (quartet), quint (quintet), sept (septet), m (multiplet). The assignment is given behind the multiplicity; if necessary for clarity the assigned atoms are underlined. ^nJ (coupling constant in Hz) is fixed by the number of bonds involved n and the coupling nuclei (e.g., $^1J(^{13}C,H)$), or by listing the coupling groups (e.g., $^3J(CH_2,CH_3)$)
ESR	electron spin resonance
MS	(mass spectrum). m/e = mass/charge; M^+ = molecular ion (1+); relative intensity in parentheses
EIMS	electron impact mass spectrum

Further Abbreviations:

i-C_3H_7	isopropyl, $CH(CH_3)_2$	tos	p-toluenesulfonyl, $SO_2C_6H_4CH_3$-4
i-C_4H_9	isobutyl, $CH_2CH(CH_3)_2$	DMF	dimethylformamide, $HCO(NCH_3)_2$
s-C_4H_9	*sec*-butyl, $CH(CH_3)C_2H_5$	DMSO	dimethyl sulfoxide, $(CH_3)_2SO$
t-C_4H_9	*tert*-butyl, $C(CH_3)_3$	THF	tetrahydrofuran, C_4H_8O
c-C_6H_{11}	cyclohexyl	TMS	tetramethylsilane, $Si(CH_3)_4$

Table of Contents

2 Sulfur-Nitrogen Compounds with 2-Coordinate Sulfur (continued)

2.6 Sulfinyl Imides, Thionyl Imides, N-Sulfinylamines and Derivatives

The range of sulfinyl imides was surveyed over the period 1967 to 1981 by Bussas, R.; Kresze, G.; Münsterer, H.; Schwöbel, A. (Sulfur Rept. **2** [1983] 215/364) preferentially considering organic derivatives with at least one carbon atom.

Several review articles appeared before 1983 that describe progress in the chemistry of organic N-sulfinylamines, e.g., Kresze, G.; Maschke, A.; Albrecht, R.; Bederke, K.; Patzschke, H. P.; Smalla, H.; Trede, A. (Angew. Chem. **74** [1962] 135/44; Angew. Chem. Intern. Ed. Engl. **1** [1962] 89/98); Kresze, G.; Wucherpfennig, W. (Angew. Chem. **79** [1967] 109/27; Angew. Chem. Intern. Ed. Engl. **6** [1967] 149/67); Beckert, R.; Mayer, R. (Z. Chem. [Leipzig] **22** [1982] 237/45); Dolars, A. (Houben-Weyl Methoden Org. Chem. 4th Ed. **11/2** [1958] 738/44).

A short review that deals with the preparation and reactions of N-sulfinylamines using the newer literature up to 1984 was given by Beckert, R.; Mayer, R. (Houben-Weyl Methoden Org. Chem. 4th Ed. **E 11** [1985] 558/71).

A short survey on the preparation and molecular structures of sulfinyl imides obtained with the aid of representative examples here precedes the detailed description.

The reactions of the class of N-sulfinylamines with organic compounds, radicals, carbenes, ylides, and carbanions, organometallic compounds and metal complexes (to give complexes of N-sulfinylamines) are summarized in the Sections 2.6.23, pp. 186/241, 2.6.24, pp. 241/3, 2.6.25, pp. 243/51, and 2.7, pp. 251/88, respectively, whereas the reactions with inorganic compounds are described in detail for the corresponding N-sulfinylamine species.

2.6.1 Survey

2.6.1.1 Preparation

From Compounds with Primary Amino Groups

The oldest method of preparation of O=S=NR compounds, and still a standard method is the reaction of primary amines with $SOCl_2$:

$$RNH_2 + SOCl_2 \rightarrow O{=}S{=}NR + 2\,HCl$$

R = OR′ (see p. 28), SO_2R' (see p. 46), NR^1R^2 (see pp. 54/5, 57, 61, 63), $P(O)(OR')_2$ (see p. 76), $P(O)R'_2$ (see p. 79), alkyl (see pp. 95, 96, 97, 99, 100, 102, 103), aryl (see pp. 113, 128, 131), C(O)R′ (see p. 172), C(O)(OR′) (see p. 173). The HCl produced during the reaction is trapped by excess amine or by addition of an auxiliary base such as pyridine or triethylamine. The hydrochlorides of the amines were also used as educts.

The use of dialkyl sulfites (e.g. $SO(OC_4H_9)_2$, see p. 113) was published in 1939. For a few years, N, N′-sulfinyl-bisimidazole (I) or N-(chlorosulfinyl)imidazole (II) have been employed with success as sulfinylation agents, see, e.g., pp. 40, 102, 113, 136, 147.

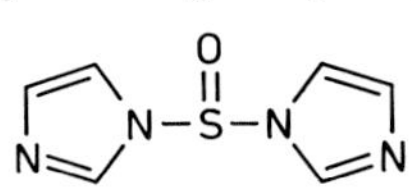

I

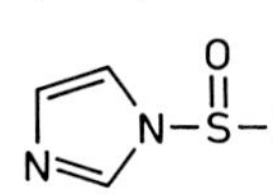

II

An important preparative reaction for sensitive unstable N-sulfinylamines is the "transsulfinylation" method, based on the equilibrium:

$$R'NH_2 + O{=}S{=}NR \rightleftharpoons RNH_2 + O{=}S{=}NR'$$

The less basic the $R'NH_2$ reactant corresponding to the starting O=S=NR, the more the equilibrium is shifted to the right. Therefore, $O{=}S{=}NSO_2C_6H_4CH_3$-4 and related compounds are most frequently used. $O{=}S{=}NC_6H_5$ is generally employed. The N-sulfinyl derivatives of hydrazines (see pp. 54/5, 57), hydrazides (see p. 61), and hydrazones (see p. 70) are accessible by this method.

From Monosubstituted Amino Compounds

Since the direct sulfinylation of amides with $SOCl_2$ fails in some cases, it is often advantageous to use the monotrimethylsilyl derivatives as educts.

$$RNHSi(CH_3)_3 + SOCl_2 \rightarrow O{=}S{=}NR + (CH_3)_3SiCl + HCl$$

R = S(O)R′ (see p. 46), $P(O)(OR')_2$ (see p. 76).

The preparation of N-sulfinylamines proceeds also by sulfinylation of lithiated amines or N-trimethylsilylamines in THF with SO_2.

$$RNHLi \xrightarrow{SO_2} RNHS(O)OLi \xrightarrow[-n\text{-}C_4H_{10}]{+LiC_4H_9\text{-}n} RLiN{-}S(O)OLi \xrightarrow[-Li_2SO_3]{SO_2} O{=}S{=}NR$$

R = alkyl (see pp. 96, 97, 102), aryl (see pp. 113, 135, 136, 144).

$$RNHSi(CH_3)_3 \xrightarrow[-n\text{-}C_4H_{10}]{+LiC_4H_9\text{-}n} RLiN{-}Si(CH_3)_3 \xrightarrow[-LiOSi(CH_3)_3]{+SO_2} O{=}S{=}NR$$

R = C_6H_{11}-c (see p. 102), aryl (see pp. 147, 156/9).

2-Chloro-N-sulfinylalkanamines can be prepared from aziridines and $SOCl_2$:

$$R^1R^2\overline{C{-}CH_2{-}N}H + SOCl_2 \xrightarrow[-HCl]{} [R^1R^2\overline{C{-}CH_2{-}N}{-}S(O)Cl] \rightarrow Cl{-}CR^1R^2{-}CH_2{-}N{=}S{=}O$$

R^1 = H, CH_3; R^2 = H, CH_3 (see Table 38, Nos. 19, 21, 23, pp. 106/7).

N-Sulfinylsulfenamides O=S=NSR were obtained by the reaction of disulfenylamines $(RS)_2NH$ with $SOCl_2$, R = CCl_3 (see p. 40), R = C_6F_5 (see pp. 41/2).

From N,N-Disubstituted Amino Compounds

The reaction of N,N-bis(trimethylsilyl) derivatives of amines or amides with $SOCl_2$ give the corresponding N-sulfinylamino compounds.

$$R{-}N(Si(CH_3)_3)_2 + SOCl_2 \rightarrow O{=}S{=}NR + 2(CH_3)_3SiCl$$

R = Hal (see pp. 23/4), SR′ (see pp. 38, 41), S(O)R′ (see p. 46), $Si(CH_3)_3$ (see p. 81), C(O)R′ (see p. 172).

Treatment of N,N-dichloro-1-cyanoalkanamines, N,N-dichloro carboxylic acid amides, sulfonamides, or carbamic esters with $SOCl_2$ yields the corresponding N-sulfinylamines.

$$R{-}NCl_2 + SOCl_2 \rightarrow O{=}S{=}NR + 2Cl_2$$

R = SO_2Ar (see p. 46), $CH(CN)CH(CH_3)_2$ (see Table 38, No. 27, p. 108), C(O)Ar (see Table 44, p. 172), C(O)OAlk (see Table 45, p. 173).

From Imino Compounds

Phenyl-trifluoromethyl-ketimine reacts with $SOCl_2$ to yield the corresponding α-chlorinated N-sulfinylamino compound

$$R^1R^2C{=}NH + SOCl_2 \xrightarrow[-HCl]{} [R^1R^2C{=}N{-}S(O)Cl] \rightarrow O{=}S{=}N{-}CClR^1R^2$$

R^1 = CF_3, R^2 = C_6H_5 (see Table 38, No. 38, p. 109).

N-Lithiated imines also yield N-sulfinylamino compounds with $SOCl_2$, which, in turn, attack the lithio derivative with elimination of LiCl.

$$R_2C{=}NLi + SOCl_2 \rightarrow O{=}S{=}NCClR_2 + LiCl$$

$$O{=}S{=}NCClR_2 + R_2C{=}NLi \xrightarrow[-LiCl]{} O{=}S{=}NCR_2N{=}CR_2$$

$R = C_6H_5$ (see Table 38, No. 40, p. 110), CF_3 (see p. 185).

N-Sulfinyl-carbamic acid esters are synthesized from dialkylesters of N-chloroimido carboxylic acids with $SOCl_2$.

$$(RO)_2C{=}NCl + SOCl_2 \rightarrow O{=}S{=}NC(O)OR + RCl + Cl_2$$

$R = CH_3$, C_2H_5 (see Table 45, p. 173).

N-Substituted iminosulfur dichlorides $Cl_2S{=}NR$ react with formic and acetic acid to yield $O{=}S{=}NR$, $R = SO_2C_6H_5$ (see p. 48), $R = C(CN)Alk_2$ (see Table 38, Nos. 25, 26, p. 107).

N-Sulfinyl-perfluoroalkanamines are obtained by treatment of $Cl_2S{=}NC_nF_{n+1}$ with Ag_2O, $AgO(O)CF_3$, or HgO, see pp. 181, 183, 184.

From Azides

Arylazides react with thiirane 1-oxides prepared in situ to give N-sulfinylanilines.

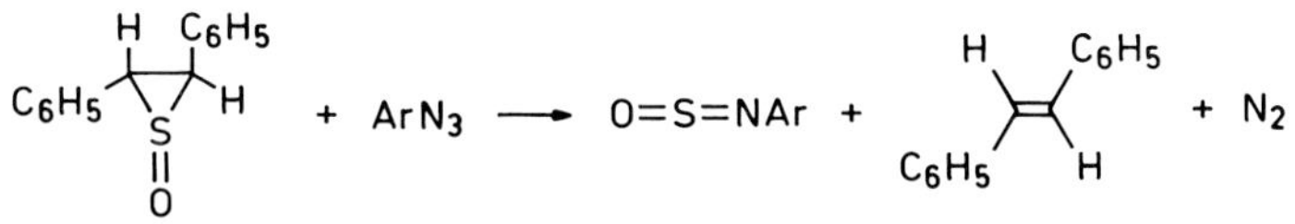

$Ar = C_6H_5$ (see p. 114), C_6H_4Cl-4 (see p. 131), $C_6H_4OCH_3$-4 (see p. 136), $C_6H_4CH_3$-4 (see p. 147).

From Isocyanates

Phenylisocyanate is converted with SO_2 to N-sulfinylaniline by use of $Fe(CO)_5$ as a catalyst, see p. 113.

$Hg(NCO)_2$ reacts with $SOCl_2$ in the following manner (see p. 180):

$$Hg(NCO)_2 + SOCl_2 \rightarrow [O{=}S(Cl)NCO] \rightarrow [O{=}S{=}N{-}C(O)Cl] \rightarrow O{=}S{=}N{-}C(O){-}N{=}C{=}O$$

From Sulfur Diimides

Treatment of $H_3CN{=}S{=}NCH_3$ and t-$C_4H_9N{=}S{=}NC_4H_9$-t with O_3 yields the corresponding N-sulfinylamines, see pp. 91 and 100, respectively.

From Other N-Sulfinylamino Compounds

In most of these syntheses, $O{=}S{=}NSi(CH_3)_3$ was the starting material which reacts readily with a multitude of halo derivatives.

$$O{=}S{=}NSi(CH_3)_3 + R{-}X \rightarrow O{=}S{=}NR + (CH_3)_3SiX$$

X = F, R = F; X = Cl, R = Cl; X = Br, R = Br (see pp. 23/4); X = Cl, $R = C_6H_5S$ (see p. 40); X = Cl, $R = C_6F_5S$ (see p. 41); X = Cl, $R = ((CH_3)_2N)_2B$ (see p. 80); X = F, R = FHg (see p. 87).

$K^+[O{=}S{=}N]^-$ was also used as a starting material for the preparation of $O{=}S{=}NR$, e.g., $R = As(C_6H_5)_2$ (see p. 80), $Ge(C_2H_5)_3$ (see p. 86), $Sn(CH_3)_3$ (see p. 86).

Isothiocyanates react in an exchange with N-sulfinylamines via an intermediate 1,2-cycloadduct.

$$O{=}S{=}N{-}R + S{=}C{=}N{-}R' \rightleftharpoons \left[\begin{array}{l} R{-}N{-}S{=}O \\ S{=}C{-}N{-}R' \end{array}\right] \rightleftharpoons O{=}S{=}N{-}R' + S{=}C{=}N{-}R$$

R = Ar, R′ = C_6H_5 (see pp. 113/4).

Higher yields are obtained by a similar reaction of carbodiimides with N-sulfinylsulfonamides.

$$O{=}S{=}N{-}SO_2R + R'N{=}C{=}NR' \rightleftharpoons RSO_2N{=}C{=}NR' + O{=}S{=}NR'$$

R = CH_3, $C_6H_4CH_3$-4; R′ = C_6H_5 (see p. 114).

2.6.1.2 Molecular Structure

Sulfinyl imides, O=S=N–R, have a nonlinear O=S=N system like all SO_2 analogs. Two stereoisomeric forms may be considered:

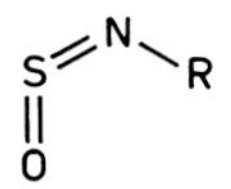

Z
(cis, syn)

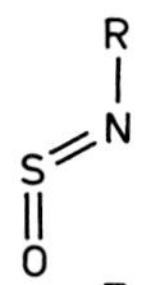

E
(trans, anti)

In the gas phase, sulfinyl imides have the (Z)-[cis] configuration. It was shown from microwave spectral or electron-diffraction analyses that O=S=N–H (see p. 6) and O=S=N–CH_3 (see pp. 92/3) have a planar cis configuration with respect to the S=N bond, whereas for O=S=N–Cl (see p. 24) and O=S=N–SiH_3 (see p. 81) nonplanar cis configurations with dihedral angles of $35.5° \pm 3.6°$ and $30° \pm 3°$, respectively, were observed.

X-ray and neutron-diffraction analyses show that in the crystalline phase the studied sulfinyl imides have a planar or roughly planar (Z)-[cis] configuration of the O=S=N–X unit (dihedral angle ~0°):

X = S in $(O{=}S{=}N)_2S$ (see pp. 31/2), O=S=N–S–N=C$(C_6H_5)_2$ (see p. 44), and O=S=N–$SO_2C_6H_5$ (see p. 48);

X = N in O=S=N–NHC_6H_5 (see p. 55), O=S=N–N$(C_3H_7\text{-}i)_2$ (see p. 66), and O=S=N–N$(CH_3)C_6H_5$ (see p. 66);

X = P in $(CO)_5Mo((C_6H_5)_2P{-}N{=}S{=}O)$ (see pp. 88/90);

X = As in $(CO)_5Cr((C_6H_5)_2As{-}N{=}S{=}O)$ (see pp. 89/90);

X = C in O=S=NC_6H_4Br-4 (see p. 133), O=S=$NC_6H_5NO_2$-4 (see p. 141), O=S=$NC_6H_2(C_4H_9\text{-t})_3$-2,4,6 (see p. 160), and $(O{=}S{=}NC_6H_2(C_4H_9\text{-t})_2\text{-3,5-S-2-})_2$ (see pp. 161/2).

The structure of sulfinyl imides in the liquid state or in solution is not known with certainty. In earlier publications a cis ⇌ trans equilibrium was proposed for N-sulfinylaniline in solutions, based on an evaluation of dipole moment data. The possible existence of two rotational isomers was also concluded from IR studies in solvents. However, in most of the recent papers, a planar cis configuration is assumed to exist in O=S=NC_6H_5 (see p. 115) and other aromatic derivatives. The difference in stability of stereoisomeric forms, however, may be small so that in solution the existence of a conformational equilibrium cannot be excluded.

The existence of *trans*-O=S=NH, *trans*-O=S=ND, and *trans*-O=S=^{15}NH in Ar matrices could be established from IR studies. The trans isomers were generated from the corresponding cis isomers by low-energy photolysis in Ar matrices, see p. 18.

2.6.2 Sulfinyl Imide, Thionyl Imide, O=S=NH

Earlier work on O=S=NH has been compiled in "Schwefel" B 3, 1963, pp. 1580/1, and in Goehring, M., Ergebnisse und Probleme der Chemie der Schwefel-Stickstoffverbindungen, Akademie Verlag, Berlin 1957, pp. 129/35.

2.6.2.1 *cis*-Sulfinyl Imide, *cis*-Thionyl Imide, *cis*-O=S=NH (and isotopic isomers)

2.6.2.1.1 Formation. Preparation

The standard method for the preparation of the gaseous O=S=NH is the gas-phase reaction of $SOCl_2$ with NH_3 [1, 2]. Low pressures of the reactants (12 Torr of NH_3 and 4 Torr of $SOCl_2$) increase the yield of O=S=NH and decrease the amount of the by-product SO_2 [3]. At higher pressure of ~210 Torr, ClS≡N was detected (by IR and UV) besides SO_2 as by-product, whereas at low pressures (~10 Torr) only small amounts of SO_2 and no ClS≡N could be observed. It was suggested that the by-products are formed via a side reaction of O=S=NH with $SOCl_2$ [4].

O=S=NH is also formed rapidly with evolution of heat in the reaction of gaseous NH_3 with SO_2 under completely anhydrous conditions. The simultaneously produced solid is $(NH_4)_2S_2O_5$, identified by IR spectroscopy [5, 6, 7]. The reaction of SO_2 with excess NH_3 was investigated in the rare-gas matrix. O=S=NH was detected IR spectroscopically at low temperatures (~40 K), however on warming, it was converted to $NH_4^+[O{=}S{=}N]^-$, immediately [8].

The reaction of H_2N^- (generated by electron impact on NH_3) with SO_2 in the gas phase by means of the flowing afterglow and selected ion-flow tube technique also produces O=S=NH [9].

The title compound was observed as an intermediate in the hydrolysis of FS≡N. Under the experimental conditions used the final products were SO_3^{2-}, F^-, and NH_4^+ [10, 11]. Later, hydrolysis of FS≡N was used for preparation of O=S=NH in the microwave absorption cell [3]. Obviously due to hydrolysis, O=S=NH was formed in small amounts (detected IR spectroscopically) by the preparation of gaseous ClS≡N via heating $S_3N_3Cl_3$ to 110°C under high vacuum [12].

Depolymerization of poly(sulfinyl imide) $(OSNH)_x$ at ca. 80°C in vacuum produces O=S=NH in ca. 40% yield [13].

Finally, O=S=NH was detected by IR spectroscopy (besides S_2N_2) in an inert matrix at 20 K (trapped vapor of S_4N_4 which had been passed over heated Ag_2Se at ~160°C) [14].

O=S=NH with various isotopes was prepared by analogous pathways. O=S=ND was produced by the reaction of ND_3 with $SOCl_2$ and by hydrolysis of FS≡N with D_2O. O=S=^{15}NH was obtained by the reaction of $^{15}NH_3$ with $SOCl_2$ and ^{18}O=S=NH by the hydrolysis of FS≡N with $H_2{}^{18}O$ [3].

References:

[1] Schenk, P. W. (Ber. Deut. Chem. Ges. **75** [1942] 94/9).
[2] Steudel, R.; Schenk, P. W. (in: Brauer, G., Handbuch der präparativen anorganischen Chemie, 3rd Ed., Vol. 1, Stuttgart 1975, pp. 484/6).
[3] Kirchhoff, W. H. (J. Am. Chem. Soc. **91** [1969] 2437/42).
[4] DeKock, R. L.; Haddad, M. S. (Inorg. Chem. **16** [1977] 216/7).
[5] Hata, T.; Asano, T. (Tohoku Daigaku Hisui Yoeki Kagaku Kenkyusho Hokoku **9** [1960] 79).
[6] Hata, T.; Kiunmaki, S. (Nature **203** [1964] 1378/9).
[7] Hata, T. (Tohoku Daigaku Hisui Yoeki Kagaku Kenkyusho Hokoku **14** [1964] 5/14).

[8] Meyer, B.; Mulliken, B.; Weeks, H. (Phosphorus Sulfur **8** [1980] 281/90).
[9] Bierbaum, V. M.; Grabowski, J. J.; DePuy, C. H. (J. Phys. Chem. **88** [1984] 1389/93).
[10] Glemser, O.; Richert, H. (Z. Anorg. Allgem. Chem. **307** [1961] 313/27, 315).

[11] Roesky, H. W; Glemser, O.; Hoff, A. (Chem. Ber. **101** [1968] 1215/22, 1216).
[12] Müller, A.; Nagarajan, G.; Glemser, O.; Cyvin, S. F.; Wegener, J. (Spectrochim. Acta A **23** [1967] 2683/9).
[13] Schenk, P. W.; Krone, E.; Kartono-Soeratman, H. (Monatsh Chem. **95** [1964] 710/5).
[14] Almond, M. J.; Downs, A. J.; Jefferey, T. L. (Polyhedron **7** [1988] 629/34).

2.6.2.1.2 Molecular Properties. Spectra

Geometrical Structure

The geometrical structure of OSNH was investigated by a microwave spectrum analysis of $H^{14}N^{32}S^{16}O$ and the isotopically substituted species, $D^{14}N^{32}S^{16}O$, $H^{15}N^{32}S^{16}O$, $H^{14}N^{34}S^{16}O$, and $H^{14}N^{32}S^{18}O$, and of $H^{14}N^{32}S^{16}O$ in the $v_5=1$ state. The microwave spectrum was assigned to the planar *cis*-O=S=NH isomer. The HOSN isomer could be excluded by chemical arguments, the mass spectral cracking pattern, and microwave information [1]. **Fig. 1**, taken from [2], presents the geometrical arrangement along with bond lengths and bond angles, obtained from I_b and I_c of 5 isotopic species [1].

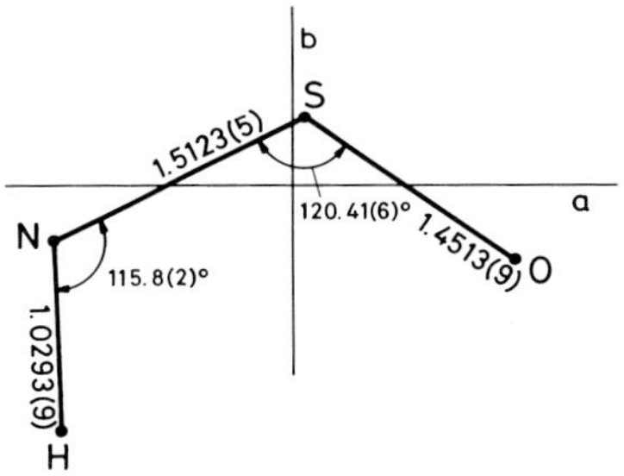

Fig. 1. Geometrical structure of *cis*-O=S=NH [1, 2]; bond lengths in Å.

For the N–H bond distance, an r_s value of 1.0272 ± 0.0001 Å was derived from another microwave spectral analysis [3]. The geometrical parameters of *cis*-O=S=NH have also been determined by several quantum mechanical methods: ab initio: SCF+CPF (coupled pair functional) [4], SCF (HF+MP) [5], SCF (HF) [6] and semiempirical methods: CNDO/2 [7, 8] and MNDO [9]. Optimized geometrical data obtained by a recently performed SCF calculation are: H–N = 1.008 Å, N=S = 1.497 Å, S=O = 1.443 Å, ∢HNS = 117.0°, ∢NSO = 120.1° [10]. These data are in qualitative agreement with the experimental data and those from other calculations [4 to 7].

References:

[1] Kirchhoff, W. H. (J. Am. Chem. Soc. **91** [1969] 2437/42).
[2] Carlotti, M.; Di Lonardo, G.; Galloni, G.; Trombetti, A. (J. Mol. Spectrosc. **84** [1980] 155/61).
[3] Dal Borgo, A.; Di Lonardo, G.; Scappini, F.; Trombetti, A. (Chem. Phys. Letters **63** [1979] 115/8).
[4] Ehrhardt, C.; Ahlrichs, R. (Chem. Phys. **108** [1986] 417/28).

[5] Raghavachari, K. (J. Chem. Phys. **76** [1982] 3668/72).
[6] Turner, A. G. (Inorg. Chim. Acta **84** [1984] 85/7).
[7] Grunwell, J. R.; Danison, W. C., Jr. (Tetrahedron **27** [1971] 5315/21).
[8] Oberhammer, H. (Theor. Chim. Acta **26** [1972] 79/86).
[9] Oleksik, J. J.; Turner, A. G. (Inorg. Chim. Acta **59** [1982] 165/9).
[10] Nonella, M.; Huber, J. R.; Ha, T.-K. (J. Phys. Chem. **91** [1987] 5203/9).

Quantum Mechanical Calculations on Stability and Electronic Structure

Most workers (with the exception of [1, 2]) agree that the *cis*-O=S=NH isomer is more stable than the trans isomer. Energy differences ΔE from several ab initio SCF calculations were ~37 [3], 21 [4], and 20 kJ/mol [5]. The energy barrier for cis-trans isomerization was determined to be 59.8 kJ/mol (5000 cm^{-1}) [5]. A consensus exists also that O=S=NH is more stable than HOSN with ΔE = 39.6 (ab initio SCF; in a former paper the reverse was concluded using a semiempirical (MNDO) calculation [7]) [6], 42 [4], and ~42 to 45 kJ/mol [5]. In addition, it was found that O=S=NH is more stable than HSNO and HONS [5]. A rough determination of the barrier to internal rotation at S–N using a PRDDO (partial retention of double differential overlap) calculation in the rigid rotator approximation yielded a value of 84 kJ/mol [8].

The **electronic structure** of *cis*-OSNH was characterized using the results of population analyses based upon several SCF calculations [1, 3, 6, 9, 10]. Orbital and overlap populations are listed in [1, 10]. Shared electron number, hypervalency population, and Mulliken occupation number for nitrogen are tabulated in [4]. Atomic charges are reported by [4, 7, 9, 10], e.g.: +0.13 for H, −0.56 for N, +1.14 for S, and −0.71 for O [4].

The total energy for the *cis*-O=S=NH molecule was computed [3, 5, 6, 9, 10]. A recent SCF calculation yielded −527.35136 a.u. as the deepest value [5]. The lowest electronically excited states (S_1, S_2, S_3, T_1, T_2, T_3) were calculated using MRCI (multireference configuration interaction) and MCSCF-CI methods. Results in terms of excitation energies and oscillator strengths are tabulated [5].

References:

[1] Grunwell, J. R.; Danison, W. C., Jr. (Tetrahedron **27** [1971] 5315/21).
[2] Oberhammer, H. (Theor. Chim. Acta **26** [1972] 79/86).
[3] Bak, B.; Svanholt, H.; Larsen, C. (J. Mol. Struct. **36** [1977] 55/9).
[4] Ehrhardt, C.; Ahlrichs, R. (Chem. Phys. **108** [1986] 417/28).
[5] Nonella, M.; Huber, J. R.; Ha, T.-K. (J. Phys. Chem. **91** [1987] 5203/9).
[6] Turner, A. G. (Inorg. Chim. Acta **84** [1984] 85/7).
[7] Oleksik, J. J.; Turner, A. G. (Inorg. Chim. Acta **59** [1982] 165/9).
[8] Zyubin, A. S.; Charkin, O. P. (Zh. Neorgan. Khim. **27** [1982] 565/70; Russ. J. Inorg. Chem. **27** [1982] 319/22).
[9] Solouki, B.; Rosmus, P.; Bock, H. (Angew. Chem. **88** [1976] 381/2).
[10] Cassoux, P.; Serafini, A.; Fong, G.; Kuczkowski, R. L. (Inorg. Chem. **17** [1978] 1204/7).

Bond Order

The bond order, derived from force constants (by a method of Siebert [1]) for the SN bond in *cis*-O=S=NH, is 1.66 and in *cis*-O=S=ND is 1.58. The SO bond order is 1.76 and 1.78, respectively [2]. In an earlier paper the SN bond order in O=S=NH was given as 1.9 and the SO bond order as 1.8 [3].

References:

[1] Siebert, H. (Anwendungen der Schwingungsspektroskopie in der Anorganischen Chemie, Springer, Berlin 1966, p. 36).
[2] Eysel, H. H. (J. Mol. Struct. **5** [1970] 275/82).
[3] Richert, H. (Z. Anorg. Allgem. Chem. **309** [1961] 171/80, 178).

Ionization Potentials E_i. Photoelectron Spectrum

The HeI photoelectron (PE) spectrum was recorded in the range ~10 to ~20 eV. The observed peaks were assigned by a comparison with the PE spectra of SO_2 and S_2O and also with the aid of an ab initio SCF calculation. The result is given below [1]:

E_i (in eV)

observed	11.60	—	12.52	15.10	15.88
calculated	12.18	12.72	13.38	16.88	17.32
assignment	$\pi_{nb}(3a'')$	$n_S - n_O(13a')$	$n_O - n_N(12a')$	$\pi_b(2a'')$	$n^+_{NH}(11a')$
observed	16.72	20.15			
calculated	18.84	23.57			
assignment	$\sigma^-_{NSO}(10a')$	$\sigma^+_{NSO}(9a')$			

Other workers compared the HeI PE spectrum of OSNH with that of O=S=NCl [2], making a slightly different assignment (3a″, 13a′ assigned to 11.60 eV) in the spectrum of O=S=NH from that given in [1].

References:

[1] Solouki, B.; Rosmus, P.; Bock, H. (Angew. Chem. **88** [1976] 381/2).
[2] Nachbaur, E.; Prossegger, P. (Monatsh. Chem. **116** [1985] 1385/93).

Dipole Moment

The dipole moment of *cis*-O=S=NH was determined by microwave spectrum analysis (MW) [1, 2] and by quantum mechanical calculation methods [2, 3, 4]. The results are listed in Table 1.

For $OS^{15}NH$ the dipole moment was determined from the Stark effect of the M = 0 components of the $1_{0,1} \rightarrow 2_{0,2}$ and $1_{1,1} \rightarrow 2_{1,2}$ transitions: $\mu_a = 0.893 \pm 0.003$ D, $\mu_b = 0.181 \pm 0.005$ D, $\mu_{tot} = 0.911 \pm 0.003$ D [1].

Table 1
Dipole Moment Values of *cis*-OSNH.

μ_a[D]	μ_b[D]	μ_{tot}[D]	method	remark	Ref.
±0.89	±0.19	0.91	MW	a)	[1, 2]
−0.94	−0.03	0.94	ab initio SCF	b)	[2]
0.915	−0.03	0.916	ab initio SCF		[3]
—	—	1.29	CNDO/2		[4]

a) The dipole moment was derived from the measurements of the M = 0 and M = 1 components of the $1_{0,1} \rightarrow 2_{0,2}$ transition. Because of quadrupole broadening, the measurements were less accurate than for $OS^{15}NH$. The sign of μ_a and μ_b was not determined [1, 2]. – b) The dipole moment lies approximately along the a inertial axis with negative end toward nitrogen [2].

The dipole moments of the lower-lying electronic states were calculated with the MRCI (multireference configuration interaction) method [5].

References:

[1] Kirchhoff, W. H. (J. Am. Chem. Soc. **91** [1969] 2437/42).
[2] Cassoux, P.; Serafini, A.; Fong, G.; Kuczkowski, R. L. (Inorg. Chem. **17** [1978] 1204/7).
[3] Bak, B.; Svanholt, H.; Larsen, C. (J. Mol. Struct. **36** [1977] 55/9).
[4] Grunwell, J. R.; Danison, W. C., Jr. (Tetrahedron **27** [1971] 5315/21).
[5] Nonella, M.; Huber, J. R.; Ha, T.-K. (J. Phys. Chem. **91** [1987] 5203/9).

^{14}N Nuclear Electric Quadrupole Coupling Constants

The quadrupole coupling constants $eq_{aa}Q$, $eq_{bb}Q$, and $eq_{cc}Q$ (in MHz) of OSNH, OSND, and $^{18}O^{32}S^{14}NH$ were obtained from an analysis of the hyperfine splitting of the rotational transitions. Use was made of the relation $eq_{aa}Q + eq_{bb}Q + eq_{cc}Q = 0$ and of the assumption that $eq_{cc}Q$ should not change on isotopic substitution H↔D [1]:

eqQ in MHz	O=S=NH	O=S=ND	^{18}O=S=NH
$eq_{aa}Q$	-1.50 ± 0.07	-1.30 ± 0.09	-1.62 ± 0.10
$eq_{bb}Q$	-0.07 ± 0.17	-0.27	0.08 ± 0.19
$eq_{cc}Q$	1.57 ± 0.14	1.57	1.54 ± 0.16

The coupling constants $eq_{xx}Q$, $eq_{yy}Q$, and $eq_{zz}Q$ of OSNH along the principal axes of the quadrupole tensor were based upon [1] and were also theoretically (ab initio SCF) determined: $eq_{xx}Q = -3.21 \pm 0.50$ and -3.09, $eq_{yy}Q = 1.64 \pm 0.75$ and 1.61, $eq_{zz}Q = 1.57 \pm 0.14$ and 1.48 MHz, respectively. The angle between the principal inertial axes (a, b frame) and the principal field gradient axes (x, y frame) was found by comparison with $DNSOF_2$ to be $36.4° \pm 4.0°$ (theoretical value 39.29°) [2].

References:

[1] Kirchhoff, W. H. (J. Am. Chem. Soc. **91** [1969] 2437/42).
[2] Cassoux, R.; Serafini, A.; Fong, G.; Kuczkowski, R. L. (Inorg. Chem. **17** [1978] 1204/7).

Rotational Constants. Centrifugal Distortion Constants. Moments of Inertia

cis-O=S=NH is a planar, asymmetric rotor close to the prolate symmetric top limit, $\varkappa = -0.919$ [1] (for OSND: $\varkappa = -0.897$ [2]). The rotational constants for the ground state were obtained from analysis of the microwave (MW) spectrum [3], from a simultaneous least squares fit to the ground state combination differences derived from the IR measurements of ν_1 and $2\nu_1$ (together with data of [3]) [1], and also from an MW analysis, including measured lines from [3] (two different methods by Dowling [4] and Watson [5] were used, leading to practically coincident molecular constants) [6]. The results are compiled in Table 2.

Table 2
Rotational Constants (A, B, C in MHz) for *cis*-O=S=NH and Isotopic Species (ground electronic state[a)]).

molecule	A	B	C	Ref.
$H^{14}N^{32}S^{16}O$	49316.00 ± 0.1	9869.77 ± 0.04	8205.19 ± 0.04	[3]
$H^{14}N^{32}S^{16}O$	49315.80 ± 0.09	9869.69 ± 0.02	8205.18 ± 0.02	[6]
$H^{14}N^{32}S^{16}O$	49315.746 ± 0.033	9869.629 ± 0.006	8205.233 ± 0.006	[1][b)]
$D^{14}N^{32}S^{16}O$	42299 ± 96[c)]	9540.86 ± 0.04	7765.92 ± 0.04	[3]
$D^{14}N^{32}S^{16}O$	42257.49 ± 0.10	9540.86 ± 0.02	7765.89 ± 0.02	[2, 6]
$H^{15}N^{32}S^{16}O$	48868.19 ± 0.1	9542.89 ± 0.03	7966.14 ± 0.04	[3]
$H^{14}N^{34}S^{16}O$	48051 ± 140	9870.00 ± 0.04	8168.64 ± 0.04	[3]
$H^{14}N^{32}S^{18}O$	48006 ± 176	9325.98 ± 0.04	7791.59 ± 0.04	[3]

[a)] For $v_5=1$ (of $H^{14}N^{32}S^{16}O$): A = 49966 ± 159, B = 9869.04 ± 0.04, and C = 8192.87 ± 0.04 [3]. – [b)] cm^{-1} converted to MHz using $c = 2.9979246 \times 10^{10}$ cm/s. – [c)] Misprinted in [3].

Rotational constants (in MHz) for vibrationally excited states of O=S=NH and O=S=ND were also determined (OSNH ν_1 band: A = 48984, B + C = 18069, B – C = 1622; $2\nu_1$ band: A = 48632, B + C = 18042; $3\nu_1$ band: A = 48414) [1] (rounded values).

Centrifugal distortion constants for the ground state of O=S=NH and O=S=ND obtained from an analysis of the microwave spectrum by Dowling's [4] and Watson's [5] formalism are given in Table 3.

Centrifugal distortion constants of OSNH (and OSND) were earlier presented for the ground state and vibrationally excited states [1, 7]. Calculated centrifugal distortion constants from force constants are given in [8].

The sextic constant H_K of O=S=NH for the ground state is $-3.842(43) \times 10^{-9}$ cm^{-1} (for ν_1: $-7.0(5) \times 10^{-9}$; for $2\nu_1$: $-24(5) \times 10^{-9}$ cm^{-1}) [1].

The moments of inertia I_a, I_b, I_c (in uÅ^2) of OSNH and its isotopic species were determined from the MW spectrum analysis. They are tabulated together with the inertial defect Δ ($= I_c - I_a - I_b$) (in uÅ^2) in Table 4. The conversion factor from MHz to uÅ^2 was 505376 [3].

Table 3

Centrifugal Distortion Constants of O=S=NH and O=S=ND (in MHz) [6].

molecule[a]	τ_{aaaa}	τ_{bbbb}	τ_{aabb}	τ_{abab}
cis-O=S=NH	-5.535 ± 0.009	-0.0413 ± 0.0009	0.340 ± 0.006	-0.051 ± 0.001
cis-O=S=ND	-3.395 ± 0.003	-0.0470 ± 0.0008	0.264 ± 0.005	-0.059 ± 0.001

molecule[b]	d_J	d_{JK}	d_K	d_{WJ}	d_{WK}
cis-O=S=NH	-0.0351 ± 0.0004	-0.72 ± 0.01	-0.51 ± 0.06	$(0.463 \pm 0.008) \times 10^{-5}$	$(0.49 \pm 0.02) \times 10^{-4}$
cis-O=S=ND	-0.0340 ± 0.0004	-0.69 ± 0.01	-0.89 ± 0.06	$(0.479 \pm 0.006) \times 10^{-5}$	$(0.54 \pm 0.02) \times 10^{-4}$

[a] Dowling's method [4]. – [b] Watson's method [5].

Table 4

Moments of Inertia (I in uÅ^2) and Inertial Defect (Δ in uÅ^2) [3].

	$H^{14}N^{32}S^{16}O$	$D^{14}N^{32}S^{16}O$	$H^{15}N^{32}S^{16}O$	$H^{14}N^{34}S^{16}O$	$H^{14}N^{32}S^{18}O$
I_a	10.24771 ± 0.00002	11.948 ± 0.027	10.34161 ± 0.00002	10.517 ± 0.031	10.527 ± 0.039
I_b	51.20445 ± 0.00021	52.96967 ± 0.00022	52.95837 ± 0.00016	51.20326 ± 0.00021	54.19010 ± 0.00023
I_c	61.59222 ± 0.00030	65.07616 ± 0.00034	63.44048 ± 0.00032	61.86781 ± 0.00030	64.86170 ± 0.00033
Δ*)	0.14006 ± 0.00046	0.158 ± 0.027	0.14050 ± 0.00044	0.148 ± 0.031	0.145 ± 0.039

*) From MW and IR analysis: $\Delta = 0.13899(2)$ uÅ^2 [1]; calculated from vibrational frequencies and Coriolis coupling constants: 0.1436 uÅ^2 [8].

References:

[1] Carlotti, M.; Di Lonardo, G.; Galloni, G.; Trombetti, A. (J. Mol. Spectrosc. **60** [1976] 388/99).
[2] Carlotti, M.; Di Lonardo, G.; Galloni, G.; Trombetti, A. (J. Mol. Spectrosc. **80** [1980] 291/300).
[3] Kirchhoff, W. H. (J. Am. Chem. Soc. **91** [1969] 2437/42).
[4] Dowling, J. M. (J. Mol. Spectrosc. **6** [1961] 550/3).
[5] Watson, J. K. G. (J. Chem. Phys. **45** [1966] 1360/1, **48** [1968] 181/5).
[6] Dal Borgo, A.; Di Lonardo, G.; Scappini, F.; Trombetti, A. (Chem. Phys. Letters **63** [1979] 115/8).
[7] Carlotti, M.; Di Lonardo, G.; Galloni, G.; Trombetti, A. (J. Mol. Spectrosc. **84** [1980] 155/61).
[8] Namasivayam, R.; Kannan, K.; Sakkarapani, S. (Indian J. Phys. B **58** [1984] 369/75).

Molecular Vibrations. Infrared Spectrum

The IR spectral data of *cis*-O=S=NH were recorded from the gas phase in the ranges 4000 to 420 cm^{-1} [1] and 4100 to 300 cm^{-1} [2] and from Ar [3, 4] and N_2 matrices [3] below 1500 and between 3500 and 3000 cm^{-1}. Since *cis*-O=S=NH is a four atom molecule of C_s symmetry, six fundamental vibrations are expected: five of a′ type and one of a″ type. The observed IR absorptions ν (in cm^{-1}) together with assignments are listed in the following table (see also [5]):

type	gas phase [1]	gas phase [2]	Ar matrix [3]	Ar matrix [4]	assignment
ν_1 a′	3345(w)	3308.50	3308.5	3309	NH stretch
ν_2 a′	1261(s)	1257.1	1248.7	1249	SO stretch
ν_3 a′	1090(w)	1086.1	1082.7	1083	NS stretch
ν_4 a′	911(m)	906.3	900.4	900	HNS bend
ν_5 a′	453(m)	447.7	447.4	447	NSO bend
ν_6 a″	759(s)	757.0	754.7	754	torsion

The NH stretch, ν_1, and the torsion, ν_6, were observed as doublets in Ar matrices [3]. This fact was ascribed to the occupancy of different sites within the Ar lattice. In comparison with the spectrum [1] of gaseous O=S=NH, the vibrations in Ar matrices were observed at lower frequencies. In N_2 matrices both positive and negative shifts were found. The average absolute shift is slightly less in Ar (13 cm^{-1}) than in N_2 (16 cm^{-1}). Fermi resonance between ν_4 and $2\nu_5$ was observed. The effect is stronger in *cis*-$OS^{15}NH$ (showing a doublet with components of nearly equal intensity) [3]. The higher resolution of the IR spectrum of gaseous *cis*-OSNH, ranging from 0.2 cm^{-1} at 700 cm^{-1} to 0.6 cm^{-1} at 3500 cm^{-1}, permitted the measurement and assignment of overtones and combination bands (compiled in a table) [2]. The ground-state progression of the ν_1 band (N–H stretching fundamental) was studied earlier [6].

IR spectra of the isotopic species of OSNH were also recorded. Gas-phase spectra of OSND were recorded in the range 4000 to 420 cm^{-1} [1] and at higher resolution from 4100 to 300 cm^{-1} [2]. The six fundamental vibrations of OSND are (in cm^{-1}): 2480(w), 1257(s), 1055(w), 757(m), 594(s), ~410(m) [1], 2454.76, 1254.4, 1052.3, 755.0, 595.7, 400.2 [2], see also [5]. The IR spectrum of OSND was studied in the region of the N–D stretching fundamental vibration and of the first overtone with an even higher resolution (0.05 cm^{-1}) [7]. In the range 4100 to 300 cm^{-1}, a spectrum of gaseous $OS^{15}NH$ was recorded. Overtones and combination bands were also presented and assigned [2]. Ar matrix IR spectra showed ν_1 through ν_6 for OSND [3, 4] and for $OS^{15}NH$ and $OS^{15}ND$ (~95% enriched in ^{15}N) [3]. Only ν_2 [3, 4] and ν_4 [3] were observed for $O^{34}SNH$ and $O^{34}SND$, only ν_2 [3] for $O^{34}S^{15}NH$ (^{34}S in natural abundance).

Harmonic vibration frequencies, ω_1 to ω_6 for OSNH (3462.1, 1271.7, 1089.9, 934.1, 449.1, and 765.1 cm^{-1}), $OS^{15}NH$, OSND, and $OS^{15}ND$, ω_2 and ω_4 for $O^{34}SNH$ and $O^{34}SND$, were derived by [2] from experimental data. The data of OSNH were compared with computed data from Hartree-Fock and Møller-Plesset methods. The inclusion of electron correlation yielded frequencies within 5% of the experimental values [8]. The vibrational frequencies, which were estimated by using force constants from an SCF calculation, are higher than the experimental ones [9].

The **anharmonicity constant** x_{11} for OSNH was determined from the ν_1, $2\nu_1$, and $3\nu_1$ frequencies to be $-76.76(2)$ cm^{-1} [6] (x_{11}(OSND) = $-40.87(1)$ cm^{-1} [7]).

Vibrational mean amplitudes were calculated [10, 11] using Cyvin's relation [12].

Coriolis coupling constants were calculated from the force constants [11] by the relation of Meal and Polo [13].

References:

[1] Richert, H. (Z. Anorg. Allgem. Chem. **309** [1961] 171/80).
[2] Carlotti, M.; Di Lonardo, G.; Galloni, G.; Trombetti, A. (J. Mol. Spectrosc. **84** [1980] 155/61).
[3] Tchir, P. O.; Spratley, R. D. (Can. J. Chem. **53** [1975] 2311/7).
[4] Allegretti, J. M.; Merer, A. J. (Can. J. Phys. **50** [1972] 404/7).
[5] Jacox, M. E. (J. Phys. Chem. Ref. Data **13** [1984] 945/1068, 996).
[6] Carlotti, M.; Di Lonardo, G.; Galloni, G.; Trombetti, A. (J. Mol. Spectrosc. **60** [1976] 388/99).
[7] Carlotti, M.; Di Lonardo, G.; Galloni, G.; Trombetti, A. (J. Mol. Spectrosc. **80** [1980] 291/300).
[8] Raghavachari, K. (J. Chem. Phys. **76** [1982] 3668/72).
[9] Nonella, M.; Huber, J. R.; Ha, T.-K. (J. Phys. Chem. **91** [1987] 5203/9).
[10] Sebestyen, A.; Vizi, B. (Magy. Kem. Folyoirat **84** [1978] 282/8; C.A. **89** [1978] No. 16396).
[11] Namasivayam, R.; Kannan, K.; Sakkarapani, S. (Indian J. Phys. B **58** [1984] 369/75).
[12] Cyvin, S. J. (Acta Chem. Scand. **13** [1959] 2135/6).
[13] Meal, J. H.; Polo, S. R. (J. Chem. Phys. **24** [1956] 1119/25).

Force Constants

The first calculation of the SO and SN valence force constants of *cis*-OSNH was performed on the basis of a simple three-mass model, yielding $f_{NS} = 8.2(9)$, $f_{SO} = 8.6(6)$, and the deformation constant d = 0.60(3) mdyn/Å [1]. Seven force constants, computed for a general valence force field, were in accordance with the cis structure of the molecule [2].

A force field with nine constants was derived from an IR study of various isotopic species (using fundamental frequencies) [3] (see Table 5, p. 14). A simple and a modified valence force field with 5 and 12 force constants, respectively, was determined (see Table 5, p. 14) [4]. A force field with eleven constants was derived from an IR study of the fundamental vibrations, overtones and combination bands of gaseous OSNH in combination with frequencies of isotopic species and with centrifugal distortion constants of OSNH and OSND (see Table 5, p. 14) [5]. Force constants were also calculated theoretically [6, 7]. The results of various calculations at the Hartree-Fock and Møller-Plesset levels were compared with experimentally derived data from [3] and [5] (see Table 5, p. 14). Differences between experimental and theoretical values, for example of $f_{NS,HNS}$, $f_{SO,NSO}$, and $f_{HN,HNS}$ were found; the experimental ones were believed to be in error [6]. The Green's function formalism was also used to calculate force constants. Slight deviations from experimental results were found [7].

Table 5
Force Constants of *cis*-O=S=NH Derived from IR Studies [3, 4, 5] and Ab Initio Calculations [6].

force constant	[3][a)]	[4][b)]	[5][a)]	[6][a), c)]
f_{HN}	6.05	6.20	6.6225	6.127
f_{NS}	6.47	6.80	6.99	6.557
f_{SO}	9.01	9.90	9.30	8.906
f_{HNS}	0.457	0.40	0.5041	0.525
f_{NSO}	1.66	0.60	1.578	1.544
f_{HNSO}	0.225	0.152	0.23091	0.206
$f_{HN,NS}$	—	0.01	−0.110	−0.23
$f_{HN,SO}$	—	—	—	0.02
$f_{HN,HNS}$	—	0.05 to 0.10	—	0.34
$f_{HN,NSO}$	—	—	—	−0.13
$f_{NS,SO}$	—	0.40	0.108	−0.22
$f_{NS,HNS}$	−0.15	—	−0.094	0.64
$f_{NS,NSO}$	—	0.20	—	−0.11
$f_{SO,HNS}$	−0.14	—	—	−0.04
$f_{SO,NSO}$	—	0.125	0.73	0.10
$f_{HNS,NSO}$	−0.18	0.10 to 0.15	−0.195	−0.24

[a)] Stretching, bending, and stretch-bend interaction constants in mdyn/Å, mdyn·Å/rad^2, and mdyn/rad, respectively. – [b)] In mdyn/Å. – [c)] MP2/3-21G* level of theory.

References:

[1] Richert, H. (Z. Anorg. Allgem. Chem. **309** [1961] 171/80, 177).
[2] Eysel, H. H. (J. Mol. Struct. **5** [1970] 275/82).
[3] Tchir, P. O.; Spratley, R. D. (Can. J. Chem. **53** [1975] 2311/7).
[4] Sebestyen, A.; Vizi, B. (Magy. Kem. Folyoirat **84** [1978] 282/8; C.A. **89** [1978] No. 16396).
[5] Carlotti, M.; Di Lonardo, G.; Galloni, G.; Trombetti, A. (J. Mol. Spectrosc. **84** [1980] 155/61).
[6] Raghavachari, K. (J. Chem. Phys. **76** [1982] 3668/72).
[7] Namasivayam, R.; Kannan, K.; Sakkarapani, S. (Indian J. Phys. B **58** [1984] 369/75).

Microwave Spectrum

The microwave (MW) spectrum of $^{16}O^{32}S^{14}NH$ was measured in the range 16185.66 to 37815.27 MHz (17 lines) [1] and in the range 9647.37 to 37814.51 MHz (33 lines) [2]. The $5_{1,4}\rightarrow 4_{2,3}$ transition line in [1] was corrected to 18298.80 MHz in [2]. The MW spectrum consists of strong a-type and weak b-type transitions. Low J transitions were identified by their Stark effects. Many of the transitions consist of multiplets that were caused by the coupling of the spin of the nitrogen nucleus to the overall rotation of the molecular framework [1]. The MW spectrum was used for a structure determination of O=S=NH, see p. 6.

The MW spectrum of OSND was recorded in the range 17306.42 to 36389.18 (10 lines) [1] and 8375.52 to 36388.49 MHz (30 lines) [2]. The MW spectra of $^{18}O^{32}S^{14}NH$, $^{16}O^{34}S^{14}NH$, and $^{16}O^{32}S^{15}NH$ were also presented [1].

References:

[1] Kirchhoff, W. H. (J. Am. Chem. Soc. **91** [1969] 2437/42).
[2] Dal Borgo, A.; Di Lonardo, G.; Scappini, F.; Trombetti, A. (Chem. Phys. Letters **63** [1979] 115/8).

UV Spectrum

The UV spectrum of gaseous *cis*-OSNH was recorded in the range 210 to 350 nm and showed two absorption regions: an extremely weak continuous band system from 325 to 350 nm and a long series of diffuse bands between 238.7 and 269 nm (17 bands ranging from 238.68 to 268.92 nm and 3 temperature-sensitive bands at 258.38, 260.39, and 262.05 nm were tabulated) [1]. Based on their MRCI and MCSCF-CI calculations, other workers attributed the measured absorptions between 238.7 and 269 nm to an $S_0 \rightarrow S_1$ ($n \rightarrow \pi^*$) transition and that between 325 and 350 nm to an $S_0 \rightarrow T_1$ ($n \rightarrow \pi^*$) transition (probably in combination with an $S_0 \rightarrow T_2$ ($\pi \rightarrow \pi^*$) transition) [2]. A $\pi \rightarrow \pi^*$ transition was calculated (using CNDO/2) to be 6.9716 and 7.8474 eV (178 and 158 nm) excluding and including d orbitals, respectively [3].

The UV spectrum of OSND was recorded in the range 240 to 320 nm (15 bands ranging from 242.2 to 263.5 nm were observed). The vibrational pattern of OSNH in the region 238.7 to 269 nm was not changed on deuteration, indicating that the electronic excitation is centered at the sulfur atom [1].

References:

[1] Allegretti, J. M.; Merer, A. J. (Can. J. Phys. **50** [1972] 404/7).
[2] Nonella, M.; Huber, J. R.; Ha, T.-K. (J. Phys. Chem. **91** [1987] 5203/9).
[3] Grunwell, J. R.; Baker, H. J. (J. Chem. Soc. Perkin Trans. II **1973** 1542/4).

2.6.2.1.3 Heat of Formation. Thermodynamic Functions

The heat of formation $\Delta H^\circ_{f,298} \leq -46$ kJ/mol was determined from the enthalpy for the reaction $H_2N^- + SO_2 \rightarrow HO^- + O{=}S{=}NH$ in the gas phase, $\Delta H \leq 0$ [1]. The value $\Delta H^\circ_f = 174.5$ kJ/mol was calculated by using the semiempirical MNDO molecular orbital method [2].

Thermodynamic functions for *cis*-O=S=NH were calculated from vibrational frequencies (taken from the argon matrix study) and shown in Table 6 [3].

Table 6
Selected Values of Thermodynamic Functions for O=S=NH (ideal gas) [3].

T in K	S°	C°_p	$-(G^\circ - H^\circ_0)/T$	$(H^\circ - H^\circ_0)/T$
		in $cal \cdot mol^{-1} \cdot K^{-1}$		
273.15	65.95	10.57	57.22	8.72
298.10	66.89	11.01	57.99	8.90
200.00	62.86	9.30	54.57	8.29
300.00	66.96	11.04	58.05	8.91
400.00	70.37	12.65	60.71	9.65
500.00	73.33	13.90	62.95	10.38
1000.00	84.11	16.95	71.06	13.05
1500.00	91.24	18.17	76.66	14.58

References:

[1] Bierbaum, V. M.; Grabowski, J. J.; DePuy, C. H. (J. Phys. Chem. **88** [1984] 1389/93).
[2] Oleksik, J. J.; Turner, A. G. (Inorg. Chim. Acta **59** [1982] 165/9).
[3] Tchir, P. O.; Spratley, R. D. (Can. J. Chem. **53** [1975] 2311/7, 2316).

2.6.2.1.4 Stability and Solubility

For earlier information see "Schwefel" B 3, 1963, pp. 1580/1.

O=S=NH can be stored in the gas phase for weeks [1].

It is soluble without color in absolute methanol, acetone, ether, and pyridine. Apparently, O=S=NH is stabilized via adduct formation. However, the brown polymer ($(OSNH)_x$) separates in the ether solution at −80°C and at −40°C in the pyridine solution. The methanol solution becomes red to violet and S_4N_4 separates. 0.1 to 0.2% H_2O in the solvents lead to a complete hydrolysis of O=S=NH [2].

References:

[1] Meyer, B.; Mulliken, B.; Weeks, H. (Phosphorus Sulfur **8** [1980] 281/90).
[2] Schenk, P. W.; Krone, E.; Kartono-Soeratman, H. (Monatsh. Chem. **95** [1964] 710/5).

2.6.2.1.5 Chemical Reactions

Reactions with transition metal complexes to give OSNH complexes are described on pp. 251/2.

Dimerization. Polymerization

In an N_2 matrix study of O=S=NH, a dimer $(OSNH)_2$ was detected at 35 K based on IR absorptions at 3200, 1245, 1080, 945, 852, and 478 cm^{-1} [1]. The result was confirmed by a matrix isolation study of the S_4N_4 vapor species (produced by the passage of S_4N_4 vapor over heated silver selenide) showing that OSNH and $(OSNH)_2$ were present. A hydrogen-bonded eight-membered ring was proposed for the geometrical structure. The dimeric species is assumed to be the first stage in the polymerization of OSNH [2]. Sulfinyl imide polymerizes above −70°C to yield the brown poly(sulfinyl imide) $(OSNH)_x$. The reaction is accomplished after standing for 1 h at room temperature. The polymer separates also from an ether or pyridine solution at −80 and −40°C, respectively [3]; see above.

References:

[1] Tchir, P. O.; Spratley, R. D. (Can. J. Chem. **53** [1975] 2311/7, 2313).
[2] Almond, M. J.; Downs, A. J.; Jeffery, T. L. (Polyhedron **7** [1988] 629/34).
[3] Schenk, P. W.; Krone, E.; Kartono-Soeratman, H. (Monatsh. Chem. **95** [1964] 710/5).

Photolysis of Matrix-Isolated *cis*-O=S=NH

The selective photolysis of *cis*-O=S=NH in an argon matrix (T = 12 K) permits the formation of the following isomers: *cis*-HOSN [1, 2], *cis*- and *trans*-HSNO [1, 2], *trans*-O=S=NH [3], *trans*-HONS [4] and the radicals $SNO^•$ [1] and $NSO^•$ [3]. The various transformations with irradiation wavelengths are summarized in **Fig. 2**.

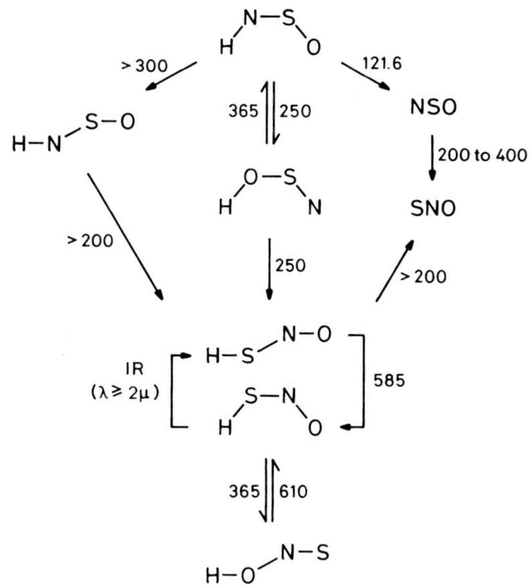

Fig. 2. Transformations of matrix-isolated *cis*-O=S=NH by photolysis. The irradiation wavelengths are given in nm [4, 5].

A freshly prepared sample of *cis*-O=S=NH is converted to a mixture of *cis*-HSNO and *trans*-HSNO via *cis*-HOSN when it is irradiated with light of $\lambda = 250$ nm [1, 2, 4]. Further photolysis with light of $\lambda = 365$ nm leads to the formation of *trans*-HONS. A portion of the initial *cis*-O=S=NH is reformed [4]. The photolysis of matrix-isolated *cis*-O=S=NH with light limited to $\lambda > 300$ nm yields *trans*-O=S=NH [3]. Using vacuum UV ($\lambda = 121.6$ nm) produces the radical $NSO^{\bullet}$ (further photolysis in the 200 to 400 nm region results in the disappearance of $NSO^{\bullet}$ and probable in the appearance of $SNO^{\bullet}$) [3]. The various products were identified by IR spectroscopy (use of D and ^{15}N isotopes).

References:

[1] Tchir, P. O.; Spratley, R. D. (Can. J. Chem. **53** [1975] 2318/30).
[2] Müller, R. P.; Nonella, M.; Russegger, P.; Huber, J. R. (Chem. Phys. **87** [1984] 351/61).
[3] Tchir, P. O.; Spratley, R. D. (Can. J. Chem. **53** [1975] 2331/6).
[4] Nonella, M.; Huber, J. R.; Ha, T.-K. (J. Phys. Chem. **91** [1987] 5203/9).
[5] Perutz, R. N. (Chem. Rev. **85** [1985] 97/127, 108).

Reaction with NH_3

Condensing an excess of NH_3 onto colorless OSNH leads to the formation of an adduct with the composition $OSNH \cdot NH_3$. A conductometric titration with KNH_2 in liquid NH_3 showed it to be $O{=}S(NH_2)_2$.

Reference:

Schenk, P. W.; Krone, E.; Kartono-Soeratman, H. (Monatsh. Chem. **95** [1964] 710/5).

2.6.2.2 *trans*-Sulfinyl Imide, *trans*-Thionyl Imide, *trans*-O=S=NH

trans-O=S=NH was generated by low-energy photolysis (λ>300 nm, using a medium-pressure mercury lamp with a filter) of *cis*-OSNH in an Ar matrix and was detected by IR spectroscopy. *trans*-OSND and *trans*-$OS^{15}NH$ were generated and detected under analogous conditions [1].

For a comparison of the thermodynamic stability between *trans*-OSNH and *cis*-OSNH see p. 7.

The following structure parameters of *trans*-OSNH were calculated:

method	bond lengths in Å			bond angles in °		Ref.
	N–H	N–S	S–O	∢HNS	∢NSO	
ab initio SCF	1.004	1.50	1.432	109.6	114.8	[3]
ab initio SCF	1.007	1.503	1.436	109.6	114.8	[4]
MNDO	1.00	1.54	1.48	110	105	[2]

A population analysis was based on an ab initio SCF calculation. Atomic charges were: 0.12 at H, −0.53 at N, 1.11 at S, and −0.68 at O. Shared electron number, hypervalency population, and Mulliken occupation number for nitrogen were listed, together with values of the cis isomer and a hypothetical linear isomer of OSNH (for which the highest total energy of the three isomers was estimated) [3].

The bond orders of the SN and SO bonds were calculated from the force constants using the method of Siebert [5] to be 1.49 (OSND: 1.50) and 1.82 (OSND: 1.82), respectively [6].

The dipole moment of *trans*-OSNH was calculated by an ab initio method: $\mu_{tot}=3.88$, $\mu_a=2.62$, and $\mu_b=-2.86$ D [7]. $\mu_{tot}=4.25$ D was computed by the CNDO/2 method [8].

The following four IR bands (in cm^{-1}) were attributed to *trans*-OSNH: 1382 (S=O stretch), 986 (S=N stretch), 881 (HN=S bend), and 651 (torsion). HN and N=S=O bands were not observed. Four frequencies (in cm^{-1}) of *trans*-$OS^{15}NH$ (1379, 963, 868, and 650) and two frequencies of *trans*-OSND (1380 and 951) were presented [1, 9]. But the absorption band at 951 cm^{-1} was later attributed to the SN stretching vibration of *trans*-DONS [4]. The IR frequencies of *trans*-OSNH were also calculated by an ab initio SCF method [4]. A normal coordinate analysis was made, using the structure of *cis*-OSNH and taking the dihedral angle at 180°. Force constants of *trans*-OSNH were listed and calculated frequencies of *trans*-OSNH, *trans*-$OS^{15}NH$, and *trans*-OSND were compared with the observed ones [1]. Force constants of *trans*-OSNH were also computed for a general force field (based on [10]) [6].

The heat of formation was calculated by the semiempirical MNDO method to be 187.0 kJ/mol [2].

References:

[1] Tchir, P. O.; Spratley, R. D. (Can. J. Chem. **53** [1975] 2331/6).
[2] Oleksik, J. J.; Turner, A. G. (Inorg. Chim. Acta **59** [1982] 165/9).
[3] Ehrhardt, C.; Ahlrichs, R. (Chem. Phys. **108** [1986] 417/28).
[4] Nonella, M.; Huber, J. R.; Ha, T.-K. (J. Phys. Chem. **91** [1987] 5203/9).
[5] Siebert, H. (Anwendungen der Schwingungsspektroskopie in der Anorganischen Chemie, Springer, Berlin 1966, p. 36).
[6] Eysel, H. H. (J. Mol. Struct. **5** [1970] 275/82).
[7] Bak, B.; Svanholt, H.; Larsen, C. (J. Mol. Struct. **36** [1977] 55/9).

[8] Grunwell, J. R.; Danison, W. C., Jr. (Tetrahedron **27** [1971] 5315/21).
[9] Jacox, M. E. (J. Phys. Chem. Ref. Data **13** [1984] 945/1068, 996).
[10] Richert, H. (Z. Anorg. Allgem. Chem. **309** [1961] 171/80, 178).

2.6.3 Sulfinyl Imide Ion (1−), Thionyl Imide Ion (1−), $[O{=}S{=}N]^-$

Earlier work on $[O{=}S{=}N]^-$ is summarized in "Schwefel" B 3, 1963, p. 1581, see also Goehring, M., Ergebnisse und Probleme der Chemie der Schwefel-Stickstoffverbindungen, Akademie Verlag, Berlin 1957, pp. 131/8.

2.6.3.1 Sulfinyl Imide Ion (1−), $[O{=}S{=}N]^-$ in the Gas Phase and in Solution

2.6.3.1.1 Formation

The anion OSN^- has been recently synthesized in the gas phase at 298 K in a flowing afterglow system by the reaction of H_2N^- and SO_2. Experimental rate coefficients, reaction efficiencies, and further details are presented [1]. The ion was also formed by electrochemical reduction of $O{=}S{=}N{-}S{-}N{=}C(C_6H_5)_2$ at −1.6 V (Hg pool) or −1.75 V (Pt gauze) and by electrolysis of a solution of $(O{=}S{=}N)_2S$ in $(C_2H_5)_4N^+ClO_4^-{-}CH_3CN$ on Pt gauze at −2.0 V (reference electrode: Ag/0.1M $AgClO_4{-}CH_3CN$) [2].

2.6.3.1.2 Molecular Properties. Spectra

Molecular Geometry

The equilibrium geometry of OSN^- has been computed by [3] and was recalculated by [4]. An optimized bond angle was obtained by the MNDO method (values of S–N and S–O bond lengths were taken from related compounds) [5].

method	bond lengths in Å		bond angle in °	Ref.
	N–S	S–O	∢NSO	
ab initio SCF	1.467	1.495	121.4	[3]
ab initio SCF	1.462	1.489	122.4	[4]*)
MNDO	—	—	111	[5]

*) Larger basis values.

Quantum Mechanical Calculations on Electronic Structure and Stability

Energies ε_i of nine occupied molecular orbitals and the first virtual orbital of OSN^- were calculated by an HFS (Hartree-Fock-Slater)-SCF-LCAO method [5]; energies of three occupied MO's (different numbering) were also calculated by an ab initio SCF method [3]:

orbital	1a′	2a′	3a′	4a′	1a″	5a′
ε_i in a.u. [5]	−0.65	−0.43	−0.20	−0.09	−0.08	−0.03
ε_i in a.u. [3]	—	—	—	—	−0.3358	—
description [5]	σ	σ	σ*	σ	π	σ*(N–S) σ (S–O)

(continued)

orbital	$2a''$	$6a'$	$7a'$	$(3a'')$[a]
ε_i in a.u. [5]	0.06	0.07	0.12	0.20
ε_i in a.u. [3]	−0.1565	—	−0.1335[b]	—
description [5]	π (N–S) π* (S–O)	σ*	σ (N–S) σ* (S–O)	π*

[a] Virtual orbital. – [b] Antibonding [3].

Total overlap population studies [3, 4, 5] show that OSN^- has a strong polar NS double bond and a very strong SO single bond ($^-N{=}S^+{-}O^-$ [3]). Bond orders of 1.5 and 1 for S–N and S–O, respectively, were obtained [5]. Atomic charges were computed; the sulfur atom is positively charged: 0.93 [3], 0.467 [4], 0.24 [5] at S; −0.93 [3], −0.736 [4], −0.58 [5] at N; −1.00 [3], −0.730 [4], −0.66 [5] at O.

Recent quantum mechanical calculations including correlation effects show that OSN^- is more stable than SNO^- by ~70 kJ/mol (CPF (coupled pair functional) method) [3] or 37.8 and 65.7 kJ/mol (MP method, using two different basis sets) [4]. The MNDO method gave the opposite result (SNO^- more stable than OSN^- by about 418 kJ/mol) [5].

Electronic Spectrum

One absorption band at 379 nm ($\varepsilon = 70\ L \cdot cm^{-1} \cdot mol^{-1}$) was observed in the UV-visible spectrum, measured from a solution of $((CH_3)_2N)_3S^+OSN^-$ in CH_3CN [5].

An HFS calculation using Slater's transition state method [6] yielded two transitions at 300 and 379 nm which were attributed to a $7a' \to 3a''$ ($\sigma/\sigma^* \to \pi^*$) and a $2a'' \to 3a''$ ($\pi/\pi^* \to \pi^*$) transition, respectively [5].

IR Spectrum

The vibrational frequencies of OSN^- were computed by an SCF method to be 574, 1139, and 1443 cm^{-1} (with the intensities 56, 201, and 156 $km \cdot mol^{-1}$, respectively). Scaling the calculated frequencies by a factor 0.89 yielded reasonable agreement with experimentally observed frequencies from K^+OSN^-, see p. 21 [4].

References:

[1] Bierbaum, V. M.; Grabowski, J. J.; DePuy, C. H. (J. Phys. Chem. **88** [1984] 1389/93).
[2] Chivers, T.; Hojo, M. (Inorg. Chem. **23** [1984] 4088/93).
[3] Ehrhardt, C.; Ahlrichs, R. (Chem. Phys. **108** [1986] 417/28).
[4] Suk Ping So (Inorg. Chem. **28** [1989] 2888/90).
[5] Chivers, T.; Da Silva, A. B. F.; Treu, O., Jr.; Trsic, M. (J. Mol. Struct. **162** [1987] 351/7).
[6] Slater, J. C. (Advan. Quantum Chem. **6** [1972] 1/92).

2.6.3.1.3 Heat of Formation

The value $\Delta H^\circ_{f,298} \leq -142.25$ kJ/mol was determined from the equation: $\Delta H^\circ_{f,298}(OSN^-) = \Delta H^\circ_{acid}(OSNH) - \Delta H^\circ_{f,298}(H^+) + \Delta H^\circ_{f,298}(OSNH)$ with $\Delta H^\circ_{acid} = 1439 \pm 21$, $\Delta H^\circ_{f,298}(H^+) = -1536.4$, and $\Delta H^\circ_{f,298}(OSNH) \leq -46$ kJ/mol [1]. The value $\Delta H^\circ_f = 266.5$ kJ/mol was calculated by the semiempirical MNDO method [2].

References:

[1] Bierbaum, V. M.; Grabowski, J. J.; DePuy, C. H. (J. Phys. Chem. **88** [1984] 1389/93).
[2] Chivers, T.; Da Silva, A. B. F.; Treu, O., Jr.; Trsic, M. (J. Mol. Struct. **162** [1987] 351/7).

2.6.3.1.4 Chemical Reactions

The isomerization energy for the $OSN^- \rightarrow SNO^-$ process was calculated to be ~25 kJ/mol based on MNDO method [1].

The electrolytically produced OSN^- anion reacts on addition of S_4N_4 to yield $S_3N_3O^-$. With elemental sulfur the red species $[OSNS_x]^-$ (x = 1 or 2) is presumably formed [2], see also below. In order to determine gas-phase proton affinity of OSN^- the compound was allowed to react with acids of known gas-phase acidity. The anion did not react with CH_3COOH, reacted slowly with HCOOH, moderately with $ClCH_2CH_2COOH$ and rapidly with HCl. The gas-phase proton affinity of OSN^- was determined to be 1459.5 ± 20.9 kJ/mol [3].

References:

[1] Chivers, T.; Da Silva, A. B. F.; Treu, O., Jr.; Trsic, M. (J. Mol. Struct. **162** [1987] 351/7).
[2] Chivers, T.; Hojo, M. (Inorg. Chem. **23** [1984] 4088/93).
[3] Bierbaum, V. M.; Grabowski, J. J.; DePuy, C. H. (J. Phys. Chem. **88** [1984] 1389/93).

2.6.3.2 Salts of the Sulfinyl Imide Ion (1−), $[O{=}S{=}N]^-$

In earlier work the salts $NH_4^+OSN^-$, Li^+OSN^-, Na^+OSN^-, and Ag^+OSN^- were described, but not unambiguously characterized, see "Schwefel" B 3, 1963, pp. 1580/1.

2.6.3.2.1 $K^+[O{=}S{=}N]^-$

The title compound was prepared in a good yield by the reaction of $K^+OC_4H_9\text{-t}^-$ and O=S=N-$Si(CH_3)_3$ in THF [1]. Other workers found that a successful preparation of K^+OSN^- requires the sublimation of commercial $K^+OC_4H_9\text{-t}^-$ before use and that this reaction be carried out in THF at 50°C [2].

Caution: K^+OSN^- explodes on ignition and ignites on contact with water.

The compound darkens on heating at about 135°C and melts at 190°C to a bubbling orange liquid [1].

IR spectrum (in cm^{-1}): 1270, ν_{as}(O=S=N); 990, ν_s(O=S=N); 515, δ(O=S=N) [1].

The salt is soluble in a benzene solution of 18-crown-6, forming the crown complex in good yield as a white crystalline solid with melting point 150 to 153°C (dec.) [1].

Addition of elemental sulfur to a slurry of K^+OSN^- in CH_3CN gave an unstable red anion $[OSNS_x]^-$ (x = 1, 2) [5] (see also above). The title compound reacts with $(CH_3)_3SiCl$, $(CH_3)_2SiCl_2$, and $(C_2H_5)_3GeCl$ to produce the corresponding sulfinyl imides in low yield. Reaction with $(CH_3)_3SnCl$ yields the sulfinyl imide in a good yield when heated under reflux for 6 h in THF:

$K^+OSN^- + R_3MCl \rightarrow O{=}S{=}NMR_3 + KCl$ (M = Si, Ge, Sn; R = CH_3, C_2H_5). The yield was increased in the case of the reactions of $(CH_3)_3SiCl$ and $(C_2H_5)_3GeCl$ by adding 18-crown-6 [1]. A similar reaction with $(C_6H_5)_2AsCl$ in acetonitrile (23°C, 2 h) led to the formation of $O{=}S{=}NAs(C_6H_5)_2$ [2].

The title compound reacts with the transition metal complexes $(CO)_5Mo((C_6H_5)_2PCl)$ [3, 4, 6], $(CO)_5Mo((CH_3)_2PCl)$, $(CO)_5Cr((C_6H_5)_2PCl)$, and $(CO)_5Cr((CH_3)_2PCl)$ [4, 6], in acetonitrile to yield the corresponding $(CO)_5M(R_2PNSO)$ complexes (M = Mo, Cr; R = CH_3, C_6H_5). When the reaction mixture was stirred at room temperature for 1 d (instead of 30 min) a ~10% conversion of $(CO)_5Mo((C_6H_5)_2PNSO)$ to $(CO)_5MoP(C_6H_5)_2NSN(C_6H_5)_2PMo(CO)_5$ took place [3]. Similarly, a small excess of K^+OSN^- in the reaction with $(CO)_5M(R_2PCl)$ leads to the formation of small amounts of $(CO)_5M(R_2PNSNPR_2)M(CO)_5$ (M = Mo, Cr; R = CH_3, C_6H_5) [4]. The reaction of $Cr(CO)_5(RPCl_2)$ with K^+OSN^- leads to the formation of $(CO)_5Cr[RP(NSN)_2PR]Cr(CO)_5$ (R = $(i\text{-}C_3H_7)_2N$, 3% yield (trans); R = $t\text{-}C_4H_9$, 21% yield (cis)) [7, 8].

References:

[1] Armitage, D. A.; Brand, J. C. (J. Chem. Soc. Chem. Commun. **1979** 1078/9).

[2] Chivers, T.; Dhathathreyan, K. S.; Lensink, C.; Richardson, J. F. (Inorg. Chem. **27** [1988] 1570/4).

[3] Chivers, T.; Lensink, C.; Richardson, J. F. (Organometallics **5** [1986] 819/21).

[4] Chivers, T.; Lensink, C.; Richardson, J. F. (Organometallics **6** [1987] 1904/9).

[5] Chivers, T.; Edelmann, F.; Schmidt, K. J. (unpublished results in: Chivers, T., Sulfur Rept. **7** [1986] 89/107, 91).

[6] Chivers, T.; Lensink, C.; Richardson, J. F. (Phosphorus Sulfur **30** [1987] 189/92).

[7] Chivers, T.; Dhathathreyan, K. S.; Lensink, C.; Liblong, S. W. (Phosphorus Sulfur Silicon Relat. Elem. **41** [1989] 85/95, 86/7).

[8] Chivers, T.; Dhathathreyan, K. S.; Lensink, C.; Meetsma, A.; van de Grampel, J. C.; de Boer, J. L. (Inorg. Chem. **28** [1989] 4150/4).

2.6.3.2.2 $NH_4^+[O{=}S{=}N]^-$

The compound forms in the reaction of SO_2 with excess ammonia in a rare gas matrix (Ar or Kr) in the temperature range between 40 to 90 K. The molecule was identified by six IR absorptions.

Reference:

Meyer, B.; Mulliken, B.; Weeks, H. (Phosphorus Sulfur **8** [1980] 281/90).

2.6.3.2.3 $((CH_3)_2N)_3S^+[O{=}S{=}N]^-$

The salt was prepared by the reaction of $((CH_3)_2N)_3S^+[Si(CH_3)_3F_2]^-$ with $O{=}S{=}NSi(CH_3)_3$ in CH_3CN at −40°C (57% yield) [1] or in THF at 23°C [2].

^1H NMR (CD_3CN/TMS): $\delta = 2.96$ ppm [1].

IR (Nujol): 19 bands ranging from 1262 to 378 cm^{-1}; 1262 vs, $\nu_{as}(N{=}S{=}O)$; 991 to 980 s, $\nu_s(N{=}S{=}O)$; 498 m, $\delta(N{=}S{=}O)$ [1]. Similar values were obtained by [2]: 1261 s, $\nu_{as}(N{=}S{=}O)$; 991 s, $\nu_s(N{=}S{=}O)$; 493 m, $\delta(N{=}S{=}O)$.

The compound is a moisture-sensitive yellow solid [2]. Decomposition occurs at ~180°C [1].

It reacts with t-$C_4H_9AsCl_2$ and $C_6H_5AsCl_2$ in CH_3CN at 23°C (16 h) to give $(O{=}S{=}N)_2AsC_4H_9$-t and $(O{=}S{=}N)_2AsC_6H_5$, respectively [2]. The reaction with $(CO)_5Cr(RPCl_2)$ (R = $(i\text{-}C_3H_7)_2N$, t-C_4H_9) in CH_3CN at 23°C (mole ratio 2:1) leads to the formation of *trans*-$(CO)_5Cr(C_3H_7\text{-}i)_2NP$-$(NSN)_2PN(C_3H_7\text{-}i)_2Cr(CO)_5$ (yield 3%) [3, 4] and *cis*-$(CO)_5Cr(t\text{-}C_4H_9P(NSN)_2PC_4H_9\text{-}t)Cr(CO)_5$ [4].

References:

[1] Heilemann, W.; Mews, R. (Chem. Ber. **121** [1988] 461/3).
[2] Chivers, T.; Dhathathreyan, K. S.; Lensink, C.; Richardson, J. F. (Inorg. Chem. **27** [1988] 1570/4).
[3] Chivers, T.; Lensink, C.; Meetsma, A.; van de Grampel, J. C.; de Boer, J. L. (J. Chem. Soc. Chem. Commun. **1988** 335/6).
[4] Chivers, T.; Dhathathreyan, K. S.; Lensink, C.; Meetsma, A.; van de Grampel, J. C.; de Boer, J. L. (Inorg. Chem. **28** [1989] 4150/4).

2.6.4 Sulfinyl Imide Radical, Thionyl Imide Radical, $[O{=}S{=}N]^{\bullet}$

The $OSN^{\bullet}$ radical forms by photolysis with hydrogen 1216 Å emission of *cis*-O=S=NH in an Ar matrix. Arguments for the formation of $OSN^{\bullet}$ are based on an IR absorption band at 1195 cm^{-1}, which was assigned to ν(SO). Further photolysis of $OSN^{\bullet}$ in the region 200 to 400 nm has been observed to lead probably to the $SNO^{\bullet}$ radical.

Reference:

Tchir, P. O.; Spratley, R. D. (Can. J. Chem. **53** [1975] 2331/6).

2.6.5 Sulfinyl Halogenoimides, Thionyl Halogenoimides, O=S=NX, X = F, Cl, Br, I

Sulfinyl halogenoimides are colorless substances [1, 2]. Their odor is similar to that of halogenoamines and of SO_2 [2]. For O=S=NCl a suffocating odor is reported [1].

2.6.5.1 Preparation

O=S=NF was obtained in poor yield when fluorine diluted with nitrogen was passed into a solution of $O{=}S{=}NSi(CH_3)_3$ in $CFCl_3$ at −80°C: $O{=}S{=}NSi(CH_3)_3 + F_2 \rightarrow O{=}S{=}NF + (CH_3)_3SiF$. It was separated by preparative gas chromatography but tends to explode on crystallization (at −196°C) [1].

O=S=NCl forms analogously, but in nearly quantitative yield in a smooth reaction when liquid chlorine was added dropwise to an equimolar amount of $O{=}S{=}NSi(CH_3)_3$ with stirring at −76°C [1]. It formed in only 25% yield in the reaction of $SOCl_2$ with $((CH_3)_3Si)_2NCl$ at room temperature [2]: $((CH_3)_3Si)_2NCl + SOCl_2 \rightarrow O{=}S{=}NCl + 2(CH_3)_3SiCl$. Separation from $(CH_3)_3SiCl$ is possible only by gas chromatography [1, 2, 7]. Alternatively, $(CH_3)_3SiCl$ may be transformed to $(CH_3)_3SiF$ by halogen exchange with HgF_2, and then separated by fractional distillation [1].

In a more recent method of preparation the difficulty of separation is avoided: N-trichloroisocyanuric acid was added with stirring into $O{=}S{=}NSi(CH_3)_3$; when heated slowly the reaction mixture dissolved spontaneously and reacted with evolution of heat. $(ClNCO)_3 + 3\,O{=}S{=}NSi(CH_3)_3 \rightarrow 3\,O{=}S{=}NCl + ((CH_3)_3SiOCN)_3$. The quantitatively formed O=S=NCl was separated by distillation with a Vigreux column [3]. O=S=NCl formation by thermal decomposition of $Cl_3P{=}NP(Cl)_2{=}NS(O)_2Cl$ or $Cl_3P{=}NS(O)_2Cl$ is described in two patents [4, 5].

The compound must be handled with caution. It is reported that on melting an ampule containing ca. 3 mL O=S=NCl a violent detonation occurred. Similar treatment in the past, as well as distillation at atmospheric pressure or shock proof had not previously indicated the dangerous nature of the compound [6].

O=S=NBr was obtained in quantitative yield on treatment of $(O{=}S{=}N)_2Hg$ with stoichiometric amounts of bromine in $CFCl_3$ at room temparature [1]: $(O{=}S{=}N)_2Hg + 2Br_2 \rightarrow 2O{=}S{=}NBr + HgBr_2$. It formed in the reaction of $SOCl_2$ with $((CH_3)_3Si)_2NBr$ (analogously to the corresponding iodide, but without solvent). The products may be separated by vacuum distillation at low temperature [2]. The reaction of $O{=}S{=}NSi(CH_3)_3$ with bromine gave a difficultly separable mixture containing O=S=NBr [1].

O=S=NI was prepared in 70% yield in the exothermic reaction of $SOCl_2$ and $((CH_3)_3Si)_2NI$ in $CFCl_3$ or ClF_2CCF_2Cl at $-10°C$. Solvent and $(CH_3)_3SiCl$ which formed in the reaction were removed in vacuum and O=S=NI was sublimed twice in high vacuum [2].

2.6.5.2 Molecular Properties. Spectra

The sulfinyl halogenoimides are expected to possess planar cis conformation because of their 24 valency electrons [3]. In agreement with this postulate the Raman spectra of the liquid compounds O=S=NX (X = F, Cl, Br, I) show five polarized lines and one depolarized line as evidence for C_s symmetry; this means all four atoms of the molecule are situated in one plane. In addition, the IR spectra of gaseous O=S=NF, O=S=NCl, and O=S=NBr indicate that the halogeno substituents are in a cis position. O=S=NBr only behaves as an almost linear molecule [8]. In agreement with the IR and Raman spectra, ab initio LCGO-MO Hartree-Fock calculations [9] and semiempirical CNDO/2 calculations for O=S=NCl also indicate a planar cis structure [10], but the latter results of the calculations agree only poorly with the experimental values for geometric parameters [10]. Calculations with the semiempirical CNDO/INDO method predict a three-membered ring structure for O=S=NF as the most stable geometry. But the authors themselves state, that for sulfur compounds, no satisfactory molecular parameters are obtained with this method [11].

The only direct experiments to elucidate the structure were performed with gaseous O=S=NCl by means of electron diffraction [12]. Bond distances and bond angles were determined: $r(N\text{–}Cl) = 1.696 \pm 0.003$ Å, $r(S\text{–}N) = 1.559 \pm 0.004$ Å, $r(S\text{–}O) = 1.445 \pm 0.004$ Å, $\sphericalangle ClNS = 116.3° \pm 0.4°$, $\sphericalangle NSO = 116.0° \pm 0.6°$. In contrast with all suggestions given above, the Cl atom is found in the cis position, but projected out of the OSN plane with a dihedral angle of $\varrho = 35.5° \pm 3.6°$ [12]. Another indication for a nonplanar structure comes from the vibration progression in the ultraviolet photoelectron spectrum showing a torsion vibration ($359\ cm^{-1}$) which is not allowed to occur during ionization of a planar structured molecule [13].

The ^{19}F NMR spectrum of O=S=NF at room temperature shows a triplet, $\delta = -123.3$ ppm referred to $CFCl_3$ as internal standard, with $^1J(^{14}N, ^{19}F) = 112$ Hz [1].

The HeI photoelectron spectrum of O=S=NCl (consisting of two double bands and four single bands) was measured and interpreted in terms of ionization from the eight highest filled valence molecular orbitals. The analysis was made by comparison with the known spectra of the parent compounds O=S=NH, SO_2, and the related molecules O=C=NCl and O=C=NH [13], as well as with the aid of ab initio SCF calculations [9, 13]. The results are given in Table 7 [13].

Table 7

Vertical Ionization Energies, E_i, Progression of Vibration, νM^+ [13], and Ab Initio Orbital Energies, ε_j^{SCF} [13] (method of [9]) of O=S=NCl.

band No.	orbital	E_i in eV	νM^+ in cm^{-1}	$-\varepsilon_j^{SCF}$ in eV
1	5a″	11.8	780	11.24
	19a′			12.18
2	18a′	12.17	660, 360, 180	13.43
3	4a″	13.57	—	14.70
	17a′	13.62		14.88
4	3a″	15.61	—	17.54
5	16a′	16.57	—	18.15
6	15a′	17.25	—	18.75

The vibrational structure of the first band was assigned to a reduced S=N vibration. That of the second band was assigned to the bending modes δ(O=S=N) and δ(S=N–Cl), and to the torsion vibration τ(O=S=N–Cl) [13].

The IR and Raman spectra show the six bands expected for acyclic tetra-atomic molecules with C_s symmetry. The bands with their assignment are listed in Table 8. Assignments were made by depolarization measurements and by comparison with the related compounds O=S=NH and O=S=ND [8].

Table 8

IR and Raman Spectra (frequencies in cm^{-1}) of Sulfinyl Halogenoimides, O=S=NX (X = F, Cl, Br, I).

compound	spectrum	state	ν(S=O)	ν(S=N)	δ(O=S=N)	ν(NX)	γ(O=S=N)	δ(S=N–X)	Ref.
O=S=NF	IR	gas	1246	1007	606	834	394	228	[1]
	IR	gas	1246	1006	606	834	394	—	[8]
	Ra	liquid	1230	995	600	825	395	228	[8]
O=S=NCl	IR	gas	1248	995	676 [a]	532 [b]	359	187	[1]
	IR	gas	1243	995	673	531	358	—	[8]
	Ra	liquid	1221	989	672	526	359	187	[8]
O=S=NBr	IR	gas	1245	1003	621 [a]	456 [b]	335	161	[1]
	IR	gas	1243	1002	620	457	336	161	[2]
	IR	gas	1243	1003	628	457	335	—	[8]
	Ra	liquid	1214	1000	624	451	342	161	[8]
O=S=NI	IR	?	1205	1037	588	378	317	154	[2]
	IR	solid	1210	1037	588	378	330	—	[8]
	Ra	liquid	1247	1028	602	372	330	154	[8]

[a] Tentatively assigned to ν(NX) in the original paper. – [b] Tentatively assigned to δ(O=S=N) in the original paper.

Seven force constants, computed for a general valence force field, were in accordance with the cis structure of the molecules [8]. Five force constants were determined for a simple force field [16]. A modified valence force field with twelve force constants computed by [16] is given in Table 9.

Table 9
Force Constants f (in mdyn/Å) of Sulfinyl Halogenoimides O=S=NX (X = F, Cl, Br, I) [16].

	O=S=NF	O=S=NCl	O=S=NBr	O=S=NI
f_{NX}	3.76	2.50	2.20	1.60
f_{NS}	5.68	5.62	5.74	6.07
f_{SO}	9.51	9.37	9.26	9.77
f_{SNX}	0.30	0.33	0.30	0.30
f_{NSO}	1.40	1.40	1.04	0.954
f_{XNSO}	0.163	0.134	0.119	0.1076
$f_{XN,NS}$	0.50	0.30	0.30	0.30
$f_{XN,XNS}$	0.25	0.20	0.20	0.20
$f_{NS,SO}$	0.50	0.40	0.40	0.40
$f_{NS,NSO}$	0.20	0.20	0.20	0.20
$f_{SO,NSO}$	0.25	0.25	0.25	0.25
$f_{XNS,NSO}$	0.30	0.20	0.20	0.20

Bond orders were calculated from force constants [8] using the method of [14]:

compound	O=S=NF	O=S=NCl	O=S=NBr	O=S=NI
S–N bond	1.24	1.32	1.40	1.56
S–O bond	1.80	1.79	1.78	1.78

Melting points and boiling points reported for the compounds are listed in the following table:

compound	O=S=NCl	O=S=NBr	O=S=NI	Ref.
m.p. in °C	−80	−48	—	[1]
	—	—	32 (dec.)	[2]
b.p. in °C/Torr	65.5*)	12/12	—	[1]
	65	6/10	—	[2]

*) The vapor-pressure curve for O=S=NCl is represented by log (p/Torr) = −1784/T + 8.15. Trouton's constant $\Delta H_v/T_v = 24.1$ cal·mol^{-1}·K^{-1} [1].

In the mass spectra of the four halides, the molecule ion appears, proving the existence of the compounds:

compound	m/e; assignment	Ref.
O=S=NF	81 $OSNF^+$; 48 SO^+; 46 SN^+; 33 NF^+; 32 S^+	[1]
O=S=NCl	97 $OSN^{35}Cl^+$; 49 $N^{35}Cl^+$; 48 SO^+; 46 SN^+; 35 Cl^+; 32 S^+	[1]
O=S=NBr	141, 143 $OSNBr^+$; 93, 95 NBr^+; 79, 81 Br^+; 48 SO^+; 46 SN^+; 32 S^+	[1,2]

2.6.5.3 Chemical Reactions

The sulfinyl halogenoimides are relatively stable up to room temperature [2].

O=S=NF, however, tends to explode on crystallization at −196°C [1]. Gaseous O=S=NF is not stable enough in a pure state for the measurement of its IR spectrum; it must be diluted with N_2 or $CFCl_3$ [8].

O=S=NCl can be distilled under an inert gas at atmospheric pressure without decomposition [1]. When heated in air it deflagrates with a bluish white flame [1]. It reacts explosively with water [1, 2], and vigorously with mercury [1]. With polar halogenating reagents, cleavage at the S=N bond is the preferred reaction. Thus, the reaction of gaseous O=S=NCl with ClF (mole ratio 1:2) for 5 h in helium atmosphere resulted in the formation of SOF_2, Cl_2, and N_2. The same reaction in the mole ratio 1:1 yielded SOF_2, Cl_2, and a highly explosive material, which was strongly oxidizing. When O=S=NCl was condensed onto $ClOSO_2F$ (mole ratio 1:2) at −196°C, and the mixture allowed to warm to 25°C during 4 h, the main products were $SO(OSO_2F)_2$, Cl_2, and N_2. With the mole ratio 1:1 the same reaction yielded mainly Cl_2, SOF_2, $SO(OSO_2F)_2$, N_2, $S_2O_5F_2$, SO_2, $ClN(SO_2F)_2$, Cl_2NSO_2F. When O=S=NCl was condensed into an equimolar amount of PCl_5 at −83.5°C, and the vessel allowed to warm to 25°C in 24 h, the products were $SOCl_2$, Cl_2, and a mixture of phosphonitrilic chlorides. The mechanism of the reactions and suggested intermediate products are discussed [7]. The reaction with $NPCl_2$ affording cyclic S–N–P compounds is claimed in patents [4, 5]. When a solution of O=S=NCl and $Hg(CF_3)_2$ in CH_2Cl_2 was UV-irradiated the formation of $O{=}S{=}NCF_3$ was observed [15].

O=S=NBr, a colorless liquid, becomes yellow at room temperature owing to loss of bromine [1, 2]. It reacts explosively with water [2].

O=S=NI decomposes on melting at 32°C. It is sensitive toward oxygen and reacts explosively with water [2]. Irradiation of a solution of O=S=NI and CF_3I in CH_2Cl_2 gives I_2 and C_2F_6 [15].

References:

[1] Verbeek, W.; Sundermeyer, W. (Angew. Chem. **81** [1969] 331; Angew. Chem. Intern. Ed. Engl. **8** [1969] 376).

[2] Seppelt, K.; Sundermeyer, W. (Naturwissenschaften **56** [1969] 281/2).

[3] Nachbaur, E.; Kosmus, W.; Krannich, H. J.; Sundermeyer, W. (Monatsh. Chem. **109** [1978] 1211/6).

[4] Otsuka Chemical Co., Ltd. (Japan Kokai Tokkyo Koho 83-194721 [1982/83] 1/7 from C.A. **100** [1984] No. 88135).

[5] Otsuka Chemical Co., Ltd. (Japan Kokai Tokkyo Koho 83-194722 [1982/83] 1/7 from C.A. **100** [1984] No. 177261).

[6] Anonym (Nachr. Chem. Tech. **18** [1970] 318); Leleu, M. J. (Cahiers Notes Doc. No. 94 [1979] 127/32, 128).

[7] Kuta, G. S.; Noftle, R. E. (Intern. J. Sulfur Chem. **8** [1973] 335/40).

[8] Eysel, H. H. (J. Mol. Struct. **5** [1970] 275/82).

[9] Faegri, K., Jr.; Kosmus, W. (J. Mol. Struct. **66** [1980] 217/20).

[10] Oberhammer, H. (Theor. Chim. Acta **26** [1972] 79/86).

[11] Mathies, P.; Sladky, F. O.; Rode, B. M. (J. Mol. Struct. **90** [1982] 335/40 [THEOCHEM 7]).

[12] Oberhammer, H. (Z. Naturforsch. **25a** [1970] 1497/501).

[13] Nachbaur, E.; Prossegger, P. (Monatsh. Chem. **116** [1985] 1385/93).

[14] Siebert, H. (Anwendungen der Schwingungsspektroskopie in der anorganischen Chemie, Springer, Berlin 1966, p. 36 from Ref. [8]).

[15] Leidinger, W.; Sundermeyer, W. (Chem. Ber. **115** [1982] 2892/7).
[16] Sebestyen, A.; Vizi, B. (Magy. Kem. Folyoirat **84** [1978] 282/8; C. A. **89** [1978] No. 169396).

2.6.6 Sulfinyl Organyloxyimides, N-Sulfinyl-O-organyl-hydroxylamines, O=S=NOR, R = CH_3, $CH_2C_6H_5$, $Si(CH_3)_3$

Only a few compounds of this type are described.

O=S=NOCH$_3$ was prepared in 49% yield from CH_3ONH_2 and $ClS(O)OC_2H_5$ (mole ratio 2 : 1) in absolute ether. The colorless liquid with a penetrating odor decomposes with inflammation on heating in a test glass. B.p. 36 to 38°C/53 Torr; $n_D^{20} = 1.4602$ [1].

O=S=NOCH$_2$C$_6$H$_5$ was prepared in 1893 from $C_6H_5CH_2ONH_2$ and $SOCl_2$ in absolute ether in an exothermic reaction. It was purified by vacuum distillation, for it decomposes on distillation at ambient pressure. The light yellow liquid has a weak odor resembling that of $C_6H_5CH_2ONH_2$. B.p. 153 to 154°C/50 Torr. On standing in moist air or in moist ether, it decomposed to give a compound of the composition $2(C_6H_5ONH_2) \cdot SO_2$. On boiling with H_2O it decomposed gradually with formation of H_2SO_3. It readily decomposed by reaction with NaOH solution to form $C_6H_5CH_2ONH_2$ and Na_2SO_3 [2].

O=S=NOSi(CH$_3$)$_3$ was obtained in 53% yield when $(CH_3)_3SiON(Si(CH_3)_3)_2$ reacted with $SOCl_2$ or $ClS(O)OC_2H_5$ at 10 to 20°C. It was separated by vacuum distillation. B.p. 30.5°C/11 Torr. IR (film or KBr): ν (in cm^{-1}) = 2960 m, ν_{as}(CH); 2900 w, ν_s(CH); 2230 vw; 1270 sh; 1257 s, δ_s(CH); 1210 s, ν_{as}(N=S=O); 1033 vs, ν(SiO); 875 s; 855 vs, ϱ_{as}(CH); 777 w; 750 m, ϱ_s(CH); 630 w; 530 m, δ(N=S=O); 432 w.

MS: m/e (rel. int. in %) = 151 (1.3) M^+, 136 (13.3) $C_2H_6SiONSO^+$, 121 (26.6) $CH_3SiONSO^+$, 106 (32.9) $SiONSO^+$, 73 (100) $C_3H_9Si^+$, 59 (5.05) CH_3SiO^+, 46 (16.4) NS^+.

O=S=NOSi$(CH_3)_3$ is a clear liquid with an odor of SO_2. It is very sensitive towards hydrolysis but stable when moisture is excluded [3].

References:

[1] Zinner, G.; Ritter, W. (Arch. Pharm. **296** [1963] 681/8, 682, 685).
[2] Michaelis, A.; Schröter, G. (Ber. Deut. Chem. Ges. **26** [1893] 2155/7).
[3] Lidy, W.; Sundermeyer, W. (Z. Naturforsch. **29 b** [1974] 276).

2.6.7 N-Sulfinyl-sulfenamides

2.6.7.1 Sulfinyl Mercaptoimide, N-Sulfinyl-thiohydroxylamine, O=S=NSH

The compound has not been synthesized to date, but semiempirical MNDO calculations for the electronic structure were made. It was taken as a model substance for a comparison with another model compound, S=S=NSH, see p. 299. Atomic charges and bond orders for the planar model substance are shown in **Fig. 3**.

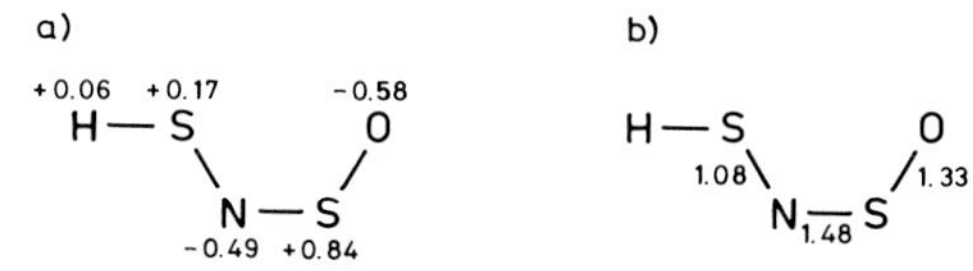

Fig. 3. Atomic charges (a) and bond orders (b) for planar O=S=NSH.

The π orbital energies (in eV) of O=S=NSH are π_1 (HOMO) −15.706; π_2 −12.547; π_3 −9.707; π_4 (LUMO) −1.825. For S=S=NSH the HOMO energy is similar (−15.588 eV), but the LUMO energy is lower (−2.557 eV). Therefore the π(HOMO)→π*(LUMO) transition is expected to occur at higher energy for compounds O=S=NSR compared to S=S=NSR. Indeed, experimental UV absorptions for O=S=NSR were found at about 370 nm (for instance see p. 43), for S=S=NSR above 400 nm.

Reference:

Chivers, T.; Oakley, R. T.; Pieters, R.; Richardson, J. F. (Can. J. Chem. **63** [1985] 1063/7).

2.6.7.2 Disulfur Nitride Oxide Ion (1−), $[O\overset{..}{-}S\overset{..}{-}N\overset{..}{-}S]^-$, or Trisulfur Nitride Oxide Ion (1−), $[O\overset{..}{-}S\overset{..}{-}N\overset{..}{-}S\overset{..}{-}S]^-$

The unstable red anion forms when the electrochemically produced OSN^- anion (see p. 19) [1] or a slurry of K^+OSN^- [2] reacts with elemental sulfur in CH_3CN. When a solution of the OSN^- anion is titrated with a solution of elemental sulfur in CH_3CN the absorption band at 496 nm reaches a maximum for an OSN^- : S ratio of 1:1, and the polarogram of the solutions shows the presence of unreacted sulfur for a 1:2 mole ratio, suggesting that the red species is $OSNS^-$. But there are some doubts about the UV-visible absorption of $OSNS^-$ at 496 nm. The possibility cannot be ruled out that the 496 nm species is $OSNSS^-$ formed by disproportionation of $OSNS^-$ [1].

References:

[1] Chivers, T.; Hojo, M. (Inorg. Chem. **23** [1984] 4088/93, 4092).
[2] Chivers, T.; Edelmann, F.; Schmidt, K. J. (unpublished results in: Chivers, T., Sulfur Rept. **7** [1986] 89/107, 91).

2.6.7.3 Bis(sulfinyl-amino)sulfane, Bis(sulfinyl)-sulfoxylic Diamide, $(O{=}S{=}N)_2S$

The literature on preparation, properties, and reactions of $(O{=}S{=}N)_2S$ through 1960 can be found in "Schwefel" B 3, 1963, pp. 1562/4; see also Goehring, M. (Ergebnisse und Probleme der Chemie der Schwefelstickstoffverbindungen, Akad.-Verlag, Berlin 1957, pp. 1/173, 80/4).

2.6.7.3.1 Preparation. Formation

In 1952 the preparation of $(O{=}S{=}N)_2S$ was described for the first time [1]. It was obtained from S_4N_4 and SO_2 in $SOCl_2$ [1]. It formed also by reacting $SOCl_2$ with dry NH_3 gas in petroleum ether [2]. Later publications described the preparation from S_4N_4 [3, 4, 5] or NH_4Cl [6, 7], and $SOCl_2$. Similarly the reaction of S_4N_4 with $SOCl_2$ and SO_2 [4], or of NH_4Cl with S_2Cl_2 and $SOCl_2$ [6] yielded $(O{=}S{=}N)_2S$. Much higher yields (68%) were attained when (in an atmosphere of dry nitrogen) SCl_2 (0.145 mol) was added slowly to $O{=}S{=}NSi(CH_3)_3$ (0.307 mol) with stirring. The reaction was exothermic, and fine yellow crystals separated. The mixture was heated under reflux for 48 h to ensure complete reaction. Long yellow needles formed on cooling: $2O{=}S{=}N\text{-}Si(CH_3)_3 + SCl_2 \rightarrow (O{=}S{=}N)_2S + 2(CH_3)_3SiCl$. $(CH_3)_3SiCl$ is an excellent recrystallization solvent for $(O{=}S{=}N)_2S$ [8]. When $(CH_3)_3SiN{=}S{=}NSi(CH_3)_3$ was reacted with $SOCl_2$ [9, 10] in C_6H_6 at 60°C for 3 h $(O{=}S{=}N)_2S$ was obtained in 83% yield [10]. Formation of $(O{=}S{=}N)_2S$ was also observed in the reaction of $O{=}S{=}NSi(CH_3)_3$ with $SOCl_2$ [9] or with BrCOF, and of $Hg(N{=}S{=}O)_2$ with ClFSO [16]. The compound was obtained in 66% yield when $((CH_3)_3Si)_2NSO_2N(Si(CH_3)_3)_2$ was dropped into $SOCl_2$ (mole ratio 1:2) in absolute ether at −78°C. After the solution was warmed to room temperature, the solvent was removed in vacuum [9]. The compound formed in 50% yield by reaction of $Cs^+S_3N_3^-$ with $SOCl_2$ (mole ratio ~1:1) in CH_3CN at 23°C for 7 h. Other products are S_4N_4, $S_4N_3^+Cl^-$, and CsCl [11]. It formed from $((C_6H_5)_3P{=})_2N^+S_3N_3^-$ and $SOCl_2$ [12], and was obtained, along with S_3N_2O, S_4N_4, and S_5N_6, in the reaction of $Na^+S_4N_5^-$ with $SOCl_2$ (mole ratio ~1:1) in CH_2Cl_2 at 0°C [13]. The compound was prepared in 67% yield by reacting a mixture of $(BrS)_2N^+AsF_6^-$ and CsF (mole ratio 1:1.8) with an excess of SO_2 at room temperature for 2 d [14]. The formation of small amounts of $(O{=}S{=}N)_2S$ in the reaction of NH_3 with excess SO_2 in sealed tubes is concluded from the appearance of a Raman line at 980 cm^{-1} in the spectrum of the red solid product [15].

$(O{=}S{=}N)_2S$ is generally purified by sublimation [2, 3, 9, 10, 13, 14]. When the compound was held at 45°C under a dynamic vacuum, crystals were formed above the heated zone, which were used for structure determinations [14].

It can be recrystallized from benzene [5, 6] or from $(CH_3)_3SiCl$ [8] to give large yellow rhombic crystals [8].

References:

[1] Goehring, M.; Schuster, D. (Angew. Chem. **64** [1952] 617).
[2] Goehring, M.; Heinke, J. (Z. Anorg. Allgem. Chem. **272** [1953] 297/302).
[3] Goehring, M.; Heinke, J. (Z. Anorg. Allgem. Chem. **278** [1955] 53/7).
[4] Becke-Goehring, M. (Progr. Inorg. Chem. **1** [1959] 207/34, 227).
[5] Banister, A. J.; Padley, J. S. (J. Chem. Soc. A **1967** 1437/9).
[6] Jolly, W. L.; Becke-Goehring, M. (Inorg. Chem. **1** [1962] 76/8).
[7] Becke-Goehring, M.; Magin, G. (Z. Anorg. Allgem. Chem. **340** [1965] 126/32).
[8] Armitage, D. A.; Sinden, A. W. (Inorg. Chem. **11** [1972] 1151/2).
[9] Appel, R.; Montenarh, M. (Chem. Ber. **108** [1975] 2340/8, 2342, 2347).
[10] Zibarev, A. V.; Gurin, G. G.; Yakobson, G. G. (Izv. Akad. Nauk SSSR Ser. Khim. **1985** 2774/80; Bull. Acad. Sci. USSR Div. Chem. Sci. **1985** 2571/6, 2574/5).

[11] Chivers, T.; Rao, M. N. S. (Can. J. Chem. **61** [1983] 1957/62).
[12] Chivers, T.; Hojo, M. (Inorg. Chem. **23** [1984] 4088/93).
[13] Chivers, T.; Proctor, J. (Can. J. Chem. **57** [1979] 1286/93, 1288).
[14] MacLean, G.; Passmore, J.; White, P. S.; Banister, A.; Durrant, J. A. (Can. J. Chem. **59** [1981] 187/90).
[15] Meyer, B.; Mulliken, B.; Weeks, H. (Phosphorus Sulfur **8** [1980] 291/9, 294).
[16] Krannich, J. H.; Sundermeyer, W. (Z. Anorg. Allgem. Chem. **450** [1979] 45/9).

2.6.7.3.2 Lattice Structure. Molecular Structure

$(O{=}S{=}N)_2S$ crystals are described as long yellow needles [1], large rhombic crystals [1], yellow plate-like crystals [5], or longish plates with a flat pyramide on the surface [2].

According to X-ray diffraction studies, $(O{=}S{=}N)_2S$ crystallizes in a monoclinic unit cell, space group C2/c–C^6_{2h} (No. 15). The unit cell contains four $(O{=}S{=}N)_2S$ molecules [3, 4, 5]. Lattice parameters and X-ray density are listed in Table 10, p. 32. Positional parameters [3, 4, 5] and thermal parameters [4] are given in the original papers. The packing of the $(O{=}S{=}N)_2S$ molecules is shown in **Fig. 4**. The molecules are arranged in layers, the twofold rotating axes parallel in the same layer, and antiparallel with respect to the neighbored layers. The intermolecular atomic distances are equal to or longer than the sum of van-der-Waals radii of the concerned atoms [4, 5].

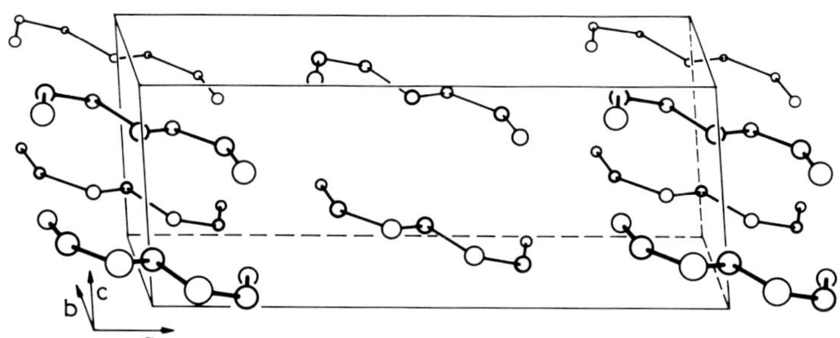

Fig. 4. Unit cell of $(O{=}S{=}N)_2S$ [4].

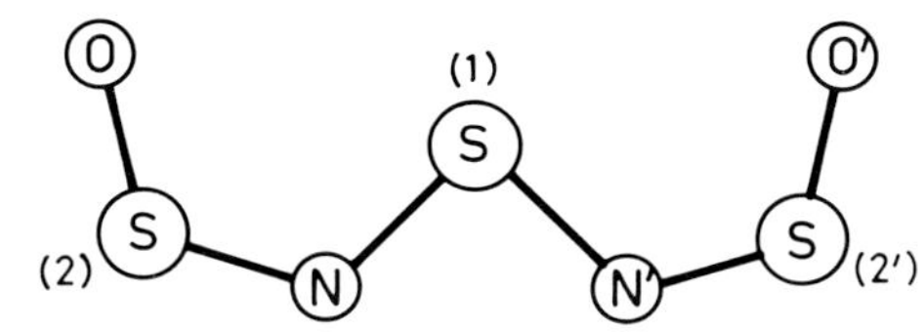

Fig. 5. Molecular structure of $(O{=}S{=}N)_2S$ [4].

The $(O{=}S{=}N)_2S$ molecule is almost planar with a twofold rotating axis, according to C_{2v} symmetry. The N=S=O groups are in cis position with respect to the central S atom, as shown in **Fig. 5** [3, 4, 5]. Atomic distances and bond angles are listed in Table 11, p. 32. The oldest published S–O distance [3] appeared to be too small when compared with similar S–N compounds [6], and was corrected by structure re-investigations [4, 5]. The overall bonding may, to a first order of approximation, be depicted by valence bond structures 1, 2, and 3 [5]:

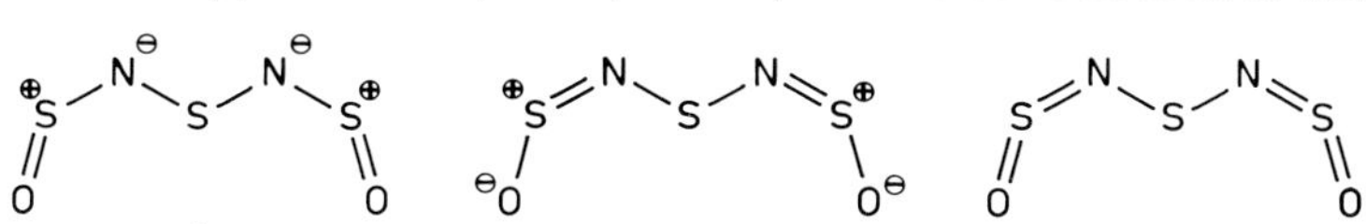

1 2 3

The central S–N bond distance is somewhat less than that of other formally single S–N bonds, and suggests that there is some delocalization of the π system across the central sulfur atom, thus accounting for the planarity of the molecule. Using Pauling's approach, the S(1)–N, N–S(2), and S(2)–O bonds have bond orders 1.37, 1.88, and 1.93 respectively. Planarity and a cis, cis configuration may also result in part as a consequence of weak bonding or electrostatic attraction between the central sulfur atom and terminal oxygen atoms [5]. MNDO calculations give an almost identical minimum of energy for the planar cis, trans configuration. Therefore, the cis, cis configuration observed experimentally to be the only one existing in the solid state, is supposed to be due to packing effects. In solution or in the gas phase the cis, trans configuration should exist also [7]. But in benzene solution the b_2 torsion vibration (SNSO) is the same as in solid $(O{=}S{=}N)_2S$, which means there is no change of conformation on dissolution [4].

Table 10

Lattice Parameters and X-Ray Density of $(O{=}S{=}N)_2S$ (monoclinic symmetry, space group C2/c–C^6_{2h} (No. 15), Z = 4).

a in Å	16.389(3)	16.329(8)	16.52
b in Å	4.5719(6)	4.518(2)	4.56
c in Å	6.790(1)	6.629(4)	6.84
β in °	96.64(2)	97.31(4)	97
V in A^3	505.3	485.1(4)	—
D_x in g/cm³	2.053	2.14	2.01
R	0.057	0.024	0.185
R_w	0.076	—	—
No. of observed reflexes ($I \geq 2\sigma$)	550	408	—
temperature	r.t.	−105°C	r.t.
Ref.	[5]	[4]	[3]

Table 11

Atomic Distances and Bond Angles of $(O{=}S{=}N)_2S$.

atomic distance in Å			
S(1)–N	1.647(3)	1.657(2)	1.69
N–S(2)	1.535(3)	1.534(2)	1.58
S(2)–O	1.446(4)	1.465(2)	1.37
S(1)–S(2)	2.8132(9)	2.813(1)	2.83
S(1)–O	3.142(4)	3.129(2)	—
bond angle in °			
O–S(2)–N	118.0(2)	117.5(1)	115.3
S(1)–N–S(2)	124.3(2)	123.6(2)	120
N–S(1)–N	97.3(2)	97.2(1)	95.3
temperature	r.t.	−105°C	r.t.
Ref.	[5]	[4]	[3]

References:

[1] Armitage, D. A.; Sinden, A. W. (Inorg. Chem. **11** [1972] 1151/2).
[2] Goehring, M.; Heinke, J. (Z. Anorg. Allgem. Chem. **272** [1953] 297/302).
[3] Weiss, J. (Z. Naturforsch. **16b** [1961] 477).
[4] Steudel, R.; Steidel, J.; Rautenberg, N. (Z. Naturforsch. **35b** [1980] 792/6).
[5] MacLean, G.; Passmore, J.; White, P. S.; Banister, A.; Durrant, J. A. (Can. J. Chem. **59** [1981] 187/90).

[6] Banister, A. J.; Durrant, J. A. (J. Chem. Res. S **1978** 150/1; J. Chem. Res. M **1978** 1912/30, 1929).
[7] Gleiter, R.; Bartetzko, R. (Z. Naturforsch. **36b** [1981] 492/7).

2.6.7.3.3 Nuclear Magnetic Resonance

^{14}N NMR (neat liquid/aq. NH_4^+, 100°C): $\delta = 282$ ppm (referred to the CH_3NO_2 scale: $\delta = -77.5$ ppm).

Reference:

Armitage, D. A.; Mason, J.; Vinter, J. G. (Inorg. Chem. **17** [1978] 776/7).

2.6.7.3.4 Molecular Vibrations

The $(O{=}S{=}N)_2S$ molecule possesses C_{2v} symmetry and therefore 15 fundamental vibrations are expected, with the symmetry classes: $6a_1$, $2a_2$, $5b_1$, $2b_2$. All of these vibrations are Raman-allowed, but only a_1, b_1, and b_2 vibrations are IR-active.

One group of authors made an assignment of the measured vibrational spectra by comparison with the wavenumbers calculated from Urey-Bradley force constants [1], see Table 12. In the range 120 to 300 cm^{-1} the three in-plane bending modes of the central part of the molecule appear; the assignment to special vibrations is difficult. In the range <120 cm^{-1}, lattice vibrations should exist in addition to torsional vibrations, and the assignment was not possible [1].

Independently an IR absorption at 1185 cm^{-1} (Nujol) was assigned to ν(SO), and a Raman line at 1039 cm^{-1} to ν(SN) [2]; other IR bands (e.g. at 1036 and 993 cm^{-1}, and nine further bands) and Raman lines (e.g. at 1199 and 996 cm^{-1}, and 18 further lines) are listed in the paper [2]. Previously an IR spectrum (Nujol) with eight bands (e.g. at 1184, 1040, and 998 cm^{-1}) [3] and IR absorptions (in CS_2) at 1184 and 988 cm^{-1} were reported [4].

Table 12
IR and Raman Spectra of $(O{=}S{=}N)_2S$ (frequencies in cm^{-1}; relative Raman intensities in parentheses) [1] (ν = stretching, δ = bending, τ = torsion vibration).

Raman (−100°C) crystal	in Nujol	IR (25°C) in CS_2	in c-C_6H_{12}	assignment
40 (50)				lattice and torsional vibrations
66 (20)				
94 (6)				
113 (4)				in-plane deformation vibrations
145 (1)				
172 (1)				
238 (14)				
		292vw		
	372sh			

Table 12 (continued)

Raman (−100°C) crystal	in Nujol	IR (25°C) in CS_2	in c-C_6H_{12}	assignment
385 (0.5)	377 s	380 sh	380 m	b_2, τ(SNSO)
		508 vvw		
514 (5)	504 w	510 vvw		a_1 } δ(NSO) and ν(SN)
554 (0.5)	556 w	552 m		b_1 } δ(NSO) and ν(SN)
661 (0.5)	655 w	665 vw, sh		b_1 } δ(NSO) and ν(SN)
692 (8)	685 m	691 m	695 w	a_1 } δ(NSO) and ν(SN)
		715 ww		
725 (1) br		733 vvw		
998 (1)	998 w	992 w to m		b_1, ν(S=N)
	1020			
1040 (100)	1042 m	1038 vw		a_1, ν(S=N)
1165 (1)		1190 vs	1196 sh } 1200 s }	b_1, ν(SO)
1199 (8)	1193 vs, br	1208 s	1214 m	a_1, ν(SO)

References:

[1] Steudel, R.; Steidel, J.; Rautenberg, N. (Z. Naturforsch. **35b** [1980] 792/6).
[2] MacLean, G.; Passmore, J.; White, P. S.; Banister, A.; Durrant, J. A. (Can. J. Chem. **59** [1981] 187/90).
[3] Chivers, T.; Proctor, J. (Can. J. Chem. **57** [1979] 1286/93, 1288).
[4] Armitage, D. A.; Sinden, A. W. (Inorg. Chem. **11** [1972] 1151/2).

2.6.7.3.5 Electronic Structure

Core Levels. The ionization energies, E_i, of N 1s and S 2p electrons of a thin film of $(O{=}S{=}N)_2S$ on Au were determined from X-PES obtained by use of Al Kα radiation. The results were compared with theoretical CNDO predictions. The experimental E_i values of the core electrons are given below [1]:

E_i N 1s	E_i S(2)*⁾ 2p	E_i S(1)*⁾ 2p
399.0 eV	164.3 eV	162.6 eV

*⁾ Enumeration of the atoms as on p. 31.

The ionization spectra of S 2p orbitals give evidence for two sets of inequivalent sulfur atoms with area ratios 1:2. The band of N 1s ionization has a width of only 1.7 eV, characteristic of the presence of only one type of nitrogen atoms [1], in agreement with the molecular structure.

Valence Orbitals. The UV photoelectron spectrum (UPS) of gaseous $(O{=}S{=}N)_2S$ was measured by use of HeI radiation by two groups of authors with different assignments, see Table 13. One UPS was interpreted by comparison with the known and assigned PES of O=S=NH, as well as with CNDO/2 calculations of the sequence of valence orbital data for O=S=NH and $(O{=}S{=}N)_2S$ [1]. The assignment of the other UPS was done by comparison with calculated (MNDO) orbital energies [2].

Table 13

Experimental Ionization Energies, E_i, from UPS for $(O{=}S{=}N)_2S$, Molecular Symmetry C_{2v}, Assignments from MO Calculations.

E_i a) in eV	assignment [1]	E_i in eV	assignment [2]
9.41	$b_2(p_zS(1))$ b)	9.25	$b_1(\pi)$
11.11	$b_1(\sigma_{SN})$	11.08	$a_2(\pi)$
11.26	$a_1(p_yS(1), \pi_{SN})$	11.38	$a_1(n)$
11.50	$a_2(\pi_{NSO})$	11.60	$b_2(n)$
11.78	b_1(mixed)	—	
12.66	$a_1(\sigma$ mixed)	12.51	$a_1(n)$
13.66	$b_2(\pi_{SN})$		
14.56	a_1(mixed)		
14.85	b_1(mixed)		
16.35	$a_2(\pi, n_O)$		
~18	$b_2(\pi_{NSO})$		

a) Five calculated (CNDO) higher E_i values in the paper [1]. – b) 3p-(out-of-plane) orbital of the S(1) atom.

Excited Levels. $(O{=}S{=}N)_2S$ in CH_3CN solution (1.0 mM) exhibits a UV absorption $\lambda_{max} = 365$ nm [3]. The position of the UV maximum depends on the solvent. From a diagram, the maximum may be estimated to be $\lambda_{max} \approx 380$ nm in heptane and dioxane and $\lambda_{max} \approx 350$ nm in ether, $\log \varepsilon = 4$ [4].

References:

[1] Andreocci, M. V.; Bossa, M.; DiCastro, V.; Furlani, C.; Mattogno, G.; Roesky, H. W. (Z. Phys. Chem. [N.F.] **118** [1979] 137/50, 139, 142).
[2] Gleiter, R.; Bartetzko, R. (Z. Naturforsch. **36b** [1981] 492/7, 496).
[3] Chivers, T.; Hojo, M. (Inorg. Chem. **23** [1984] 4088/93, 4092).
[4] Goehring, M.; Heinke, J. (Z. Anorg. Allgem. Chem. **272** [1953] 297/302, 301).

2.6.7.3.6 Thermal Properties. Enthalpy of Formation

The melting point is 100.8°C [1], 100.7°C [2, 3], 100 to 101°C [4], 99 to 100°C [5, 6], 99°C [7], and at somewhat lower temperature, 95 to 96°C [8].

Sublimation was performed at 35°C/0.01 Torr [2] and at 25 to 30°C/0.01 Torr [8].

The enthalpy of formation from atoms of gaseous $(O{=}S{=}N)_2S$ $\Delta H_f = 534.5$ kcal/mol was calculated from the measured heat of combustion [9].

References:

[1] Banister, A. J.; Padley, J. S. (J. Chem. Soc. A **1967** 1437/9).
[2] Goehring, M.; Heinke, J. (Z. Anorg. Allgem. Chem. **272** [1953] 297/302).
[3] Becke-Goehring, M. (Progr. Inorg. Chem. **1** [1959] 207/34, 228).
[4] Armitage, D. A.; Sinden, A. W. (Inorg. Chem. **11** [1972] 1151/2).
[5] Zibarev, A. V.; Gurin, G. G.; Yakobson, G. G. (Izv. Akad. Nauk SSSR Ser. Khim. **1985** 2774/80; Bull. Acad. Sci. USSR Div. Chem. Sci. **1985** 2571/6).
[6] Jolly, W. L.; Becke-Goehring, M. (Inorg. Chem. **1** [1962] 76/8).
[7] Appel, R.; Montenarh, M. (Chem. Ber. **108** [1975] 2340/8, 2342).
[8] Chivers, T.; Proctor, J. (Can. J. Chem. **57** [1979] 1286/93, 1288).
[9] Fleig, H.; Becke-Goehring, M. (Z. Anorg. Allgem. Chem. **375** [1970] 8/14, 11).

2.6.7.3.7 Solubility

$(O{=}S{=}N)_2S$ is soluble in organic solvents such as benzene, nitrobenzene, heptane, petroleum ether, alcohols. The solutions are stable if the solvent ist completely dry [1]. Other solvents used in reactions with $(O{=}S{=}N)_2S$ are: dichloromethane [2, 3], 1,2-dichloroethane, nitromethane, acetonitrile [4], toluene [5, 6].

References:

[1] Goehring, M.; Heinke, J. (Z. Anorg. Allgem. Chem. **272** [1953] 297/302).
[2] Roesky, H. W.; Schaper, W. (Chem. Ber. **107** [1974] 3451/3).
[3] Roesky, H. W.; Anhaus, J.; Sheldrick, W. S. (Inorg. Chem. **23** [1984] 75/9).
[4] Becke-Goehring, M.; Magin, G. (Z. Anorg. Allgem. Chem. **340** [1965] 126/32).
[5] Roesky, H. W.; Thomas, M.; Jones, P. G.; Pinkert, W.; Sheldrick, G. M. (J. Chem. Soc. Dalton Trans. **1983** 1211/3).
[6] Roesky, H. W.; Thomas, M.; Bats, J. W.; Fuess, H. (Inorg. Chem. **22** [1983] 2342/3).

2.6.7.3.8 Electrochemical Behavior

The voltammogram of $(O{=}S{=}N)_2S$ (0.21 mM) in 0.1M $(C_2H_5)_4NClO_4$–CH_3CN on a dropping mercury electrode gave three cathodic waves. The first wave was poorly defined, but the second and third waves at $E_{1/2} = -1.50$ and -2.90 V (vs. Ag|0.1M $AgClO_4$–CH_3CN) were well behaved. On the rotating platinum electrode two waves were observed, at −1.21 and −1.62 V.

Electrolysis of $(O{=}S{=}N)_2S$ (0.20 mM) on a platinum-gauze electrode at the first reduction step (−1.4 V) resulted in the uptake of one electron/molecule. The electrolyzed solution exhibited a strong absorption band at 330 nm. (The same species was also produced when a solution of $(O{=}S{=}N)_2S$ in CH_3CN was allowed to stand at 23°C for 3 to 4 h.) Coulometric reduction at −2.0 V on a platinum-gauze electrode consumed more than two electrons/molecule $(O{=}S{=}N)_2S$, and produced $[O{=}S{=}N]^-$.

Reference:

Chivers, T.; Hojo, M. (Inorg. Chem. **23** [1984] 4088/93, 4092).

2.6.7.3.9 Chemical Behavior

$(O{=}S{=}N)_2S$ is stable at room temperature when kept in dry air, dry nitrogen, or in an atmosphere of dry SO_2 [1]. On heating to 80°C it turns red. On heating above the melting point the substance appeared to boil [1, 2, 3], and then a yellow vapor evolved which ignited at 300°C [1]. Under ultrahigh vacuum $(O{=}S{=}N)_2S$ does not decompose when heated up to 400°C, and is deposited unchanged in the cooling trap [4].

The heat of combustion in O_2 according to $(O{=}S{=}N)_2S + 7/2\,O_2 + aq \rightarrow 3\,H_2SO_4 \cdot aq + N_2$ was measured in an O_2-filled calorimeter (30 atm O_2): $Q_v = 415 \pm 3.5$ and $Q_p = 416.5$ kcal/mol [5]. An exothermic reaction takes place with liquid chlorine with evolution of SO_2. In addition to $SOCl_2$, the oxocyclotrithiazenes $[(NSCl)_2(NSOCl)]$ [11, 12], and in small amounts $[(NSCl)(NSOCl)_2]$ and $[NSOCl]_3$ form [12] (the compounds are described in "Schwefel-Stickstoff-Verbindungen" Teil 1, 1977, pp. 16, 22). The reports on the behavior of $(O{=}S{=}N)_2S$ towards hydrolysis are contradictory. Thus, it was reported that $(O{=}S{=}N)_2S$ is not wetted by water [1, 2], and hydrolyzed only on long contact [1]. But when exposed to moist air or moist nitrogen, decomposition to S_4N_4 and SO_2 occurs [1, 2], and even sensitivity to traces of moisture is reported [3]. Caustic soda causes rapid hydrolysis according to $(O{=}S{=}N)_2S + 4\,H_2O \rightarrow (NH_4^+)_2S_3O_6^{2-}$ [1, 2].

HI dissolved in dry HCOOH is oxidized by $(O{=}S{=}N)_2S$ to form 3 I_2, 2 NH_3, 2 S, and SO_3 [1].

S_2Cl_2 in CH_3CN or in CH_3NO_2 reacts with $(O{=}S{=}N)_2S$ with formation of small green-black crystals of $(S_3N_2Cl)_2$. When a mixture of $(O{=}S{=}N)_2S$ and S_2Cl_2 is refluxed in CH_3CN, yellow $S_4N_3^+Cl^-$ precipitates on cooling [6]. Condensation of SO_3 on solid $(O{=}S{=}N)_2S$ yields a black product of the composition $(O{=}S{=}N)_2S \cdot SO_3$ [14] which decomposes with oxidation to $S_3N_2O_5$ [2, 14].

It reacts with PCl_5 in CCl_4 at 35 to 38°C to form the heterocyclic compound $SN_3P_2Cl_5$ [13] ($(NSCl)(NPCl_2)_2$, described in "Sulfur-Nitrogen Compounds" Part 3, 1987, p. 24).

In the reaction with $(CH_3)_2BSCH_3$ the SN group attacks the B–S bond [15]: $(O{=}S{=}N)_2S + 6(CH_3)_2BSCH_3 \rightarrow O(N(SSCH_3)(B(CH_3)_2))_2 + 2\ CH_3SSCH_3 + 2((CH_3)_2B)_2O$.

The reaction of $(O{=}S{=}N)_2S$ with N-(trimethylsilyl)-dialkyl amines, $(CH_3)_3SiNR_2$ ($R = CH_3$, C_2H_5), in dry boiling CH_2Cl_2 can be used for the preparation of N-sulfinyl compounds: $(O{=}S{=}N)_2S + (CH_3)_3SiNR_2 \rightarrow O{=}S{=}NSNR_2 + O{=}S{=}NSi(CH_3)_3$. By-products are S_4N_4, $(CH_3)_3SiOSi(CH_3)_3$, and $(CH_3)_3SiN{=}S{=}NSNR_2$ [7].

The reactions with the Lewis acids $SbCl_5$ or $TiCl_4$ yield adducts of the cyclic SN compounds S_4N_4 and S_2N_2. Thus, $(O{=}S{=}N)_2S$ reacts with $SbCl_5$ or $TiCl_4$ in dry 1,2-dichloroethane to form the adducts $S_4N_4 \cdot SbCl_5$ and $S_4N_4 \cdot 2TiCl_4$, respectively (see "Sulfur-Nitrogen Compounds" Part 2, 1985, pp. 212, 213, 217) [6]. When $(O{=}S{=}N)_2S$ is reacted with $TiCl_4$ in CH_2Cl_2 another adduct, $S_2N_2 \cdot TiCl_4$ is obtained [4] (see "Sulfur-Nitrogen Compounds" Part 2, 1985, p. 21). Under identical conditions the reaction of $(O{=}S{=}N)_2S$ with $SeCl_4$, as well as with $SnCl_4$, did not result in the analogous S_2N_2 complexes [4]. $(O{=}S{=}N)_2S$ reacts with $((C_6H_5)_3P)_2Pt(\eta^2\text{-}C_2H_4)$ (mole ratio 1:1.5) in toluene at room temperature under evolution of gas (SO_2 and C_2H_4) and formation of a yellow precipitate [8]:

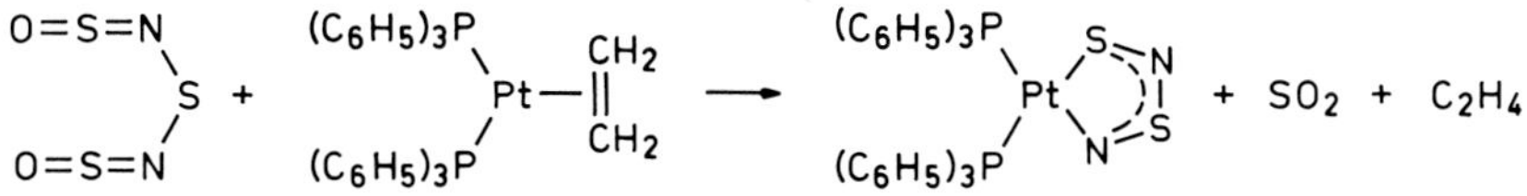

When $(O{=}S{=}N)_2S$ reacts with $[Zn(SO_2)_2]^{2+}(AsF_6^-)_2$ (mole ratio 2:1) in liquid SO_2 in a pressure flask at room temperature $[Zn((O{=}S{=}N)_2S)_2]^{2+}(AsF_6^-)_2 \cdot 2SO_2$ is formed [9] (see p. 264). With $AgAsF_6$ under similar conditions the Ag(I) complex $[Ag_4((O{=}S{=}N)_2S)_9]^{4+}(AsF_6^-)_4 \cdot SO_2$ forms [10] (see p. 266).

References:

[1] Goehring, M.; Heinke, J. (Z. Anorg. Allgem. Chem. **272** [1953] 297/302).
[2] Becke-Goehring, M. (Progr. Inorg. Chem. **1** [1959] 207/34, 228/9).
[3] Armitage, D. A.; Sinden, A. W. (Inorg. Chem. **11** [1972] 1151/2).
[4] Roesky, H. W.; Anhaus, J.; Sheldrick, W. S. (Inorg. Chem. **23** [1984] 75/9).
[5] Fleig, H.; Becke-Goehring, M. (Z. Anorg. Allgem. Chem. **375** [1970] 8/14, 10).
[6] Becke-Goehring, M.; Magin, G. (Z. Anorg. Allgem. Chem. **340** [1965] 126/32).
[7] Roesky, H. W.; Schaper, W. (Chem. Ber. **107** [1974] 3451/3).
[8] Chivers, T.; Edelmann, F.; Behrens, U.; Drews, R. (Inorg. Chim. Acta **116** [1986] 145/51, 146, 149).
[9] Roesky, H. W.; Thomas, M.; Bats, J. W.; Fuess, H. (Inorg. Chem. **22** [1983] 2342/3).
[10] Roesky, H. W.; Thomas, M.; Jones, P. G.; Pinkert, W.; Sheldrick, G. M. (J. Chem. Soc. Dalton Trans. **1983** 1211/3).

[11] Schläfer, D.; Becke-Goehring, M. (Z. Anorg. Allgem. Chem. **362** [1968] 1/6).
[12] Weiss, J.; Mews, R.; Glemser, O. (Inorg. Nucl. Chem. Herbert H. Hyman Mem. Vol. **1976** 213).
[13] Roesky, H. W. (Angew. Chem. **84** [1972] 685; Angew. Chem. Intern. Ed. Engl. **11** [1972] 642/3).
[14] Goehring, M.; Hohenschutz, H.; Appel, R. (Z. Naturforsch. **9b** [1954] 678/81).
[15] Meller, A.; Maringgele, W.; Fetzer, H. (Chem. Ber. **113** [1980] 1950/61, 1951/2).

2.6.7.4 N-Sulfinyl-methanesulfenamide, Sulfinyl Methanethioimide, $O{=}S{=}NSCH_3$

The compound was prepared in 69% yield by slowly dropping $SOCl_2$ into a solution of $H_3CSN(Si(CH_3)_3)_2$ in ether at −60°C (mole ratio ~1:1).

$$H_3CSN(Si(CH_3)_3)_2 + SOCl_2 \rightarrow O{=}S{=}NSCH_3 + 2(CH_3)_3SiCl$$

The yellow oil can be distilled at 54 to 55°C/12 Torr; m.p. −50°C. ^{1}H NMR (10% CCl_4 solution/internal TMS): $\delta = 2.89$ ppm(s) [1] or 2.88 ppm [2]; ^{13}C NMR (solvent?/TMS): $\delta = 24.80$ ppm [2]. IR (film, absorption bands were tentatively assigned): ν (in cm^{-1}) = 2990w and 2920w, ν(CH); 1430m, $\delta_{as}(CH_3)$; 1315m, $\delta_s(CH_3)$; 1185s, ν_{as}(N=S=O); 1010s, ν_s(N=S=O); 950s, $\varrho(CH_3)$; 720m, ν(SN); 675s, ν(SC); 540s, δ(N=S=O) [1].

The yellow liquid discolors when exposed to light. It is exceptionally stable towards hydrolysis. With 2,3-dimethylbutadiene, no addition reaction occurs, even at raised temperature [1]. It is alkylated to form $O{=}S{=}NS(CH_3)_2^+\,BF_4^-$ by $(CH_3)_3O^+BF_4^-$ in the absence of a solvent (see below) [3]. It has an irritating effect on the eyes, even in high dilution [1].

References:

[1] Scherer, O. J.; Schmitt, R. (Chem. Ber. **101** [1968] 3302/12, 3307/8, 3312).
[2] Schwöbel, A. (unpublished results in: Bussas, R.; Kresze, G.; Münsterer, H.; Schwöbel, A., Sulfur Rept. **2** [1983] 215/377, 236).
[3] Rössert, M.; Kraus, W.; Kresze, G. (Tetrahedron Letters **1978** 4669/70).

2.6.7.5 Sulfinylamino-dimethylsulfonium Tetrafluoroborate, $O{=}S{=}NS(CH_3)_2^+$ BF_4^-

The salt was formed by alkylation of N-sulfinylmethane-sulfenamide, $O{=}S{=}NSCH_3$, with $(CH_3)_3O^+BF_4^-$ without solvent.

Yield 80%; solid, m.p. 49.5 to 50.5°C (from CH_2Cl_2–CH_3CN).

The compound reacts (fast and slightly exothermically) with the dienes, $CH_2{=}CR^1CR^2{=}CHR^3$ ($R^1 = R^2 = H$, $R^3 = CH_3$; $R^1 = CH_3$, $R^2 = R^3 = H$; $R^1 = R^2 = CH_3$, $R^3 = H$) in CH_3CN to give the corresponding N-dimethylsulfonio-3,6-dihydro-1-oxo-1,2-thiazine tetrafluoroborates (I).

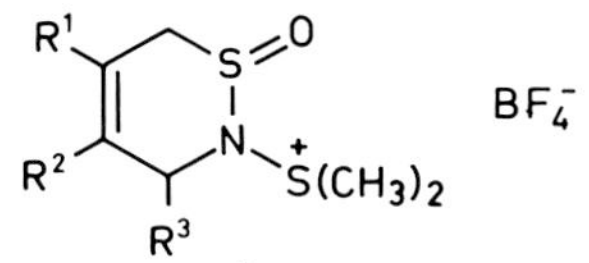

I

The compound undergoes "quasi-Wittig" reactions with sulfinyl compounds $R^1R^2S{=}O$ ($R^1 = R^2 = CH_3$; $R^1 = CH_3$, $R^2 = N(CH_3)_2$; $R^1 = R^2 = N(CH_3)_2$) to give (in CH_3CN at room temperature) the corresponding N-dimethylsulfonio-sulfimine tetrafluoroborates $[R^1R^2S{=}NS(CH_3)_2]^+BF_4^-$ in a yield of 75 to 82%. With C_6H_5CHO in CH_3CN, $[C_6H_5CH{=}NS(CH_3)_2]^+BF_4^-$ is formed.

Reference:

Rössert, M.; Kraus, W.; Kresze, G. (Tetrahedron Letters **1978** 4669/70).

2.6.7.6 N-Sulfinyl-trifluoromethanesulfenamide, $O{=}S{=}NSCF_3$

CF_3SNH_2 (98.4 mmol) was condensed onto NaF (381.0 mmol) in an autoclave, mixed by briefly thawing and shaking, and then reacted with SOF_2 (116.2 mmol) at 130°C for 6 h. Distillation into a trap cooled by liquid air, followed by fractional distillation to remove CF_3SSCF_3 and SOF_2, yielded colorless $O{=}S{=}NSCF_3$ in the −70°C fraction; yield 68%. The compound also formed slowly along with other products, when $CF_3SN{=}SF_2$ was allowed to stand in a sealed glass tube by reaction with SiO_2 of the glass at 20°C: $2CF_3SN{=}SF_2 + SiO_2 \rightarrow 2O{=}S{=}NSCF_3 + SiF_4$.

The colorless liquid vaporizes easily. M.p. −72°C. ^{19}F NMR ($CFCl_3$ solvent and internal standard): $\delta = -51.5$ ppm at 30°C. IR (capillary): $\nu_{as}(O{=}S{=}N) = 1129\ cm^{-1}$, $\nu_s(O{=}S{=}N) = 1025\ cm^{-1}$, $\nu(C{-}F) = 1095\ cm^{-1}$; seven other unassigned frequencies are given in the paper.

$O{=}S{=}NSCF_3$ decomposes at 20°C, to form a yellow unidentified solid, vaporous products, and CF_3SSCF_3 as the only identified substance.

Reference:

Haas, A.; Schott, P. (Chem. Ber. **101** [1968] 3407/18, 3411, 3416/7).

2.6.7.7 N-Sulfinyl-heptafluoro-2-propanesulfenamide, $O{=}S{=}NSC_3F_7$-i

The compound formed along with other products when FSN was heated with an equimolar amount of $CF_3CF{=}CF_2$ in the presence of CsF at 130°C for 16 h. It resulted from the secondary reaction of i-$C_3F_7SN{=}SF_2$ with SiO_2 of the glass vessel or from hydrolysis. It was separated by gas chromatography.

B.p. 112°C/750 Torr; ^{19}F NMR ($CFCl_3$ internal standard): $\delta = -74.0$ ppm $(CF_3)_2$, -173.5 ppm (CF), $^3J(CF_3, CF) = 10.0$ Hz. The IR spectrum (KBr) with 14 bands between 1290 and 480 cm^{-1} is listed in the original paper.

MS: m/e (rel. int. in %) = 263 (100) M^+, 113 (16) CF_3CS^+, 100 (15) CF_3CF^+, 94 (90) $SNSO^+$, 78 (25) SNS^+, 69 (56) CF_3^+, 63 (23) CFS^+, 48 (35) SO^+, 46 (70) SN^+.

Reference:

Bludssus, W.; Mews, R.; Glemser, O.; Alange, G. G. (Israel J. Chem. **17** [1978] 137/40).

2.6.7.8 N-Sulfinyl-trichloromethanesulfenamide, O=S=NSCCl$_3$

The compound was obtained in 65% yield by reaction of $HN(SCCl_3)_2$, with an excess of $SOCl_2$ in the presence of catalytic quantities of $(C_4H_9)_4NCl$ under reflux for 18 h with evolution of HCl. Distillation in vacuum gave pure $O{=}S{=}NSCCl_3$ as a yellow liquid.

B.p. 57°C/1 Torr; $n_D^{20} = 1.5958$; IR (NaCl): 1205 vs and 1025 cm^{-1} w, ν(S=O); ^{13}C NMR ($CDCl_3$/TMS): $\delta = 97.0$ ppm. In the mass spectrum the molecule ion appears at m/e = 210.851.

Reference:

Trompen, W.P.; Geevers, J. (Recl. Trav. Chim. **95** [1976] 106/8).

2.6.7.9 N-Sulfinyl-pentachloroethanesulfenamide, O=S=NSC$_2$Cl$_5$

The compound was obtained in 59% yield by treating a solution of N-(chlorosulfinyl)-imidazole in CH_2Cl_2 at −40°C with an equimolar solution of $C_2Cl_5SNH_2$ in CH_2Cl_2. After 30 min stirring at 20°C the compound was separated by distillation:

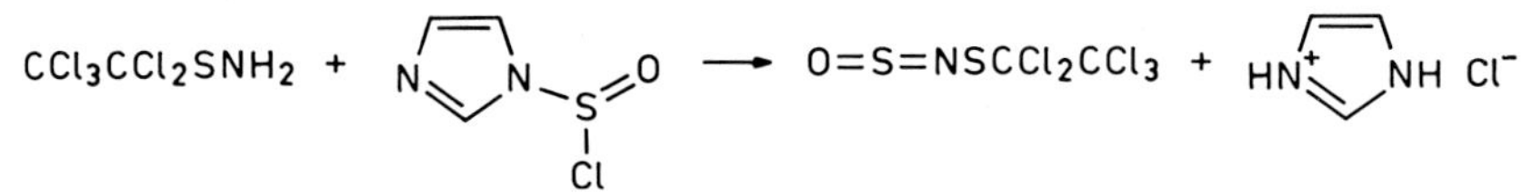

Yellow oil, b.p. 80 to 86°C/0.4 Torr; solidifies in the refrigerator. ^{13}C NMR ($CDCl_3$/Cr(CH_3C-(O)CHC(O)CH_3)$_3$/TMS): δ (in ppm) = 102.36, 103.62 (first cis, trans isomer?); 102.54, 104.39 (second cis, trans isomer?). UV (c-C_6H_{12}): λ_{max} = 250 and 320 nm. IR (neat): ν (in cm^{-1}) = 1190, 780, 755, 665.

MS (70 eV): m/e = 293 M^+, 199 $C_2Cl_5^+$, 176 $CCl_2NOS_2^+$, 164 $C_2Cl_4^+$, 129 $C_2Cl_3^+$, 117 CCl_3^+, 94 NOS_2^+, 78 NS_2^+, 46 NS^+.

Reference:

Boureghda, A.; Abdal-Megeed, M. F.; Ghattas, A.-B. A. G.; Jensen, B.; Senning, A. (Sulfur Letters **5** [1987] 159/63).

2.6.7.10 N-Sulfinyl-benzenesulfenamide, O=S=NSC$_6$H$_5$

The compound was obtained in 85% yield when C_6H_5SCl was added slowly to an equimolar amount of $O{=}S{=}NSi(CH_3)_3$ with cooling, and then the mixture heated to 130°C. The by-product $(CH_3)_3SiCl$ was isolated, and vacuum distillation of the residual red oil gave $O{=}S{=}NSC_6H_5$ as an orange liquid [1]. The reaction of $O{=}S{=}NSi(CH_3)_3$ with $C_6H_5SOCH_3$ (mole ratio 1:1) gave the title compound in 66% yield [2]. Another method of preparation is the reaction of $SOCl_2$ with $((CH_3)_3Si)_2NSC_6H_5$ [1].

B.p. 80°C/0.2 Torr [1] or 72 to 74°C/0.05 Torr [2].

^{14}N NMR (neat/NH_4^+ aq.): $\delta = 296(4)$ ppm [3], referred to the CH_3NO_2 scale: $\delta = -63.5$ ppm. ^{13}C NMR ($CDCl_3$/TMS): δ (in ppm) = 135.6 (C-1); 125.5 (ortho-C), ^1J (^{13}C, ^1H) = 268.1 Hz; 129.4 (meta-C), ^1J (^{13}C, ^1H) = 257.9 Hz; 128.5 (para-C), ^1J(^{13}C, ^1H) = 257.3 Hz [4].

IR: 1195 cm^{-1} ν_{as}(O=S=N) and 1000 cm^{-1} ν_s(O=S=N) [1].

Mass spectrum (70 eV): the molecule ion (m/e = 171) and 33 other peaks are observed. A fragmentation scheme including cyclization to O–S–N–S–C rings is suggested [5].

$O{=}S{=}NSC_6H_5$ reacts exothermically with an equimolar amount of $(CH_3)_3SiSC_6H_5$ giving $O{=}S{=}NSi(CH_3)_3$ and $(C_6H_5S)_2$. Analogously, it reacts with $(C_2H_5)_3SiSC_6H_5$ on heating at 170°C, and with $(CH_3)_3GeSC_6H_5$ on heating at 90°C to form $O{=}S{=}NSi(C_2H_5)_3$ and $O{=}S{=}NGe(CH_3)_3$, respectively, and $(C_6H_5S)_2$ [1].

References:

[1] Armitage, D. A.; Clark, M. J. (J. Organometal. Chem. **24** [1970] 623/31).
[2] Armitage, D. A.; Clark, M. J.; Kinsey, A. C. (J. Chem. Soc. C **1971** 3867/9).
[3] Armitage, D. A.; Mason, J.; Vinter, J. G. (Inorg. Chem. **17** [1978] 776/7).
[4] Brands, G.; Golloch, A. (Z. Naturforsch. **38b** [1983] 326/31).
[5] Brands, G.; Golloch, A. (Phosphorus Sulfur **13** [1982] 211/34, 224, 232).

2.6.7.11 N-Sulfinyl-2-nitro-benzenesulfenamide, $O{=}S{=}NSC_6H_4NO_2$-2

The compound was prepared in 90% yield by sulfinylation of the corresponding amine. Thus, equimolar amounts of $2\text{-}NO_2C_6H_4SNH_2$ and $O{=}S{=}NSO_2C_6H_4CH_3$-4 were refluxed in benzene. Removal of $4\text{-}CH_3C_6H_4SO_2NH_2$ and of the solvent yielded the raw compound, which was recrystallized twice from acetone.

The yellow crystals melt at 127 to 128°C.

The compound is relatively insensitive to moisture. It hydrolyzes in the presence of acid with evolution of SO_2 [1]. $O{=}S{=}NSC_6H_4NO_2$-2 reacts with PCl_5 (mole ratio ~1:1) when refluxed in CCl_4 for 30 min with exchange of oxygen to give unstable $2\text{-}NO_2C_6H_4SN{=}SCl_2$ and $POCl_3$. An exchange reaction occurs with $S({=}NSO_2C_6H_4CH_3\text{-}4)_2$ in dry benzene in the presence of pyridine (refluxing for 6 h, 10% yield) [2]: $O{=}S{=}N{-}S{-}R + R'{-}SO_2{-}N{=}S{=}N{-}SO_2{-}R' \rightarrow O{=}S{=}N{-}SO_2{-}R' + R'{-}SO_2{-}N{=}S{=}N{-}S{-}R$; $R = C_6H_4NO_2$-2, $R' = C_6H_4CH_3$-4.

References:

[1] Kresze, G.; Maschke, A.; Albrecht, R.; Bederke, K.; Patzschke, H. P.; Smalla, H.; Trede, A. (Angew. Chem. **74** [1962] 135/44, 144; Angew. Chem. Intern. Ed. Engl. **1** [1962] 89/98, 97).
[2] Buchholt, H. C.; Senning, A.; Kelly, P. (Acta Chem. Scand. **23** [1969] 1279/87).

2.6.7.12 N-Sulfinyl-pentafluorobenzenesulfenamide, $O{=}S{=}NSC_6F_5$

For preparation of the title compound, C_6F_5SCl was dropped onto warmed $O{=}S{=}NSi(CH_3)_3$ (mole ratio 1:1.3; bath temperature 100 to 120°C). During the reaction $(CH_3)_3SiCl$ was separated by distillation and the residue was distilled in vacuum; yield 80%. In another method of preparation a mixture of $(C_6F_5S)_2NH$ and pyridine (mole ratio 2:3) was reacted with 1.5 mole equivalents of $SOCl_2$ in ether with ice-cooling for 2 h. The compound formed quantitatively in accordance with the following equation: $2(C_6F_5S)_2NH + SOCl_2 + 2C_5H_5N \rightarrow O{=}S{=}NSC_6F_5 + (C_6F_5S)_3N + 2C_5H_5N \cdot HCl$. Or $C_6F_5SNH_2$ was reacted with $SOCl_2$ (mole ratio 1.6:1) at 0°C and then at room temperature for 30 min; yield 45%: $2C_6F_5SNH_2 + SOCl_2 \rightarrow O{=}S{=}NSC_6F_5 + C_6F_5SCl + NH_4Cl$. The residue was fractionated in vacuum. Addition of pyridine to the reaction mixture does not raise the yield. The preparation by the reaction of $C_6F_5SONH_2$ with $SOCl_2$ in the presence of pyridine in ether at room temperature is not feasible because the products $O{=}S{=}NSC_6F_5$ and C_6F_5SOCl cannot be separated by distillation [1].

$O{=}S{=}NSC_6F_5$ is a yellow oily liquid with pungent odor. B.p. 58°C/10^{-1} Torr or 42 to 44°C/10^{-2} Torr. Refraction index $n_D^{20} = 1.5438$ [1].

^{13}C NMR ($CDCl_3$/TMS): $\delta = 110.8$ ppm (C-1); 146.28 ppm (ortho-C; $^1J(^{13}C,^{19}F) = 261.3$ Hz); 138.23 ppm (meta-C; $^1J(^{13}C,^{19}F) = 259.3$ Hz); 144.20 ppm (para-C; $^1J(^{13}C,^{19}F) = 267.5$ Hz) [2].

^{19}F NMR ($CCl_4/C_4Cl_4F_4$ [1] or C_6F_6 [2], referred to $CFCl_3$): δ (in ppm) = −133.2 (ortho-F), −162.0 (meta-F), −149.1 (para-F) [1, 2].

IR (capillary) absorption bands at 1640, 1505, and 1485 cm^{-1} are characteristic of the perfluorated ring; those at 1090 and 974 cm^{-1} are ascribed to the C–F bond. The absorption band at 1190 cm^{-1} is assigned to the O=S=N group [1].

MS (70 eV): m/e (rel. int. in %) = 261 (69) M^+, 213 (15) $C_6F_5SN^+$, 199 (100) $C_6F_5S^+$, 155 (64) $C_5F_5^+$, and 16 other peaks are observed [1]; see also [3]. A fragmentation scheme including cyclization to five-membered C–S–N–S–C rings is proposed by [3].

$O{=}S{=}NSC_6F_5$ is thermally quite stable. There is no reaction with Lewis bases such as pyridine. The attempt to chlorinate the OSN group with PCl_5 at room temperature to form $C_6F_5S{-}N{=}SCl_2$ resulted in the splitting of the S–N bond: $O{=}S{=}NSC_6F_5 + PCl_5 \rightarrow C_6F_5SCl$ + unidentified products [1].

References:

[1] Glander, I.; Golloch, A. (J. Fluorine Chem. **14** [1979] 403/20, 405/7, 410, 412/5).
[2] Brands, G.; Golloch, A. (Z. Naturforsch. **38b** [1983] 326/31, 327).
[3] Brands, G.; Golloch, A. (Phosphorus Sulfur **13** [1982] 221/34, 225, 232).

2.6.8 Sulfinyl-amino-diorganylamino-sulfanes, $O{=}S{=}NSNR_2$,

$NR_2 = N(CH_3)_2$, $N(C_2H_5)_2$, $N(C_3H_7\text{-}n)_2$, $N(CH_3)CH_2C_6H_5$, $N{=}C(C_6H_5)_2$, NC_4H_8 (pyrrolidyl)

Preparation of these compounds was accomplished by the reaction of a thionyl (S^{IV}) compound with an N-trimethyl-silyl-substituted compound. Most of the products are liquid and they were purified by vacuum distillation. Starting compounds, conditions for preparation, yields, and bulk properties are listed in Table 14, NMR spectra in Table 16, p. 45. Unassigned IR spectra of $O{=}S{=}NSN(CH_3)_2$ [1], $O{=}S{=}NSN(C_2H_5)_2$ [1], $O{=}S{=}NSN(CH_3)CH_2C_6H_5$ [3], $O{=}S{=}NS{-}N{=}C(C_6H_5)_2$ [4], and $O{=}S{=}NSNC_4H_8$ [3] are listed in the original papers.

The structure of the only solid in this group of compounds, $O{=}S{=}NSN{=}C(C_6H_5)_2$, was studied by X-ray analysis. The crystals belong to the triclinic space group $P\bar{1}$–C_i^1 (No. 2) with the lattice parameters a = 9.9078(8), b = 10.0967(9), c = 15.1682(14) Å, α = 78.646(7)°, β = 71.065(7)°, γ = 63.449(7)°, V = 1281.5(5) Å^3; Z = 4; D_x = 1.422 g/cm^3 (21°C). R = 0.033, R_w = 0.027. Positional parameters are listed in the paper [4]. Bond lengths and bond angles are listed in Table 15, p. 44. The structure of molecule A is shown in **Fig. 6**, p. 44.

Table 14
Preparation, Yields, and Bulk Porperties of Sulfinyl-amino-diorganylamino-sulfanes, O=S=NSNR$_2$.

compound*)	preparation	conditions	yield in %	properties	Ref.
O=S=NSN$(CH_3)_2$	$(O{=}S{=}N)_2S + (CH_3)_3SiN(CH_3)_2$	mole ratio 1:1, in CH_2Cl_2, at reflux	38	yellow liquid, b.p. 48 to 50°C/12 Torr	[1]
	$O{=}S{=}NSi(CH_3)_3 + (CH_3)_2NS(O)Cl$	mole ratio 1:1, neat, gently heating	21	b.p. 39°C/7 Torr, $n_D^{25} = 1.5405$	[2]
O=S=NSN$(C_2H_5)_2$	$(O{=}S{=}N)_2S + (CH_3)_3SiN(C_2H_5)_2$	mole ratio 1:1, in CH_2Cl_2, at reflux	34	yellow liquid, b.p. 28 to 30°C/0.01 Torr	[1]
O=S=NSN$(CH_3)CH_2C_6H_5$	$C_6H_5CH_2(CH_3)NSN(Si(CH_3)_3)_2 + SOCl_2$	mole ratio 1:1, in n-pentane, 12 h with cooling	52	red liquid, b.p. 91 to 93°C/0.5 Torr	[3]
O=S=NSN=C$(C_6H_5)_2$	$(C_6H_5)_2C{=}NSi(CH_3)_3 + SCl_2 + O{=}S{=}NSi(CH_3)_3$	mole ratio 1:1, in CH_2Cl_2, 23°C for 3d	37	orange-red crystals, m.p. 95 to 96°C, λ_{max} (CH_3CN) = 370 nm, $\varepsilon = 2 \times 10^4$ $L \cdot mol^{-1} \cdot cm^{-1}$	[4]
O=S=NSNC_4H_8	$C_4H_8NSN(Si(CH_3)_3)_2 + SOCl_2$	mole ratio 1:1, in CH_2Cl_2, at room temperature for 1 h	54	red liquid, b.p. 26.5 to 27.5°C/0.15 Torr	[3]

*) The formation of O=S=NSNR$_2$ compounds by reaction of O=S=NSi$(CH_3)_3$ with R_2NSCl ($R = CH_3$, C_2H_5, n-C_3H_7) is reported in [7]. The compounds are yellow liquids which do not readily decompose on warming nor in sunlight. Information on preparation, spectra, and further properties is not given [7].

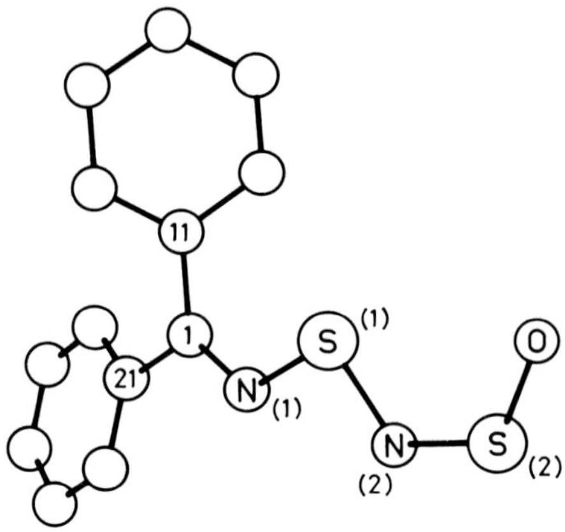

Fig. 6. Molecular structure of $O{=}S{=}NSN{=}C(C_6H_5)_2$ [4].

Table 15
Selected Bond Lengths and Bond Angles of $O{=}S{=}NSN{=}C(C_6H_5)_2$ (molecule A and B) [4]*).

bond length	in Å		bond angle	in °	
	A	B		A	B
S(1)–N(1)	1.679(2)	1.681(3)	C(1)–N(1)–S(1)	117.8(2)	120.8(2)
S(1)–N(2)	1.641(2)	1.646(2)	N(1)–S(1)–N(2)	97.5(1)	96.3(1)
S(2)–N(2)	1.533(2)	1.525(2)	S(1)–N(2)–S(2)	121.6(1)	123.9(2)
S(2)–O	1.455(2)	1.451(3)	N(2)–S(2)–O	116.9(1)	117.5(1)
C(1)–N(1)	1.281(2)	1.292(3)	N(1)–C(1)–C(11)	124.4(2)	115.5(2)
			N(1)–C(1)–C(21)	117.9(2)	124.7(3)

*) For molecule B, C(11) in Fig. 6 is C(21).

The O=S=NSNC unit is an essentially planar chain, the largest deviations from the best plane being 0.067 Å for S(1) in molecule A, and −0.087 Å for S(1) in molecule B. The short C(1)–N bond length and the alternation in the S–N bond lengths involving N(2) suggest the valence bond structure $O{=}S{=}N{-}S{-}N{=}C(C_6H_5)_2$. The delocalization of π electrons across S(1) is obviously not well developed. The dihedral angle between the planes of the two phenyl groups attached to C(1) is 77.4° in molecule A and 66.7° in molecule B. The dihedral angles between the OSNSNC plane and the individual phenyl groups are 59.6° and 22.9° for A, vs. 56.0° and 26.3° for B [4].

The compounds are easily soluble in organic solvents. CH_2Cl_2, $(C_2H_5)_2O$, and CH_3COCH_3 were used as solvents for $O{=}S{=}NSN(CH_3)_2$ and $O{=}S{=}NSN(C_2H_5)_2$ [1], and CH_3CN for $O{=}S{=}NSN{=}C(C_6H_5)_2$ [6].

Electrochemical studies were performed with $O{=}S{=}NSN{=}C(C_6H_5)_2$. At the dropping mercury electrode (0.40 mM in CH_3CN – 0.1 M $(C_2H_5)_4NClO_4$ at 23°C) four cathodic waves were observed: at −1.39, −1.86, −2.55, and −2.95 V vs. Ag|0.1 M $AgClO_4$. The second wave was under kinetic control and disappeared at −15°C. The other three waves were diffusion-controlled. The height of the fourth wave at −15°C became smaller relative to the first and third. At the rotating platinum electrode under similar experimental conditions, a single cathodic wave was observed at −1.53 V, and an anodic wave at +1.33 V. Cyclic voltammograms on a stationary Pt electrode showed the cathodic process to be irreversible.

Coulometric reductions of 0.27 mM $O{=}S{=}NSN{=}C(C_6H_5)_2$ in CH_3CN on Hg-pool or Pt-gauze electrodes at −1.6 or −1.75 V, respectively, both gave n values of 1.0. Polarograms and UV spectra of the resulting solutions proved the quantitative formation of $(C_6H_5)_2C{=}NSS{-}N{=}C(C_6H_5)_2$ and the $[O{=}S{=}N]^-$ ion (see p. 19) [6].

The liquids are distillable in vacuum without decomposition (see Table 14, p. 43) [1, 2, 3]. $O{=}S{=}NSN{=}C(C_6H_5)_2$ decomposes at about 200°C in the solid state to give S_4N_4 and $(C_6H_5)_2CO$ [4].

Table 16

NMR Spectra of Sulfinyl-amino-diorganylamino-sulfanes, $O{=}S{=}NSNR_2$.
(^1H NMR: standard TMS. ^{14}N NMR: neat compound, standard NH_4^+ in saturated aqueous solution; δ in ppm; figures in parentheses are the estimated probable errors in the last digit.)

$-NR_2$	^1H NMR (solvent)	Ref.	^{14}N NMR	Ref.
$-N(CH_3)_2$	2.83 (in CH_2Cl_2)	[1]	18(5) ($N(CH_3)_2$); 322(2) ($O{=}S{=}N$)[a]	[5]
$-N(C_2H_5)_2$	1.21 (CH_3); 3.17 (CH_2); $^3J(H,H)$ = 7.2 Hz (in CH_2Cl_2)	[1]	—	
$-N(C_3H_7\text{-}n)_2$	—		30(10) ($N(C_3H_7\text{-}n)_2$); 303(2) ($O{=}S{=}N$)[b]	[5]
$-N(CH_3)CH_2C_6H_5$	2.80 (CH_3); 4.15(CH_2); 7.27 (C_6H_5) (in CCl_4)	[3]	—	
$-NC_4H_8$	1.88 (CH_2CH_2); 3.15 (CH_2NCH_2) (in $CDCl_3$)	[3]	—	

[a] Referred to the CH_3NO_2 scale: δ = −341.5 and −37.5 ppm, respectively. – [b] Referred to the CH_3NO_2 scale: δ = −329 and −56 ppm, respectively.

The mass spectrum was studied for five compounds:

$O{=}S{=}NSN(CH_3)_2$ (70 eV): m/e (rel. int. in %) = 138 (30) M^+, 76 (77) $SN(CH_3)_2^+$, 75 (100) $NSNCH_3^+$, 44 (88) $N(CH_3)_2^+$, 42 (80) $NC_2H_4^+$ [1].

$O{=}S{=}NSN(C_2H_5)_2$ (70 eV): m/e (rel. int. in %) = 166 (62) M^+, 151 (39) $OSNSNC_3H_7^+$, 104 (84) $SN(C_2H_5)_2^+$, 76 (71) $SN(CH_3)_2^+$, 72 (85) $N(C_2H_5)_2^+$, 56 (100) $NC_3H_6^+$ [1].

$O{=}S{=}NSN{=}(CH_3)CH_2C_6H_5$ (100 eV): m/e (rel. int. in %) = 214 (6.4) M^+, 152 (2.0) $M^+ - NSO$ [3].

$O{=}S{=}NSN{=}C(C_6H_5)_2$ (70 eV; source temperature 100°C): m/e (rel. int. in %) = 274 (100) M^+, 212 (62), 180 (76), 109 (41), 77 (87), 51 (36); (source temperature 90°C): 274 (19) M^+, 212 (29), 180 (53), 109 (39), 104 (28), 77 (100) [4].

$O{=}S{=}NSNC_4H_8$ (100 eV): m/e = 164 M^+ (40.1) [3].

$O{=}S{=}NSN(CH_3)_2$ and $O{=}S{=}NSN(C_2H_5)_2$ are only slightly sensitive to hydrolysis. They react with trimethylsilylamines to form the corresponding sulfur diamines:

$$O{=}S{=}NSNR_2 + (CH_3)_3SiNR_2 \rightarrow O{=}S{=}NSi(CH_3)_3 + S(NR_2)_2 \quad (R = CH_3,\ C_2H_5)\ [1].$$

References:

[1] Roesky, H. W.; Schaper, W. (Chem. Ber. **107** [1974] 3451/3).
[2] Armitage, D. A.; Sinden, A. W. (J. Inorg. Nucl. Chem. **36** [1974] 993/5).
[3] Neidlein, R.; Lenhard, Th. (Chem. Ber. **116** [1983] 3133/40, 3139).
[4] Chivers, T.; Oakley, R. T.; Pieters, R.; Richardson, J. F. (Can. J. Chem. **63** [1985] 1063/7).
[5] Armitage, D. A.; Mason, J.; Vinter, J. G. (Inorg. Chem. **17** [1978] 776/7).
[6] Chivers, T.; Hojo, M. (Inorg. Chem. **23** [1984] 4088/93, 4092).
[7] Armitage, D. A.; Tso, C. C. (J. Chem. Soc. D **1971** 1413/4).

2.6.9 N-Sulfinyl-sulfinamides, O=S=NS(O)R, R = CF_3, C_6H_5

O=S=NS(O)CF_3 was prepared by slowly adding $SOCl_2$ to an equimolar amount of $(CH_3)_3$-SiNHS(O)CF_3 or $((CH_3)_3Si)_2$NS(O)CF_3 at 15°C, under a nitrogen atmosphere in well-dried vessels. The yellow liquid was purified by distillation (b.p. 48°C/15 Torr). Yield 52%.

$(CH_3)_3$SiNRS(O)CF_3 + $SOCl_2$ → O=S=NS(O)CF_3 + $(CH_3)_3$SiCl + RCl, R = H or $(CH_3)_3$Si.

^{19}F NMR (neat/external $CFCl_3$): $\delta = -76.35$ ppm. IR (capillary film): ν (in cm^{-1}) = 1243 sh, 1200 vs, 1160 s, 1110 vs, 1093 sh, 755 w [1].

O=S=NS(O)C_6H_5. Apparently the title compound is only mentioned once in literature. It was prepared from $C_6H_5N(Si(CH_3)_3)_2$ and $SOCl_2$ [2].

References:

[1] Roesky, H. W.; Holtschneider, G. (J. Fluorine Chem. **7** [1976] 77/84, 78, 81).
[2] Scherer, O. J.; Schmitt, R. (Chem. Ber. **101** [1968] 3302/12, 3302).

2.6.10 N-Sulfinylsulfonamides, O=S=NSO_2R, R = CH_3, CCl_3, aryl

Perfluorinated O=S=NSO_2R_f compounds are described in "Schwefel-Stickstoff Verbindungen" Teil 1, 1977.

Preparation of N-sulfinylsulfonamides is generally accomplished by the reaction of sulfonamides, N,N-dichloro-arylsulfonamides, and the sodium salt of N-chloro-arylsulfonamides with $SOCl_2$, see Table 17, p. 47.

The reaction with sulfonamides occurs readily when they are refluxed for some hours with excess $SOCl_2$ without solvent. In benzene solution refluxing for several days is necessary to obtain good yields: RSO_2NH_2 + $SOCl_2$ → O=S=NSO_2R + 2HCl [1, 2, 3].

N,N-dichloroarylsulfonamides and excess $SOCl_2$ are heated at 80 to 100°C until Cl_2 evolution ceases (2 to 4 h): RSO_2NCl_2 + $SOCl_2$ → O=S=NSO_2R + 2Cl_2 [4].

When excess $SOCl_2$ is slowly added to the sodium salt of N-chlorosulfonamide in C_6H_6 or CCl_4, the reaction mixture warms slowly and NaCl precipitates: RSO_2NClNa + $SOCl_2$ → O=S=NSO_2R + Cl_2 + NaCl [4].

Normally, after distillation of the excess $SOCl_2$, a yellow oil remains, and this crystallizes on cooling [1 to 4].

Table 17

Preparation, Melting Points, and Boiling Points of N-Sulfinylsulfonamides, $O{=}S{=}NSO_2R$.

R	starting sulfonamide + $SOCl_2$	yield in %	m.p. in °C	b.p. in °C/10^{-4} Torr	Ref.
CH_3	RSO_2NH_2 (in C_6H_6)	86	oil	80 to 81	[2, 3]
CCl_3	RSO_2NH_2	not isolated*)			[6]
C_6H_5	RSO_2NH_2 (in C_6H_6)	75 to 80	70 to 71	115 to 118	[1, 3]
	RSO_2NCl_2 RSO_2NClNa (in C_6H_6 or CCl_4)	~100 82	68 to 70	—	[4]
4-FC_6H_4	RSO_2NH_2	69	59.5	—	[3]
4-ClC_6H_4	RSO_2NH_2 (in C_6H_6)	88	58 to 60	—	[3]
	RSO_2NH_2 (in C_6H_6)	98	52	122 to 125	[1, 2]
	RSO_2NCl_2 RSO_2NClNa (in C_6H_6 or CCl_4)	~100 92	58 to 60	—	[4]
4-BrC_6H_4	RSO_2NH_2 (in C_6H_6)	83	60 to 61	—	[3]
	RSO_2NH_2 (in C_6H_6)	—	57	133 to 136	[1]
	RSO_2NCl_2 RSO_2NClNa (in C_6H_6 or CCl_4)	~100 87	60 to 61	—	[4]
2-$NO_2C_6H_4$	RSO_2NCl_2 RSO_2NClNa (in C_6H_6 or CCl_4)	~100 99	112 to 113	—	[4]
3-$NO_2C_6H_4$	RSO_2NCl_2 RSO_2NClNa (in C_6H_6 or CCl_4)	~100 73	79 to 80	—	[4]
4-$NO_2C_6H_4$	RSO_2NCl_2 RSO_2NClNa (in C_6H_6 or CCl_4)	~100 96	115 to 117	—	[4]
4-$CH_3OC_6H_4$	RSO_2NH_2	75	30	—	[3]
4-$CH_3C_6H_4$	RSO_2NH_2	80	53	—	[3]
	RSO_2NH_2 (in C_6H_6)	93	53	108 to 110	[1]
	RSO_2NH_2 (in C_6H_6)	22	49.5 to 51.0	—	[7]

*) The raw product was used directly in the reaction with DMSO, see p. 221.

The compounds O=S=NSO_2R (R = C_6H_5, 4-BrC_6H_4, 4-$NO_2C_6H_4$) were also obtained in high yield by acidolysis of RSO_2N=SCl_2 with formic or acetic acid (mole ratio 1:1, in benzene, heating): RSO_2N=SCl_2 + CH_3COOH → O=S=NSO_2R + HCl + CH_3COCl.

O=S=$NSO_2C_6H_5$ forms in 85% yield by the reaction of $C_6H_5SO_2N$=SCl_2 with CH_3ONa (mole ratio 1:1, in benzene, room temperature): $C_6H_5SO_2N$=SCl_2 + CH_3ONa → O=S=$NSO_2C_6H_5$ + NaCl + CH_3Cl [5].

Recrystallization from benzene, petroleum ether (pure or mixed), or benzene–hexane yields colorless prisms or needles [4]. Methods of preparation, yields, melting points, and boiling points of the individual compounds are compiled in Table 17, p. 47; densities, dielectric constants, and dipole moments in Table 18. Bond moments were estimated to be μ(S=N) = 0.45 D and μ(S=O) = 3.5 D [12].

Table 18
Densities, d, Dielectric Constants, ε, and Dipole Moments, μ, of O=S=NSO_2R [12].

R	C_6H_5	4-ClC_6H_4	4-BrC_6H_4	4-$NO_2C_6H_4$	4-$CH_3C_6H_4$
d (l) (in g/cm³)	1.45 (85°C)	1.49 (25°C)	1.82 (60°C)	1.56 (122°C)	1.41 (50°C), 1.35 (90°C)
$d_4^{-190°C}$ (in g/cm³)	1.57	—	1.95	—	1.50
ε (l)	29.0 (85°C)	17.4 (25°C)	16.3 (60°C)	8.9 (122°C)	27.2 (50°C), 24.4 (90°C)
ε (−190°C)	3.06	—	3.18	—	3.06
μ (in D)	4.28	3.12	3.12	2.29	4.2

An X-ray structure analysis was made for O=S=$NSO_2C_6H_5$. It crystallizes in the monoclinic space group $P2_1/c$–C_{2h}^5 (No. 14) with a = 5.454(3), b = 16.600(5), c = 9.448(4) Å, β = 122.16(9)°, V = 820.5 Å³; Z = 4, D_x = 1.64 g/cm³, R = 0.038. The O=S=$NSO_2C_6H_5$ molecule has cis structure, as shown in **Fig. 7**. Bond distances and bond angles are also given there. The distances between unbonded S and O atoms in the same molecule, and also between neighboring molecules, are short. The crystal packing shows a weak stacking of aromatic rings [8].

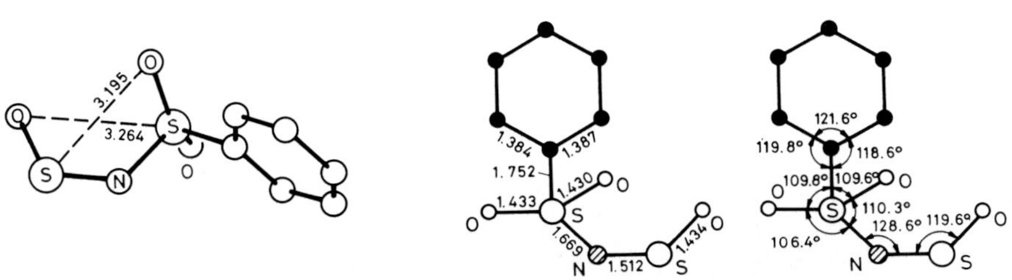

Fig. 7. Molecular structure of O=S=$NSO_2C_6H_5$ (bond lengths in Å, bond angles in °) [8].

¹H NMR ($CDCl_3$/TMS) of O=S=$NSO_2C_6H_4CH_3$-4: δ (in ppm) = 7.68 (A_2B_2 quartet, aromatic H); 2.47 (s, CH_3) [7].

Tentatively assigned IR spectra are listed in Table 19. The IR spectra show two ν(SO_2) bands at 1175 and 1375 cm^{-1} and two other strong bands at 1090 and 1250 cm^{-1} [2]. For O=S=$NSO_2C_6H_4CH_3$-4 in $CHCl_3$ intense peaks at 1364, 1245, 1170, and 1089 cm^{-1} (7.33, 8.03, and 9.18 μ in the original paper) are reported [7].

Table 19
IR Spectra of Compounds O=S=NSO$_2$R (ν in cm^{-1}).

R	state	ν_s(N=S=O)	ν_{as}(N=S=O)	ν_s(SO$_2$)	ν_{as}(SO$_2$)	Ref.
CH_3	neat	1108	1237	1173	—	[12]
C_6H_5	solid film	1090	1243	1172	1362	[9]
	in Nujol	—	—	1160	1344	[9]
	in CS_2	1090	1219	1145	1343	[9]
	in CCl_4	1104	1250	1179	1375	[12]
4-ClC$_6$H$_4$	in CCl_4	1093	1250	1177	1367	[12]
4-NO$_2$C$_6$H$_4$	in CCl_4	1096	—	—	—	[12]
4-CH$_3$C$_6$H$_4$	in CCl_4	1095	1242	1170	—	[12]

N-Sulfinylsulfonamides are soluble in most inert organic solvents [2]; readily soluble in benzene, dichlorethane, and CH_2Cl_2, less soluble in ether and CCl_4, and sparingly soluble in petroleum ether and hexane [4].

The compounds are stable up to ~165°C: on heating the pure substances SO_2 evolution is observed and sulfur diimides form:

$$2\,O{=}S{=}NSO_2R \rightarrow RSO_2N{=}S{=}NSO_2R + SO_2$$

R = CH_3, C_6H_5, 4-FC$_6$H$_4$, 4-ClC$_6$H$_4$, 4-BrC$_6$H$_4$, 4-CH$_3$OC$_6$H$_4$, 4-CH$_3$C$_6$H$_4$.

At higher temperatures decomposition predominates. The presence of Lewis acids, such as $AlCl_3$, or better of bases in an inert solvent (e.g., C_6H_5Cl) catalyzes the formation of diimides. When an appreciable amount of dry $AlCl_3$ (1 to 7 mol%) is added, the reaction temperature can be lowered to 125 to 100°C. Tertiary amines catalyze the reaction even at room temperature. Strong basic amines, e.g. $N(C_2H_5)_3$, cause a violent reaction. The primary product is a brown-red complex which slowly loses SO_2 with formation of the diimide. For the preparation of diimides a weaker base, such as pyridine, is preferable [13].

The compounds are very sensitive towards water and moist air, and decompose readily and quantitatively into SO_2 and arylsulfonamides [4]: $O{=}S{=}NSO_2R + H_2O \rightarrow RSO_2NH_2 + SO_2$. (For alcoholysis see pp. 205/8.) Hydrogen chloride reverses the formation reaction: $O{=}S{=}NSO_2R + 2\,HCl \rightarrow RSO_2NH_2 + SOCl_2$. In this case, however, the N-sulfinylsulfonamides O=S=NSO$_2$R are more stable than the N-sulfinylamines O=S=NR, R = organyl (in agreement with the slower formation reaction, see p. 46) [3]. They react exothermically and quantitatively with PCl_5 [4]:

$$O{=}S{=}NSO_2C_6H_5 + PCl_5 \rightarrow O{=}S{=}NC_6H_5 + POCl_3 + SOCl_2.$$

O=S=NSO$_2$C$_6$H$_4$CH$_3$-4 forms no adduct with BF_3 in toluene at −23°C [10]. With $COCl_2$ in benzene solution in the presence of pyridine or N,N-dimethylformamide forms the corresponding isocyanate [11]:

$$O{=}S{=}NSO_2C_6H_4CH_3\text{-}4 + COCl_2 \rightarrow 4\text{-}CH_3C_6H_4SO_2N{=}C{=}O + SOCl_2.$$

When O=S=NSO$_2$CH$_3$ reacts with the five-membered SN ring compound, S_3N_2O, only elemental sulfur, cyclic S_4N_4, and volatile reaction products form [14].

N-Sulfinylsulfonamides, O=S=NSO$_2$R, react with sulfur dichloride imides of the R'SO$_2$-N=SCl$_2$ type (about 150°C for 50 min, mole ratio 1:1) to give symmetrically or asymmetrically substituted sulfur diimides. Formation of an intermediate adduct (which then split off $SOCl_2$) was assumed [15]:

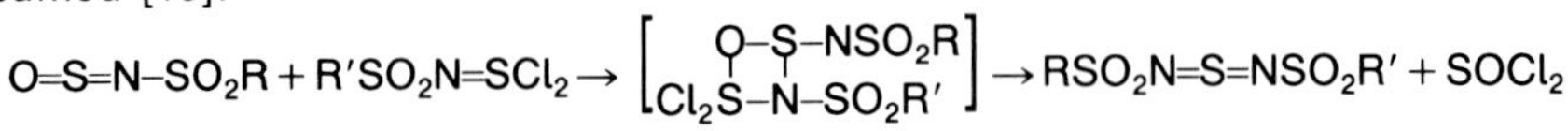

R, R' = C_6H_5, 4-$CH_3C_6H_4$.

$O{=}S{=}NSO_2C_6H_5$, when heated with an equimolar amount of $((CH_3)_3Si)_2NSO_2C_6H_5$ at 100 to 120°C in an oil bath under the vacuum of a water-jet pump for 10 to 12 h, yields $(C_6H_5SO_2N{=})_2S$ and $((CH_3)_3Si)_2O$ [16].

With diaryl-selenones, $(R'C_6H_4)_2SeO_2$ (R' = H, CH_3), $O{=}S{=}NSO_2C_6H_4R$ (R = H, Cl, NO_2, CH_3), when heated in benzene at 70 to 80°C for 2 h, evolves SO_2 and forms $(R'C_6H_4)_2Se(O){=}NSO_2$-$C_6H_4R$, quantitatively [17, 18].

N-Sulfinylsulfonamides in CH_2Cl_2 react with $(CH_3)_2BSCH_3$ at room temperature, quantitatively inserting the NS group into the B–S bond [19]:

$$O{=}S{=}NSO_2R + 3(CH_3)_2BSCH_3 \rightarrow RSO_2N\langle^{S-SCH_3}_{B(CH_3)_2} + (CH_3)_2S + ((CH_3)_2B)_2O$$

R = CH_3, C_6H_5, 4-$CH_3C_6H_4$.

In cyclic BS compounds the S–S bond is split (24 h reflux in CCl_4) [20]:

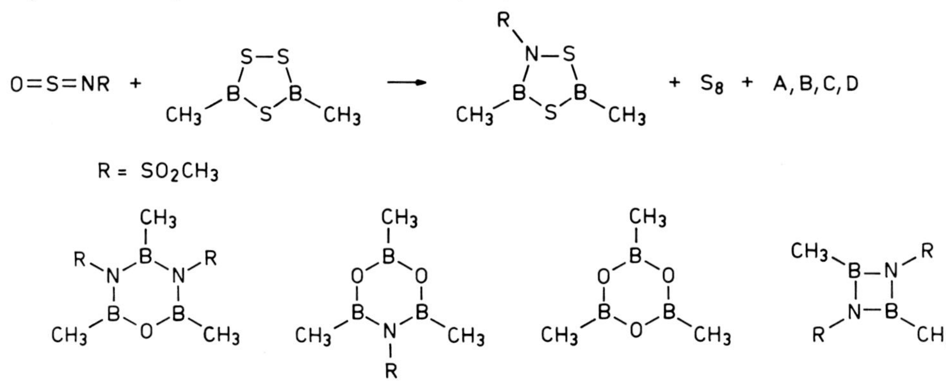

A B C D

Phosphorus(III) compounds are oxidized to phosphorus(V) compounds. $O{=}S{=}NSO_2CH_3$ and $O{=}S{=}NSO_2C_6H_4CH_3$-4, when mixed with $P(C_6H_5)_3$ in C_6H_6, react exothermically. Refluxing gives $(C_6H_5)_3P{=}NSO_2R$ (R = CH_3, $C_6H_4CH_3$-4) in high yield [21, 22]. When $O{=}S{=}NSO_2CH_3$ and $P(OC_2H_5)_3$ are refluxed in benzene for 7 h, $(C_2H_5O)_3P{=}S$ and $(C_2H_5O)_3P{=}NSO_2CH_3$ form [21]. With $(C_6H_5)_3P{=}O$, $(C_6H_5)_3P{=}S$, and $(C_6H_5)_3P{=}NC_6H_5$ exchange takes place on reflux in benzene [21, 22]:

$$O{=}S{=}NSO_2R + (C_6H_5)_3P{=}O \rightarrow (C_6H_5)_3P{=}NSO_2R + SO_2$$

R = CH_3, 4-$CH_3C_6H_4$;

$$O{=}S{=}NSO_2R + (C_6H_5)_3P{=}S \rightarrow (C_6H_5)_3P{=}NSO_2R + [SSO]$$

$$O{=}S{=}NSO_2R + (C_6H_5)_3P{=}NC_6H_5 \rightarrow (C_6H_5)_3P{=}NSO_2R + O{=}S{=}NC_6H_5$$

R = 4-$CH_3C_6H_4$.

With $(C_2H_5O)_3P{=}S$, $O{=}S{=}NSO_2C_6H_4CH_3$-4 reacts similarly to give SO_2 and $(C_2H_5O)_3P{=}NSO_2$-$C_6H_4CH_3$-4 in 10% yield [21, 22].

$(C_6H_5)_3As$ and $(C_6H_5)_3Sb$, when refluxed with $O{=}S{=}NSO_2R$ in benzene for several hours, afford the corresponding imine [21]:

$$O{=}S{=}NSO_2R + (C_6H_5)_3M \rightarrow (C_6H_5)_3M{=}NSO_2R + [SO]$$

$R = CH_3$; $M = Sb$; yield 38%;
$R = C_6H_5$; $M = As$; yield 67%.

For reactions of N-sulfinylsulfonamides with organic compounds, organometallic compounds, and metal complexes, see Sections 2.6.23, pp. 186/241, 2.6.25, pp. 243/51, and 2.7.4, pp. 268/9, respectively.

References:

[1] Kresze, G.; Maschke, A. (Ger. 1117566 [1960/61] from C.A. **57** [1962] 11110).
[2] Kresze, G.; Maschke, A.; Albrecht, R.; Bederke, K.; Patzschke, H. P.; Smalla, J.; Trede, A. (Angew. Chem. **74** [1962] 135/44, 141/4; Angew. Chem. Intern. Ed. Engl. **1** [1962] 89/98).
[3] Kresze, G.; Wucherpfennig, W. (Angew. Chem. **79** [1967] 109/27, 111, 120; Angew. Chem. Intern. Ed. Engl. **6** [1967] 149/67, 151, 157/60).
[4] Levchenko, E. S.; Kirsanov, A. V. (Zh. Obshch. Khim. **32** [1962] 161/5; J. Gen. Chem. [USSR] **32** [1962] 157/61).
[5] Markovskii, L. N.; Fedyuk, G. S.; Levchenko, E. S. (Zh. Org. Khim. **6** [1970] 2426/31; J. Org. Chem. [USSR] **6** [1970] 2438/42).
[6] Buchholt, H. C.; Senning, A. (Acta Chem. Scand. **24** [1970] 2255/6).
[7] Cram, D. J.; Day, J.; Rayner, D. R.; von Schriltz, D. M.; Duchamp, D. J.; Garwood, D. C. (J. Am. Chem. Soc. **92** [1970] 7369/84, 7371/2, 7382).
[8] Déléris, G.; Courseille, C.; Kowalski, J.; Dunoguès, J. (J. Chem. Res. S **1979** 122/3).
[9] Lutskii, A. E.; Shevchenko, Z. A.; Kochergina, L. A.; Levchenko, E. S.; Uzarov, B. N.; Men'shova, I. I.; Bal'on, Ya. G. (Zh. Obshch. Khim. **37** [1967] 2095/2101; J. Gen. Chem. [USSR] **37** [1967] 1986/91).
[10] Severson, R. G.; Leung, T. W.; Wojcicki, A. (Inorg. Chem. **19** [1980] 915/23, 916).

[11] Ulrich, H. ; Tucker, B.; Sayigh, A. A. R. (J. Org. Chem. **34** [1969] 3200/1).
[12] Borovikov, Yu. Ya.; Borovikova, G. S.; Koval, N. V.; Borovik, E. I. (Ukr. Khim. Zh. **52** [1986] 875/9; Soviet Progr. Chem. **52** No. 8 [1986] 96/9).
[13] Wucherpfennig, W.; Kresze, G. (Tetrahedron Letters **1966** 1671/5).
[14] Roesky, H. W.; Holtschneider, G.; Wiezer, H.; Krebs, G. (Chem. Ber. **109** [1976] 1358/61).
[15] Markovskii, L. N.; Fedyuk, G. S.; Levchenko, E. S. (Zh. Org. Khim. **8** [1972] 286/91; J. Org. Chem. [USSR] **8** [1972] 287/91).
[16] Derkach, N. Ya.; Barashenkov, G. G.; Levchenko, E. S. (Zh. Org. Khim. **18** [1982] 964/7; J. Org. Chem. [USSR] **18** [1982] 837/9).
[17] Derkach, N. Ya.; Pasmurtseva, N. A.; Levchenko, E. S. (Zh. Obshch. Khim. **42** [1972] 1652/3; J. Gen. Chem. [USSR] **42** [1972] 1646).
[18] Derkach, N. Ya.; Lyapina, T. V.; Pasmurtseva, N. A. (Zh. Org. Khim. **10** [1974] 807/10; J. Org. Chem. [USSR] **10** [1974] 810/2).
[19] Meller, A.; Maringgele, W.; Fetzer, H. (Chem. Ber. **113** [1980] 1950/61, 1952).
[20] Meller, A.; Habben, C. (Monatsh. Chem. **113** [1982] 139/53, 142/3, 149).

[21] Senning, A. (Acta Chem. Scand. **19** [1965] 1755/9).
[22] Senning, A. (Angew. Chem. **77** [1965] 379; Angew. Chem. Intern. Ed. Engl. **4** [1965] 357).

2.6.11 Selenium-Substituted Sulfinyl Imides

2.6.11.1 N-Sulfinyl-trifluoromethaneselenamide, $O{=}S{=}NSeCF_3$

$O{=}S{=}NSeCF_3$ was prepared in 22% yield by reaction of CF_3SeNH_2 with SO_2 (mole ratio ~1:4) in a glass bomb, containing a molecular sieve A_4, with ice cooling for 3 h. Fractional condensation afforded O=S=NSeCF$_3$ in the $-80°C$ trap. It was produced in 98% yield by heating $((CF_3Se)_2N)_2SO$ in a glass bomb at 40 to 50°C for 3 h.

Yellow, light-sensitive liquid, m.p. $-61.5°C$; b.p. 126.85°C (extrapolated from $\log(p/Torr) = 7.35 - 1756/T$). The vapor pressure increases in the temperature range of -23.5 to 18°C from 1.9 to 21.8 Torr; $\Delta H_v = -8034.8$ cal/mol, Trouton's constant $\Delta H_v/T_v = 20.1\ cal \cdot mol^{-1} \cdot K^{-1}$ [1].

^{19}F NMR (solvent?/$CFCl_3$): $\delta = -41.4$ ppm; ^{77}Se NMR (solvent?/$(CH_3)_2Se$): $\delta = 1091$ ppm [2], see also [1]. IR (gas, 5 Torr) and Raman (liquid): 17 and 11 frequencies between 2281 and 464 cm^{-1} and 1215 and 146 cm^{-1}, respectively.

MS: m/e (rel. int. in %) = 298 (2) $(CF_3Se)_2^+$, 229 (3) $CF_3Se_2^+$, 211 (8) M^+, 160 (5) Se_2^+, 149 (6) CF_3Se^+, 142 (8) $SeNSO^+$, 94 (21) SeN^+, 80 (16) Se^+, 69 (100) CF_3^+, 50 (9) CF_2^+, 48 (20) SO^+, 46 (55) SN^+. The compound decomposes in sealed capillaries after 24 h. It reacts with H_2O to give CF_3SeOH and $(HNSO)_x$ [1].

References:

[1] Darmadi, A. (Diss. Univ. Bochum 1980).
[2] Gombler, W. (Z. Naturforsch. **36b** [1981] 535/43).

2.6.11.2 N-Sulfinyl-trifluoromethaneseleninamide, O=S=NSe(O)CF$_3$

O=S=NSe(O)CF$_3$ was prepared in 69% yield by reaction of $CF_3Se(O)NH_2$ with a slight excess of $SOCl_2$ at 25°C for 2 h.

The colorless, moisture-sensitive needles melt at 39.5°C. The vapor pressure at 23°C is 0.7 ± 0.1 Torr. ^{19}F NMR (20% ether solution/$CFCl_3$): $\delta = -66.95$ ppm; IR (gas, ~1.5 Torr, 50°C): ν (in cm^{-1}) = 1123s; 1185vs; 1106vs; 1076w,sh; 948s; 736m.

MS: m/e (rel. int. in %) = 227 (2) M^+, 208 (3) $M^+ - F$, 181 (9) $CF_3SeO_2^+$, 179 (5) CF_3SeON^+, 165 (5) CF_3SeO^+, 149 (15) CF_3Se^+, 96 (11) SeO^+, 80 (18) Se^+, 69 (100) CF_3^+, 50 (9) CF_2^+.

Reference:

Darmadi, A. (Diss. Univ. Bochum 1980).

2.6.11.3 Bis(sulfinyl-amino)selane, (O=S=N)$_2$Se

The title compound was prepared in 69.2% yield by stirring a mixture of O=S=NSi(CH$_3$)$_3$ and Se_2Cl_2 (mole ratio ~2:1) in CH_2Cl_2 in a glass bomb with a teflon valve at 22°C for 24 h. After separation of the solvent and $(CH_3)_3SiCl$, the residue was sublimed at 22°C/10^{-3} Torr. Using $((CH_3)_3Si)_2O$ as solvent, the yield increased to 80 to 85%. Lower yields were obtained, when $SeOCl_2$ or $SeCl_4$ were treated with O=S=NSi(CH$_3$)$_3$ [1].

An X-ray structure analysis shows the compound to be planar, angled, with an atomic distance d(Se–N) = 1.827 Å and a bond angle ∢(N–Se–N) = 92.0° [2].

The compound, a yellow, crystalline solid, melts at 122 to 123°C.

It is easily soluble in polar organic solvents, like acetone, CH_2Cl_2 or $CHCl_3$ and decomposes spontaneously in air [1].

$(O{=}S{=}N)_2Se$ reacts with equimolar amounts of MF_5 (M = As, Sb, Nb) at 22°C in SO_2 to give $[SN_2Se_2]_2^{2+}(MF_6^-)_2$, Ia to Ic:

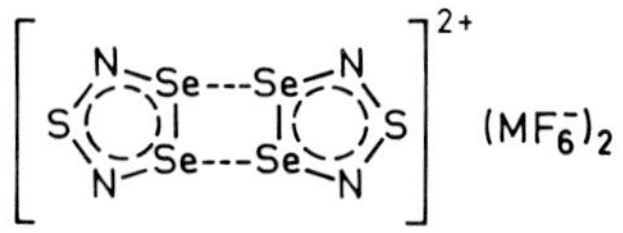

Ia to Ic

M	yield in %
Ia As	39.6
Ib Sb	38.2
Ic Nb	47.7

Similarly, $(O{=}S{=}N)_2Se$ reacted with PCl_5 in CH_2Cl_2 to give $[SN_2Se_2Cl]^+Cl^-$ (IIa). With $SbCl_5$ (mole ratio 1:0.5 to 0.7) in CH_2Cl_2 solution $[SN_2Se_2Cl]^+SbCl_6^-$ (IIb) was analogously obtained. The reaction of $(O{=}S{=}N)_2Se$ with an excess of AsF_5 (mole ratio 1:3) yields $[SN_2Se_2]^{2+}(AsF_6^-)_2$ (III) [3].

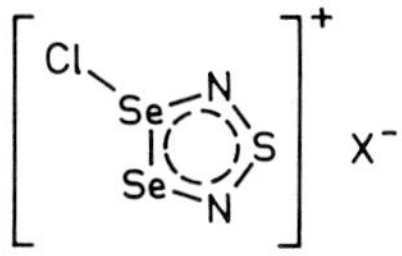

IIa (X = Cl), IIb (X = $SbCl_6$)

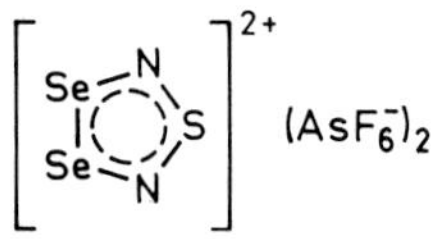

III

References:

[1] Haas, A.; Kasprowski, J. (Chimia [Switz.] **41** [1987] 340).
[2] Angermund, K.; Krüger, C. (unpublished results cited in [1]).
[3] Haas, A.; Kasprowski, J. (Chimia [Switz.] **42** [1988] 343/4).

2.6.12 Tellurium-Substituted Sulfinyl Imides

2.6.12.1 N-Sulfinyl-aminopentafluoro-λ^6-tellane, $O{=}S{=}NTeF_5$

The title compound was obtained in 68% yield by reaction of $F_5TeN{=}SCl_2$ with $CF_3C(O)OAg$ (mole ratio 1:1.6) at room temperature and subsequently warming to 45°C for 1 h. $O{=}S{=}NTeF_5$ was separated by trap-to-trap distillation in the −45°C trap [1]. It was formed in 29% yield by treating $B(OTeF_5)_3$ at −196°C with FSN (mole ratio 1.3:4), followed by stirring at −35°C for 1 h. The compound was isolated by fractional condensation. It formed also by reaction of $(CH_3)_3M$-$OTeF_5$ (M = Si, Sn) with FSN or of $(CH_3)_3SiOTeF_5$ with $S_3N_3Cl_3$ or by spontaneous rearrangement of $N{\equiv}SOTeF_5$ at low temperature [2].

The colorless, foul smelling, moisture-sensitive liquid solidifies at −24.6°C. It turns yellow on warming or on standing in the presence of light [1]. ^{19}F NMR (solvent?/$CFCl_3$): AB_4 type, $\delta(F_A) = -41.2$ ppm, $\delta(F_B) = -33.4$ ppm; $^2J(F_A, F_B) = 171.7$ Hz [1, 2], $^1J(^{125}Te, F_A) = 3459$ Hz, $^1J(^{125}Te, F_B) = 3854$ Hz. IR (gas) and Raman (liquid): 9 and 15 absorptions between 1291 and 654 cm^{-1} and 1261 and 110 cm^{-1}, respectively [1]; 1291, $\nu_{as}(O{=}S{=}N)$; 1138 cm^{-1}, $\nu_s(O{=}S{=}N)$ [2].

In the mass spectrum (70 eV) the following molecular fragments are observed: $OTeF_5^+$, TeF_5^+, TeF_4^+, TeF_3^+, TeF_2^+, TeF^+, Te^+, SO_2^+, SO^+ [1]. The BF_3 catalyzed elimination of SO_2 afforded $F_5TeN{=}S{=}NTeF_5$ [2].

References:

[1] Hartl, H.; Huppmann, P.; Lentz, D.; Seppelt, K. (Inorg. Chem. **22** [1983] 2183/7).
[2] Hoppenheit, R.; Mews, R. (Chem. Ber. **118** [1985] 4276/80).

2.6.12.2 Bis(sulfinyl-amino)tellane, $(O{=}S{=}N)_2Te$

The title compound was prepared by condensing O=S=NCl to a solution of $(t\text{-}C_4H_9(CH_3)_2\text{-}Si)_2Te$ (mole ratio 2:1) in CH_2Cl_2. After stirring of the reaction mixture at 22°C for 48 h the solvent and $t\text{-}C_4H_9(CH_3)_2SiCl$ were removed and the side-product $(O{=}S{=}N)_2S$ was sublimed at 22°C/10^{-3} Torr. The compound was isolated by sublimation at 50°C/10^{-3} Torr in 36% yield.

The X-ray structure analysis showed that $(O{=}S{=}N)_2Te$ is isomorphous with $(O{=}S{=}N)_2Se$ (see p. 52). Bond distances (in Å): Te–N = 2.039(7), N–S = 1.492(7), S–O = 1.460(7); bond angles (in °): N–Te–N = 87.3(3), N–S–O = 116.6(4).

The compound, a red crystalline solid, decomposes at 132°C with formation of a grey-black solid, which exploded on further heating.

^{17}O NMR ($CDCl_3$/external D_2O): δ = 461 ppm (s). ^{125}Te NMR ($CDCl_3$/external $(C_6H_5)_2Te_2$ in $CDCl_3$): δ = 1273 ppm (s). IR (KBr): ν (in cm^{-1}) = 1180 vs, 1039 s, 595 s, 413 s. Mass spectrum (70 eV): in addition to the M^+ peak, the fragments $OSNTe^+$ (100%) and NTe^+ (43%) were observed. The spectroscopic data are consistent with a planar chain structure.

Reference:

Haas, A.; Pohl, R. (Chimia [Switz.] **43** [1989] 261/2).

2.6.13 N′-Sulfinyl-hydrazines

2.6.13.1 N′-Sulfinyl-N-organyl-hydrazines

2.6.13.1.1 N′-Sulfinyl-p-toluenesulfonohydrazide, $O{=}S{=}NNHSO_2C_6H_4CH_3\text{-}4$

The compound formed in 56% yield by treating a solution of O=S=Ntos (tos = $SO_2C_6H_4CH_3$-4) in $CHCl_3$ with $tosNHNH_2$ (mole ratio 1:1) at room temperature. After 5 h the precipitate was removed, the solvent reduced in volume and O=S=NNHtos was precipitated with petroleum ether; m.p. 80 to 85°C with decomposition.

On slight heating the compound decomposes with evolution of N_2 to give $4\text{-}CH_3C_6H_4SO_3H$ and $4\text{-}CH_3C_6H_4SO_2S\text{–}S\text{–}S\text{–}SO_2C_6H_4CH_3\text{-}4$.

Reference:

Kresze, G.; Horn, A.; Philippson, R.; Trede, A. (Chem. Ber. **98** [1965] 3401/9).

2.6.13.1.2 N′-Sulfinyl-N-benzyl-hydrazine, $O{=}S{=}NNHCH_2C_6H_5$ and N′-Sulfinyl-N-(2-phenylethyl)-hydrazine, $O{=}S{=}NNHCH_2CH_2C_6H_5$

Both compounds were prepared by treating a solution of $C_6H_5CH_2NHNH_2$ or $C_6H_5CH_2CH_2\text{-}NHNH_2$ and pyridine in $CHCl_3$ with a solution of $SOCl_2$ in $CHCl_3$ (mole ratio 2:2:1) at 0°C for 1 h. The products were formed as colorless solids after work-up in 41 and 42% yield, respectively. Careful crystallization from C_2H_5OH gave $O{=}S{=}NNHCH_2C_6H_5$, with m.p. 30 to 30.5°C, and $O{=}S{=}NNHCH_2CH_2C_6H_5$, with m.p. 27.5 to 29°C.

Reference:

Pearce, L. B.; Feingold, M. H.; Cerny, K. F.; Anselme, J.-P. (J. Org. Chem. **44** [1979] 1881/3).

2.6.13.1.3 N′-Sulfinyl-N-aryl-hydrazines

2.6.13.1.3.1 N′-Sulfinyl-N-phenyl-hydrazine, $O{=}S{=}NNHC_6H_5$

$O{=}S{=}NNHC_6H_5$ was obtained in ~50% yield by passing a stream of SO_2 into a solution of $C_6H_5NHNH_2$ in C_6H_6 at ~75°C for ~2 h. It also formed by heating the addition product $(C_6H_5NHNH_2)_2 \cdot SO_2$, which is produced in the reaction of $C_6H_5NHNH_2$ and SO_2 in C_6H_6 at room temperature [1].

$O{=}S{=}NNHC_6H_5$ was first prepared in 1889 by Michaelis [2] by reaction of $C_6H_5NHNH_2$ with $SOCl_2$ (mole ratio 3:1) in ether (no yield given). More than 70 years later it was reported to form in ~82% yield by reaction of $C_6H_5NHNH_2$ with $SOCl_2$ and pyridine (mole ratio 1:1:2) in $CHCl_3$ with ice cooling for 2 h and 1 h at room temperature [3]. More recent attempts to duplicate the reported yield failed. The best yield obtained by the latter method was 34%. The yield increased up to 40% when $SOCl_2$ was added to a twofold excess of $C_6H_5NHNH_2$ and of pyridine in $CHCl_3$ at 0°C [4]. The compound formed nearly quantitatively by transthionylation of $C_6H_5NHNH_2$, preferentially in CH_3COOH with an equimolar amount of $O{=}S{=}NC_6H_5$ [5, 6], see also [4] or with $O{=}S{=}NOCH_2C_6H_5$ [7] both in C_2H_5OH.

The crystral structure was determined by X-ray diffraction at room temperature [9] and by neutron diffraction at 120 K [10]. The compound crystallizes in the monoclinic space group $P2_1/a{-}C^5_{2h}$ (No. 14), Z=8 [9, 10] with a=10.50(5), b=15.75(1), c=8.45(9) Å and β=94.4°, D_x= 1.47, D_m=1.45 g/cm³, R=0.042 (2642 independent reflections used) [9]; a=10.451(10), b= 15.561(18), c=8.346(10) Å and β=95.44°, V=1351.2 Å³, D_x=1.516 g/cm³, R=0.056 and R_w= 0.055 (1789 unique reflections used) [10]. The asymmetric unit contains two identical molecules which form a dimer, see **Fig. 8**. They are linked by two N–H···O=S hydrogen bonds. The hydrogen bonds are bifurcated, forming a five-membered chelate ring with an intramolecular branch. The distances in the neutron study at 120 K are larger than the corresponding X-ray values, with one exception. The S–O bond is 0.007(8) Å shorter. The largest differences occur in the inter- and intramolecular O···H distances which are on average 0.13(2) Å longer in the neutron study [9, 10]. Bond lengths and bond angles are given in Fig. 8, selected distances from the X-ray diffraction and neutron diffraction study are compared in Table 20, p. 56, for atomic coordinates and isotropic temperature factors, see [10].

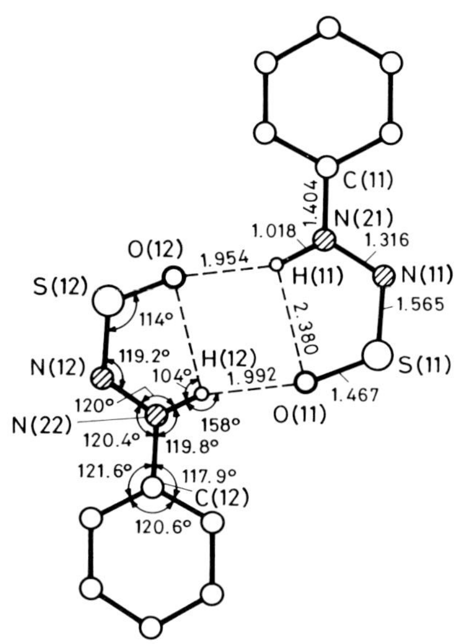

Fig. 8. Molecular structure of $O{=}S{=}NNHC_6H_5$. Bond lengths (in Å) and bond angles (in °) averaged over the two molecules of the asymmetric unit. The mean e.s.d.'s of the bond lengths and bond angles (not including S and H) are 0.005 Å and 0.3°, respectively. For bonds including S and H atoms the means are 0.008 Å and 0.5° [10].

Table 20
Atomic Distances of $O{=}S{=}NNHC_6H_5$ (in Å).

	X-ray data [9] at room temperature	neutron data [10] at 120 K
C(11)–N(21)	1.400	1.404
N(21)–N(11)	1.312	1.316
N(11)–S(11)	1.560	1.565
S(11)–O(11)	1.474	1.467
N(21)–H(11)	0.87	1.018
H(11)···O(11)	2.43	2.380
H(11)···O(12)	2.13	1.954
H(12)···O(11)	—	1.992

The yellow prisms melt at 105°C (from C_2H_5OH) [2, 3, 6]; melting points between 103 and 105.5°C are given in [4].

UV (cyclohexane): λ_{max} (in nm) (log ε) = 356 (4.22); a figure and additional data pertaining to it is given in the paper [8].

EIMS (75 eV): m/e (rel. int. in %) = 154 (5) M^+, 105 (16), 78 (16), 77 (100), 76 (6), 74 (8), 65 (8), 63 (5), 52 (10), 51 (47), 50 (22), 49 (9), 48 (10), 39 (12), 38 (6) [11].

$O{=}S{=}NNHC_6H_5$ is readily soluble in ether, $CHCl_3$, CS_2, C_6H_6, and hot C_2H_5OH [2].

Thermal decomposition of $O{=}S{=}NNHC_6H_5$ afforded H_2O, SO_2, N_2, $(C_6H_5)_2S$, and $(C_6H_5)_2S_2$ [2, 6]. The thermolysis of $O{=}S{=}NNHC_6H_5$ in a number of dry solvents (C_6H_5Cl, C_6H_5Br, $C_6H_5OCH_3$, and $C_6H_5NO_2$) stirred at 135°C under nitrogen for 72 h led to the formation of o-, m-, and p-substituted biphenyl products, C_6H_6, $(C_6H_5S)_2$, $(C_6H_5)_2S$, $C_6H_5SSO_2C_6H_5$, sulfur, and SO_2 [12, 13]. The photolysis of $O{=}S{=}NNHC_6H_5$ in dry solvents (C_6H_5Cl, C_6H_5Br, and $C_6H_5OCH_3$) at room temperature for 24 h with a high-pressure mercury lamp afforded a mixture of o-, m-, and p-substituted biphenyls, sulfur, and SO_2 [14]. Treatment of the compound with Br_2 gave $C_6H_5N{=}NBr_3$, H_2SO_4, and HBr. It was not attacked by H_2O and was only slowly attacked by diluted acids. With concentrated HCl the hydrazinium chloride and H_2SO_3 was formed. It reacted rapidly with diluted NaOH solution to give $C_6H_5NHNH_2$ and Na_2SO_3 [2, 6]. Oxidation of $O{=}S{=}NNHC_6H_5$ with H_2O_2 (38%) (mole ratio 1:4) in acetone for several days gives $(C_6H_5)_2SO_2$ (42%) [3]. When it was reacted with $SOCl_2$ in ether solution $C_6H_5N{=}NCl$, SO_2, HCl, and sulfur were formed [6]. Treatment of a 1:1 mixture of the compound and pyridine in $CHCl_3$ with an equimolar amount of $SOCl_2$ at 0°C for 1.5 h resulted in the evolution of SO_2 and the precipitation of elemental sulfur [4]. Treatment of the compound with PCl_3 in ether led to $C_6H_5N{=}NCl$ and $P(S)Cl_3$ [6]. It reacted with C_6H_5NCO (mole ratio 1:1) in C_6H_6 for several days to give $(C_6H_5NH)_2CO$ [3]. The reaction of equimolar amounts of $O{=}S{=}NNHC_6H_5$ and $(C_6H_5CO_2)_2$ in boiling dry benzene for 72 h under N_2 afforded $(C_6H_5)_2$, $(C_6H_5)_2S$, $(C_6H_5S)_2$, $(C_6H_5)_2SO$, and $C_6H_5SSO_2C_6H_5$ [15, 16]. When equimolar amounts of $O{=}S{=}NNHC_6H_5$ and styrene oxide, $C_6H_5\overline{CH{-}CH_2{-}O}$, were heated in the presence of $(C_2H_5)_4N^+Br^-$ in refluxing C_6H_6 for 16 h $(C_6H_5S)_2$, $(C_6H_5)_2S$, $(C_6H_5)_2$, and $C_6H_5CH(CH_2OH)SO_2C_6H_5$ were obtained [17]. It reacted with $CH_3C(O)Cl$ to give $C_6H_5N{=}NCl$, CH_3COOH, and sulfur [6]. $O{=}S{=}NNHC_6H_5$, gently warmed with Na powder in dry toluene, gave C_6H_5SH with N_2 evolution [18]. When $O{=}S{=}NNHC_6H_5$ was added to a solution of $NaOC_4H_9$-t in C_6H_6 (mole ratio 2:1) at room temperature, evolution of N_2 was observed after 10 to 15 min, and after 48 h a mixture of products, $C_6H_5SO_2H$, C_6H_5SH, $(C_6H_5S)_2$, $(C_6H_5)_2S$, $(C_6H_5)_2$, $C_6H_5N_2C_6H_5$, and $C_6H_5NHNH_2$ was isolated [19]. Treatment of $O{=}S{=}NNHC_6H_5$ with C_6H_5MgBr (mole ratio

1:1.3 or 1:2) yielded $MgBr^+[OSNNC_6H_5]^-$ [3]. When the compound was reacted with $Fe_2(CO)_9$ (mole ratio ~1:1) in C_6H_6 at room temperature for 7 h $(Fe(CO)_3(SC_6H_5))_2$ and C_6H_5SH were produced [20]. The oxidation of $O{=}S{=}NNHC_6H_5$ with an equimolar amount of Ag_2O in C_6H_6 at room temperature for 48 h afforded a mixture of $(C_6H_5)_2S$, $(C_6H_5S)_2$, C_6H_5SH, $C_6H_5SO_2H$, $C_6H_5SO_3H$, $C_6H_5SSO_2C_6H_5$, $(C_6H_5)_2SO$, $(C_6H_5)_2SO_2$, $(C_6H_5)_2$, and $C_6H_5N_2C_6H_5$ [21].

References:

[1] Michaelis, A.; Ruhl, J. (Chem. Ber. **23** [1890] 474/7).
[2] Michaelis, A. (Chem. Ber. **22** [1889] 2228/33).
[3] Klamann, D.; Krämer, U.; Weyerstahl, P. (Chem. Ber. **95** [1962] 2694/9).
[4] Pearce, L. B.; Feingold, M. H.; Cerny, K. F.; Anselme, J.-P. (J. Org. Chem. **44** [1979] 1881/3).
[5] Michaelis, A. (Chem. Ber. **24** [1891] 745/57).
[6] Michaelis, A.; Ruhl, J. (Liebigs Ann. Chem. **270** [1892] 114/22).
[7] Michaelis, A.; Schröter, G. (Ber. Deut. Chem. Ges. **26** [1893] 2155/7).
[8] Leandri, G.; Mangini, A. (Boll. Sci. Fac. Chim. Ind. Bologna **17** [1959] 15/8).
[9] Gieren, A.; Dederer, B. (Angew. Chem. **89** [1977] 190/1; Angew. Chem. Intern. Ed. Engl. **16** [1977] 179).
[10] Schanda, F.; Gieren, A.; Filhol, A. (Acta Cryst. C **40** [1984] 306/8).

[11] Bowie, J. H.; Larsson, F. C. V.; Schroll, G.; Lawesson, S.-O.; Cooks, R. G. (Tetrahedron **23** [1967] 3743/52).
[12] De Luca, G.; Renzi, G.; Cipollini, R.; Pizzabiocca, A. (J. Chem. Soc. Perkin Trans. I **1980** 1901/3).
[13] De Luca, G.; Renzi, G.; Pizzabiocca, A. (Chim. Ind. [Milan] **61** [1979] 73).
[14] De Luca, G.; Renzi, G.; Pizzabiocca, A. (Chem. Ind. [London] **1979** 899).
[15] De Luca, G.; Renzi, G.; Pizzabiocca, A. (Chem. Ind. [London] **1982** 328/9).
[16] De Luca, G.; Renzi, G.; Pizzabiocca, A. (Chim. Ind. [Milan] **61** [1979] 695/6).
[17] Tsuge, O.; Mataka, S. (Bull. Chem. Soc. Japan **44** [1971] 1896/900).
[18] Leandri, G.; Spinelli, D. (Experientia **15** [1959] 9).
[19] Carpanelli, C.; Leandri, G.; Poluzzi Corallo, G. (Gazz. Chim. Ital. **102** [1972] 539/45).
[20] Glass, W. K.; McBreen, J. O. (J. Organometal. Chem. **198** [1980] 71/2).

[21] Carpanelli, C.; Leandri, G.; Poluzzi Corallo, G. (Gazz. Chim. Ital. **103** [1973] 469/75).

2.6.13.1.3.2 Other N′-Sulfinyl-N-aryl-hydrazines, O=S=NNHAr, Ar is given in Table 21

N′-Sulfinyl-N-aryl-hydrazines were prepared analogously to N′-sulfinyl-N-phenyl-hydrazine (see p. 55) either by passing a stream of SO_2 into a solution of $ArNHNH_2$ in boiling C_6H_6 (method I), by reaction of $ArNHNH_2$ with $SOCl_2$ in ether or $CHCl_3$ (method II), or by reaction of $ArNHNH_2$ (or $ArNHNH_2 \cdot 2HCl$) with $O{=}S{=}NC_6H_5$ (and CH_3COONa) in alcohol–CH_3COOH (method III). The compounds are summarized in Table 21, p. 58.

N′-Sulfinyl-N-aryl-hydrazines, O=S=NNHAr, reacted similarly to N′-sulfinyl-N-phenyl-hydrazine, see pp. 56/7. The compounds were not attacked by H_2O, but reacted with aqueous NaOH solution to give the respective hydrazides, $ArNHNH_2$, and Na_2SO_3 [2, 4, 5]. Treatment with Br_2 in ether or heating with $SOCl_2$ generally yields the corresponding diazo compounds, $ArN{=}NBr_3$ and ArN=NCl, respectively [2].

Table 21
N′-Sulfinyl-N-aryl-hydrazines, O=S=NNHAr.

O=S=NNHAr Ar	method of preparation	appearance, properties	Ref.
2-ClC_6H_4	II	yellow needles, m.p. 42°C (from aqueous C_2H_5OH); UV*): λ_{max} (in nm) (log ε) = 248 (3.89), 355 (4.27)	[1]
4-ClC_6H_4	III	greenish yellow needles, m.p. 159°C (from C_2H_5OH)	[2]
4-BrC_6H_4	II [1], III [2]	yellow needles, m.p. 168°C (from C_2H_5OH) [2], 176°C (from C_2H_5OH) [1]; UV*): λ_{max} (in nm) (log ε) = 239 (3.93), 360 (4.33); a figure is given in the paper [1]; readily soluble in ether, C_6H_6, $CHCl_3$, and hot C_2H_5OH [2]	[1, 2]
3-$CF_3C_6H_4$	III	yield 64%; m.p. 100°C	[3]
4-$CF_3C_6H_4$	III	m.p. 140°C	[3]
4-$CF_3OC_6H_4$	III	m.p. 96 to 97°C	[3]
2-$NO_2C_6H_4$	II [1], III [2]	small red crystal leaves [2], yellow-orange crystals [1], m.p. 128°C (from C_2H_5OH or C_6H_6) [1, 2]; UV: λ_{max} (in nm) (log ε) = 217 (3.99), 268 (3.88), 337 (4.01), 386 (4.04); a figure is given in the paper [1]; readily soluble in C_6H_6 and hot C_2H_5OH; difficultly soluble in cold C_2H_5OH, ether, and ligroine [2]	[1, 2]
3-$NO_2C_6H_4$	II [1], III [2]	pale yellow needles, m.p. 185°C (from benzene) [1, 2]; UV*): λ_{max} (in nm) (log ε) = 252 (4.23), 346 (4.02); a figure is given in the paper [1]; readily soluble in C_6H_6 and hot C_2H_5OH; difficultly soluble in cold C_2H_5OH, ether, and ligroine [2]	[1, 2]
4-$NO_2C_6H_4$	II	rust-brown needles, m.p. 209°C (from C_6H_6); UV*): λ_{max} = 367 to 368 nm, a figure is given in the paper	[1]
4-$C_6H_5SO_2C_6H_4$	II	yellow crystals, m.p. 213°C (from C_2H_5OH); UV*): λ_{max} = 356 nm; a figure is given in the paper	[1]
2-$CH_3OC_6H_4$	II	gold-colored prisms, m.p. 49°C (from aqueous C_2H_5OH); UV*): λ_{max} (in nm) (log ε) = 250 (3.80), 368 (4.26); a figure is given in the paper	[1]

Table 21 (continued)

O=S=NNHAr Ar	method of preparation	appearance, properties	Ref.
4-$CH_3OC_6H_4$	II	gold-yellow-colored plates, m.p. 125°C (from ligroine); UV*): λ_{max} (in nm) (log ε) = 239 (3.82), 370 (4.25); a figure is given in the paper	[1]
2-$HOOCC_6H_4$	III	pale yellow needles, m.p. 152°C (from C_2H_5OH–H_2O (1:2)); dec. at ~155°C explosively	[2]
3-$HOOCC_6H_4$	III	yellow lustrous crystal leaves, m.p 231°C (from C_2H_5OH); almost insoluble in C_6H_6, petroleum ether, and C_2H_5OH	[2]
4-$HOOCC_6H_4$	III	yellowish crystals, m.p. 258°C (from C_2H_5OH); slowly soluble in C_2H_5OH	[2]
4-$C_6H_5C(O)C_6H_4$	II	yellow plates, m.p. 197°C (from C_2H_5OH); UV*): λ_{max} (in nm) (log ε) = 248 (4.14), 362 (4.19); a figure is given in the paper	[1]
2-$CH_3C_6H_4$	I	yellow oil with an aromatic odor	[4, 5]
3-$CH_3C_6H_4$	II	pale yellow tabular crystals, m.p. 90°C (from ligroine); UV*): λ_{max} (in nm) (log ε) = 358 to 359 (4.26)	[1]
4-$CH_3C_6H_4$	I [4, 5], II, III [5]	yellow, silky lustrous needles (from C_2H_5OH), m.p. 112°C	[4, 5]
4-$C_6H_5C_6H_4$	II	yellow prisms, m.p. 187°C (from C_6H_6); UV: λ_{max} (in nm) (log ε) = 258 (4.07), 368 to 369 (4.25); a figure is given in the paper	[1]
2-Cl-4-$CF_3C_6H_3$	III	m.p. 46 to 47°C	[3]
3-Cl-4-$CF_3C_6H_3$	III	m.p. 158°C	[3]
4-Cl-3-$CF_3C_6H_3$	III	m.p. 148 to 149°C	[3]
$Br_2C_6H_3$	III	transparent yellow needles, m.p. 99°C (from C_2H_5OH); sublimes on heating slightly	[2]
4-Br-2-$NO_2C_6H_3$	III	pale yellow crystals, m.p. 157°C (from C_2H_5OH)	[2]
α-$C_{10}H_7$	II [1, 5], III [1]	yellow-red plates (from C_2H_5OH) [5] or pink needles (from ligroine) [1], m.p. 100°C [1, 5]; UV: λ_{max} (in nm) (log ε) = 248 (4.41), 329 to 337 (3.92); a figure is given in the paper [1]	[1, 5]
β-$C_{10}H_7$	II, III	yellow-red needles, m.p. 136 to 139°C (from C_2H_5OH) with decomposition	[5]

*) Additional data are given in the paper [1].

O=S=NNHAr (Ar = 4-ClC_6H_4, 4-BrC_6H_4, and 4-$CH_3OC_6H_4$) reacted with an equimolar amount of $(C_6H_5CO_2)_2$ in boiling dry benzene for 72 h to give ArC_6H_5, $(C_6H_5)_2S$, $(C_6H_5S)_2$, $(C_6H_5)_2SO$, $C_6H_5SSO_2C_6H_5$, and $(C_6H_5)_2$ [6, 7]. When O=S=$NNHC_6H_4CH_3$-4 was heated with an equimolar amount of styrene oxide, $C_6H_5\text{–}\overline{CH\text{–}CH_2\text{–}O}$, in the presence of $(C_2H_5)_4N^+Br^-$ in C_6H_6 for 16 h (4-$CH_3C_6H_4S)_2$, (4-$CH_3C_6H_4)_2S$, and 4-$CH_3C_6H_4C_6H_5$ were obtained [8]. O=S=NNHAr (Ar = $CH_3C_6H_4$, $CH_3OC_6H_4$, and $HalC_6H_4$; in ortho, meta, or para position), gently heated with Na powder in dry toluene, afforded ArSH along with N_2 [9]. When O=S=NNHAr (Ar = 4-$CH_3C_6H_4$, 4-BrC_6H_4, and α-$C_{10}H_7$) was added to a solution of $NaOC_4H_9$-t in C_6H_6 (mole ratio 2:1) at room temperature, evolution of N_2 was observed after 10 to 15 min and after 48 h a mixture of products, $ArSO_2H$, ArSH, $(ArS)_2$, $(Ar)_2S$, ArH, ArC_6H_5, ArN=NAr, and $ArNHNH_2$ was isolated [10]. The oxidation of O=S=NNHAr (Ar = 4-$CH_3C_6H_4$, 4-BrC_6H_4, α-$C_{10}H_7$, β-$C_{10}H_7$, and 2-$C_6H_5C_6H_4$) with an equimolar amount of Ag_2O in C_6H_6 at room temperature for 48 h yielded a mixture of ArSH, $ArSO_2H$, $ArSO_3H$, $(ArS)_2$, $(Ar)_2S$, $ArSSO_2Ar$, $(Ar)_2SO$, $(Ar)_2SO_2$, ArH, ArC_6H_5, and ArN=NAr [11].

The use of O=S=NNHAr derivatives as fungicides [12, 13], acaricides, and ovicides [13] is claimed in patents.

References:

[1] Leandri, G.; Mangini, A. (Boll. Sci. Fac. Chim. Ind. Bologna **17** [1959] 15/8).
[2] Klieeisen, J. (Chem. Ber. **27** [1894] 2549/56).
[3] Kühle, E.; Klauke, E.; Frohberger, P.-E. (Ger. Offen. 2244616 [1972/74] 1/22; C.A. **80** [1974] No. 145781).
[4] Michaelis, A.; Ruhl, J. (Chem. Ber. **23** [1890] 474/7).
[5] Michaelis, A.; Ruhl, J. (Liebigs Ann. Chem. **270** [1892] 114/22).
[6] De Luca, G.; Renzi, G.; Pizzabiocca, A. (Chem. Ind. [London] **1982** 328/9).
[7] De Luca, G.; Renzi, G.; Pizzabiocca, A. (Chim. Ind. [Milan] **61** [1979] 695/6).
[8] Tsuge, O.; Mataka, S. (Bull. Chem. Soc. Japan **44** [1971] 1896/1900).
[9] Leandri, G.; Spinelli, D. (Experientia **15** [1959] 9).
[10] Carpanelli, C.; Leandri, G.; Poluzzi Corallo, G. (Gazz. Chim. Ital. **102** [1972] 539/45).

[11] Carpanelli, C.; Leandri, G.; Poluzzi Corallo, G. (Gazz. Chim. Ital. **103** [1973] 469/75).
[12] Brandes, W.; Frohberger, P.-E.; Scheinpflug, H.; Krämer, W. (Ger. Offen. 2552967 [1975/77] 1/37; C.A. **87** [1977] No. 64060).
[13] Wegler, R.; Unterstenhöfer, G. (U.S. 2898265 [1959]; C.A. **1960** 6020).

2.6.13.1.4 2-Sulfinyl-N-phenyl-hydrazinecarboxamide, 1-Sulfinyl-4-phenyl-semicarbazide, O=S=NNHC(O)NHC_6H_5

The compound was prepared nearly quantitatively by treatment of a solution of O=S=NSO_2-$C_6H_4CH_3$-4 in $CHCl_3$ with an equimolar amount of $C_6H_5NHC(O)NHNH_2$ at room temperature. After 5 h the precipitate was filtered, the solvent reduced in volume, and O=S=NNHC(O)NHC_6H_5 precipitated with petroleum ether; m.p. 137 to 139°C with decomposition (from benzene).

On heating slightly the compound decomposed with evolution of N_2 to give $C_6H_5NH_2$ and CO_2.

Reference:

Kresze, G.; Horn, A.; Philippson, R.; Trede, A. (Chem. Ber. **98** [1965] 3401/9).

2.6.13.1.5 N′-Sulfinyl-carbohydrazides

2.6.13.1.5.1 N′-Sulfinyl-alkanohydrazides, O=S=NNHC(O)R, R = CH_3, $CH_3(CH_2)_4$, $CH_3(CH_2)_6$, $C_6H_5CH_2$, α-$C_{10}H_7CH_2$

The compounds were prepared by boiling $SOCl_2$ and the corresponding hydrazide, RC(O)-$NHNH_2$ (mole ratio 4:1) under reflux in $CHCl_3$ for 1 h (R = CH_3, $CH_3(CH_2)_4$, $CH_3(CH_2)_6$), 2 h (R = α-$C_{10}H_7CH_2$), or 4 h (R = $C_6H_5CH_2$). Evaporation of the solvent and recrystallization of the residue from light petroleum (b.p. 40 to 60°C) or C_6H_6 gave the sulfinyl-alkanohydrazides. Yields and properties of the compounds are compiled in Table 22, p. 62.

The chemical reactivity of the compounds may be influenced by a tautomeric equilibrium O=S=NNHC(O)R ↔ O=S=NN=C(OH)R. On heating, the compounds decompose with evolution of N_2 to give the corresponding carboxylic acids, RCOOH, and elemental sulfur. Hydrolysis with H_2O yields, with one exception, the corresponding hydrazides, RC(O)$NHNH_2$ and SO_2. O=S=NNHC(O)$CH_2C_6H_5$ is not hydrolyzed. The rates of hydrolysis depend on R; generally they are faster with smaller R (see the paper). When the sulfinyl-hydrazides are boiled under reflux with $SOCl_2$ in $CHCl_3$ for 24 h they are converted into the corresponding acid chlorides RC(O)Cl and/or carboxylic acids RCOOH, also in dependence of R.

Reference:

Hope, P.; Wiles, L. A. (J. Chem. Soc. **1965** 5386/90).

2.6.13.1.5.2 N′-Sulfinyl-benzohydrazides, O=S=NNHC(O)C_6H_4R, R = H, 2-NH_2, 4-CH_3

O=S=NNHC(O)C_6H_5 and O=S=NNHC(O)$C_6H_4CH_3$-4 were prepared in 81 and 83.5% yield, respectively, by boiling $C_6H_5C(O)NHNH_2$ or 4-$CH_3C_6H_4C(O)NHNH_2$ and $SOCl_2$ (mole ratio 1:4) under reflux in $CHCl_3$ for 2 h [1]. O=S=NNHC(O)C_6H_5 and O=S=NNHC(O)$C_6H_4NH_2$-2 were obtained in 77% [2] and 79% [3] yield, respectively, by reacting $C_6H_5C(O)NHNH_2$ or 2-$NH_2C_6H_4$-C(O)$NHNH_2$ with equimolar amounts of O=S=$NSO_2C_6H_4CH_3$-4 in $CHCl_3$ at room temperature. Melting points and spectral data of the compounds are compiled in Table 23.

Table 23
Melting Points and Spectra of O=S=NNHC(O)C_6H_4R.

R	m.p. in °C	spectra	Ref.
H	50 to 55 (dec.) [2], 82 (dec.) [1]	IR (Nujol or CCl_4): ν (in cm^{-1}) = 1720 ν(CO); 1470, 1290 ν(NH/CN); 1090 ν_s(O=S=N) [1]	[1, 2]
2-NH_2	97 (dec.) (orange crystals)	^{1}H NMR ($(CD_3)_2CO$/TMS): δ (in ppm) = 5.90 (s, broad, 3H), 6.85 (m, 2H), 7.45 (m, 1H), 7.90 (m, 1H) IR (KBr): ν (in cm^{-1}) = 1250 ν_{as}(O=S=N); 1080 ν_s(O=S=N)	[3]
4-CH_3	105 (dec.)	IR (Nujol or CCl_4): ν (in cm^{-1}) = 3125 ν(NH); 1670 ν(CO); 1460, 1270 ν(NH/CN); 1210 ν_{as}(O=S=N); 1100 ν_s(O=S=N)	[1]

Table 22
Yields, Properties, and IR Spectra of O=S=NNHC(O)R (reference on p. 61).

R	yield in %	appearance m.p. in °C	IR (Nujol or CCl_4) ν in cm^{-1}	remarks
CH_3	81.7	pale yellow solid 65 to 69	—	fumes in moist air, decomposes even on drying in vacuum
$CH_3(CH_2)_4$	65	light brown crystals 33 to 34	3160 ν(NH); 1685 ν(CO); 1460, 1280 ν(NH/CN); 1220 $ν_{as}$(O=S=N); 1100 $ν_s$(O=S=N)	acidification gives an oil; it decomposes readily in 5 N NaOH solution with evolution of N_2
$CH_3(CH_2)_6$	80.5	pink-colored plates 50	3185 ν(NH); 1695 ν(CO); 1460, 1280 ν(NH/CN); 1215 $ν_{as}$(O=S=N); 1100 $ν_s$(O=S=N)	it decomposes slowly already at room temperature; hydrolysis by hot 5 N NaOH solution gives $CH_3(CH_2)_6C(O)NHNH_2$
$C_6H_5CH_2$	92	white plates 129	3155 ν(NH); 1680 ν(CO); 1460 ν(NH/CN); 1215 $ν_{as}$(O=S=N); 1100 $ν_s$(O=S=N)	stable at room temperature and against H_2O; it is readily hydrolyzed by 5 N NaOH solution with N_2 evolution; boiling under reflux in $CHCl_3$ in the presence of pyridine gives N_2, sulfur, and $C_6H_5CH_2COOH$
α-$C_{10}H_7CH_2$	70	162 (dec.)	3105 ν(NH); 1710 ν(CO); 1455, 1270 ν(NH/CN); 1215 $ν_{as}$(O=S=N); 1100 $ν_s$(O=S=N)	—

Chemical reactivity of the compounds was assumed to be influenced by the presence of a tautomeric equilibrium $O{=}S{=}NNHC(O)C_6H_4R \leftrightarrow O{=}S{=}NN{=}C(OH)C_6H_4R$. Thus, decomposition of the compounds with R = H and R = 4-CH_3 by heating at 65 and 100°C, respectively, led to N_2, sulfur and the corresponding carboxylic acids, RC_6H_4COOH, presumably via the enol form [1]. The compounds (R = H, 2-NH_2, and 4-CH_3) were readily hydrolyzed with H_2O to give SO_2 and the corresponding hydrazides, $RC_6H_4C(O)NHNH_2$ [1, 2, 3]. When the compounds with R = H and R = 4-CH_3 were boiled under reflux with $SOCl_2$ in $CHCl_3$ for 24 h the corresponding acid chlorides, $RC_6H_4C(O)Cl$, were formed in 77 and 52% yield, respectively with evolution of HCl, N_2, and SO_2 [1].

References:

[1] Hope, P.; Wiles, L. A. (J. Chem. Soc. **1965** 5386/90).
[2] Kresze, G.; Horn. A.; Philippson, R.; Trede, A. (Chem. Ber. **98** [1965] 3401/9).
[3] Grill, H.; Kresze, G. (Liebigs Ann. Chem. **749** [1971] 171/7).

2.6.13.1.6 2-Sulfinyl-hydrazinecarbothioamide, 1-Sulfinyl-thiosemicarbazide, $O{=}S{=}NNHC(S)NH_2$

The compound was obtained in 71% yield by reacting $O{=}S{=}NSO_2C_6H_4CH_3$-4 with an equimolar amount of $H_2NC(S)NHNH_2$ in ethyl acetate. The reaction mixture was allowed to stand overnight at −20°C and the compound was separated by filtration; decomposition point 104°C.

On heating, the compound decomposed violently to give H_2O, N_2, SO_2, and sulfur. Alkaline hydrolysis afforded $H_2NC(S)NHNH_2$.

Reference:

Kresze, G.; Horn, A.; Philippson, R.; Trede, A. (Chem. Ber. **98** [1965] 3401/9).

2.6.13.2 N'-Sulfinyl-N,N-diorganyl-hydrazines and Derivatives

2.6.13.2.1 N'-Sulfinyl-N,N-dialkyl-, -N-alkyl-N-phenyl-, and -N,N-diphenyl-hydrazines, $O{=}S{=}NNR^1R^2$, R^1 and R^2 are listed in Table 24

Preparation. Properties

Method Ia: The compounds were prepared by reacting ethereal solutions of $R^1R^2NNH_2$ with $SOCl_2$ and $(C_2H_5)_3N$ (mole ratio 1:1:2) with cooling for 2 to 3 h. After filtration of $(C_2H_5)_3N \cdot HCl$, the solvent was removed and the resulting liquid distilled in vacuum.

Method Ib: $O{=}S{=}NN(CH_3)_2$ formed by reaction of solutions of $(CH_3)_2NNH_2$ and $SOCl_2$ (mole ratio 3.3:1) in benzene, followed by refluxing of the reaction mixture for 30 min. After separation of the hydrochloride, the compound was distilled in vacuum.

Method Ic: The compounds were obtained by treating a mixture of hydroxyalkylhydrazines, $[HO(CH_2)_n]_2NNH_2$ or $HO(CH_2)_nNCH_3NH_2$, and pyridine at 0°C with a solution of $SOCl_2$ in $CHCl_3$ (mole ratio 1:4:3). After 2 h the reaction mixture was extracted with Skellysolve A and the products isolated by distillation.

Method II: $O{=}S{=}NN(CH_3)_2$ was prepared by dissolving N,N′-bis(sulfinyl)-1,4-phenylenediamine, $O{=}S{=}NC_6H_4N{=}S{=}O$-4, slowly in $(CH_3)_2NNH_2$ (mole ratio ~1:1.2) followed by distillation under reduced pressure.

Method III: $O{=}S{=}NN(C_6H_5)_2$ was prepared by adding $(C_6H_5)_2NNH_2$ to a solution of $O{=}S{=}NSO_2$-$C_6H_4CH_3$-4 in CCl_4 (mole ratio 1:1). After stirring for 12 h at room temperature the solvent was reduced and the compound precipitated.

Yields, properties, and spectra of the compounds are summarized in Table 24.

Table 24
Yields, Properties, and Spectra of $O{=}S{=}NNR^1R^2$.

$O{=}S{=}NNR^1R^2$ R^1	R^2	method of preparation (yield in %)	properties, spectral data
CH_3	CH_3	Ia (58) [1], Ia [2], Ib (70) [3], II (91) [4]	yellow liquid, b.p. 159°C [3], 88°C/50 Torr [1], yellow oil 69°C/20 Torr [2], 48 to 49°C/10 Torr [4] n_D^{25} = 1.5407 [4] ^1H NMR (CCl_4/TMS): δ = 3.42 ppm [1]; (neat/TMS): δ = 3.37 ppm [4] IR (capillary film): ν (in cm^{-1}) = 1185, 1130(?) ν_{as}(O=S=N); 1060 ν_s(O=S=N); additionally 9 unassigned bands between 2920 and 740 cm^{-1} recorded [1] UV: λ_{max} (in nm) (log ε) = 234 (3.15), 312 (4.05), a figure is given in the paper [2]
C_2H_5	C_2H_5	Ia	b.p. 60 to 62°C/11 Torr ^1H NMR ($CDCl_3$/TMS): δ = 1.30 (t, 6H), 3.70 ppm (q, 4H) ^{13}C NMR: δ = 13.0 (CH_3), 52.3 ppm (CH_2) IR: ν (in cm^{-1}) = 1160 ν_{as}(O=S=N); 1065 ν_s(O=S=N) MS: m/e = 134 M^+ [5]
i-C_3H_7	i-C_3H_7	Ia (62)	m.p. 43 to 45°C ^1H NMR ($CDCl_3$/TMS): δ = 1.28 (d, 12H), 4.48 ppm (sept, 2H) IR: ν (in cm^{-1}) = 1140 ν_{as}(O=S=N); 1065 ν_s(O=S=N) [5]
$ClCH_2CH_2$	$ClCH_2CH_2$	Ic (11)	boiling range 107 to 112°C/0.5 Torr [6]
$CH_2{=}CClCH_2$	$CH_2{=}CClCH_2$	Ic (24)	boiling range 78 to 88°C/0.1 Torr [6]

Table 24 (continued)

$O{=}S{=}NNR^1R^2$ R^1	R^2	method of preparation (yield in %)	properties, spectral data
cis-$ClCH{=}CHCH_2$	*cis*-$ClCH{=}CHCH_2$	Ic (36)	boiling range 80 to 95°C/0.1 Torr [6]
trans-$ClCH{=}CHCH_2$	*trans*-$ClCH{=}CHCH_2$	Ic (16)	boiling range 75 to 91°C/0.1 Torr [6]
$C_6H_5CHClCH_2$	$C_6H_5CHClCH_2$	Ic (2)	m.p. 94 to 96°C [6]
CH_3	$ClCH_2CH_2$	Ic (7)	boiling range 93 to 98°C/0.2 Torr [6]
CH_3	$CH_3CHClCH_2$	Ic (6)	boiling range 72 to 75°C/0.1 Torr [6]
CH_3	$CH_3CH_2CHClCH_2$	Ic (23)	boiling range 70 to 94°C/0.2 Torr [6]
CH_3	C_6H_5	Ia (72.5) [7]	yellow prisms, m.p. 77°C (from ligroine) [2], m.p. 76.5°C [7], yellow needles, m.p. 77°C (from C_2H_5OH) [8] UV: λ_{max} (in nm) (log ε) = 238 (3.97), 349 (4.25), a figure with additional data is given in the paper [2]
C_2H_5	C_6H_5	Ia	yellow oil with an aromatic smell [8, 9]
i-C_4H_9	C_6H_5	Ia	yellow oil with a penetrating smell [8]
$C_6H_5CH_2$	C_6H_5	Ia (84.3) [7]	yellow prisms, m.p. 61.5°C [7], 65°C (from ligroine) [8]
C_6H_5	C_6H_5	Ia (79.8) [7], see also [5], III (93) [10]	yellow needles, m.p. 45°C [10], 46 to 47°C (from C_2H_5OH) [7], 48 to 50°C [5] ^{13}C NMR ($CDCl_3$/TMS): δ (in ppm) = 121.6 (4C, ortho), 128.9 (4C, meta), 126.6 (2C, para), 143.8 (2C) [5] IR: ν = 1160, 1115 cm^{-1} [5], IR (KBr): ν (in cm^{-1}) = 1590, 1480, 1290, 1162, 1113 [10] MS: m/e = 230 M^+ [5]

Structure

X-ray diffraction analyses have been performed for $O{=}S{=}NN(C_3H_7\text{-}i)_2$ [5] and $O{=}S{=}NN(CH_3)C_6H_5$ (for atomic coordinates, see the paper) [11]. The crystallographic data are listed in Table 25. **Fig. 9** shows that in the solid state the OSN(1)N(2)C(1)C(2) moieties of both compounds are nearly planar (see also pp. 55/6). The methine hydrogen bonded to C(1) in $O{=}S{=}NN(C_3H_7\text{-}i)_2$ (Fig. 9a) and the position of H(31) in $O{=}S{=}NN(CH_3)C_6H_5$ (Fig. 9b) are very close to this plane. The distance C(1)H···O of 2.3 Å supports the existence of a strong C(1)H···O interaction in $O{=}S{=}NN(C_3H_7\text{-}i)_2$ [5]. However, it is deduced that the position of H(31) in $O{=}S{=}NN(CH_3)C_6H_5$ is determined by steric restrictions rather than by an H···O interaction [11].

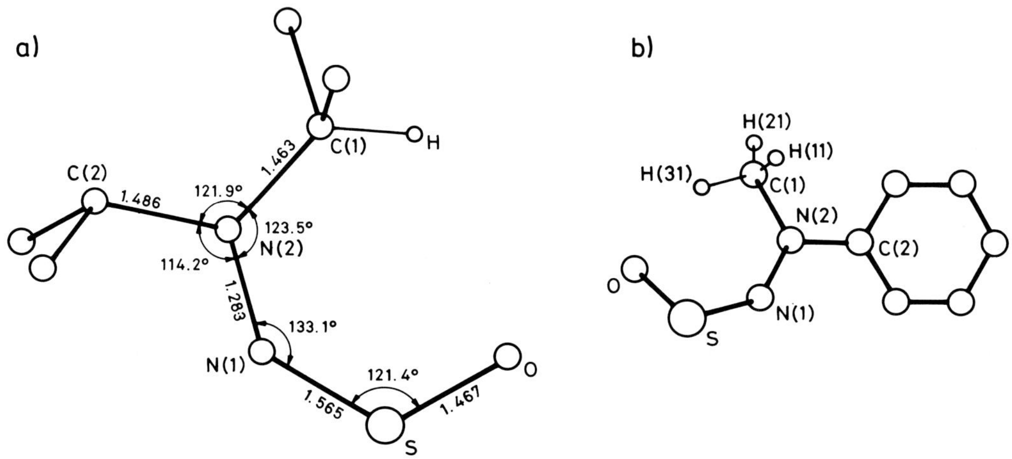

Fig. 9. a) Molecular structure of $O{=}S{=}NN(C_3H_7\text{-}i)_2$ [5].
b) Molecular structure of $O{=}S{=}NN(CH_3)C_6H_5$ [11].

Table 25
Crystal Data of $O{=}S{=}NN(C_3H_7\text{-}i)_2$ [5] and $O{=}S{=}NN(CH_3)C_6H_5$ [11].

	$O{=}S{=}NN(C_3H_7\text{-}i)_2$	$O{=}S{=}NN(CH_3)C_6H_5$
symmetry	monoclinic	orthorhombic
space group	$P2_1/n{-}C^5_{2h}$ (No. 14)	$P2_12_12_1{-}D^4_2{=}V^4$ (No. 19)
a in Å	12.63(2)	10.07(1)
b in Å	7.93(1)	10.92(1)
c in Å	9.28(1)	7.26(1)
β in °	95.4(2)	—
V in Å^3	925	—
Z	4	4
D_x in g/cm^3	1.16	1.40
D_m in g/cm^3	—	1.38
R	0.095	0.051
refined reflections	1578	670
Ref.	[5]	[11]

Table 26

Selected Bond Distances and Bond Angles of $O{=}S{=}NN(C_3H_7\text{-}i)_2$ [5] and $O{=}S{=}NN(CH_3)C_6H_5$ [11].

	$O{=}S{=}NN(C_3H_7\text{-}i)_2$	$O{=}S{=}NN(CH_3)C_6H_5$
bond distance in Å		
S–O	1.467(5)	1.455(4)
S–N(1)	1.565(5)	1.542(4)
N(1)–N(2)	1.283(5)	1.303(5)
N(2)–C(1)	1.463(6)	1.478(6)
N(2)–C(2)	1.486(6)	1.414(6)
bond angle in °		
O–S–N(1)	121.4(2)	121.2(2)
S–N(1)–N(2)	133.1(3)	133.2(3)
N(1)–N(2)–C(1)	123.5(3)	122.6(4)
N(1)–N(2)–C(2)	114.2(4)	116.7(4)
C(1)–N(2)–C(2)	121.9(3)	120.7(4)

^1H and ^{13}C NMR data (see Table 27) at low temperatures indicate that $O{=}S{=}NN(C_3H_7\text{-}i)_2$ exists in solution in two conformations owing to a "correlated orientation" (see the paper) [5].

Table 27

^{13}C and ^1H Chemical Shifts (in CHF_2Cl/TMS, δ in ppm) of $O{=}S{=}NN(C_3H_7\text{-}i)_2$ at Low Temperatures [5].

	CH	CH	CH_3	CH_3
^{13}C				
−60°C	59.6	52.7(5)	22.3	20.3(4)
−100°C	59.8	52.0(8)	22.4	20.1(5)
^1H				
−60°C	5.49	3.50	1.24	1.13
−100°C	5.56	3.51	1.23	1.13

Free energies of activation, $\Delta G^{\neq}$, of the N–N rotation for the derivatives of $O{=}S{=}NNR^1R^2$ with $R^1 = R^2 = CH_3$, C_2H_5, $i\text{-}C_3H_7$, and C_6H_5 are given in the table below. The ^1H frequency is 100 MHz, the ^{13}C frequency 25.16 MHz [5]:

$R^1 = R^2$	solvent	$\Delta G^{\neq}$ in kcal/mol	nucleus observed	signals	coalescence t in °C
CH_3	$CDCl_3$	10.5 (0.1)	^1H	CH_3	−54.5
C_2H_5	$CDCl_3$	11.0 (0.1)	^{13}C	CH_2	−42
$i\text{-}C_3H_7$	CHF_2Cl	11.6 (0.1)	^{13}C	CH_3	−35
		11.7 (0.1)	^1H	CH_3	a)

$R^1 = R^2$	solvent	ΔG^{+} in kcal/mol	nucleus observed	signals	coalescence t in °C
C_6H_5	$(CH_3)_2O$	9.7 (0.15)	^{13}C	C3, C5	−83
		9.9 (0.1)	^{13}C	C2, C6	b)
		9.9 (0.1)	^{13}C	C4	b)

a) Line shape analysis in the range −50 to −55°C. – b) Line shape analysis in the range −78 to −83°C.

Reactions

The reduction of a solution of $O{=}S{=}NN(CH_3)_2$ in ether with a 1M solution of $LiAlH_4$ in ether (mole ratio 4:1) at −75°C for 1 h, followed by stirring at room temperature for 1 h yielded $(CH_3)_2$-NN=S [4, 12]. The compound reacted with $(CH_3)_3O^+BF_4^-$ without solvent to give $O{=}S{=}NN(CH_3)_3{}^+$ BF_4^- [13]. In the reaction of $O{=}S{=}NN(CH_3)_2$ with CH_3I in ether at reflux for 18 h, the starting materials were recovered [1]. $O{=}S{=}NN(R)C_6H_5$ ($R = CH_3$, C_2H_5) reacted with NaOH to give C_6H_5-$NRNH_2$ and Na_2SO_3 [8, 9]. Treatment of $O{=}S{=}NN(CH_3)C_6H_5$ with $LiCH_3$ at room temperature yielded $C_6H_5(CH_3)NNHS(O)CH_3$ in 87% yield. When the same reaction was carried out at reflux in ether, the yield decreased to 38% [14].

References:

[1] Kanamueller, J. M. (J. Chem. Eng. Data **15** [1970] 356/8).
[2] Leandri, G.; Mangini, A. (Boll. Sci. Fac. Chim. Ind. Bologna **17** [1959] 15/8).
[3] Meller, A.; Maringgele, W.; Fetzer, H. (Chem. Ber. **113** [1980] 1950/61, 1954/5).
[4] Middleton, W. J. (J. Am. Chem. Soc. **88** [1966] 3842/4).
[5] Cerioni, G.; Piras, P.; Marongiu, G.; Macciantelli, D.; Lunazzi, L. (J. Chem. Soc. Perkin Trans. II **1981** 1449/53).
[6] Smith, W. T., Jr.; Chen, W. Y. (J. Med. Chem. **11** [1968] 504/5).
[7] Klamann, D.; Krämer, U.; Weyerstahl, P. (Chem. Ber. **95** [1962] 2694/9).
[8] Michaelis, A.; Ruhl, J. (Liebigs Ann. Chem. **270** [1892] 114/22).
[9] Michaelis, A. (Chem. Ber. **22** [1889] 2228/33).
[10] Augart, K.-D.; Kresze, G.; Schönberger, N. (Liebigs Ann. Chem. **1973** 1457/66).

[11] Cannas, M.; Marongiu, G.; Cerioni, G. (Cryst. Struct. Commun. **11** [1982] 1115/8).
[12] Middleton, W. J. (U.S. 3344135 [1964/67] 1/3; C.A. **67** [1967] No. 116823).
[13] Rössert, M.; Kraus, W.; Kresze, G. (Tetrahedron Letters **1978** 4669/70).
[14] Mataka, S.; Anselme, J.-P. (J. Chem. Soc. Chem. Commun. **1974** 554/5).

2.6.13.2.2 N′-Sulfinyl-N,N,N-trimethylhydrazinium Tetrafluoroborate, $O{=}S{=}NN(CH_3)_3^+BF_4^-$

The title compound was prepared in 35% yield by reaction of $O{=}S{=}NN(CH_3)_2$ with $(CH_3)_3O^+$ BF_4^- without solvent; m.p. 194 to 195°C with decomposition (from CH_2Cl_2–CH_3CN).

Treatment with $C_6H_5CH{=}O$ in CH_3CN yielded $C_6H_5CH{=}NN(CH_3)_3^+BF_4^-$. It underwent cycloaddition with $C_6H_5CH{=}\overset{+}{N}(O^-)C_6H_5$ in CH_3CN at room temperature to give the unstable five-membered heterocyclus I, which lost SO_2 to give $(C_6H_5)_2NCH{=}NN(CH_3)_3^+BF_4^-$. It reacted with $CH_2{=}C(CH_3)CH{=}CH_2$ or $CH_2{=}C(CH_3)C(CH_3){=}CH_2$ in CH_3CN for 7 to 14 d to give the cyclic compounds IIa and IIb.

I 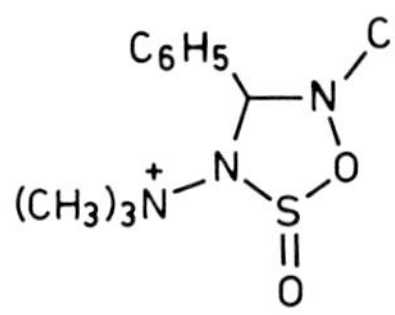 BF_4^-

IIa: R = H
IIb: R = CH_3 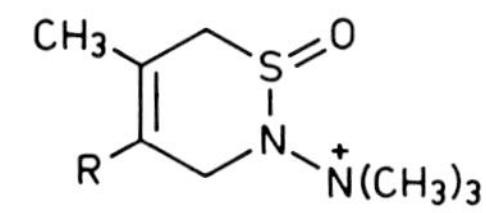 BF_4^-

Reference:

Rössert, M.; Kraus, W.; Kresze, G. (Tetrahedron Letters **1978** 4669/70).

2.6.13.2.3 Magnesium Bromide N′-Sulfinyl-N-phenylhydrazide, $MgBr^+[O \dot{=} S \dot{=} N \dot{=} NC_6H_5]^-$

The salt was obtained by treating $O{=}S{=}NNHC_6H_5$ with an excess of C_6H_5MgBr (mole ratio 1:1.3 to 2) in ether. The anion is described by the mesomeric structures $|\overset{\ominus}{\underline{\bar{O}}}{-}\underline{\bar{S}}{-}\bar{N}{=}\bar{N}{-}C_6H_5 \leftrightarrow |\overset{\ominus}{\underline{O}}{-}\overset{\oplus}{\bar{S}}{=}\bar{N}{-}\overset{\ominus}{\underline{N}}{-}C_6H_5$.

On hydrolysis with H_2O or aqueous NH_4Cl the starting compound was recovered. It reacted with $C_6H_5CH_2Cl$ in DMF for 1 h at 100°C to give $C_6H_5S(O)CH_2C_6H_5$. Solutions of the salt in DMF, THF, or pyridine were decomposed to give C_6H_5COOH even at −18°C when treated with C_6H_5-COCl; with $NaNH_2$ the corresponding $Na^+[OSNNC_6H_5]^-$ can be obtained (not isolated).

Reference:

Klamann, D.; Krämer, U.; Weyerstahl, P. (Chem. Ber. **95** [1962] 2694/9).

2.6.13.2.4 Sulfinyl-hydrazones of Heterocyclic Ketones

I 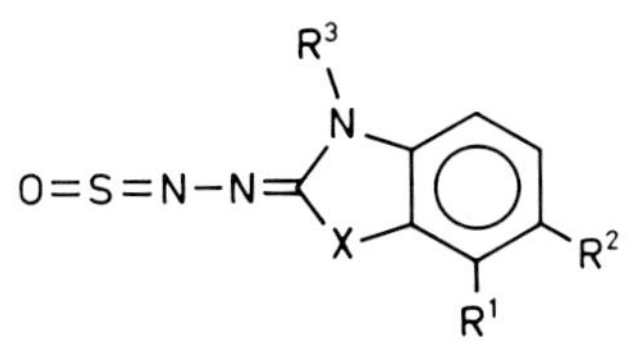

IIa: R = CH_3
IIb: R = C_2H_5 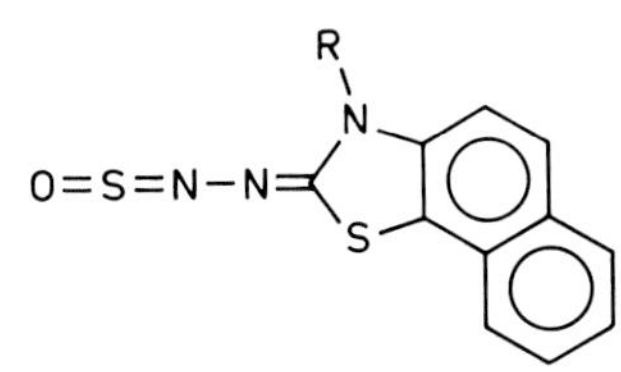

R^1, R^2, R^3, and X are compiled in Table 28.

The title compounds I summarized in Table 28, p. 70, were obtained by reaction of the hydrazone III

III 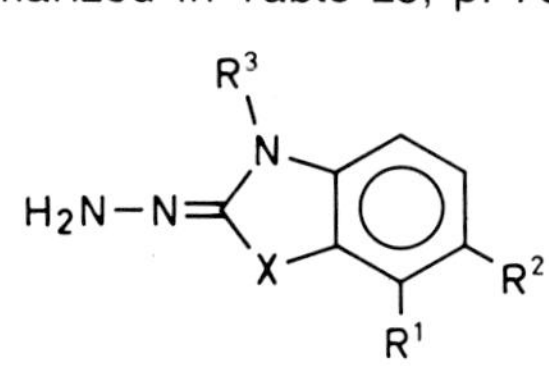

R^1, R^2, R^3, and X are compiled in Table 28.

a) with SO_2 in benzene. While passing SO_2 into the reaction mixture, it is refluxed at a water separator for 10 h [2];
b) with $SOCl_2$ in benzene in the presence of $(C_2H_5)_3N$ (mole ratio 1:1:2) at 0°C and subsequently stirring for 30 min [1, 2];

c) with O=S=NC_6H_5 (mole ratio 1:1.1) in C_2H_5OH in the presence of glacial acetic acid or only in glacial acetic acid [1, 2];

d) with O=S=$NSO_2C_6H_4CH_3$-4 (mole ratio 1:1) in $CHCl_3$. The reaction mixture is allowed to stand for 6 h at room temperature [1, 2].

e) Hydrazone hydrochloride, III·HCl, suspended in C_2H_5OH is treated with an equimolar amount of $(C_2H_5)_3N$ and subsequently reacted with a 10% excess of O=S=NC_6H_5 [1, 2].

Table 28

Methods of Preparation, Yields, and Melting Points of Sulfinyl-hydrazones of Heterocyclic Ketones (I).

I

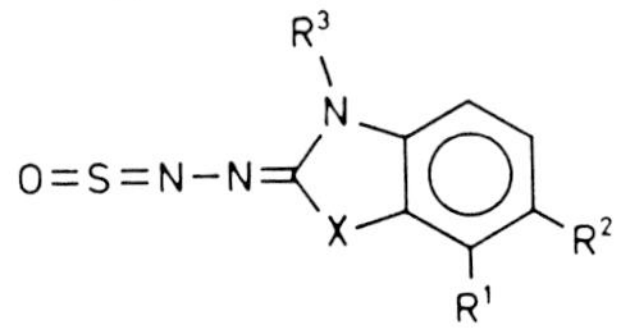

R^1	R^2	R^3	X	method of preparation (yield in %)	m.p. in °C	appearance, remark	Ref.
H	H	CH_3	NCH_3	e) (69)	137 to 138/dec. (from C_2H_5OH)		[1, 2]
H	H	CH_3	S	a) (53) b) (75) c) (89) d) (93) e) (65)	128.5/dec. (from C_2H_5OH)	yellow needles with tendency to decomposition or deflagration near m.p.	[1, 2]
H	OCH_3	CH_3	S	c) (86)	135.5 to 139/dec. (from C_2H_5OH)	yellow needles	[1, 2]
H	H	CH_3	Se	preparation not described			[3]

The title compounds IIa and IIb were synthesized by reaction of the hydrazone IVa and IVb with O=S=NC_6H_5 (mole ratio 1:1.1) in C_2H_5OH in the presence of glacial acetic acid or only in glacial acetic acid in 87 and 68% yield, respectively [1, 2].

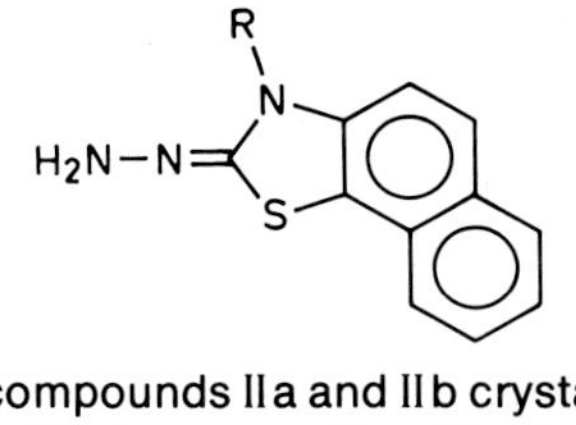

IVa: R = CH_3
IVb: R = C_2H_5

The yellow needles of compounds IIa and IIb crystallized from C_2H_5OH melt at 143.4 to 144.5°C (IIa) and 127.5°C (IIb) with decomposition [1, 2].

The compounds I and II show characteristic absorptions in the UV spectra between 395 and 420 nm (log ε = 4.40 to 4.55) [1].

3-Methyl-benzthiazolon-N-sulfinylhydrazone (I, $R^1 = R^2 = H$, $R^3 = CH_3$, X = S) reacted with RMgX (R = i-C_4H_9, C_6H_5, 4-$CH_3C_6H_4$, X = Br; R = CH_3, X = I) (mole ratio 1:1.5) in diethyl ether at room temperature to give the corresponding alkyl- or aryl-substituted sulfinic acid hydrazides.

$R = i\text{-}C_4H_9$, C_6H_5, $4\text{-}CH_3C_6H_4$, $X = Br$; $R = CH_3$, $X = I$

Analogously, the reaction of 3-methylnaphtho[2,1-d]-thiazol-(2)-on-N-sulfinyl-hydrazone (IIa) and 3-methyl-benzselenazol-(2)-on-N-sulfinyl-hydrazone (I, $R^1 = R^2 = H$, $R^3 = CH_3$, X = Se) with C_6H_5MgBr led to the corresponding phenyl-substituted sulfinic acid hydrazides [3].

References:

[1] Mayer, R.; Bleisch, S.; Heller, V. (Z. Chem. [Leipzig] **20** [1980] 185).
[2] Bleisch, S.; Mayer, R.; Domschke, G.; Beckert, R. (Ger. [East] 149215 [1981] 1/7; C.A. **96** [1982] No. 52294).
[3] Heller, V.; Mayer, R.; Beckert, R. (Ger. [East] 204085 [1982/83] 1/8; C.A. **100** [1984] No. 209800).

2.6.13.2.5 N-Sulfinyl-azacycloalkan-N-amines

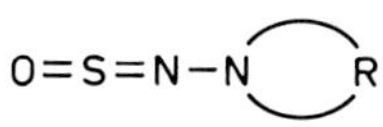

$R = -(CH_2)_4-$, $-(CH_2)_5-$, $-CHCH_3(CH_2)_3CHCH_3-$, $-C(CH_3)_2(CH_2)_3C(CH_3)_2-$, $-(CH_2)_6-$

The title compounds were prepared by adding a solution of $SOCl_2$ in ether to solutions of N-aminopyrrolidine [1], N-aminopiperidine [2], N-amino-2,6-*cis*-dimethyl-, and N-amino-2,2,6,6-tetramethylpiperidine [1] or N-aminohomopiperidine [2], and $(C_2H_5)_3N$ each (mole ratio 1:1:2) in ether with ice cooling for 3 h. After the reaction mixture reached room temperature, the hydrochloride was filtered and the ether removed [2]. N-Sulfinyl-1-piperidin-amine was also obtained by distilling a mixture of equimolar amounts of N-aminopiperidine and N-sulfinyl-aniline through a column under reduced pressure [3].

Table 29
Yields, Boiling Points, and Spectral Data of N-Sulfinyl-azacycloalkan-N-amines,

R (yield in %)	b.p. in °C/Torr (color)	properties and spectral data
$-(CH_2)_4-$	—	^{13}C NMR*): $\delta = 55.0$ (C-2,5), 23.3 ppm (C-3,4); free energy of activation for the N–N rotation (−28°C, in $CDCl_3$): $\Delta H^{\neq} = 11.7 \pm 0.1$ kcal/mol IR: 1185 to 1170 cm^{-1} ν_{as}(O=S=N), 1090 to 1080 cm^{-1} ν_s(O=S=N) MS: m/e = 132 M^+ [1]
$-(CH_2)_5-$ (58) [2], (89) [3]	48 to 50/0.2 (yellow) [2]; 60 to 61/0.2 (light yellow) [3]	$n_D^{25} = 1.5603$ [3] ^1H NMR ($CDCl_3$/TMS): δ (in ppm) = 1.70 (m, CCH_2), 3.90 (m, NCH_2) [2] ^1H NMR (CCl_4?/TMS?): δ (in ppm) = 1.65 (m, CCH_2), 3.85 (m, NCH_2) [3]

Table 29 (continued)

R (yield in %)	b.p. in °C/Torr (color)	properties and spectral data
$-(CH_2)_5-$ (continued)		^{13}C NMR*): δ = 57.8 (C-2,6), 26.0 (C-3,5), 23.7 ppm (C-4); free energy of activation for the N–N rotation (−63 and −77°C, in CHF_2Cl): $\Delta H^{\neq}$ = 9.9 ± 0.1 kcal/mol [1] IR (capillary film): 1140 cm^{-1} ν_{as}(O=S=N), 1080 cm^{-1} ν_s(O=S=N); additionally 15 unassigned absorption bands between 2930 and 642 cm^{-1} recorded [2]
$-CHCH_3(CH_2)_3$-$CHCH_3-$	—	^{13}C NMR*): δ = 59.9 (C-2,6), 30.0 (C-3,5), 19.3 (CH_3), 13.0 ppm (C-4); free energy of activation for the N–N rotation (−102 and −106°C, in CHF_2Cl): $\Delta H^{\neq}$ = 8.0 ± 0.1 kcal/mol IR: 1185 to 1170 cm^{-1} ν_{as}(O=S=N), 1090 to 1080 cm^{-1} ν_s(O=S=N) MS: m/e = 174 M^+ [1]
$-C(CH_3)_2(CH_2)_3$-$C(CH_3)_2-$	—	^{13}C NMR: δ = 64.0 (C-2,6), 40.6 (C-3,5), 29.1 (CH_3), 16.9 ppm (C-4); free energy of activation for the N–N rotation (−97 and −101°C, in CHF_2Cl–$CHFCl_2$): $\Delta H^{\neq}$ = 8.2 ± 0.1 kcal/mol IR: 1185 to 1170 cm^{-1} ν_{as}(O=S=N); 1090 to 1080 cm^{-1} ν_s(O=S=N) MS: m/e = 202 M^+ [1]
$-(CH_2)_6-$ (54)	62/0.1 (yellow)	^1H NMR ($CDCl_3$/TMS): δ (in ppm) = 1.71 (m, CCH_2), 3.98 (m, NCH_2) IR (capillary film): 1180 cm^{-1} ν_{as}(O=S=N); 1080 cm^{-1} ν_s(O=S=N); additionally 18 unassigned absorption bands recorded [2]

*) The ^{13}C NMR spectra display, at temperatures < −30°C, two different signals for the carbons which do not lie along the N–N axis.

When a solution of N-sulfinyl-1-piperidin-amine, $O{=}S{=}NNC_5H_{10}$, was reacted with a 1 M ethereal solution of $LiAlH_4$ (mole ratio ~4:1) at −50°C N-thionitrosopiperidine, $C_5H_{10}NN{=}S$, $Al(OH)_3$ and LiOH were formed [3, 4].

References:

[1] Lunazzi, L.; Macciantelli, D.; Cerioni, G. (J. Org. Chem. **47** [1982] 4579/81).
[2] Kanamueller, J. M. (J. Chem. Eng. Data **15** [1970] 356/8).
[3] Middleton, W. J. (J. Am. Chem. Soc. **88** [1966] 3842/4).
[4] Middleton, W. J. (U.S. 3344135 [1964/67]; C.A. **67** [1967] No. 116823).

2.6.13.2.6 N-Sulfinyl-4-morpholinamine

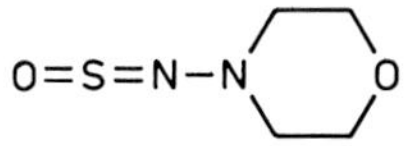

The compound was obtained in 96% yield by adding, over a period of 3 h, an ethereal solution of $SOCl_2$ to a mixture of N-aminomorpholine and triethylamine (mole ratio 1:1:2) in ether with ice cooling.

The pale yellow crystals melt at 50.5 to 52.5°C (from cyclohexane). ^1H NMR ($CDCl_3$/TMS): $\delta = 3.93$ ppm; IR: 1185(?), 1115 cm^{-1} ν_{as}(O=S=N); 1085 cm^{-1} ν_s(O=S=N); additionally 17 unassigned absorption bands between 2975 and 665 cm^{-1} recorded.

Reference:

Kanamueller, J. M. (J. Chem. Eng. Data **15** [1970] 356/8).

2.6.13.2.7 N-Sulfinyl-4H-1,2,4-triazol-4-amine

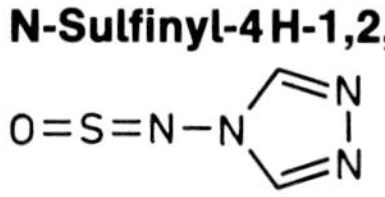

The title compound was prepared by treatment of suspensions of either 4-amino-4H-1,2,4-triazole and pyridine (mole ratio 1:2) in CH_2Cl_2 at 0°C or bis(trimethylsilyl)amino-4H-1,2,4-triazole in CH_2Cl_2 at −78°C with the stoichiometric amount of $SOCl_2$. After stirring at room temperature for 30 min to 1 h the compound was obtained in 55 and 77% yields, respectively.

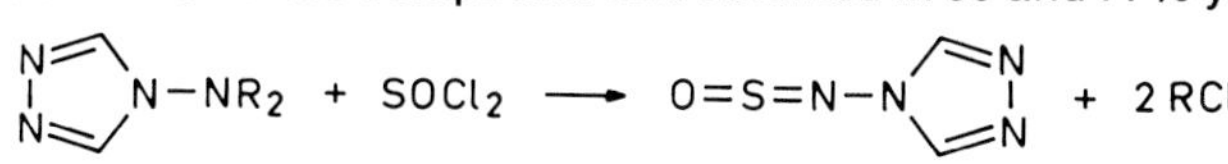

R = H or $Si(CH_3)_3$

The yellow, moisture-sensitive crystals melt at 90 to 91°C with decomposition from toluene–cyclohexane. ^1H NMR (CD_3CN/TMS): $\delta = 9.0$ ppm (s, 2H); (DMSO/TMS): $\delta = 8.46$ ppm; ^{13}C NMR (CD_3CN/TMS): $\delta = 141.6$ ppm (d, 2C).

It reacted with an excess of $(CH_3)_3O^+BF_4^-$ in CH_2Cl_2 for 15 h to yield the salt I with evolution of $(CH_3)_2O$. The compound underwent Diels-Alder reactions with 2,3-dimethyl-1,3-butadiene or 1,3-cyclohexadiene in CH_3CN at room temperature for 14 and 10 d to give the cycloadducts II and III, respectively.

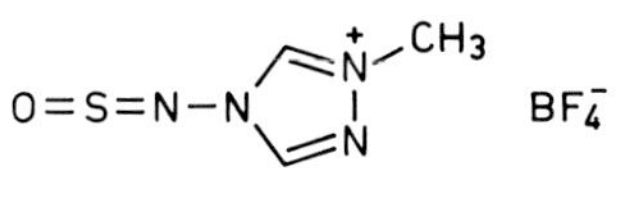

I

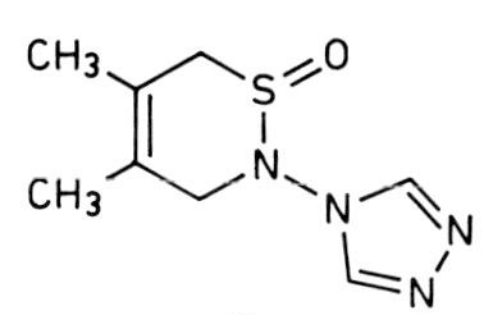

II

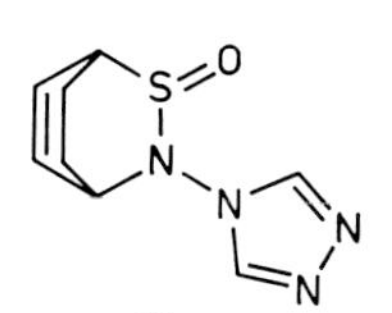

III

Reference:

Malach, H.-P.; Bussas, R.; Kresze, G. (Liebigs Ann. Chem. **1982** 1384/6).

2.6.13.2.8 4-Sulfinylamino-1-methyl-1H⁺, 4H-1,2,4-triazol-1-ium Tetrafluoroborate

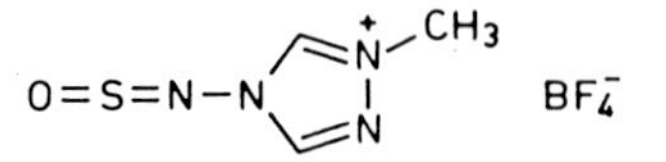

The salt was obtained by treatment of 4-sulfinylamino-4H-1,2,4-triazole in CH_2Cl_2 with an excess of $(CH_3)_3O^+BF_4^-$ and the reaction mixture subsequently allowed to stand at room temperature for 15 h. Removal of the volatile components yielded a yellow, extremely moisture-sensitive, sticky solid. ^{1}H NMR (CH_3CN/TMS): δ (in ppm) = 4.25 (s, CH_3), 9.35, 10.4 (2 s, CH).

It reacted with an excess of 2,3-dimethyl-1,3-butadiene in CH_3CN at 20°C for 10 d to give the Diels-Alder adduct I.

I

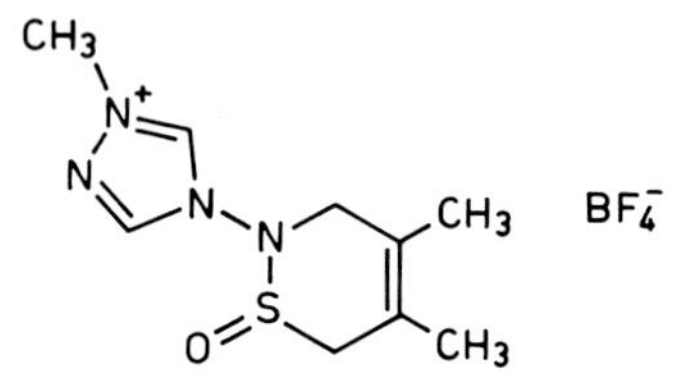

Reference:

Malach, H.-P.; Bussas, R.; Kresze, G. (Liebigs Ann. Chem. **1982** 1384/6).

2.6.13.2.9 N′,N″-Bis(sulfinyl)-ethylenebis(N-phenyl-hydrazine), $O{=}S{=}NN(C_6H_5)CH_2CH_2(C_6H_5)NN{=}S{=}O$

The compound precipitated by shaking $O{=}S{=}NC_6H_5$ with a mixture of $H_2NN(C_6H_5)CH_2CH_2N(C_6H_5)NH_2 \cdot 2HCl$ and CH_3COONa in H_2O for several minutes [1]; it was also formed from the reaction of $H_2NN(C_6H_5)CH_2CH_2N(C_6H_5)NH_2$ with $O{=}S{=}NC_6H_5$ or $SOCl_2$ as yellow shiny flakes, m.p. 121 to 123°C (from C_2H_5OH) [2].

References:

[1] Michaelis, A. (Chem. Ber. **24** [1891] 745/57).
[2] Michaelis, A.; Ruhl, J. (Liebigs Ann. Chem. **270** [1892] 114/22).

2.6.13.3 Bis(sulfinyl)-hydrazine, $O{=}S{=}N{-}N{=}S{=}O$

Early attempts of Michaelis in 1891 to synthesize the compound by reaction of hydrazine with $SOCl_2$ failed [1], but later the compound was prepared by adding $SOCl_2$ for 1 h at room temperature to a solution of $H_2NNH_2 \cdot H_2O$ (mole ratio ~2:1) in C_6H_6. $O{=}S{=}N{-}N{=}S{=}O$ separated from the solution in 70% yield.

Solid, m.p. 139°C. IR: ν (in cm^{-1}) = 2090 (?), 1495 ν(S=N), 1100 ν(O=S). The compound is insoluble in common solvents [2]. The polycondensation of 9,10-phenanthrenequinone, 9,10-anthraquinone, 1- and 2-chloroanthraquinone [3], 1,2- and 1,4-naphthoquinone [2] with O=S=N–N=S=O, producing the corresponding polyazines, was carried out in pyridine acting both as a solvent and as a catalyst.

References:

[1] Michaelis, A. (Chem. Ber. **24** [1891] 745/57, 752).
[2] Păstrăvanu, M.; Dumitriu, M.; Lixandru, T. (Eur. Polym. J. **22** [1986] 565/7).
[3] Păstrăvanu, M.; Dumitriu, M.; Lixandru, T. (Eur. Polym. J. **19** [1983] 633/7).

2.6.14 Phosphorus-Substituted Sulfinyl Imides

2.6.14.1 N-Sulfinyl-(cyclotriphosphazen)amine. N-Sulfinyl- and N,N′-Bis(sulfinyl)-(cyclotetraphosphazen)amine

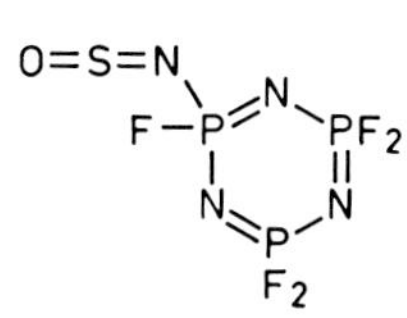

$O{=}S{=}NP_3N_3F_5$ (I)

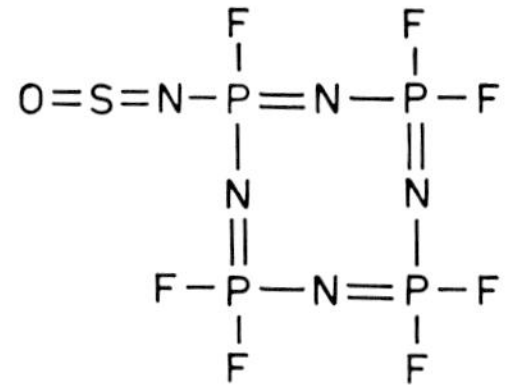

$O{=}S{=}NP_4N_4F_7$ (II)

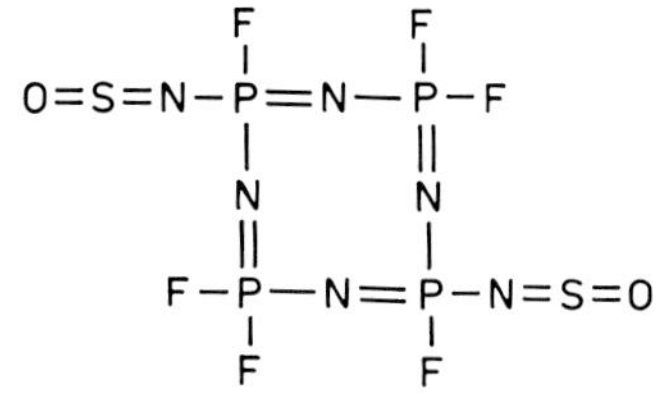

$(O{=}S{=}N)_2P_4N_4F_6$ (III)

$O{=}S{=}NP_3N_3F_5$ (I) was obtained in 82% yield by boiling $P_3N_3F_5NH_2$ with excess $SOCl_2$ until the solution became clear. Removal of excess $SOCl_2$ and fractionated distillation afforded the compound as a pale yellow liquid.

B.p. 143°C/760 Torr; ^{19}F NMR (solvent?/external $CFCl_3$): δ (in ppm) = −70 (PF_2), −57 (PFNSO), $^1J(^{31}P,^{19}F)_{PF_2}$ = 890 Hz, $^1J(^{31}P,^{19}F)_{PFNSO}$ = 920 Hz. IR (capillary): 20 absorptions between 1279 and 409 cm^{-1} recorded; 1279 ν(P=NP); 1250 ν_{as}(O=S=N); 1169 ν_s(O=S=N); 750 cm^{-1} ν(NP). MS: m/e (rel. int. in %) = 292 (32.2) M^+, 230 (100) $P_3N_3F_3^+$.

The compound is nearly insensitive to H_2O at 20°C. With PCl_5 the corresponding sulfur dichloride imide, $N_3F_5N{=}SCl_2$, and $P(O)Cl_3$ were formed [1]. $O{=}S{=}NP_3N_3F_5$ reacted with $(COCl)_2$ in the mole ratio 1:1 [2] or with $COCl_2$ in C_6H_5Cl solution in the mole ratio 2.5:1 [3] in the presence of catalytic amounts of pyridine to give $P_3N_3F_5NCO$ in 8 and 85% yield, respectively. Heating a mixture of the compound and pyridine at 100°C for 20 min yielded $F_5N_3P_3N{=}S{=}NP_3N_3F_5$ and SO_2 [1]. When 1:1 mixtures of $O{=}S{=}NP_3N_3F_5$ and carboxylic acids, RCOOH (R = CH_3, C_2H_5, and n-C_3H_7) were heated at 80°C for 5 to 6 h SO_2 evolved and $P_3N_3F_5NHC(O)R$ was obtained [2].

$O{=}S{=}NP_4N_4F_7$ (II) was prepared in 70% yield by refluxing $P_4N_4F_7NH_2$ in excess $SOCl_2$ for 7 d. After removal of excess $SOCl_2$ the residue was distilled under reduced pressure, giving a pale yellow liquid.

B.p. 51°C/13 Torr [4]. ^{19}F NMR (neat/$CFCl_3$): δ (in ppm) = ca. −60 (PF), ca. −71 (PF_2). ^{31}P NMR (neat/standard?): δ = ca. −17 ppm (PF_2) [5]. IR (capillary film): ν (in cm^{-1}) = 1430 ν_{as}(PNP); 1270 ν_{as}(O=S=N); 1170 ν_s(O=S=N); 1090 ν(PN); 1000, 960, 900, 800, 755, 690, 595 ν(PN), ν(PF), ν_s(P=NP); 545 δ(P=NP) [4].

$(O{=}S{=}N)_2P_4N_4F_6$ (III) was obtained in 81% yield by refluxing of $P_4N_4F_6(NH_2)_2$ with an excess of $SOCl_2$ for about one week. Subsequently the excess $SOCl_2$ was removed and the compound distilled in vacuum.

Yellowish liquid of b.p. 75°C/0.01 Torr. ^{19}F NMR (neat/$CFCl_3$): δ (in ppm) = −57.4 (PF), −68.5 (PF_2); $^1J(P,F)_{PF_2}$ = 890 Hz, $^1J(P,F)_{PF}$ = 920 Hz. ^{31}P NMR (neat/external 85% H_3PO_4): δ (in ppm) = ca. −22 (PF_2), ca. −33 (PF). IR (capillary film): 11 absorptions between 1400 and 665 cm^{-1}. MS: m/e (rel. int. in %) = 418 (5) M^+, 372 (9) $P_4N_4F_6NSONH_2^+$, 356 (5) $P_4N_4F_6NSO^+$, 326 (94) $P_4N_4F_6(NH_2)_2^+$, 314 (2) $P_4N_4F_7H^+$, 313 (1) $P_4N_4F_7^+$, 312 (4) $P_4N_4F_6NH_2H_2^+$, 311 (100) $P_4N_4F_6NH_2H^+$, 308 (16) $P_4N_4F_6N^+$, 291 (29) $P_4N_4F_5NH_2^+$, and characteristic phosphazene fragments [6].

Treatment with catalytic amounts of pyridine and heating in a water bath for 2 d afforded the sulfur diimide bridged cyclophosphazene IV [7].

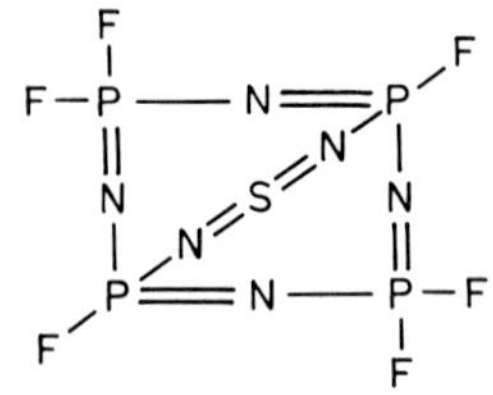

IV

References:

[1] Niecke, E.; Glemser, O.; Thamm, H. (Chem. Ber. **103** [1970] 2864/7).
[2] Roesky, H. W.; Janssen, E. (Z. Naturforsch. **26b** [1971] 679/83).
[3] Adolphi, H.; Wunsch, G.; Kiener, V. (Ger. Offen. 2334917 [1973/75] 1/16; C.A. **82** [1975] No. 156392).
[4] Roesky, H. W.; Grosse-Böwing, W. (Inorg. Nucl. Chem. Letters **6** [1970] 781/3).
[5] Grosse-Böwing, W. (Diss. Univ. Göttingen 1971, pp. 1/101).
[6] Janssen, E. (Diss. Univ. Frankfurt 1975, pp. 1/107).
[7] Roesky, H. W.; Janssen, E. (Angew. Chem. **88** [1976] 24/5; Angew. Chem. Intern. Ed. Engl. **15** [1976] 39).

2.6.14.2 N-Sulfinyl-phosphoramidic Acid Diorganyl Esters, O=S=NP(O)(OR)$_2$, R = CH_3, C_2H_5, n-C_3H_7, i-C_3H_7, n-C_4H_9, C_6H_5

Formation

The compounds were generally prepared by reaction of $(RO)_2P(O)NH_2$ with a slight excess of $SOCl_2$, followed by heating of the reaction mixture until HCl evolution ceased. Reaction conditions, yields, and properties are summarized in Table 30. $O{=}S{=}NP(O)(OC_2H_5)_2$ was also obtained by reaction of $(C_2H_5O)_2P(O)NHSi(CH_3)_3$, formed in situ from silylation of $(C_2H_5O)_2$-$P(O)NH_2$ with $(CH_3)_3SiN(C_2H_5)_2$, with $SOCl_2$ in benzene [1].

Mass Spectrum

The EIMS of $O{=}S{=}NP(O)(OC_2H_5)_2$ is shown in **Fig. 10**, p. 78. The spectrum shows no peak at m/e = 137 corresponding to loss of NSO, but a metastable peak at m/e = 94.3 indicates that the P–N bond is cleaved. No cleavage in the NSO group is observed [8].

Reactions

H_2O. The phosphoro(thionylamidates), $O{=}S{=}NP(O)(OR)_2$ (R = C_2H_5 [3 to 6], n-C_3H_7, n-C_4H_9 [5, 6], and C_6H_5 [2]) reacted with H_2O to give the corresponding phosphoramidates and SO_2 almost quantitatively: $O{=}S{=}NP(O)(OR)_2 + H_2O \rightarrow (RO)_2P(O)NH_2 + SO_2$.

$C_6H_5NH_2$. $(C_2H_5)_3N$. C_5H_5N. 3-$NO_2C_6H_4CH{=}NC_6H_5$. $O{=}S{=}NP(O)(OC_2H_5)_2$ reacted with $C_6H_5NH_2$ to give $(C_2H_5O)_2P(O)NH_2$ and $O{=}S{=}NC_6H_5$ [3, 4]. On heating the compound in the presence of $(C_2H_5)_3N$ or C_5H_5N, evolution of SO_2 was observed and $(C_2H_5O)_2P(O)N{=}S{=}NP(O)(OC_2H_5)_2$ was possibly formed, but could not be isolated. The reaction of excess $O{=}S{=}NP(O)(OC_2H_5)_2$ with 3-$NO_2C_6H_4CH{=}NC_6H_5$ yielded $O{=}S{=}NC_6H_5$, 3-$NO_2C_6H_4C(H)(NHP(O)(OC_2H_5)_2)_2$, and SO_2 [1].

Table 30
Formation, Yields, and Properties of $O{=}S{=}NP(O)(OR)_2$.

R	conditions	yield in %	b.p. in °C (Torr)	n_D (t in °C)	appearance, properties	Ref.
C_2H_5	in C_6H_6 at 65 to 70°C for ~1 h	75	75 to 76 (0.04)	1.4569 (22)	—	[2]
	in C_6H_6, heating for 30 min	70	72 (0.03)	—	yellow liquid	[3, 4]
	in C_6H_6 at 80°C	78	77 (1)	1.4576 (20)	yellowish liquid, odor similar to SO_2; IR: $\nu_s(O{=}S{=}N) = 1152\ cm^{-1}$	[5, 6]
	in C_6H_6 at 80°C	72	76 (0.1)	1.4578 (20)		[8]
	in C_6H_6 at 80°C for ~45 min	37	72 to 74 (0.3)	—	^{15}N labeled compound	[7]
	$(C_2H_5O)_2P(O)NHSi(CH_3)_3$ + $SOCl_2$ (in C_6H_6 at room temperature and reflux)	72	67 (0.02)	1.4582 (21)	—	[1]
n-C_3H_7	in C_6H_6 at 80°C	70	71 (0.05)	1.4562 (20)	yellowish liquid, odor similar to SO_2; IR: $\nu_s(O{=}S{=}N) = 1152\ cm^{-1}$	[5, 6]
i-C_3H_7	in C_6H_6 at 65 to 70°C for ~1 h	74	69 (0.06)	1.4482 (22)	—	[2]
n-C_4H_9	in C_6H_6 at 80°C	78	84 (0.1)	1.4552 (20)	yellowish liquid, odor similar to SO_2; IR: $\nu_s(O{=}S{=}N) = 1152\ cm^{-1}$	[5, 6]
C_6H_5	without solvent, at 80 to 85°C	81	97 to 99 (0.05)	1.5730 (20)	light yellow liquid	[2]

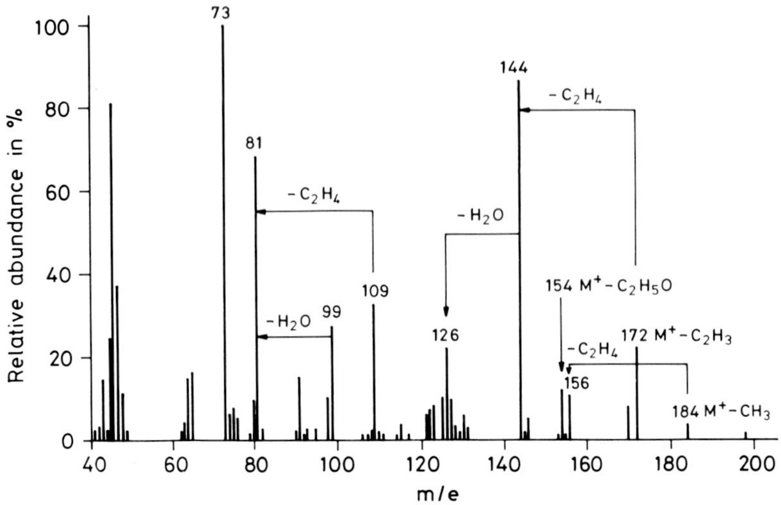

Fig. 10. Mass spectrum of $O{=}S{=}NP(O)(OC_2H_5)_2$ (m/e = 126 HNO_3PS^+, 109 $C_2H_6O_3P^+$, 99 $H_4O_4P^+$, 81 $H_2O_3P^+$, 73 $C_2H_3NS^+$) [8].

$(R'O)_3P$ ($R' = C_2H_5$, $n\text{-}C_4H_9$). Adding $O{=}S{=}NP(O)(OR)_2$ ($R = C_2H_5$, $n\text{-}C_4H_9$) [6, 9] or $O{=}S{=}^{15}N\text{-}P(O)(OC_2H_5)_2$ [7] in C_6H_6 to a solution of $(C_2H_5O)_3P$ in C_6H_6 (mole ratio 1:3) at 10 to 12°C [6, 9], or at 0°C [7], followed by stirring of the reaction mixture at room temperature for 15 min yielded $(RO)_2P(O)N{=}P(OC_2H_5)_3$ [6, 9], or $(C_2H_5O)_2P(O)^{15}N{=}P(OC_2H_5)_3$ [7], respectively, and $(C_2H_5O)_3P(O)$ and $(C_2H_5O)_3P(S)$. Analogously, the reaction with $(n\text{-}C_4H_9O)_3P$ afforded the corresponding phosphorimidic ester, phosphate, and thiophosphate. A reaction mechanism is suggested [6, 9].

$(i\text{-}C_3H_7O)_2P(S)SH$. The reaction of $O{=}S{=}NP(O)(OCH_3)_2$ (formation not described) with $(i\text{-}C_3H_7O)_2\text{-}P(S)SH$ (mole ratio 1:2) afforded $(CH_3O)_2P(O)NH_2$ and $((i\text{-}C_3H_7O)_2P(S))_2S_2$ [10].

R'OH (R' not given). Treatment of $O{=}S{=}NP(O)(OR)_2$ ($R = C_2H_5$, $n\text{-}C_3H_7$, and $n\text{-}C_4H_9$) with R'OH yielded $(RO)_2P(O)NH_2$ and R'OS(O)OR' [5,6].

R'CHO ($R' = 2\text{-}NO_2C_6H_4$, $3\text{-}NO_2C_6H_4$, $4\text{-}NO_2C_6H_4$, $4\text{-}ClC_6H_4$, and $3\text{-}NO_2\text{-}4\text{-}ClC_6H_3$). When $O{=}S{=}NP(O)(OC_2H_5)_2$ was treated with aromatic aldehydes, R'CHO, in C_6H_6 and a stream of N_2 bubbled through the reaction mixture for 5 to 8 h the corresponding aminales $((C_2H_5O)_2\text{-}P(O)NH)_2CHR'$ were obtained [1].

1,3-Dienes. When dry 1,3-butadiene was passed through a boiling benzene solution of $O{=}S{=}NP(O)(OC_2H_5)_2$ the 3,6-dihydro-2-diethoxyphosphinyl-1,2-thiazine 1-oxide was formed [3 to 6].

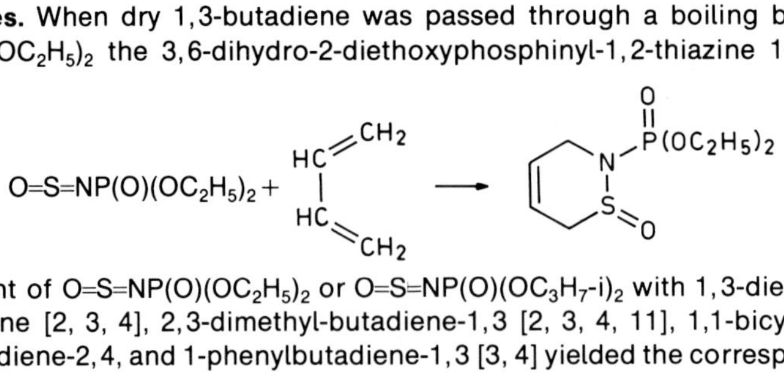

Treatment of $O{=}S{=}NP(O)(OC_2H_5)_2$ or $O{=}S{=}NP(O)(OC_3H_7\text{-}i)_2$ with 1,3-dienes such as piperylene, isoprene [2, 3, 4], 2,3-dimethyl-butadiene-1,3 [2, 3, 4, 11], 1,1-bicyclohexenyl, chloroprene, hexadiene-2,4, and 1-phenylbutadiene-1,3 [3, 4] yielded the corresponding 1,2-thiazine 1-oxides.

References:

[1] Tomaschewski, G.; Maniewski, C. (Tetrahedron Letters **1973** 561/4).
[2] Bal'on, Ya. G.; Levchenko, E. S. (Zh. Org. Khim. **3** [1967] 2232/6; J. Org. Chem. [USSR] **3** [1967] 2181/4).
[3] Kataev, E. G.; Plemenkov, V. V.; Markin, V. V. (Dokl. Akad. Nauk SSSR **165** [1965] 1313/6; Dokl. Chem. Proc. Acad. Sci. USSR **160/165** [1965] 1208/11).
[4] Kataev, E. G.; Plemenkov, V. V. (Khim. Primen. Fosfororg. Soedin. Tr. 3rd Vses. Konf., Moscow 1965 [1972], pp. 268/72; C.A. **77** [1972] No. 88415).
[5] Wieczorkowski, J. (Chem. Ind. [London] **1963** 825).
[6] Wieczorkowski, J. (Khim. Primen. Fosfororg. Soedin. Tr. 3rd Vses. Konf., Moscow 1965 [1972], pp. 262/7; C.A. **78** [1973] No. 3619).
[7] Reynolds, M. A.; Oppenheimer, N. J.; Kenyon, G. L. (J. Labelled Comp. Radiopharm. **18** [1981] 1357/63).
[8] Treppendahl, S.; Jakobsen, P.; Wieczorkowski, J. (Acta Chem. Scand. B **28** [1974] 657/61).
[9] Wieczorkowski, J. (Bull. Acad. Polon. Sci. Ser. Sci. Chim. **13** [1965] 155/61).
[10] Kutyrev, G. A.; Mavrin, V. Yu.; Cherkasov, R. A.; Pudovik, A. N. (Izv. Akad. Nauk SSSR Ser. Khim. **1980** 1402/4; C.A. **93** [1980] No. 167773).

[11] Ichimura, K.; Ichikawa, S.; Imamura, K. (Bull. Chem. Soc. Japan **49** [1976] 1157/8).

2.6.14.3 N-Sulfinyl-diphenylphosphonic Amide, $O{=}S{=}NP(O)(C_6H_5)_2$

The title compound was obtained in 90% yield by heating a mixture of $(C_6H_5)_2P(O)NH_2$ and $SOCl_2$ (mole ratio 1:1.1) at 80 to 85°C until the evolution of HCl ceased. After removal of excess $SOCl_2$, $O{=}S{=}NP(O)(C_6H_5)_2$ was distilled in vacuum; b.p. 106 to 108°C/0.06 Torr.

It hydrolyzed with H_2O to give $(C_6H_5)_2P(O)OH$. Reactions with dienes led to impure products.

Reference:

Bal'on, Ya. G.; Levchenko, E. S. (Zh. Org. Khim. **3** [1967] 2232/6; J. Org. Chem. [USSR] **3** [1967] 2181/4).

2.6.15 Arsenic-Substituted Sulfinyl Imides

2.6.15.1 N-Sulfinyl-diphenylarsenious Amide, $O{=}S{=}NAs(C_6H_5)_2$

The compound was prepared in 68% yield by adding a solution of $(C_6H_5)_2AsCl$ in CH_3CN to a suspension of $K^+[O{=}S{=}N]^-$ in CH_3CN (mole ratio 1:1) at −40°C, followed by stirring at 23°C for 2 h.

Yellow, moisture-sensitive liquid, b.p. 115 to 118°C/0.1 Torr. IR (neat): 1234 cm^{-1} $\nu_{as}(O{=}S{=}N)$; 1084 cm^{-1} $\nu_s(O{=}S{=}N)$.

It reacts with an equimolar amount of $Cr(CO)_5(THF)$ in THF at 0°C and subsequent stirring at 23°C for 16 h to yield $Cr(CO)_5((C_6H_5)_2AsN{=}S{=}O)$.

2.6.15.2 N,N′-Bis(sulfinyl)-organyl-arsonous Diamides, $(O{=}S{=}N)_2AsR$, R = t-C_4H_9, C_6H_5

$(O{=}S{=}N)_2AsC_4H_9$-t and $(O{=}S{=}N)_2AsC_6H_5$ were prepared in 77 and 73% yield, respectively, by treating a solution of $((CH_3)_2N)_3S^+$ $[O{=}S{=}N]^-$ in CH_3CN with solutions of t-$C_4H_9AsCl_2$ and $C_6H_5AsCl_2$ (mole ratio ~2:1) in CH_3CN at −40°C. After 16 h at 23°C the solvent was removed and the compounds were extracted with hexane.

$(O{=}S{=}N)_2AsC_4H_9$-t is a pale yellow, moisture-sensitive liquid, b.p. 42 to 43°C/0.05 Torr. IR (neat): 1246, 1232 cm^{-1} $\nu_{as}(O{=}S{=}N)$; 1090 cm^{-1} $\nu_s(O{=}S{=}N)$.

$(O{=}S{=}N)_2AsC_6H_5$ is a pale yellow, extremely moisture-sensitive solid, m.p. 23°C (from hexane). IR (Nujol): 1224, 1197 cm^{-1} $\nu_{as}(O{=}S{=}N)$; 1079 cm^{-1} $\nu_s(O{=}S{=}N)$.

When $(O{=}S{=}N)_2AsC_6H_5$ was heated at 100°C/0.1 Torr for 8 h the cyclic compound $S_2N_4As_2$-$(C_6H_5)_2$ (see "Sulfur-Nitrogen Compounds", Part 3, 1987, pp. 51/2) was obtained. The compound reacted with a slight excess of $(\eta^2\text{-}C_2H_4)Pt(P(C_6H_5)_3)_2$ in toluene at −17°C, followed by stirring at 23°C for 17 h to give *cis*-$Pt(P(C_6H_5)_3)_2(NSO)_2$. Reaction of $(O{=}S{=}N)_2AsC_4H_9$-t and $(O{=}S{=}N)_2AsC_6H_5$ with $Cr(CO)_5(THF)$ in THF at 0°C and subsequent stirring at 23°C for 16 h afforded $Cr(CO)_5(\text{t-}C_4H_9As(N{=}S{=}O)_2)$ and $Cr(CO)_5(C_6H_5As(N{=}S{=}O)_2)$, respectively.

Reference:

Chivers, T.; Dhathathreyan, K. S.; Lensink, C.; Richardson, J. F. (Inorg. Chem. **27** [1988] 1570/4).

2.6.16 Boron-Substituted Sulfinyl Imides

Sulfinyl-tetramethyl-boranetriamine, $O{=}S{=}NB(N(CH_3)_2)_2$ (I) and N-Sulfinyl-1,3-dimethyl-1,3,2-diazaborolidin-2-amine (II)

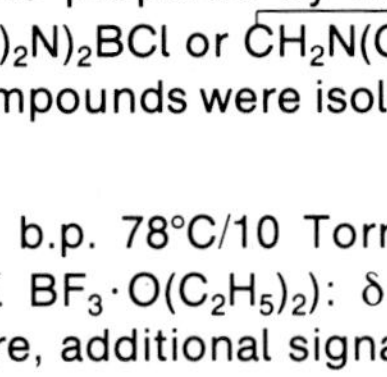

II

The title compounds I and II were prepared by heating a mixture of about equimolar amounts of $O{=}S{=}NSi(CH_3)_3$ and $((CH_3)_2N)_2BCl$ or $\overline{CH_2N(CH_3)BClN(CH_3)CH_2}$ at 100°C for ~1 h and 80°C for 2 h, respectively. The compounds were isolated by fractional distillation in 82 and 89% yields, respectively.

Compound I: Pale yellow liquid, b.p. 78°C/10 Torr. ^1H NMR ($CDCl_3$/external TMS): δ = 2.64 ppm. ^{11}B NMR ($CDCl_3$/external $BF_3 \cdot O(C_2H_5)_2$): δ = 21.5 ppm. When the solution was allowed to stand at room temperature, additional signals appeared in the ^1H NMR spectrum and the solution turned red.

Compound II: The dark red liquid, b.p. 68°C/11 Torr, is very sensitive to hydrolysis. ^1H NMR ($CDCl_3$/external TMS): δ (in ppm) = 2.55 (CH_3), 3.20 (CH_2). ^{11}B NMR ($CDCl_3$/external $BF_3 \cdot O(C_2H_5)_2$): δ = 24.6 ppm. IR (film): 25 absorptions between 2960 and 321 cm^{-1}.

Reference:

Haubold, W.; Fehlinger, H. G.; Frey, G. (Z. Naturforsch. **36b** [1981] 157/60).

2.6.17 Silicon-Substituted Sulfinyl Imides

2.6.17.1 N-Sulfinyl-silanamine, $O{=}S{=}NSiH_3$

Formation. The title compound formed along with H_3SiCl when $(H_3Si)_3N$ and $SOCl_2$ were condensed together in an ampule, fitted with a greaseless tap and allowed to warm to room temperature. Attempts to isolate pure $O{=}S{=}NSiH_3$ from the reaction mixture failed. It was prepared in 72 and 40%, respectively, by condensation of $O{=}S{=}NGe(CH_3)_3$ or $O{=}S{=}NSn(C_4H_9\text{-}n)_3$ (inserted in situ as product from the reaction of $(n\text{-}C_4H_9)_3SnOCH_3$ with excess $O{=}S{=}NSi(CH_3)_3$) with a slight excess of H_3SiBr in an ampule, fitted with a greaseless tap at 195 K. The ampule was allowed to warm to room temperature and from the volatile products $O{=}S{=}NSiH_3$ was obtained by condensation at 195 K.

Molecular Structure. Bonding. The molecular parameters from electron-diffraction scattering patterns are given in Table 31. In least-squares refinements, $O{=}S{=}NSiH_3$ was assumed to have an H_3SiN unit with local C_{3v} symmetry. The final R factor is 0.11.

Table 31
Molecular Parameters of $O{=}S{=}NSiH_3$.

bond distance	in Å	angle	in °
Si–N	1.762(6)	Si–N–S	129.9(2)
N–S	1.517(7)	N–S–O	119.2(17)
S–O	1.442(5)	H–Si–N	110.5(21)
Si–H	1.486 (fixed)	Si–N–S–O (dihedral angle)	30(3)

^{1}H NMR Spectrum (TMS solvent and standard): $\delta = 4.55$ ppm (broadened single peak with ^{29}Si satellites), $^1J(^{29}Si,^1H) = 230$ Hz; $\delta(^{29}Si) = -62$ ppm (from heteronuclear spin decoupling).

Molecular Vibrations. The IR spectrum of gaseous $O{=}S{=}NSiH_3$ at low pressure (≤ 10 Torr), the Raman spectrum of solid $O{=}S{=}NSiH_3$, condensed at low temperature on a copper block and suggested assignments are given in Table 32.

Table 32
Vibrational Spectra of $O{=}S{=}NSiH_3$.

IR (gas)	Raman (solid)	assignment	IR (gas)	Raman (solid)	assignment
ν in cm^{-1}			ν in cm^{-1}		
2195 vs	2195 vs	ν(SiH)	710 s	740 m	$\varrho(SiH_3)$
1310 vs	1290 m	$\nu_{as}(O{=}S{=}N)$	615 s	605 m	ν(SiN)
1147 s	1130 vs	$\nu_s(O{=}S{=}N)$	510 m	505 m	$\delta(O{=}S{=}N)$
	995 m	impurity?		275 w	δ(SiNS) or τ(NS)
942 vs	940 s	$\delta(SiH_3)$			

Electronic and Photoelectron Spectra. UV spectrum (vapor, up to 192 nm): λ_{max} (in nm) (ε in $L\cdot mol^{-1}\cdot cm^{-1}$) = 284 (~30) and 230 (~1000). The 284 nm band is due to $a_1 \rightarrow b_1^*$ excitation of an electron from the sulfur lone-pair orbital to the b_1 component of the 3π level. The origin of the 230 nm band is not certain, but it seems likely that several transitions are involved, possibly an excitation from the a_1 sulfur lone-pair to the antibonding a_1^* orbital.

The HeI PES (shown in a figure in the paper) exhibits 6 vertical ionization potentials, E_i, at 10.4, 10.55 (11.5?), 12.8, 13.2, 15.0, and 16.7 eV. The sulfur lone-pair band is supposed to be that at 11.5 eV. The broad band near 13 eV is due to the SiH bonding orbitals derived from Si 3p. The band at 15 eV is due to the 1π levels and the weaker band at 16.7 eV due to the mainly SiH bonding orbital derived from Si 3s.

The **vapor pressure** at 273 K was found to be 70 ± 1 Torr. The compound has a vapor density M of 95.2.

Mass Spectrum. The spectrum exhibits a weak parent peak at m/e = 93 even at low ionizing voltages. The major peaks are at m/e = 92 $M^+ - H$ and 63 $HNSO^+$.

Stability. The compound is unstable as vapor or liquid. At room temperature, decomposition is complete within 15 min, forming SiH_4, traces of H_2, NH_3, and $(H_3Si)_3N$, and an unidentified yellow polymeric residue.

Reference:

Cradock, S.; Ebsworth, E. A. V.; Meikle, G. D.; Rankin, D. W. H. (J. Chem. Soc. Dalton Trans. **1975** 805/9).

2.6.17.2 N-Sulfinyl-trimethylsilanamine, $O{=}S{=}NSi(CH_3)_3$

Preparation. Formation

The compound was prepared by reaction of SO_2 with $((CH_3)_3Si)_2NH$, pure or diluted with pentane. SO_2 was transfered efficiently into a reaction vessel (cooled with liquid N_2) containing degassed $((CH_3)_3Si)_2NH$ (mole ratio 3:1) and the reaction mixture warmed to room temperature. The reaction occurred rapidly and appeared to be complete before the system reached 0°C; yield ~22% [1]. The experimentally results indicate that SO_2 reacted with $((CH_3)_3Si)_2NH$ in a 1:1 mole ratio, according to the following equation: $4((CH_3)_3Si)_2NH + 4SO_2 \rightarrow 3O{=}S{=}N\text{-}Si(CH_3)_3 + 2((CH_3)_3Si)_2O + NH_4^+[(CH_3)_3SiOSO_2]^-$ [2]. Purification of the product was carried out gas-chromatographically at 67°C [1, 2]. $O{=}S{=}NSi(CH_3)_3$ could be isolated from the complex product mixture, which formed, when $((CH_3)_3SiN{=})_3S$ was treated with SO_2 above 0°C [3].

The compound was first synthesized in 1966 from the reaction of $((CH_3)_3Si)_3N$ with $SOCl_2$ (mole ratio 1:1) without a solvent at ~70°C in the presence of $AlCl_3$ catalyst in ~50% [4] or 42% yield [5]. Using a solvent, CH_2Cl_2, it was obtained in 70% yield after stirring for 20 h [6]. The compound was prepared in 30% yield by reacting ethereal solutions of $((CH_3)_3Si)_2NNa$ and $ClS(O)OCH_3$ (mole ratio 1:1) at −60°C for 10 min and subsequently stirring the reaction mixture at room temperature for 1 h [5]. It was obtained in 96% yield by adding $(CH_3)_3SiSC_6H_5$ to an equimolar amount of $O{=}S{=}NSC_6H_5$, followed by warming of the reaction mixture at 100°C for 1 h [7]. $O{=}S{=}NSi(CH_3)_3$ was formed by reaction of $K^+[O{=}S{=}N]^-$ [8] or $O{=}S{=}NSn(C_4H_9\text{-}n)_3$ [9] with $(CH_3)_3SiCl$ or by treatment of $O{=}S{=}NSO_2R$ (R = $(CH_3)_2N$ and $(CH_3)_3SiO$) with $HN(Si(CH_3)_3)_2$ [10].

Properties. Spectra

Colorless liquid, m.p. <−78°C [4, 5]; b.p. 104 to 106°C [6], 106°C [8], 108°C [12], and 108 to 110°C [4, 5]; D_m (23°C) = 0.978 g/cm^3 [2]; n_D^{25} = 1.4254 [11] or 1.4258 [7]. The thermal conductivity ratio for $O{=}S{=}NSi(CH_3)_3 : ((CH_3)_3Si)_2O$ = 1.29 [2].

^{1}H NMR (CCl_4/TMS): δ = 0.35 ppm [1, 2] or 0.36 ppm (s, CH_3), $^1J(^{13}C,H)$ = 120.5 Hz, $^2J(^{29}Si,H)$ = 7.2 Hz [4, 5]. ^{14}N NMR (neat/NH_4^+ aq.): δ = 309 ppm [12]; referred to CH_3NO_2: δ = −50.5 ppm. ^{29}Si NMR (CH_2Cl_2/TMS): δ = −1.8 ppm [13].

IR: ν (in cm^{-1}) = 2970, 2910 ν(CH); 1420 $\delta_{as}(CH_3)$; 1255 $\delta_s(CH_3)$; 1300 ν_{as}(O=S=N); 1125 ν_s(O=S=N); 965 ν(SiN) (?); 845, 765 $\varrho(CH_3)$; 570 δ(O=S=N) [5]; 1302 [1], 1297 [4] ν_{as}(O=S=N); 1132 [1], 1122 [4] ν_s(O=S=N); 570 m δ(O=S=N); 1258 s, 858 m, 770 m $\nu(SiCH_3)$ [1]; 12 absorptions between 2985 and 575 cm^{-1} are given in [2].

UV: λ_{max} = 237 and 295 nm in an absorbance ratio of ~10:1 [1]. The PE spectrum is shown in a figure in [35].

MS: m/e (rel. int. in %) = 120 (100) M^+-15, 90 (8) $HOSi(CH_3)_3^+$, 75 (17) $HOSi(CH_3)_2^+$, 73 (5) $(CH_3)_3Si^+$, 46 (40) NS^+ [1]; see also [2].

Reactions

The compound is sensitive to hydrolysis [4]. With elemental halogens X_2 (X = F, Cl, Br) $(CH_3)_3SiX$ is split off. When F_2, diluted with N_2 was passed into a solution of $O{=}S{=}NSi(CH_3)_3$ in $CFCl_3$ at −80°C O=S=NF was obtained [14]. When liquid Cl_2 was added at −76°C to an equimolar amount of $O{=}S{=}NSi(CH_3)_3$ a smooth reaction takes place, giving O=S=NCl. The analogous reaction with Br_2 gave a difficultly separable mixture that contained O=S=NBr.

Adding SCl_2 to $O{=}S{=}NSi(CH_3)_3$ (mole ratio ~1:2) gave $(O{=}S{=}N)_2S$ in an exothermic reaction. The reaction was complete after refluxing of the reaction mixture for 48 h [11]. The same product was obtained by reaction of the compound with SO_2Cl_2 [10] or $SOCl_2$ [15].

The reaction of $O{=}S{=}NSi(CH_3)_3$ with $HN(SO_2Cl)_2$ (mole ratio ~2:1) in CH_2Cl_2, followed by stirring at room temperature for 18 h yielded the heterocyclic compound I [16].

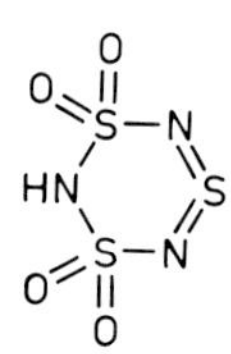

I

The compound reacted with $(CH_3)_2NSCl$ to give $O{=}S{=}NSN(CH_3)_2$ and $(CH_3)_3SiCl$ [17]. When $(CH_3)_2NS(O)Cl$ was added slowly to an equimolar amount of $O{=}S{=}NSi(CH_3)_3$ and the reaction mixture gently heated, $O{=}S{=}NSN(CH_3)_2$ was also formed, along with $(CH_3)_3SiCl$ and SO_2 [15]. When $O{=}S{=}NSi(CH_3)_3$ was reacted with $((CH_3)_2N)_3S^+(CH_3)_3SiF_2^-$ (mole ratio ~1:1) in THF at 23°C [36], or (mole ratio ~3:2) at −35°C in CH_3CN [37], see also [39], $((CH_3)_2N)_3S^+OSN^-$ and $(CH_3)_3SiF$ were formed. The compound reacted with an equimolar amount of $(C_6H_5)_2C{=}NSCl$ (in situ from $(C_6H_5)_2C{=}NSi(CH_3)_3 + SCl_2$) in CH_2Cl_2 at −78°C and subsequent standing at 23°C for 3 d to give $O{=}S{=}NSN{=}C(C_6H_5)_2$, $(C_6H_5)_2CNH_2^+\,Cl^-$, and $((C_6H_5)_2C{=}N)_2S$ [18]. Stirring a mixture of the compound and cyclic [–SCl–N=C(N(CH$_3$)$_2$)–N=SCl–N–] (mole ratio 2:1) in CCl_4 solution for 27 d at 23°C afforded the heterocyclic compound II as the major product, along with small amounts of III [20], see also [19].

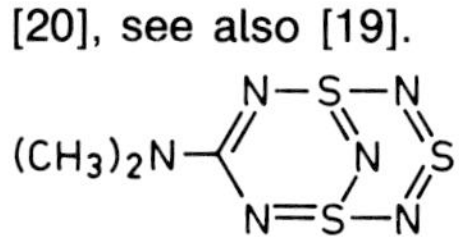

II

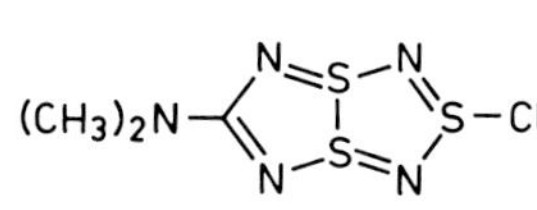

III

Treatment of $O{=}S{=}NSi(CH_3)_3$ with an equimolar amount of C_6H_5SCl [7, 21] or with C_6H_5-$SOCH_3$ [38] yielded $O{=}S{=}NSC_6H_5$ and $(CH_3)_3SiCl$ or $(CH_3)_3SiOCH_3$, respectively.

Stirring a mixture of the compound and Se_2Cl_2 (mole ratio ~2:1) in CH_2Cl_2 or $((CH_3)_3Si)_2O$ in a glass bomb at 22°C for 24 h yielded $(O{=}S{=}N)_2Se$ and $(CH_3)_3SiCl$, and elemental Se [22].

Mixing of nearly equimolar amounts of O=S=NSi(CH_3)$_3$ and ((CH_3)$_2$N)$_2$BCl and heating of the reaction mixture subsequently at 100°C for ~1 h afforded O=S=NB(N(CH_3)$_2$)$_2$ and (CH_3)$_3$SiCl [23]. O=S=NSi(CH_3)$_3$ reacted with 1,2,4-trithia-3,5-diborolane ring systems [–S–BX–S–S–BX–], X = Cl or Br (mole ratio 1:1), in refluxing CCl_4 for 3 or 24 h, respectively, to form 1,2-dithia-4-aza-3,5-diborolidine, [–N(Si(CH_3)$_3$)–BX–S–S–BX–], and sulfur [24, 25]. The reaction with an equimolar amount of [–S–B(CH_3)–S–S–B(CH_3)–] in refluxing CCl_4 for 24 h afforded [–S–B(CH_3)–N(Si(CH_3)$_3$)–S–B(CH_3)–], and sulfur [24]. No defined products were isolated from reactions with [–S–BX–S–S–BX–] (X = I, N(C_2H_5)$_2$) and with [–Se–BI–Se–Se–BI–] [25]. Heating of a ~1:1 mixture of O=S=NSi(CH_3)$_3$ and [–CH_2–N(CH_3)–BCl–N(CH_3)–CH_2–] at ~80°C for 2 h yielded [–CH_2–N(CH_3)–B(N=S=O)–N(CH_3)–CH_2–] and (CH_3)$_3$SiCl [23].

Treatment of O=S=NSi(CH_3)$_3$ with OCNC(O)Cl (mole ratio 1:1) or with (ClNCO)$_3$ (mole ratio 3:1) yielded compound IV and (ClCONSO)$_x$ or O=S=NCl, respectively [27]. Excess O=S=N-Si(CH_3)$_3$ reacted with FC(O)Br to give (O=S=N)$_2$S, (CH_3)$_3$SiF, (CH_3)$_3$SiBr and polymeric products [26]. Heating the compound and [–N($CH_2C_6H_5$)–C(=N$SO_2C_6H_4CH_3$-4)–S–N=N–] (mole ratio 1.5 to 3:1) in C_6H_6 at 60°C for 2 h and at 70°C for another short period led to compound V [28].

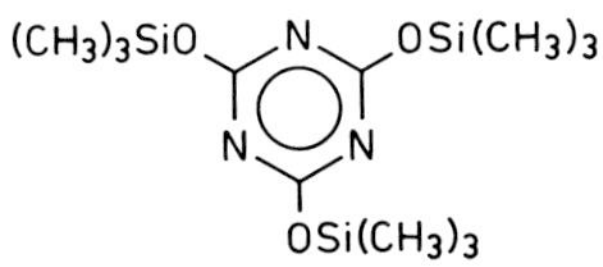

IV

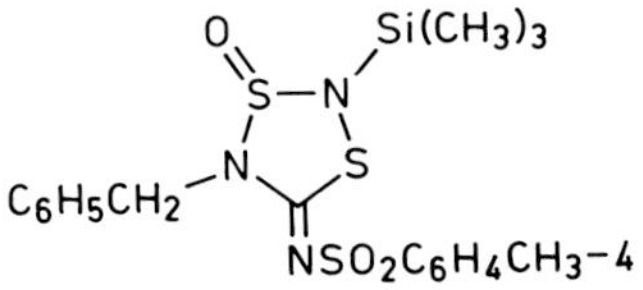

V

Adding a solution of O=S=NSi(CH_3)$_3$ in CCl_4 to an equimolar amount of (CH_3)$_3$SiOSO$_2$Cl produced O=S=NSO$_2$OSi(CH_3)$_3$ and (CH_3)$_3$SiCl [29]. Heating an ethereal solution of the compound and ((CH_3)$_3$Si)$_2$NNC (mole ratio 1:3) in a sealed tube at 90°C for 4 h led to N_2, (CH_3)$_3$SiCN, ((CH_3)$_3$Si)$_2$O, (CH_3)$_3$SiN=C=NSi(CH_3)$_3$, and the yellow adduct ((CH_3)$_3$Si)$_2$NNC ·((CH_3)$_3$Si)$_2$NN=C=S [30]. Treatment of (CH_3)$_3$SiNC$_6H_5$Li, prepared by reaction of a 1.6 M solution of n-C_4H_9Li in hexane with a solution of (CH_3)$_3$SiNHC$_6H_5$ in ether, with an ethereal solution of O=S=NSi(CH_3)$_3$ (mole ratio 1:1) at room temperature overnight gave C_6H_5N=S=N-Si(CH_3)$_3$ [31]. When a solution of O=S=NSi(CH_3)$_3$ in ether was reacted with a solution of Li(CH_3)$_3$SiN(1,4-C_6H_4)NSi(CH_3)$_3$Li in ether-hexane (mole ratio 1:1) at room temperature for 5 h and subsequently an equimolar amount of (CH_3)$_3$SiCl was added and the reaction mixture allowed to react for another 10 h, (CH_3)$_3$SiN=S=N(1,4-C_6H_4)N=S=NSi(CH_3)$_3$ was obtained [6].

Treatment of O=S=NSi(CH_3)$_3$ with (C_6H_5)$_3$P=NC$_4H_9$-t led to (C_6H_5)$_3$P=NSi(CH_3)$_3$ and O=S=N-C_4H_9-t [32].

O=S=NSi(CH_3)$_3$ reacted with an equimolar amount of $SbCl_5$ in liquid SO_2 and subsequently stirring at room temperature for 24 h to give S_4N_4·$SbCl_5$. It reacted with $SnCl_4$ (mole ratio 4:1) in CH_2Cl_2 at −40°C, followed by stirring at room temperature for 24 h to give (CH_3)$_3$SiN=S=N-Si(CH_3)$_3$·$SnCl_4$ and SO_2 [33]. The reaction with HgF_2 (mole ratio 2:1) in boiling THF for 30 min gave (O=S=N)$_2$Hg and (CH_3)$_3$SiF [34].

The compound reacted slowly with a solution of KOC$_4H_9$-t in THF to give K^+[O=S=N]$^-$ and (CH_3)$_3$SiOC$_4H_9$-t [8]. It reacted with (CH_3)$_3$GeOCH$_3$ in the presence of $AlCl_3$ catalyst to give (CH_3)$_3$SiOCH$_3$ and O=S=NGe(CH_3)$_3$ [9]. Treatment of (CH_3)$_3$SnOC$_2H_5$ [9] or (n-C_4H_9)$_3$SnOCH$_3$ [9, 35] with an excess of O=S=NSi(CH_3)$_3$ at room temperature yielded O=S=NSn(CH_3)$_3$ or O=S=NSn(C_4H_9-n)$_3$, respectively, and (CH_3)$_3$SiOR (R = CH_3, C_2H_5).

References:

[1] Davis, J. F.; Spicer, L. D. (Inorg. Chem. **19** [1980] 2191/2).
[2] Bennett, D. W.; Spicer, L. D. (Inorg. Chem. **21** [1982] 410/3).

[3] Tesky, F.-M.; Mews, R.; Krebs, B. (Z. Naturforsch. **36b** [1981] 1465/78, 1469/70).
[4] Scherer, O. J.; Hornig, P. (Angew. Chem. **78** [1966] 776/7; Angew. Chem. Intern. Ed. Engl. **5** [1966] 729/30).
[5] Scherer, O. J.; Schmitt, R. (Chem. Ber. **101** [1968] 3302/12).
[6] Scherer, O. J.; Wolmershäuser, G.; Jotter, R. (Z. Naturforsch. **37b** [1982] 432/6).
[7] Armitage, D. A.; Clark, M. J. (J. Organometal. Chem. **24** [1970] 629/31).
[8] Armitage, D. A.; Brand, J. C. (J. Chem. Soc. Chem. Commun. **1979** 1078/9).
[9] Armitage, D. A.; Sinden, A. W. (J. Organometal. Chem. **44** [1972] C43/C45).
[10] Appel, R.; Montenarh, M. (Chem. Ber. **108** [1975] 2340/8).

[11] Armitage, D. A.; Sinden, A. W. (Inorg. Chem. **11** [1972] 1151/2).
[12] Armitage, D. A.; Mason, J.; Vinter, J. G. (Inorg. Chem. **17** [1978] 776/7).
[13] Kroner, J.; Schneid, W.; Wiberg, N.; Wrackmeyer, B.; Ziegleder, G. (J. Chem. Soc. Faraday Trans. II **74** [1978] 1909/19).
[14] Verbeek, W.; Sundermeyer, W. (Angew. Chem. **81** [1969] 331; Angew. Chem. Intern. Ed. Engl. **8** [1969] 376).
[15] Armitage, D. A.; Sinden, A. W. (J. Inorg. Nucl. Chem. **36** [1974] 993/5).
[16] Buckendahl, W.; Glemser, O. (Chem. Ber. **110** [1977] 1154/8).
[17] Armitage, D. A.; Tso, C. C. (J. Chem. Soc. D **1971** 1413/4).
[18] Chivers, T.; Oakley, R. T.; Pieters, R.; Richardson, J. F. (Can. J. Chem. **63** [1985] 1063/7).
[19] Chivers, T.; Richardson, J. F.; Smith, N. R. M. (Mol. Cryst. Liquid Cryst. **125** [1985] 319/27).
[20] Chivers, T.; Richardson, J. F.; Smith, N. R. M. (Inorg. Chem. **25** [1986] 272/5).

[21] Glander, J.; Golloch, A. (J. Fluorine Chem. **14** [1979] 403/20).
[22] Haas, A.; Kasprowski, J. (Chimia [Switz.] **41** [1987] 340).
[23] Haubold, W.; Fehlinger, H. G.; Frey, G. (Z. Naturforsch. **36b** [1981] 157/60).
[24] Meller, A.; Habben, C. (Monatsh. Chem. **113** [1982] 139/53).
[25] Habben, C. (Diss. Univ. Göttingen 1981).
[26] Krannich, H.-J.; Sundermeyer, W. (Z. Anorg. Allgem. Chem. **450** [1979] 45/9).
[27] Nachbaur, E.; Kosmus, W.; Krannich, H. J.; Sundermeyer, W. (Monatsh. Chem. **109** [1978] 1211/6).
[28] L'abbe, G.; van Asch, A.; Declerq, J. P.; Germain, G.; van Meerssche, M. (Bull. Soc. Chim. Belges **87** [1978] 285/92).
[29] Buss, W.; Krannich, H.-J.; Sundermeyer, W. (Chem. Ber. **109** [1976] 1486/90).
[30] Wiberg, N.; Hübler, G. (Z. Naturforsch. **32b** [1977] 1003/9).

[31] Cordes, A. W.; Hojo, M.; Koenig, H.; Noble, M. C.; Oakley, R. T.; Pennington, W. T: (Inorg. Chem. **25** [1986] 1137/45).
[32] Ruppert, I.; Bastian, V.; Appel, R. (Chem. Ber. **108** [1975] 2329/39).
[33] Roesky, H. W.; Schmidt, H.-G.; Noltemeyer, M.; Sheldrick, G. M. (Chem. Ber. **116** [1983] 1411/4).
[34] Verbeek, W.; Sundermeyer, W. (Angew. Chem. **81** [1969] 330/1; Angew. Chem. Intern. Ed. Engl. **8** [1969] 376).
[35] Cradock, S.; Ebsworth, E. A. V.; Meikle, G. D.; Rankin, D. W. H. (J. Chem. Soc. Dalton Trans. **1975** 805/9).
[36] Chivers, T.; Dhathathreyan, K. S.; Lensink, C.; Richardson, J. F. (Inorg. Chem. **27** [1988] 1570/4).
[37] Heilemann, W.; Mews, R. (Chem. Ber. **121** [1988] 461/3).
[38] Armitage, D. A.; Clark, M. J.; Kinsey, A. C. (J. Chem. Soc. C **1971** 3867/9).
[39] Heilemann, W.; Meier, T.; Mews, R. (J. Fluorine Chem. **35** [1987] 143).

2.6.17.3 N-Sulfinyl-triethylsilanamine, $O{=}S{=}NSi(C_2H_5)_3$

The compound was prepared by mixing equimolar amounts of $O{=}S{=}NSC_6H_5$ and $(C_2H_5)_3Si$-SC_6H_5 and heating the reaction mixture at 170°C for 2 d; yield 44%. B.p. 40°C/0.2 Torr; n_D^{25} = 1.4540; IR: 1300 cm^{-1} ν_{as}(O=S=N); 1125 cm^{-1} ν_s(O=S=N).

Reference:

Armitage, D. A.; Clark, M. J. (J. Organometal. Chem. **24** [1970] 629/31).

2.6.17.4 N,N'-Bis(sulfinyl)-dimethylsilanediamine, $(O{=}S{=}N)_2Si(CH_3)_2$

The title compound was obtained in good yield (~65%) by reaction of $O{=}S{=}NSn(C_4H_9\text{-}n)_3$ with $(CH_3)_2SiCl_2$. The compound is a bright yellow liquid with a sweet sickly smell; b. p. 54°C/7 Torr, n_D^{25} = 1.4787; ^{1}H NMR: δ = 0.62 ppm; IR: intense absorptions around 1300 cm^{-1} ν_{as}(O=S=N) and 1100 cm^{-1} ν_s(O=S=N).

Reference:

Armitage, D. A.; Sinden, A. W. (J. Organometal. Chem. **44** [1972] C43/C45).

2.6.18 Germanium-Substituted Sulfinyl Imides
N-Sulfinyl-trialkylgermanamines, $O{=}S{=}NGeR_3$, R = CH_3, C_2H_5

$O{=}S{=}NGe(CH_3)_3$ was prepared in 83% yield by heating a mixture of $(CH_3)_3GeSC_6H_5$ and O=S=N-SC_6H_5 (mole ratio 1:1) at 90°C [1]. The compound also formed by reaction of $(CH_3)_3GeOCH_3$ with $O{=}S{=}NSi(CH_3)_3$ in the presence of $AlCl_3$ catalyst [2].

B.p. 131°C; n_D^{25} = 1.4555; ^{1}H NMR: δ = 0.60 ppm; IR: 1270 cm^{-1} ν_{as}(O=S=N), 1100 cm^{-1} ν_s(O=S=N) [1]. The PE spectrum is shown as a figure in [3].

It reacts with H_3SiBr (mole ratio 1.7:2) in an ampule for 5 min to give $O{=}S{=}NSiH_3$ [3].

$O{=}S{=}NGe(C_2H_5)_3$ is obtained in low yield by reaction of $(C_2H_5)_3GeCl$ with $K^+[O{=}S{=}N]^-$. The addition of 18-crown-6 increases the yield of the compound considerably [4].

References:

[1] Armitage, D. A.; Clark, M. J. (J. Organometal. Chem. **24** [1970] 629/31).
[2] Armitage, D. A.; Sinden, A. W. (J. Organometal. Chem. **44** [1972] C43/C45).
[3] Cradock, S.; Ebsworth, E. A. V.; Meikle, G. D.; Rankin, D. W. H. (J. Chem. Soc. Dalton Trans. **1975** 805/9).
[4] Armitage, D. A.; Brand, J. C. (J. Chem. Soc. Chem. Commun. **1979** 1078/9).

2.6.19 Tin-Substituted Sulfinyl Imides
N-Sulfinyl-trialkylstannanamines, $O{=}S{=}NSnR_3$, R = CH_3, n-C_4H_9

$O{=}S{=}NSn(CH_3)_3$ was prepared in good yields by heating $(CH_3)_3SnCl$ and $K^+[O{=}S{=}N]^-$ in THF at reflux temperature for 6 h [1] or by treatment of $(CH_3)_3SnOC_2H_5$ with $O{=}S{=}NSi(CH_3)_3$ [2]. It was produced in 88% yield by adding $O{=}S{=}NC_6F_5$ slowly to an equimolar amount of $((CH_3)_3Sn)_3N$ in CH_2Cl_2. After stirring overnight, the solvent was removed and the mixture distilled in vacuum [3].

Pale yellow liquid, b. p. 40°C/1 Torr [3] or 62°C/10 Torr; $n_D^{25} = 1.5141$; ^{1}H NMR (solvent?/TMS): $\delta = 0.37$ ppm, $^2J(^{117}Sn,H) = 60$ Hz, $^2J(^{119}Sn,H) = 62.7$ Hz [2].

$O{=}S{=}NSn(C_4H_9\text{-}n)_3$ was obtained in high yield by reacting $(n\text{-}C_4H_9)_3SnOCH_3$ with an excess of $O{=}S{=}NSi(CH_3)_3$ at room temperature for 30 min [2, 4].

The product is a high-boiling, pale yellow liquid; IR: 1260 cm^{-1} $\nu_{as}(O{=}S{=}N)$, 1090 cm^{-1} $\nu_s(O{=}S{=}N)$ [2].

It reacted with an excess of H_3SiBr at 195 K for 30 min to give $O{=}S{=}NSiH_3$ [4]. Treatment with $(CH_3)_3SiCl$ and $(CH_3)_2SiCl_2$ yielded $O{=}S{=}NSi(CH_3)_3$ and $(O{=}S{=}N)_2Si(CH_3)_2$, respectively, and $(n\text{-}C_4H_9)_3SnCl$ [2].

References:

[1] Armitage, D. A.; Brand, J. C. (J. Chem. Soc. Chem. Commun. **1979** 1078/9).
[2] Armitage, D. A.; Sinden, A. W. (J. Organometal. Chem. **44** [1972] C43/C45).
[3] Roesky, H. W.; Ambrosius, K. (Israel J. Chem. **17** [1978] 132/6).
[4] Cradock, S.; Ebsworth, E. A. V.; Meikle, G. D.; Rankin, D. W. H. (J. Chem. Soc. Dalton Trans. **1975** 805/9).

2.6.20 Bis(N-sulfinyl imidato)mercury, $(O{=}S{=}N)_2Hg$

$(O{=}S{=}N)_2Hg$ was prepared quantitatively by refluxing a mixture of $O{=}S{=}NSi(CH_3)_3$ and HgF_2 (mole ratio ~2:1) in THF for 30 min.

The compound crystallizes as colorless needles, m.p. 128°C with decomposition. IR (Nujol): ν (in cm^{-1}) = 1250 s, 1217 s, 1078 s, 604 m, 590 m, 561 m, 313 w, 280 w. MS (with reference to the isotope ^{202}Hg): m/e = 326 M^+, 264 $HgNSO^+$, 202 Hg^+, 48 SO^+, 46 NS^+, 32 S^+.

The compound is stable to hydrolysis and alcoholysis [1]. $(O{=}S{=}N)_2Hg$ reacted with equimolar amounts of Br_2 in $CFCl_3$ at room temperature to give quantitatively $O{=}S{=}NBr$ and $HgBr_2$ [2]. Shaking a mixture of $(O{=}S{=}N)_2Hg$ and excess FC(O)Br (mole ratio 1:12.5) in a sealed ampule at 10 to 15°C for 48 h afforded $O{=}C{=}NS(O)F$ and $HgBr_2$. Condensing $ClC(O)N{=}C{=}O$ to the compound in the mole ratio 4:1 and allowing the mixture to come to room temperature yielded $O{=}S{=}NC(O)N{=}C{=}O$ and $HgCl_2$ [3]. Treatment of a suspension of $(O{=}S{=}N)_2Hg$ in CCl_4 with $(CH_3)_3SiOSO_2Cl$ (mole ratio ~1:2) at room temperature and subsequently stirring overnight afforded $O{=}S{=}NSO_2OSi(CH_3)_3$ (71% yield) [4]. $(O{=}S{=}N)_2Hg$ reacted with $(\eta^5\text{-}C_5H_5)_2TiCl_2$ or $(\eta^5\text{-}C_5H_5)_2Ti(CO)_2$ to give $(\eta^5\text{-}C_5H_5)_2Ti(N{=}S{=}O)_2$ [5, 7]. Stirring a mixture of equimolar amounts of $(O{=}S{=}N)_2Hg$ and $(\eta^2\text{-}C_2H_4)Pt(P(C_6H_5)_3)_2$ in THF for 2 h formed the trans isomer of $Pt(N{=}S{=}O)_2(P(C_6H_5)_3)_2$. When *cis*-$PtCl_2(P(C_6H_5)_3)_2$ was reacted with $(O{=}S{=}N)_2Hg$ the cis isomer of $Pt(N{=}S{=}O)_2(P(C_6H_5)_3)_2$ was assumed to form [6, 7].

References:

[1] Verbeek, W.; Sundermeyer, W. (Angew. Chem. **81** [1969] 330/1; Angew. Chem. Intern. Ed. Engl. **8** [1969] 376).
[2] Verbeek, W.; Sundermeyer, W. (Angew. Chem. **81** [1969] 331; Angew. Chem. Intern. Ed. Engl. **8** [1969] 376).
[3] Krannich, H.-J.; Sundermeyer, W. (Z. Anorg. Allgem. Chem. **450** [1979] 45/9).
[4] Buss, W.; Krannich, H.-J.; Sundermeyer, W. (Chem. Ber. **109** [1976] 1486/90).
[5] Plenio, H.; Roesky, H. W.; Noltemeyer, M.; Sheldrick, G. M. (J. Chem. Soc. Chem. Commun. **1987** 1483/4).

[6] Short, R.; Hursthouse, M. B.; Purcell, T. G.; Woollins, J. D. (J. Chem. Soc. Chem. Commun. **1987** 407/8).
[7] Kelly, P. F.; Parkin, I. P.; Woollins, J. D. (Phosphorus Sulfur Silicon Relat. Elem. **41** [1989] 223/8).

2.6.21 Metal Complexes of N-Sulfinyl-diorganylphosphinous Amides, N-Sulfinyl-diorganylarsenious Amides, and N,N′-Bis(sulfinyl)-organylarsonous Diamides

2.6.21.1 $(CO)_5M(R_2PN{=}S{=}O)$, M = Cr, Mo; R = CH_3, C_6H_5

Preparation. Properties. Spectra. The metal complexes were prepared by reaction of the corresponding chlorodimethyl- or chlorodiphenylphosphine complexes, $(CO)_5M(R_2PCl)$, with $K^+[O{=}S{=}N]^-$ in CN_3CN at −40°C. Subsequently the reaction mixture was allowed to warm to room temperature, the solvent removed in vacuum and the residue extracted with hexane [1, 5], see also [5]. Yields, properties, and spectroscopic data are summarized in Table 33.

Table 33
Yields, Properties, and Spectra of $(CO)_5M(R_2PN{=}S{=}O)$ Complexes.

No.	M	R	yield in %	appearance, properties, spectra	Ref.
1	Cr	CH_3	69	dark orange crystals, m.p. >~15°C ^1H NMR ($CDCl_3$/external TMS): δ = 2.01 ppm (d, CH_3), $^2J(H,P)$ = 6.4 Hz ^{31}P NMR ($CDCl_3$/external 85% H_3PO_4): δ = 113.3 ppm IR (Nujol): 1250 cm^{-1} $\nu_{as}(O{=}S{=}N)$, 1108 cm^{-1} $\nu_s(O{=}S{=}N)$; IR (hexane): $\nu(CO)$ = 2069, 1988, 1952, 1946 cm^{-1}	[1]
2	Mo	CH_3	77	pale orange liquid ^1H NMR ($CDCl_3$/external TMS): δ = 1.99 ppm (d, CH_3), $^2J(H,P)$ = 6.4 Hz ^{31}P NMR ($CDCl_3$/external 85% H_3PO_4): δ = 86.4 ppm IR (Nujol): 1247 cm^{-1} $\nu_{as}(O{=}S{=}N)$, 1106 cm^{-1} $\nu_s(O{=}S{=}N)$; IR (hexane): $\nu(CO)$ = 2077, 1993, 1967, 1956 cm^{-1}	[1]
3	Cr	C_6H_5	75	dark orange crystals, m.p. 82 to 84°C ^{31}P NMR ($CDCl_3$/external 85% H_3PO_4): δ = 116.5 ppm IR (Nujol): 1232 cm^{-1} $\nu_{as}(O{=}S{=}N)$, 1100, 1089 cm^{-1} $\nu_s(O{=}S{=}N)$; IR (hexane): $\nu(CO)$ = 2069, 1989, 1954, 1944 cm^{-1}	[1]
4	Mo	C_6H_5	48	pale orange crystals $^{31}P\{^1H\}$NMR ($CDCl_3$/external 85% H_3PO_4): δ = 92.4 ppm (s) IR (Nujol): 1233 cm^{-1} $\nu_{as}(O{=}S{=}N)$, 1103, 1091 cm^{-1} $\nu_s(O{=}S{=}N)$; IR (n-pentane): $\nu(CO)$ = 2074, 1961, 1954 cm^{-1}	[3]

Structure. The crystal structure of $(CO)_5Mo((C_6H_5)_2PN{=}S{=}O)$ (Table 33, No. 4) was determined by X-ray analysis. The crystal data are listed in Table 34. Positional parameters are given in the paper [1]. The complex has approximately octahedral geometry around the Mo atom.

The PNSO moiety is planar to within 0.02 Å and is tilted at an angle of 30.9° with respect to the plane through the C(1)C(3)C(4)C(5)Mo unit, which is planar to within 0.04 Å. The $(C_6H_5)_2PNSO$ ligand adopts a cis configuration with respect to the SN bond. The molecular struture is shown in **Fig. 11**; selected bond lengths and bond angles are listed in Table 35 [1], see also [2]. The molecular structure of $(CO)_5Mo((C_6H_5)_2PN{=}S{=}O)$ is very similar to that of $(CO)_5Cr((C_6H_5)_2$-$AsN{=}S{=}O)$ [4], see p. 90. It is also shown in Fig. 11.

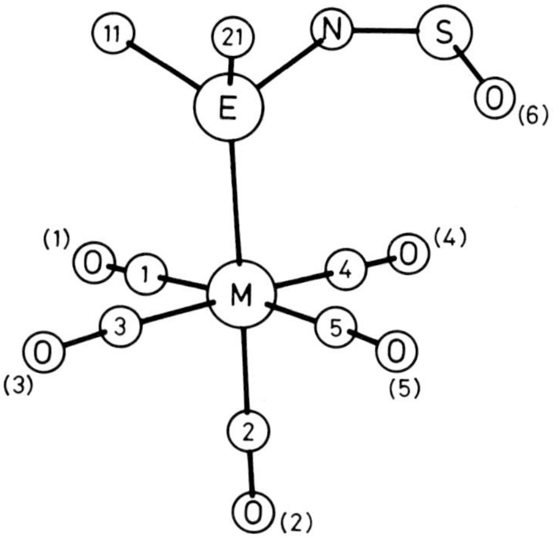

Fig. 11. Molecular structures of $(CO)_5Mo((C_6H_5)_2PN{=}S{=}O)$ (M = Mo; E = P) and $(CO)_5Cr((C_6H_5)_2AsN{=}S{=}O)$ (M = Cr; E = As).

Table 34

Crystal Data of $(CO)_5Mo((C_6H_5)_2PN{=}S{=}O)$ and $(CO)_5Cr((C_6H_5)_2AsN{=}S{=}O)$.

	$(CO)_5Mo((C_6H_5)_2PN{=}S{=}O)$ [1]	$(CO)_5Cr((C_6H_5)_2AsN{=}S{=}O)$ [4]
symmetry	monoclinic	monoclinic
space group	$C2/c{-}C_{2h}^6$ (No. 15)	$P2_1/c{-}C_{2h}^5$ (No. 14)
a in Å	30.236(3)	12.389(4)
b in Å	11.034(1)	10.293(3)
c in Å	12.056(1)	16.031(6)
β in °	96.762(4)	104.50(1)
Z	8	4
V in $Å^3$	3993.9(6)	1979.1
D_x in g/cm^3	1.607	1.622
R_w	0.030	0.057
independent reflections	1850	2178
Ref.	[1]	[4]

Table 35

Selected Bond Lengths and Bond Angles of $(CO)_5Mo((C_6H_5)_2PN{=}S{=}O)$ and $(CO)_5Cr((C_6H_5)_2As$-$N{=}S{=}O)$ (M = Mo or Cr; E = P or As).

	$(CO)_5Mo((C_6H_5)_2PN{=}S{=}O)$ [1]	$(CO)_5Cr((C_6H_5)_2AsN{=}S{=}O)$ [4]
bond length in Å		
M–E	2.484(2)	2.438(1)
E–N	1.699(5)	1.889(6)

Table 35 (continued)

	$(CO)_5Mo((C_6H_5)_2PN{=}S{=}O)$ [1]	$(CO)_5Cr((C_6H_5)_2AsN{=}S{=}O)$ [4]
N–S	1.501(6)	1.496(7)
S–O(6)	1.481(7)	1.457(9)
E–C(11)	1.803(7)	1.926(8)
E–C(21)	1.827(6)	1.926(7)
bond angle in °		
M–E–N	114.2(3)	114.5(2)
E–N–S	129.2(4)	127.5(4)
N–S–O(6)	118.6(3)	120.2(4)
N–E–C(11)	99.5(3)	95.0(3)
N–E–C(21)	103.1(3)	100.2(3)
M–E–C(11)	114.2(2)	117.1(2)
M–E–C(21)	120.8(2)	123.2(2)

Chemical Behavior. The complexes $(CO)_5M(R_2PN{=}S{=}O)$ (M = Cr or Mo; R = CH_3 or C_6H_5) (Table 33, p. 88, Nos. 1 to 4) can be handled in air in the solid state but undergo hydrolysis in solution [1]. They are readily hydrolyzed in wet solvents to the corresponding hydrolysis products $(CO)_5M(R_2PNH_2)$ and SO_2 [1, 2, 3]. The complexes reacted with KOC_4H_9-t in a 2:1 mole ratio in CH_3CN at 23°C for 17 h or in refluxing C_6H_6 for 1 h to give the corresponding homodinuclear complexes $(CO)_5M(R_2PN{=}S{=}NPR_2)M(CO)_5$ [1, 2, 5]. Treatment of the complexes with KOC_4H_9-t in a 1:1 mole ratio in C_6H_6 produced adducts of the type $K^+[(CO)_5M(R_2PNS(O)OC_4H_9\text{-t})]^-$ [1]. $(CO)_5Cr((C_6H_5)_2PN{=}S{=}O)$ reacted with the corresponding Mo complex in the presence of KOC_4H_9-t to give the heterodinuclear complex $(CO)_5Cr((C_6H_5)_2PN{=}S{=}NP(C_6H_5)_2)Mo(CO)_5$ [2]. The reaction of $(CO)_5Cr((C_6H_5)_2PN{=}S{=}O)$ with an equimolar amount of $K^+[(CO)_5Mo((C_6H_5)_2PNS(O)OC_4H_9\text{-t})]^-$ in C_6H_6 at reflux afforded the homodinuclear complexes $(CO)_5M((C_6H_5)_2PN{=}S{=}NP(C_6H_5)_2)M(CO)_5$, with M = Cr or Mo in addition to $(CO)_5Cr((C_6H_5)_2PN{=}S{=}NP(C_6H_5)_2)Mo(CO)_5$. Similarly, the reaction of $(CO)_5Cr((CH_3)_2PN{=}S{=}O)$ with $K^+[(CO)_5Cr((C_6H_5)_2PNS(O)OC_4H_9\text{-t})]^-$ in boiling C_6H_6 produced the symmetrical complexes $(CO)_5Cr(R_2PN{=}S{=}NPR_2)Cr(CO)_5$, with R = CH_3 or C_6H_5 and the unsymmetrical complex $(CO)_5Cr((CH_3)_2PN{=}S{=}NP(C_6H_5)_2)Cr(CO)_5$ [1].

2.6.21.2 $(CO)_5Cr((C_6H_5)_2AsN{=}S{=}O)$ and $(CO)_5Cr(RAs(N{=}S{=}O)_2)$, R = t-$C_4H_9$, C_6H_5

The title compounds were prepared by treatment of $(CO)_5Cr(THF)$ in THF at 0°C with $O{=}S{=}N\text{-}As(C_6H_5)_2$ and $(O{=}S{=}N)_2AsR$ (R = t-C_4H_9 and C_6H_5) (mole ratio 1:1), subsequently stirring at 23°C for 16 h, removal of the solvent in vacuum and extraction with hexane.

$(CO)_5Cr((C_6H_5)_2AsN{=}S{=}O)$. Yellow crystals, yield 69%. IR (Nujol): 1234 cm^{-1} $\nu_{as}(O{=}S{=}N)$; 1095 cm^{-1} $\nu_s(O{=}S{=}N)$; IR (CCl_4): $\nu(CO)$ = 2071, 1995, 1956 cm^{-1}. The structure of the complex was determined by X-ray structure analysis. Crystal data, selected bond lengths, and bond angles are listed in Table 34, p. 89, and Table 35, pp. 89/90, respectively, positional parameters are given in the paper and the molecular structure is shown in Fig. 11, p. 89. The structure of the compound is very similar to that of $(CO)_5Mo((C_6H_5)_2PN{=}S{=}O)$, see pp. 88/9.

$(CO)_5Cr(t\text{-}C_4H_9As(N{=}S{=}O)_2)$. Orange oil, yield 83%. IR (Nujol): 1256, 1240 cm^{-1} $\nu_{as}(O{=}S{=}N)$; 1102 cm^{-1} $\nu_s(O{=}S{=}N)$; IR (CCl_4): $\nu(CO)$ = 2075, 2003, 1957 cm^{-1}.

$(CO)_5Cr(C_6H_5As(N{=}S{=}O)_2)$. Orange oil, yield 76%. IR (Nujol): 1238 cm^{-1} $\nu_{as}(O{=}S{=}N)$; 1090 cm^{-1} $\nu_s(O{=}S{=}N)$; IR (CCl_4): $\nu(CO)$ = 2095, 2022, 1948 cm^{-1} [4].

References:

[1] Chivers, T.; Lensink, C.; Richardson, J. F. (Organometallics **6** [1987] 1904/9).
[2] Chivers, T.; Lensink, C.; Richardson, J. F. (Phosphorus Sulfur **30** [1987] 189/92).
[3] Chivers, T.; Lensink, C.; Richardson, J. F. (Organometallics **5** [1986] 819/21).
[4] Chivers, T.; Dhathathreyan, K. S.; Lensink, C.; Richardson, J. F. (Inorg. Chem. **27** [1988] 1570/4).
[5] Chivers, T.; Dhathathreyan, K. S.; Lensink, C.; Liblong, S. W. (Phosphorus Sulfur Silicon Relat. Elem. **41** [1989] 85/95).

2.6.22 N-Sulfinyl-organylamines

For reactions of N-sulfinyl-organylamines with organic compounds, radicals, carbenes, ylides, and carbanions, organometallic compounds, and metal complexes, see the Sections 2.6.23, pp. 186/241, 2.6.24, pp. 241/3, 2.6.25, pp. 243/51, 2.7.5, p. 269, and 2.7.6, pp. 270/88, respectively.

2.6.22.1 N-Sulfinyl-alkan- and -cycloalkanamines, O=S=NR, R = alkyl, cycloalkyl

2.6.22.1.1 N-Sulfinyl-methanamine, O=S=NCH_3

2.6.22.1.1.1 Formation and Preparation

The title compound was first prepared by Michaelis and Storbeck in 1893, from the reaction of O=S=NC_6H_5 and CH_3NH_2 in toluene. The reaction mixture was cooled with a freezing mixture and then allowed to warm to room temperature. It was isolated and purified by fractional distillation [1]. Later the procedure was modified by other groups. The reactants were added at dry-ice temperature [2], −50°C [3], and −40°C [4]. The mixture was allowed to warm to 0°C [2] and −5°C [3, 4] and kept at these temperatures for a period of several hours and then warmed to room temperature. Purification was achieved by fractional distillation [2, 3, 4]. The title compound was also produced by ozonolysis of a 1M solution of CH_3N=S=NCH_3 in CCl_4 at −20°C in almost quantitative yield [5]. It was formed by dropwise addition of $SOCl_2$ to $CH_3N(Si(CH_3)_3)_2$ (mole ratio 1:1) in dry petroleum at −78°C. However, it could not be separated from the by-product $(CH_3)_3SiCl$ that has nearly the same boiling point [6]. O=S=NCH_3 was obtained along with $(CH_3NH_3^+)_2S_2O_5^{2-}$ by the reaction of CH_3NH_2 and SO_2 [7] and by the exothermic reaction of CH_3N=C=NC_4H_9-t with O=S=$NSO_2C_6H_4CH_3$ (mole ratio 1:1) in 64.5% yield [8]. Finally, O=S=NCH_3 was isolated as a by-product in the reaction of $(CH_3)_3SiN(C_6H_4CF_3\text{-}3)C(O)N(CH_3)Si(CH_3)_3$ with $SOCl_2$ leading to 3-$CF_3C_6H_4N$=C=O [9]. The yield was not given in the cited papers. It should be mentioned that one group had difficulty preparing O=S=NCH_3 in a satisfactory yield following the transsulfinylation method (using O=S=NC_6H_5) or by the reaction of excess CH_3NH_2 with $SOCl_2$ [10].

References:

[1] Michaelis, A.; Storbeck, O. (Liebigs Ann. Chem. **274** [1893] 187/92, 187/8).
[2] Rao, V. M.; Yardley, J. T.; Curl, R. F., Jr. (J. Chem. Phys. **42** [1965] 284/8).
[3] Glass, W. K.; Pullin, A. D. E. (Trans. Faraday Soc. **57** [1961] 546/56).
[4] Burg, A. B.; Woodrow, H. W. (J. Am. Chem. Soc. **76** [1954] 219/20).
[5] Appel, R.; Kohnke, J. (Chem. Ber. **103** [1970] 2152/6).
[6] Abel, E. W.; Armitage, D. A. (J. Chem. Soc. **1964** 3122/6).
[7] Hata, T.; Kinumaki, S. (Nature **203** [1964] 1378/9).
[8] Ulrich, H.; Tucker, B.; Sayigh, A. A. R. (J. Am. Chem. Soc. **94** [1972] 3484/7).
[9] Roesky, H. W.; Ambrosius, K. (Z. Naturforsch. **33b** [1978] 759/62).
[10] Ruppert, I.; Bastian, V.; Appel, R. (Chem. Ber. **108** [1975] 2329/39, 2331).

2.6.22.1.1.2 Molecular Properties. Spectra

Geometrical and Electronic Structure

The molecular structure of O=S=NCH_3 was studied by microwave (MW) spectroscopy and by a gas-phase electron-diffraction analysis in combination with a vibrational analysis [2]. A complete determination with the MW spectroscopy, however, could not be performed because only one isotopic species was available. But both methods agree that O=S=NCH_3 has a cis conformation as shown in **Fig. 12**. The geometrical parameters which were obtained either by experimental methods [1, 2] or in a recent SCF calculation [3] are compiled in Table 36.

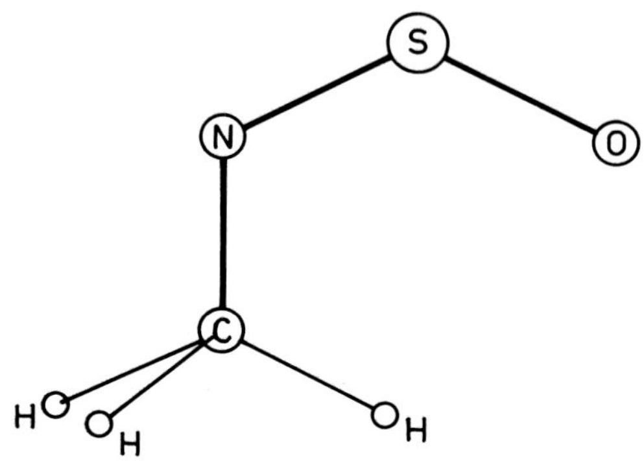

Fig. 12. Molecular Structure of N-Sulfinyl-methanamine [2].

Table 36
Bond Lengths and Bond Angles of O=S=NCH_3.

bond lengths in Å			bond angles in °		Ref.
S=N	S=O	N–C	O=S=N	S=N–C	
1.51	1.45	1.47	121	122	[1]
1.525	1.466	1.421	117	126	[2]
1.493	1.446	1.452	118.2	125.4	[3]

The equilibrium geometries obtained by the SCF calculation [3] agree better with those of [1]. Especially, the N–C distance in [2] was criticized [3].

The H–C–N angle (112°) was not determined with certainty. The O=S=NC skeleton is planar. The methyl hydrogens are eclipsed arranged [2]. Further experimental evidence for the cis form comes from a ^{14}N NMR study [4]. The existence of a trans isomer was not excluded in the MW study. At room temperature, additional lines were observed, which were absent at dry-ice temperature. These may be generated by the higher-energy trans isomer of O=S=NCH_3 [1]. It should also be mentioned that in earlier experimental works, such as mass spectroscopy [5], IR [6], and a dipole moment study [7], a trans conformation of O=S=NCH_3 was assumed.

The barrier to internal rotation of the CH_3 group was computed from MW spectra to be 335 $\pm$1.5 cal/mol [1].

The total energy of *cis*-O=S=NCH_3 and *trans*-O=S=NCH_3 was calculated by an ab initio method to be −562.76994 a.u. and −562.75310 a.u., respectively, indicating that the cis isomer is more stable than the trans isomer by ~10 kcal/mol. This result is in agreement with the most recent experimental results and structure determinations [8]. It should be mentioned that an earlier CNDO/2 calculation resulted in the opposite [9]. A population analysis (calculated at the SCF level) yielded the atomic charges: 0.01 for C, −0.50 for N, 1.11 for S, and −0.74 for O. The shared electron numbers are: $\sigma_{CN}=1.29$, $\sigma_{NS}=1.68$, and $\sigma_{SO}=1.42$ [3].

References:

[1] Rao, V. M.; Yardley, J. T.; Curl, R. F., Jr. (J. Chem. Phys. **42** [1965] 284/8).
[2] Beagley, B.; Chantrell, S. J.; Kirby, R. G.; Schmidling, D. G. (J. Mol. Struct. **25** [1975] 319/27).
[3] Ehrhardt, C.; Ahlrichs, R. (Chem. Phys. **108** [1986] 417/28).
[4] Sicińska, W.; Stefaniak, L.; Witanowski, M.; Webb, G. A. (Org. Magn. Resonance **15** [1981] 241/3).
[5] Grunwell, J. R.; Kochan, A. (J. Org. Chem. **38** [1973] 1610/1).
[6] Glass, W. K.; Pullin, A. D. E. (Trans. Faraday Soc. **57** [1961] 546/56).
[7] Egorov, Yu. P.; Borovikov, Yu. Ya.; Fedyuk, G. S.; Levchenko, E. S. (Zh. Obshch. Khim. **43** [1973] 2470/6; J. Gen. Chem. [USSR] **43** [1973] 2456/61).
[8] Bak, B.; Svanholt, H.; Larsen, C. (J. Mol. Struct. **36** [1977] 55/9).
[9] Grunwell, J. R.; Danison, W. C., Jr. (Tetrahedron **27** [1971] 5315/21).

NMR Spectra

^{1}H NMR (CCl_4/TMS): $\delta=3.58$ ppm [1]; ($CDCl_3$/TMS): $\delta=3.6$ ppm (s) [2].
^{14}N NMR (0.25 M in $(C_2H_5)_2O$/external CH_3NO_2): $\delta=-54.8$ ppm [3].
^{13}C NMR (solvent?/TMS): $\delta=34.57$ ppm [1].

References:

[1] Schwöbel, A. (unpublished results from Bussas, R.; Kresze, G.; Münsterer, H.; Schwöbel, A., Sulfur Rept. **2** [1983] 215/77, 236).
[2] Ulrich, H.; Tucker, B.; Sayigh, A. A. R. (J. Am. Chem. Soc. **94** [1972] 3484/7).
[3] Sicińska, W.; Stefaniak, L.; Witanowski, M.; Webb, G. A. (Org. Magn. Resonance **15** [1981] 241/3).

Dipole Moment

The dipole moment, from MW Stark effect measurements, is 1.70 ± 0.02 D. The components parallel and perpendicular to the a axis were estimated (based on the rotational constants given below) to be $\mu_a^2=2.74\pm0.06$ D^2 and $\mu_b^2=0.14\pm0.05$ D^2, respectively [1]. Other workers take into account the atom polarization in the gaseous state and prefer the diminished value 1.32 D (=1.70 : 1.29) [2]. The dipole moments of *cis*- and hypothetical *trans*-O=S=NCH_3 were also calculated by an ab initio method resulting in $\mu_{tot}=1.528$ D ($\mu_a=1.498$, $\mu_b=0.300$ D) and $\mu_{tot}=4.29$ D ($\mu_a=3.56$, $\mu_b=-2.4$ D), respectively [3].

References:

[1] Rao, V. M.; Yardley, J. T.; Curl, R. F., Jr. (J. Chem. Phys. **42** [1965] 284/8).
[2] Egorov, Yu. P.; Borovikov, Yu. Ya.; Fedyuk, G. S.; Levchenko, E. S. (Zh. Obshch. Khim. **43** [1973] 2470/6; J. Gen. Chem. [USSR] **43** [1973] 2456/61).
[3] Bak, B.; Svanholt, H.; Larsen, C. (J. Mol. Struct. **36** [1977] 55/9).

Rotational Spectrum

The microwave spectrum of O=S=NCH_3 was measured at dry-ice temperature with a conventional 100 kHz Stark modulation spectrometer. Only the a-type R branch transitions (J: 0 → 1 and 1→2 and some lines in the 2→3 and 3→4 region) were identified. A table presents the frequencies of all assigned lines with their assignments. In addition, the observed frequencies were compared with calculated frequencies. The Stark effect was measured for the A_0 lines and the results are given in a table. Rotational constants A=16140.00, B=5015.0, C=3892.0 Hz. The moments of inertia and some other parameters (C, λ_α, I_α, V_3, and S) were obtained by a least-squares fit from the eight lines of the ground torsional state (J: 0→1 and 1→2) and were tabulated: $I_a=31.7154$ u·Å²; $I_b=101.2087\pm0.0166$ u·Å².

Reference:

Rao, V. M.; Yardley, J. T.; Curl, R. F., Jr. (J. Chem. Phys. **42** [1965] 284/8).

Vibrational Spectrum

The IR spectrum of O=S=NCH$_3$ was recorded in the range 4000 to 365 cm^{-1} in 10% CCl$_4$ solution (except in the range 850 to 650 cm^{-1} for which n-hexane or cyclohexane was used) [1]. Vapor spectra of O=S=NCH$_3$ (range 1300 to 500 cm^{-1}) and a pure liquid spectrum were also investigated and are shown in figures. Calculated values obtained from force constants were given (estimated from an assumed geometrical structure), which do not agree very well with experimental values [1]. Calculated frequencies with better agreement are presented by other workers [2]. They are listed, together with the experimental values and suggested assignments, in Table 37.

Table 37
Experimental and Calculated Wavenumbers (in cm^{-1}) of the IR Spectrum for O=S=NCH$_3$ (C_s symmetry).

experiment [1]	calculation [2]	assignment [1, 2]
3036 m, 2980 m, 2934 s, 2860 m, 2788 w, 2740 w	—	C–H stretching fundamentals in resonance with combination tones
2962	2962.1	$\nu_{as}(CH_3)$*)
2892	2892.0	$\nu_s(CH_3)$*)
1468 sh	1472.2	$\delta_{as}(CH_3)$
1438 s	1437.4	$\delta_s(CH_3)$
1252 s	1252.0	ν_{as}(N=S=O)
1135 s	1134.8	ν_s(N=S=O)
1110 s	1110.1	$\varrho(CH_3)$
830 s	829.6	ν(CN)
577 s	577.0	δ(N=S=O)
—	423.5	CNSO torsion
—	237.4	δ(C=N=S)

*) Assignments taken from CH$_3$N$_3$ [2].

References:

[1] Glass, W. K.; Pullin, A. D. E. (Trans. Faraday Soc. **57** [1961] 546/56).
[2] Beagley, B.; Chantrell, S. J.; Kirby, R.G.; Schmidling, D. G. (J. Mol. Struct. **25** [1975] 319/27).

2.6.22.1.1.3 Thermal Properties

O=S=NCH_3 is a colorless liquid which boils at 16°C/144 Torr [1], 57 to 58°C [2], 58°C [3], 58 to 59°C [4, 5], 58 to 60°C [6]; calculated from the vapor pressure equation: b.p. 56.5°C [2].

The vapor tensions of liquid O=S=NCH_3 were found to determine the equation log (p/Torr) = 7.886 − 1650/T. The following data were obtained [2]:

t in °C	−21.0	−16.0	−13.0	−8.0	0.0	3.8
p (obs.) in Torr	22.0	29.5	35.0	46.0	71	85

References:

[1] Glass, W. K.; Pullin, A. D. E. (Trans. Faraday Soc. **57** [1961] 546/56).
[2] Burg, A. B.; Woodrow, H. H. (J. Am. Chem. Soc. **76** [1954] 219/20).
[3] Abel, E. W.; Armitage, D. A. (J. Chem. Soc. **1964** 3122/6).
[4] Michaelis, A.; Storbeck, O. (Liebigs Ann. Chem. **274** [1893] 187/92, 187/8).
[5] Ulrich, H.; Tucker, B.; Sayigh, A. A. R. (J. Am. Chem. Soc. **94** [1972] 3484/7).
[6] Rao, V. M.; Yardley, J. T.; Curl, R. F., Jr. (J. Chem. Phvs. **42** [1965] 284/8).

2.6.22.1.1.4 Chemical Reactions

O=S=NCH_3 decomposes on shaking with H_2O. It fumes in the air [1]. It forms an adduct with BF_3. The reaction was carried out in an immersible tensimeter; dissociation pressures of the adduct are listed in a table [1]. With O-alkyl esters of dithiophosphoric acid derivatives the following reaction occurs [2]: O=S=NCH_3 + 3$(RO)_2P(S)SH \rightarrow ((RO)_2P(S)S)_2 + H_3CNH_3^+(RO)_2PS_2^-$ + [SO]; R = CH_3, i-C_3H_7.

References:

[1] Burg, A. B.; Woodrow, H. W. (J. Am. Chem. Soc. **76** [1954] 219/20).
[2] Kutyrev, G. A.; Mavrin, V. Yu.; Cherkasov, R. A.; Pudovik, A. N. (Izv. Akad. Nauk SSSR Ser. Khim. **1980** 1402/4; C.A. **93** [1980] No. 167773).

2.6.22.1.2 N-Sulfinyl-ethanamine, O=S=NC_2H_5

The title compound was prepared by reactions of anhydrous $C_2H_5NH_2$ with $SOCl_2$ in dry ether (mole ratio 3:1) [1, 2], with O=S=NC_6H_5 in ether [1, 3], and with SO_2 [4], respectively. O=S=NC_2H_5 was isolated and purified by distillation. The compound is a colorless liquid which boils at 65 to 75 [5], 67 (750 Torr) [2], 73 [1], and 73 to 74°C [3]. It fumes in the air. Slow decomposition to a brown, greasy substance was observed in a sealed glass tube [1].

^{1}H NMR spectrum (CCl_4/TMS): δ (in ppm) = 4.09 (q, CH_2); 1.32 (t, CH_3); ^3J(H,H) = 7.8 Hz [5]. ^{14}N NMR spectrum: (1.9 M in $(C_2H_5)_2O$; external CH_3NO_2): δ = −37.4 ppm [6]. An IR spectrum was measured in the range 4000 to 365 cm^{-1} (CCl_4 solution, in the range 850 to 650 cm^{-1} n-hexane or cyclohexane solution). Assigned bands are: 3054 to 2787 cm^{-1} C–H stretching fundamentals in resonance with combination tones, 1464 s CH_2 "scissors", 1451 s C–CH_3 degenerate deformation, 1380 s C–CH_3 symmetric bend, 1238 s NSO pseudo-antisymmetric stretch, 1120 s NSO pseudo-symmetric stretch, and 577 m in-plane NSO bend (?) [2].

Photolysis of $O{=}S{=}NC_2H_5$ in cyclopropane yielded some insoluble white solid and gave rise to a broad featureless ESR signal ($g \approx 2.00$) which persisted when photolysis ceased [8]. Reaction with water leads to decomposition. $O{=}S{=}NC_2H_5$ reacts with $LiN(Si(CH_3)_3)_2$ in hexane to give $C_2H_5N{=}S{=}NC_2H_5$ [7].

References:

[1] Michaelis, A.; Storbeck, O. (Chem. Ber. **24** [1891] 756/7; Liebigs Ann. **274** [1893] 187/92).
[2] Glass, W. K.; Pullin, A. D. E. (Trans. Faraday Soc. **57** [1961] 546/56).
[3] Kobayashi, M.; Honda, K. (Bull. Chem. Soc. Japan **39** [1966] 1778/82).
[4] Hata, T.; Kinumaki, S. (Nature **203** [1964] 1378/9).
[5] Davies, A. G.; Kennedy, J. D. (J. Chem. Soc. C **1968** 2630/40, 2635/6).
[6] Sicińska, W.; Stefaniak, L.; Witanowski, M.; Webb, G. A. (Org. Magn. Resonance **15** [1981] 241/3).
[7] Ruppert, I.; Bastian, V.; Appel, R. (Chem. Ber. **108** [1975] 2329/39).
[8] Baban, J. A.; Roberts, B. P. (J. Chem. Soc. Perkin Trans. II **1978** 678/83).

2.6.22.1.3 N-Sulfinyl-1-propanamine, $O{=}S{=}NC_3H_7$-n

The title compound was prepared by reaction of n-$C_3H_7NH_2$ with $SOCl_2$ in ether in good yield [1, 2] or with $O{=}S{=}NC_6H_5$ ("transsulfinylation") in anhydrous ether in 46.8% yield [3]. $O{=}S{=}NC_3H_7$-n was also produced along with $((CH_3)_2N)_2SO$ by adding $(CH_3)_2NSOCl$ to n-C_3H_7N-$(Si(CH_3)_3)_2$ (mole ratio 2:1) at 0°C in ether (36% yield) [4]. Another procedure for sulfinylation of n-$C_3H_7NH_2$ was described by Sakai et al. [5]. n-$C_3H_7NH_2$ was treated with C_4H_9Li and subsequently with an equimolar amount of SO_2 in tetrahydrofuran yielding n-$C_3H_7NHS(O)OLi$. This intermediate was again allowed to react with C_4H_9Li affording n-$C_3H_7N(Li)S(O)OLi$. A second mole of SO_2 was added, forming the title compound in an exothermic reaction. Work up by direct distillation; yield 30% [5].

$O{=}S{=}NC_3H_7$-n is a colorless oil ($n_D^{25}=1.4310$ [4]) which boils at 23°C/14 Torr [2, 8], 28°C/50 Torr [4], 80 to 90°C [5], 100 to 110°C [6], and 104°C [1, 3].

^1H NMR (CCl_4/TMS): δ (in ppm) = 4.01 (t); 1.63 (sext), 1.02 (t); $^3J(H,H) \approx 7.7$ Hz [6].

^{14}N NMR: (neat/external CH_3NO_2): $\delta = -39.0$ ppm; (2.0 M in $(C_2H_5)_2O$/external CH_3NO_2): $\delta = -41.3$ ppm [7].

IR (state not given): ν (in cm^{-1}) = 1230, 1140 $\nu(N{=}S{=}O)$ [5]; 1230, 1120 $\nu(N{=}S{=}O)$ [6].

UV: two absorptions: λ_{max} = 300 to 310 nm (log ε = 1.69; in 0.4 g/L n-hexane), and λ_{max} = 235 to 236 nm (log ε = 3.63; in 10^{-2} g/L cyclohexane). The bands were tentatively assigned to a $\pi \to \pi^*$ and an $n \to \pi^*$ transition, respectively [2, 8]. Other workers observed only a single maximum at 234 nm (log ε = 3.68; in ether?) [9].

On addition of H_2O the compound decomposes slowly [1].

References:

[1] Michaelis, A.; Storbeck, O. (Liebigs Ann. **274** [1893] 187/92, 190).
[2] Leandri, G.; Mangini, A. (Spectrochim. Acta **1959** 421/31).
[3] Klamann, D.; Sass, C.; Zelenka, M. (Chem. Ber. **92** [1959] 1910/3).
[4] Armitage, D. A.; Sinden, A. W. (J. Inorg. Nucl. Chem. **36** [1974] 993/5).
[5] Sakai, S.; Fujinami, T.; Komizo, K. (J. Org. Chem. **40** [1975] 3291/2).
[6] Davies, A. G.; Kennedy, J. D. (J. Chem. Soc. C **1968** 2630/40, 2636).

[7] Sicińska, W.; Stefaniak, L.; Witanowski, M.; Webb, G. A. (Org. Magn. Resonance **15** [1981] 241/3).
[8] Leandri, G.; Mangini, A. (Ric. Sci. **27** [1957] 1920/5).
[9] Smith, W. T., Jr.; Trimnell, D.; Grinninger, L. D. (J. Org. Chem. **24** [1959] 664/6).

2.6.22.1.4 N-Sulfinyl-2-propanamine, $O{=}S{=}NC_3H_7$-i

The title compound was prepared by the standard method of Michaelis and Storbeck from i-$C_3H_7NH_2$ and $SOCl_2$ in ether at −2°C [1, 2], in dry CCl_4 at −2°C [3], or by "transsulfinylation" with $O{=}S{=}NC_6H_5$ in toluene [4]. It was also obtained by thermolysis of 2-isopropyl-3-isopropylimino-4-p-toluene-sulfonyl-1,2,4-thiadiazetidine 1-oxide (I) (synthesized by reaction of $O{=}S{=}N-SO_2C_6H_4CH_3$-4 with (i-$C_3H_7N{=})_2C$) at 80 to 100°C under reduced pressure (10 Torr) for 1 h; yield 48% [1].

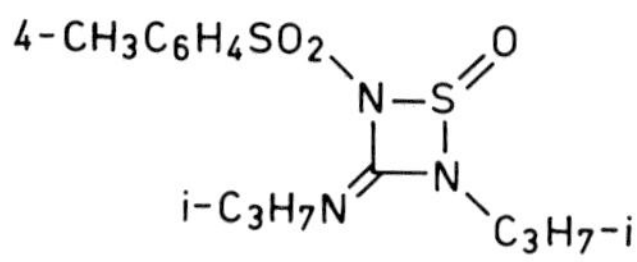

I

The yellow liquid boils at 95 to 100°C [1] or 86°C [4]. ^{1}H NMR (CCl_4/TMS): δ = 1.02 ppm (d; 6H; J = 7.0 Hz); δ = 4.90 ppm (sept; 1H; J = 6.5 Hz) [3]. ^{14}N NMR (neat/external CH_3NO_2): δ = −25.2 ppm; (1.9M in $(C_2H_5)_2O$/external CH_3NO_2): δ = −26.6 ppm [5]. The dipole moment of $O{=}S{=}NC_3H_7$-i was determined in benzene solution at 25°C to be 1.47 D. For the pure cis(syn) and trans(anti) form of $O{=}S{=}NC_3H_7$-i theoretical values of 0.85 and 2.15 D, respectively, were calculated. It was concluded that $O{=}S{=}NC_3H_7$-i exists in solution in both isomeric forms (cis:trans = 1.7) [2]. However, this is questionable (compare $O{=}S{=}NC_6H_5$, p. 115).

References:

[1] Minami, T.; Fukuda, M.; Abe, M.; Agawa, T. (Bull. Chem. Soc. Japan **46** [1973] 2156/9).
[2] Jannelli, L.; Lamanna, U.; Lumbroso, H. (Bull. Soc. Chim. France **1966** 3626/30).
[3] van Woerden, H. F.; Bijl-Vlieger, S. H. (Recl. Trav. Chim. **93** [1974] 85/90).
[4] Baban, J. A.; Roberts, B. P. (J. Chem. Soc. Perkin Trans. II **1978** 678/83).
[5] Sicińska, W.; Stefaniak, L.; Witanowski, M.; Webb, G. A. (Org. Magn. Resonance **15** [1981] 241/3).

2.6.22.1.5 N-Sulfinyl-1-butanamine, $O{=}S{=}NC_4H_9$-n

$O{=}S{=}NC_4H_9$-n was prepared by the standard method of Michaelis and Storbeck from the reaction of n-$C_4H_9NH_2$ and $SOCl_2$ in anhydrous ether with cooling [1 to 5]; yields 28.4 [1], 42.3% [3]. To reduce the consumption of primary amine, dry pyridine was used as a base [4]. For a better yield (≧70%), quinoline was found to be more favorable [5]. An analogous procedure was performed by the reaction between n-$C_4H_9NH_2$ and SOF_2 in the presence of anhydrous pyridine in ether with cooling [6]. After evaporating the ether, the remaining bright yellow oil (probably OS(F)N(H)C_4H_9-n) was boiled to yield the title compound by eliminating HF. The crude product (b.p. 35 to 36°C/20 Torr) was redistilled [6]. The reaction of ClS(O)OC_2H_5 and n-$C_4H_9NH_2$ in anhydrous ether probably led to the formation of $O{=}S{=}NC_4H_9$-n, yield 38% [7]. Adding $SOCl_2$ slowly to n-$C_4H_9N(Si(CH_3)_3)_2$ (ratio 1:1) at −78°C produced $O{=}S{=}NC_4H_9$-n in 62% yield [8]. The title compound was synthesized by adding $(CH_3)_2NSOCl$ to n-$C_4H_9N(Si(CH_3)_3)_2$ (ratio 2:1) in ether with ice cooling. The product was isolated by distillation in 70% yield [9]. It was formed by introducing $SOCl_2$ to a suspension of 1,3-di-n-butylurea (n-C_4H_9–HNC(O)NH–C_4H_9-n) ratio 1:1 in CCl_4, yield 77%. O=S=NC_4H_9-n could not be separated from the by-product n-C_4H_9N=C=O with its similar boiling point (b.p. = 115°C) [10]. Along with these methods the title compound was obtained by a sulfinylation reaction of n-$C_4H_9NH_2$ using SO_2 and C_4H_9Li in analogy to the preparation of O=S=NC_3H_7-n and O=S=NC_6H_{11}-c, yield 20% [11]. The compound can be synthesized by adding dropwise n-C_4H_9N=C=NC_4H_9-n to O=S=N$SO_2C_6H_4CH_3$-4 (ratio 1:1) in benzene. The mixture was stirred for 2 h at ambient temperature. O=S=NC_4H_9-n was separated from n-C_4H_9N=C=N$SO_2C_6H_4CH_3$-4 by distillation, yield 57% [12]. It was prepared in a similar way by the exchange reaction of O=S=NSO_2CH_3 with n-C_4H_9N=C=NC_4H_9-n; yield 47% [12]. Finally the compound was obtained in situ by sulfinylation of n-$C_4H_9NH_2$ with N-(chlorosulfinyl)-imidazole (prepared from imidazole and $SOCl_2$) in CH_2Cl_2 at 20°C for 0.3 h, yield 53% [13].

O=S=NC_4H_9-n is a colorless liquid [3, 6, 7, 8] (n_D^{20} = 1.0165 [1], 1.4367 [8], n_D^{21} = 1.438 [9]) which gradually colors yellow [3, 7] and boils at 30°C/20 Torr [3, 12], 34°C/14 Torr [9], 35°C/14 Torr [2], 35 to 36°C/20 Torr [6], 38°C/10 Torr [7], 47°C/50 Torr [4], 50°C/36 Torr [8], 64 to 65°C/100 Torr [1], 116°C/760 Torr [6, 10]. The substance decomposed on distillation at 760 Torr [13].

The statements about the stability of the compound are inconsistent. On one hand it was reported that it can be stored in dry air without decomposition [6]. Otherwise a spontaneous decomposition with elimination of NH_3 and formation of a resin was observed [3]. Decomposition of O=S=NC_4H_9-n to a greasy brown mass was also observed by others [7]. O=S=NC_4H_9-n is not hygroscopic and it is hydrolyzed after several minutes when shaking it in H_2O [6].

Based on mass spectral data studies of several N-sulfinyl-alkanamines, it was suggested that the n-butyl group is trans oriented [17]. In contrast, ^{14}N NMR data indicate that the cis form is more favorable [14].

^{14}N NMR: (neat/external CH_3NO_2): $\delta = -39.6$ ppm; (1.9 M in $(C_2H_5)_2O$/external CH_3NO_2): $\delta = -40.0$ ppm [14].

IR (neat?): ν_{as}(N=S=O) = 1235 [10], 1240 cm^{-1} [11], and ν_s(N=S=O) = 1110 [10], 1115 cm^{-1} [11].

UV: λ_{max} = 235 nm (log ε = 3.69) and a weak absorption 310 to 312 nm (log ε = 1.56) measured in n-hexane at a concentration of 10^{-2} g/L and 10^{-1} g/L, respectively. The bands were tentatively assigned to an $n \rightarrow \pi^*$ and a $\pi \rightarrow \pi^*$ transition, respectively [2, 15]. Other workers found only a single band at 234 nm (log ε = 3.72, in ether?) [16].

Mass spectrum (70 eV): m/e (rel. int. in %) = 77 (89), 76 (14), 75 (11), 71 (14), 70 (54), 55 (11), 50 (13), 43 (100), 42 (16), 41 (86), 39 (23), 30 (13), 29 (35), 27 (65). Important fragmentation pathways are the loss of propene forming an ion with m/e = 77 (McLafferty rearrangement) and loss of HSO forming an ion with m/e = 70 [17].

The reaction of O=S=NC_4H_9-n with NOCl yielded n-C_4H_9Cl by decomposition of the intermediate n-$C_4H_9N_2^+Cl^-$ [4].

References:

[1] Wichterle, O.; Roček, J. (Chem. Listy **47** [1953] 1768/80), Collection Czech. Chem. Commun. **19** [1954] 282/96, 287/8).
[2] Leandri, G.; Mangini, A. (Spectrochim. Acta **1959** 421/31).
[3] Klamann, D.; Sass, C.; Zelenka, M. (Chem. Ber. **92** [1959] 1910/3).
[4] Kobayashi, M.; Honda, K. (Bull. Chem. Soc. Japan **39** [1966] 1778/82).
[5] Schell, F. M.; Carter, J. P.; Wiaux-Zamar, C. (J. Am. Chem. Soc. **100** [1978] 2894/6).
[6] Goehring, M.; Voigt, G. (Chem. Ber. **89** [1956] 1050/4).
[7] Zinner, G. (Chem. Ber. **91** [1958] 966/71).

[8] Abel, E. W.; Armitage, D. A. (J. Chem. Soc. **1964** 3122/6).
[9] Armitage, D. A.; Sinden, A. W. (J. Inorg. Nucl. Chem. **36** [1974] 993/5).
[10] Ulrich, H.; Sayigh, A. A. R. (Angew. Chem. **78** [1966] 761/9; Angew. Chem. Intern. Ed. Engl. **5** [1966] 704/12).

[11] Sakai, S.; Fujinami, T.; Komizo, K. (J. Org. Chem. **40** [1975] 3291/2).
[12] Minami, T.; Fukuda, M.; Abe, M.; Agawa, T. (Bull. Chem. Soc. Japan **46** [1973] 2156/9).
[13] Kim, Y. H.; Shin, J. M. (Tetrahedron Letters **26** [1985] 3821/4).
[14] Sicińska, W.; Stefaniak, L.; Witanowski, M.; Webb, G. A. (Org. Magn. Resonance **15** [1981] 241/3).
[15] Leandri, G.; Mangini, A. (Ric. Sci. **27** [1957] 1920/5).
[16] Smith, W. T., Jr.; Trimnell, D.; Grinninger, L. D. (J. Org. Chem. **24** [1959] 664/6).
[17] Grunwell, J. R.; Kochan, A. (J. Org. Chem. **38** [1973] 1610/1).

2.6.22.1.6 N-Sulfinyl-2-butanamine, $O{=}S{=}NC_4H_9$-s

The compound was obtained by the method of Michaelis and Storbeck by adding $SOCl_2$ to s-$C_4H_9NH_2$ at 0°C [1].

^1H NMR (TMS): δ (in ppm) = 0.9 (3H, t; J = 7.0 Hz); 1.33 (3H, d; J = 7.0 Hz); 1.50 (2H, q; J = 7.0 Hz); 4.73 (1H, sextet; J = 7.0 Hz) [1].

^{14}N NMR (neat/external CH_3NO_2): δ = −28.6 ppm; (1.8 M in $(C_2H_5)_2O$/external CH_3NO_2): δ = −29.2 ppm [2].

MS (70 eV): m/e (rel. int. in %) = 104 (5), 91 (58), 90 (100), 89 (17), 71 (28), 70 (11), 63 (14), 60 (20), 57 (15), 56 (35), 55 (11), 44 (53), 43 (49), 42 (71), 41 (44), 39 (17), 29 (61), 27 (54) [1]. Important fragmentation pathways are the loss of ethylene to m/e = 91 (McLafferty rearrangement) and the loss of SO to m/e = 71, which subsequently fragments to m/e = 56 and 42 by cleavage of methyl and ethyl radicals, respectively [1].

References:

[1] Grunwell, J. R.; Kochan, A. (J. Org. Chem. **38** [1973] 1610/1).
[2] Sicińska, W.; Stefaniak, L.; Witanowski, M.; Webb, G. A. (Org. Magn. Resonance **15** [1981] 241/3).

2.6.22.1.7 N-Sulfinyl-2-methyl-1-propanamine, $O{=}S{=}NC_4H_9$-i

The title compound was prepared by the reaction of i-$C_4H_9NH_2$ (explicitely $(CH_3)_2CH$-CH_2NH_2) with $SOCl_2$ in ether [1]. This procedure was modified by the use of dry pyridine as a base, yield 65% [2]. It was also obtained by sulfinylation of i-$C_4H_9NH_2$ with $O{=}S{=}NC_6H_5$ in ether at room temperature (ratio 1:1), yield 24% [3]. The compound was isolated and purified by distillation. $O{=}S{=}NC_4H_9$-i is a colorless [1], yellowish [3] oil which boils at 116°C [1], 44°C/60 Torr [2], or at 45°C/50 Torr [3]. Density d_4^{25} = 0.8761 g/cm³; an orientating determination of the dipole moment yielded 1.62 D [4].

^{14}N NMR (neat/external CH_3NO_2): δ = −41.7 ppm; (1.5 M in acetone/external CH_3NO_2): δ = −42.0 ppm [5].

The compound has a sweetish, penetrating odor; it is decomposed by water on shaking [1].

The reaction of $O{=}S{=}NC_4H_9$-i with NOCl in ether gave i-$C_4H_9NH_3^+Cl^-$ and i-C_4H_9Cl. These products arose from the intermediate i-$C_4H_9N_2^+Cl^-$ by decomposition [2]. Heating $O{=}S{=}NC_4H_9$-i in

DMSO at 80°C for 3 h yielded i-C_4H_9CHO (3%) isolated by adding dimedone (5,5-dimethyl-1,3-cyclohexandione) (forming isobutylaldehyde dimethone). If O=S=NC_4H_9-i and aniline (ratio 1:1) were heated under the same conditions the yield of i-C_4H_9CHO reached 32% [3].

References:

[1] Michaelis, A.; Storbeck, O. (Liebigs Ann. Chem. **274** [1893] 187/96, 191).
[2] Kobayashi, M.; Honda, K. (Bull. Chem. Soc. Japan **39** [1966] 1778/82).
[3] Taguchi, T.; Morita, S.; Kawazoe, J. (Chem. Pharm. Bull. [Tokyo] **23** [1975] 2654/9).
[4] Jensen, K. A.; Hofman Bang, N. (Liebigs Ann. Chem. **548** [1941] 95/105).
[5] Sicińska, W.; Stefaniak, L.; Witanowski, M.; Webb, G. A. (Org. Magn. Resonance **15** [1981] 241/3).

2.6.22.1.8 N-Sulfinyl-2-methyl-2-propanamine, O=S=NC_4H_9-t

The title compound was obtained by the standard method of Michaelis and Storbeck by reacting t-$C_4H_9NH_2$ with $SOCl_2$ [1, 2, 3] in dry ether under a nitrogen inert-gas atmosphere. After evaporation ot the solvent (at standard pressure and then at 14 Torr) the residue was distilled at a pressure of 10^{-2} Torr (bath temperature of 20 to 25°C); yield 26% [2]. O=S=NC_4H_9-t was also synthesized by ozonolysis of a 1M solution of t-C_4H_9N=S=NC_4H_9-t in CCl_4 at −20°C in almost quantitative yield [4]. It was also produced along with $(CH_3)_2S$=N–$S(CH_3)_2^+BF_4^-$ in the reaction of t-C_4H_9N=S=$NS(CH_3)_2^+BF_4^-$ with $(CH_3)_2SO$ after 8 h warming at 60°C in CH_3CN [5]. O=S=NC_4H_9-t was generated as a by-product in the reactions of (i-$C_3H_7)_2NP$=NC_4H_9-t and ((i-$C_3H_7)_2NP$=N-C_4H_9-t)$Cr(CO)_5$, respectively, with SO_2, in ether [6]. O=S=NC_4H_9-t was also synthesized in situ via sulfinylation of t-$C_4H_9NH_2$ with N-(chlorosulfinyl)imidazole (prepared from imidazole ana $SOCl_2$, used without isolation) in CH_2Cl_2 at 20°C for 0.3 h; yield 63% [7]. The reactions of 2-diazo-3,3-dimethylbutane, t-C_4H_9–C(=N_2)–CH_3, or 3-diazo-2,2,4,4,-tetramethylpentane, t-C_4H_9–C(=N_2)–C_4H_9-t, with t-C_4H_9–$NHSO_2Cl$ in the presence of $N(C_2H_5)_3$ at −78°C (N_2 atmosphere) in ether led to a mixture of compounds in which O=S=NC_4H_9-t was also identified but not isolated [2].

O=S=NC_4H_9-t is a colorless liquid which boils at 98 to 99°C/760 Torr [1] or 27 to 29°C/23 Torr (decompositon was observed by distillation at 760 Torr) [7]. Density D = 1.323 g/cm^3 at 25°C, 1.59 g/cm^3 at −190°C [15]. The compound tarnished in the air [2].

^{1}H NMR (CCl_4/TMS): δ = 1.50 ppm (9H, s) [4, 8], δ = 1.51 ppm [2]. The signal did not split or broaden at −100°C [8].

^{14}N NMR (neat/external CH_3NO_2): δ = −26.0 ppm; (2.0M in $(C_2H_5)_2O$/external CH_3NO_2): δ = −27.4 ppm [11]; (neat/satd. NH_4NO_3 in H_2O): δ = 332(2) ppm [10] (referred to CH_3NO_2: δ = −27.5 ppm); (neat/referenced to liquid NH_3 via external NH_4Cl): δ = 353 ppm [12] (referred to CH_3NO_2: δ = −27.2 ppm).

^{33}S NMR (neat liquid/referenced to 2 M Cs_2SO_4 via external sulpholane): δ = 257 ppm [12].

^{13}C NMR (solvent?/TMS): δ = 62.70 ppm [9].

IR (state not given): ν (in cm^{-1}) = 1250 ν_{as}(NSO), 1095 ν_s(NSO) [1]; (without solvent): ν (in cm^{-1}) = 1370, 1260, and 1105 [2].

Dipole moment: μ = 1.46 D at 25°C [15].

Dielectric properties of the compound were determined at −190°C in the solid phase: deformation limit of dielectric constant ε_∞ = 3.08, deformation polarization P_{def} = 40.3 cm^3, and refraction R_D = 42.0 cm^3 [15].

UV (gas phase): E_{max} (in eV; ε in $L \cdot mol^{-1} \cdot cm^{-1}$) = 7.00 (1840), 6.53 (340), 5.51 (1710), 5.06 (980), 4.59 (270) [14], converted to nm: λ_{max} = 177.0, 189.8, 225.0, 245.0, and 270.1 nm. UV (in CH_3OH): λ_{max} = 233 nm (log ε = 3.10); (in c-C_6H_{12}): λ_{max} = 233 nm (log ε = 3.03) [13]; UV (no details given): λ_{max} = 305 nm (log ε = 1.6) [10]. These experimental results are in relatively good accord with excitation energies from modified CNDO CI calculation [14].

The vertical ionization energies E_i (vert), which were determined by photoelectron spectroscopy, are tabulated together with CNDO orbital energies ε and assignments [14]:

E_i (vert) in eV	10.54	10.95	12.12	12.58
−ε in eV	10.12 (7a″)	10.34 (14a′)	11.97 (13a′)	12.65 (12a′)

S $2p_{3/2}$ and N 1s core binding energies were calculated to be 167.4 and 399.3 eV, respectively [14].

MS (70 eV): m/e (rel. int. in %) = 104 (100), 74 (12), 57 (49), 56 (19), 42 (17), 41 (60), 39 (15), 29 (22), 27 (13) [8].

The photolysis of O=S=NC_4H_9-t in cyclopropane yielded an insoluble white solid and gave rise to a broad ESR signal (g ≈ 2.00) [16]. O=S=NC_4H_9-t reacted with PCl_5 (ratio 1:1) in CCl_4 at 20°C for 30 min and at 60°C for 30 min to yield t-C_4H_9N=SCl_2; yield 84%. The reaction with PBr_5 in CCl_4 (ratio 1:1) gave t-C_4H_9N=SBr_2 [17]. O=S=NC_4H_9-t reacted with $(CH_3)_2BSCH_3$ in CH_2Cl_2 (30 min refluxing, ratio 1:4) to produce the compound t-$C_4H_9N(S{-}S{-}CH_3)B(CH_3)_2$ [18]. It reacted with 3,5-dimethyl-1,2,4-trithia-3,5-diborolane (I) to form II and various cyclic byproducts [19]. Treatment of O=S=NC_4H_9-t with $LiN(Si(CH_3)_3)_2$ in n-hexane at room temperature afforded t-C_4H_9N=S=$NSi(CH_3)_3$ in 65% yield [20].

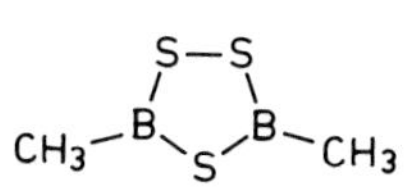

I

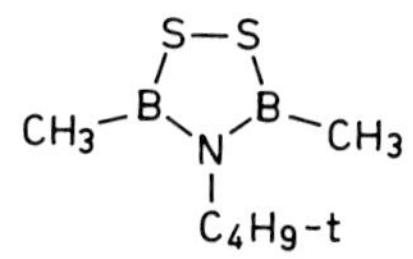

II

References:

[1] Scherer, O. J.; Hornig, P. (Angew. Chem. **78** [1966] 776/7; Angew. Chem. Intern. Ed. Engl. **5** [1966] 729/30).
[2] Quast, H.; Kees, F. (Chem. Ber. **114** [1981] 774/86).
[3] Birum, G. H.; Heniger, S. A. (U.S. 2893910 [1959]; C.A. **1960** 1434).
[4] Appel, R.; Kohnke, J. (Chem. Ber. **103** [1970] 2152/6).
[5] Schwöbel, A.; Kresze, G. (Liebigs Ann. Chem. **1985** 453/61).
[6] Niecke, E.; Engelmann, M.; Zorn, H.; Krebs, B.; Henkel, G. (Angew. Chem. **92** [1980] 737/8, 738/9).
[7] Kim, Y. H.; Shin, J. M. (Tetrahedron Letters **26** [1985] 3821/4).
[8] Grunwell, J. R.; Kochan, A. (J. Org. Chem. **38** [1973] 1610/1).
[9] Schwöbel, A. (unpublished results from Bussas, R.; Kresze, G.; Münsterer, H.; Schwöbel, A., Sulfur Rept. **2** [1983] 215/377, 236).
[10] Armitage, D.; Mason, J.; Vinter, J. G. (Inorg. Chem. **17** [1978] 776/7).

[11] Sicińska, W.; Stefaniak, L.; Witanowski, M.; Webb, G. A. (Org. Magn. Resonance **15** [1981] 241/3).
[12] Belton, P. S.; Woollins, J. D. (Magn. Resonance Chem. **24** [1986] 1080/2).
[13] Smith, W. T., Jr.; Grinninger, L. D. (J. Org. Chem. **26** [1961] 2133/4).
[14] Kroner, J.; Strack, W.; Holsboer, F.; Kosbahn, W. (Z. Naturforsch. **28b** [1973] 188/95).

[15] Egorov, Yu. P.; Borovikov, Yu. Ya.; Fedyuk, G. S.; Levchenko, E. S. (Zh. Obshch. Khim. **43** [1973] 2470/6; J. Gen. Chem. [USSR] **43** [1973] 2456/61).
[16] Baban, J. A.; Roberts, B. P. (J. Chem. Soc. Perkin Trans. II **1978** 678/83).
[17] Markovskii, L. N.; Fedyuk, G. S.; Levchenko, E. S. (Zh. Org. Khim. **8** [1972] 2057/62; J. Org. Chem. [USSR] **8** [1972] 2104/8).
[18] Meller, A.; Maringgele, W.; Fetzer, H. (Chem. Ber. **113** [1980] 1950/61, 1951).
[19] Meller, A.; Habben, C. (Monatsh. Chem. **113** [1982] 139/53, 141).
[20] Ruppert, I.; Bastian, V.; Appel, R. (Chem. Ber. **108** [1975] 2329/39).

2.6.22.1.9 N-Sulfinyl-cyclohexanamine, $O{=}S{=}NC_6H_{11}$-c

The title compound was prepared by the standard method of Michaelis and Storbeck by treatment of c-$C_6H_{11}NH_2$ dropwise with $SOCl_2$ in anhydrous benzene [1, 2] using pyridine as a base; yield 65.4% [2]. Thermolysis of the cycloadduct of the reaction of $O{=}S{=}NSO_2C_6H_4CH_3$-4 and c-$C_6H_{11}N{=}C{=}NC_6H_{11}$-c at 180°C/15 Torr for 1 h in benzene yielded $O{=}S{=}NC_6H_{11}$-c in 75% yield. It was also obtained directly from the analogous reaction of $O{=}S{=}NSO_2CH_3$ with c-$C_6H_{11}N{=}C{=}NC_6H_{11}$-c in benzene (at ambient temperature ?) in 68% yield. The similar exchange reactions of N-sulfinyl-carboxamides O=S=NCOR (R = C_6H_5, CH_3) with c-$C_6H_{11}N{=}C{=}N$-C_6H_{11}-c in dry ether (at ambient temperature ?) afforded oily products after evaporation of the solvent, which were composed of 1,2- and 1,4-cycloadducts. Decomposition took place on distillation, leading to mixtures of the title compound along with RN=C=O and RC≡N (R = CH_3, C_6H_5) and RCONH–C_6H_{11}-c (R = C_6H_5) as major products. The yields of $O{=}S{=}NC_6H_{11}$-c were 42 (R = C_6H_5) and 15% (R = CH_3) [3]. Treatment of c-$C_6H_{11}NH_2$ with C_4H_9Li and sulfinylation with SO_2 in dry THF gave c-C_6H_{11}NHS(O)OLi. This intermediate compound was allowed to react a second time with C_4H_9Li and SO_2. The resulting suspension was directly distilled in vacuum and the distillate trapped by cooling with liquid nitrogen. Purification was achieved by redistillation, yield 14% [4]. A related procedure, the sulfinylation of c-$C_6H_{11}NHSi(CH_3)_3$ by SO_2, gave a higher yield and was not so complex. c-$C_6H_{11}NH_2$ was first silylated (using n-C_4H_9Li and $(CH_3)_3SiCl$ in THF at 0°C). The solution of the anion of c-$C_6H_{11}NH$–$Si(CH_3)_3$ (obtained with n-C_4H_9Li in THF at 0°C) was added to a solution containing an excess of SO_2 in THF at −78°C. Removal of the solvent at room temperature and distillation of the residue gave the title compound in 62% yield [5]. $O{=}S{=}NC_6H_{11}$-c was also prepared in situ by sulfinylation of c-$C_6H_{11}NH_2$ with N-(chlorosulfinyl)imidazole (synthesized with imidazole and $SOCl_2$, used without isolation) in CH_2Cl_2 at 20°C for 0.3 h in 77% yield [6].

$O{=}S{=}NC_6H_{11}$-c is a colorless oil [1] which boils at 55 to 60°C/16 Torr [4], 56 to 60°C/15 Torr [5], 59 to 61°C/17 Torr [6], 74.5°C/14 Torr [1], and 78°C/12 Torr [2].

^{15}N NMR (natural-abundance ^{15}N; neat/external HNO_3): $\delta = -22.6$ ppm (referenced to CH_3NO_2: $\delta = -27.0$ ppm) [7].

IR: ν(N=S=O) = 1230 [5], 1245 [4], and 1120 cm^{-1} [4, 5] ([4]: medium ?; [5]: neat liquid).

PES: The vertical ionization energies E_i (vert), which were determined by photoelectron spectroscopy (HeI-584 Å), are tabulated together with CNDO orbital energies ε and assignments [10]:

E_i (vert) in eV	~10.0	10.38	11.16	11.61
−ε in eV	10.09 (9a″)	10.36 (17a′)	11.29 (16a′)	11.66 (8a″)

UV (gas phase): E_{max} (in eV, ε in $L \cdot mol^{-1} \cdot cm^{-1}$) = 7.75 (5500, sh), 7.25 (3200), 6.57 (2020, sh), 5.33 (3800) [10], converted to nm: λ_{max} = 160.0, 171.0, 188.7, and 232.6 nm. UV (in ether): λ_{max} =

235 nm (log ε = 3.83) [8]; (in CH_3OH): λ_{max} = 237 to 238 nm (log ε = 2.99) [8]; (10^{-2} g/L cyclohexane): λ_{max} = 239 nm (log ε = 3.70) (n → π* transition) [1, 9]; (10^{-1} g/L n-hexane): λ_{max} = 310 nm (log ε = 1.69) (π → π* transition) [1, 9]. The results are in relatively good agreement with excitation energies from modified CNDO CI calculations. A calculated band at 3.91 eV (317 nm) was not observed in the spectrum because of the low vapor pressure of the compound [10].

$O{=}S{=}NC_6H_{11}$-c afforded the 1:1 adduct $O{=}S{=}NC_6H_{11}$-c · BF_3 when treated with BF_3 in toluene at 1°C. The tensimetric titration curve is presented in a figure [11]. The title compound reacted with $COCl_2$ in refluxing benzene and with pyridine or N,N-dimethylformamide as catalyst to produce c-$C_6H_{11}N{=}C{=}O$ (yield 61%). A reaction mechanism is proposed [12]. The reaction of $O{=}S{=}NC_6H_{11}$-c and 3,5-dimethyl-1,2,4-trithia-3,5-diborolane (I) led to the formation of III and various by-products [13].

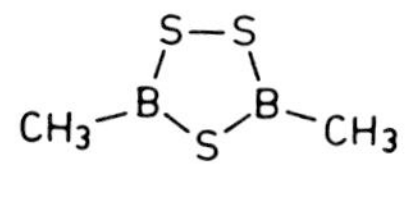

I

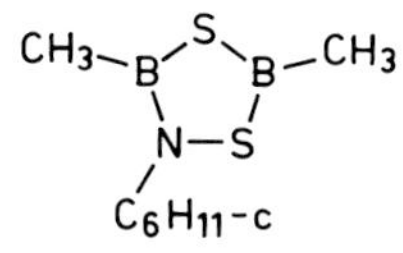

III

References:

[1] Leandri, G.; Mangini, A. (Spectrochim. Acta **1959** 421/31, 422, 429).
[2] Klamann, D.; Sass, C.; Zelenka, M. (Chem. Ber. **92** [1959] 1910/3).
[3] Minami, T.; Fukuda, M.; Abe, M.; Agawa, T. (Bull. Chem. Soc. Japan **46** [1973] 2156/9).
[4] Sakai, S.; Fujinami, T.; Komizo, K. (J. Org. Chem. **40** [1975] 3291/2).
[5] Porskamp, P. A. T. W.; Zwanenburg, B. (Synthesis **1981** 368/9).
[6] Kim, Y. H.; Shin, J. M. (Tetrahedron Letters **26** [1985] 3821/4).
[7] Yavari, I.; Staral, J. S.; Roberts, J. D. (Org. Magn. Resonance **12** [1979] 340/2).
[8] Smith, W. T., Jr.; Grinninger, L. D. (J. Org. Chem. **26** [1961] 2133/4).
[9] Leandri, G.; Mangini, A. (Ric. Sci. **27** [1957] 1920/5).
[10] Kroner, J.; Strack, W.; Holsboer, F.; Kosbahn, W. (Z. Naturforsch. **28b** [1973] 188/95).

[11] Severson, R. G.; Wai Leung, T.; Wojcicki, A. (Inorg. Chem. **19** [1980] 915/23, 916).
[12] Ulrich, H.; Tucker, B.; Sayigh, A. A. R. (J. Org. Chem. **34** [1969] 3200/1).
[13] Meller, A.; Habben, C. (Monatsh. Chem. **113** [1982] 139/53, 141/2).

2.6.22.1.10 Other N-Sulfinyl-alkanamines and Derivatives

Many N-sulfinyl-alkanamines with a C_1 to C_{18} alkyl group, including linear, cyclic, or branched alkyl groups with one or more sulfinyl moieties, were synthesized following the method of Michaelis and Storbeck [1] by direct treatment of the corresponding amine with $SOCl_2$ in a suitable solvent. The compounds were used as antiknock agents [2] or reducing agents for manganese containing deposits formed on the surfaces of jet engines from burning fuel containing organomanganese compounds [3]. However, no properties of the N-sulfinylamines were presented. Characterized substances obtained by various workers are compiled in Table 38, pp. 104/10.

References:

[1] Michaelis, A.; Storbeck, O. (Liebigs Ann. Chem. **274** [1893] 187/96, 191).
[2] Licke, G. C. (U.S. 3554712 [1968/71]; C.A. **74** [1971] No. 55933).
[3] Licke, G. C. (U.S. 3585012 [1968/71]; C.A. **75** [1971] No. 38718).

Table 38

Syntheses, Properties, and Spectra of Other N-Sulfinyl-alkanamines and Derivatives.
Further information on compounds marked with an asterisk is given at the end of the table.

No.	compound O=S=NR	synthesis	properties, spectra	Ref.
1	$O{=}S{=}N(CH_2)_4CH_3$	$RNH_2 + SOCl_2$ in ether (method of Michaelis [1])	b.p. 90°C/65 Torr	[1]
2	$O{=}S{=}NCH(CH_3)(CH_2)_2CH_3$	like No. 1	^1H NMR (solvent?/TMS): δ (in ppm) = 0.90 (3H, t, J = 7.0 Hz); 1.00 (2H, q, J = 7.0 Hz); 1.22 (3H, d, J = 7.0 Hz); 1.47 (2H, q, J = 7.0 Hz); 4.83 (1H, sext, J = 7.0 Hz) MS (70 eV): 12 m/e peaks	[2]
3	$O{=}S{=}NCH_2CH_2CH(CH_3)_2$	like No. 1	^1H NMR (solvent?/TMS): δ (in ppm) = 0.93 (6H, d, J = 6.0 Hz); 1.57 (3H, m); 3.97 (2H, t, J = 7.0 Hz) MS (70 eV): 14 m/e peaks a satisfactory analysis was not obtained	[2]
4	$O{=}S{=}NC(CH_3)_2CH_2CH_3$	like No. 1	^1H NMR (solvent?/TMS): δ (in ppm) = 0.93 (3H, t, J = 7.0 Hz); 1.47 (3H, s); 1.73 (2H, q, J = 7.0 Hz) MS (70 eV): 13 m/e peaks	[2]
5	$O{=}S{=}NCH_2C(CH_3)_3$	like No. 1	^1H NMR (solvent?/TMS): δ (in ppm) = 1.0 (9H, s); 3.80 (2H, s) MS (70 eV): 9 m/e peaks	[2]
6	$O{=}S{=}N(CH_2)_5CH_3$	like No. 1 [3, 4] quinoline as base [4]	b.p. 71.5°C/14 Torr UV (n-hexane): λ_{max} = 235 nm (log ε = 3.71); 308 nm (log ε = 1.59) [3]	[3, 4]

7	$O{=}S{=}N(CH_2)_6CH_3$	like No. 1	b.p. 87°C/14 Torr UV (n-hexane): $\lambda_{max} = 235$ nm (log $\varepsilon = 3.71$); 305 nm (log $\varepsilon = 1.63$) [3]; (ether): $\lambda_{max} = 235$ nm (log $\varepsilon = 3.67$); (CH_3OH): $\lambda_{max} = 232$ to 234 nm (log $\varepsilon = 3.24$) [5]	[3, 5]
8	$O{=}S{=}N(CH_2)_7CH_3$	like No. 1	b.p. 105°C/14 Torr UV (n-hexane): $\lambda_{max} = 235$ nm (log $\varepsilon = 3.69$); 308 nm (log $\varepsilon = 1.6$)	[3]
9	$O{=}S{=}NC_8H_{17}$-t	t-$C_8H_{17}N{=}S{=}NC_8H_{17}$-t + C_6H_5CHO at 125 to 130°C, 2 h, 30% yield or with furfural at 125°C for 9 h (small yield)	b.p. 61 to 62°C/10 Torr $n_D^{25} = 1.4513$	[6]
10*	$O{=}S{=}N(CH_2)_9CH_3$	like No. 1	b.p. 133°C/14 Torr UV (n-hexane): $\lambda_{max} = 236$ nm (log $\varepsilon = 3.70$); 310 nm (log $\varepsilon = 1.56$) [3]; (c-hexane): $\lambda_{max} = 310$ nm (log $\varepsilon = 1.60$) [3]	[3, 7]
11	$O{=}S{=}N(CH_2)_{11}CH_3$	like No. 1	b.p. 127°C/5 Torr UV (n-hexane): $\lambda_{max} = 235.5$ nm (log $\varepsilon = 3.64$); 310 nm (log $\varepsilon = 1.51$); (c-hexane): $\lambda_{max} = 308$ nm (log $\varepsilon = 1.60$)	[3]
12	$O{=}S{=}NC_{12}H_{25}$-t	like No. 1 pyridine as base	b.p. 129 to 139°C/20 Torr $n_D^{25} = 1.4597$	[8]
13	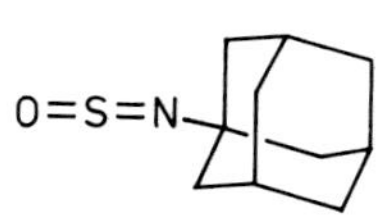	like No. 1 (from the hydrochloride) in benzene, "quantitative" yield	—	[9]

Table 38 (continued)

No.	compound O=S=NR	synthesis	properties, spectra	Ref.
14	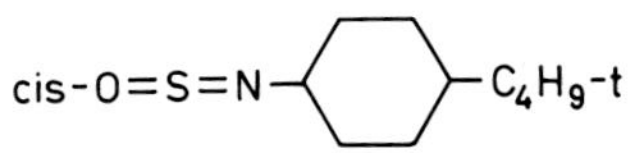	like No. 1 pyridine as base 20% yield	b.p. <75°C/0.2 Torr ^1H NMR (CCl_4): δ (in ppm) = 4.73 (br, s, 1H); 1.0 to 2.0 (br, m, 9H); 0.87 (s, 9H) IR (CCl_4): 13 bands in the range 2907 to 883 cm^{-1}	[10]
15	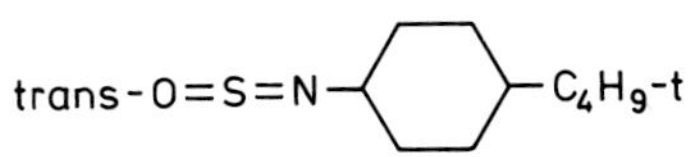	like No. 14 52% yield	colorless liquid; b.p. <70°C/0.1 Torr ^1H NMR (CCl_4): δ (in ppm) = 4.38 (br, m, 1H); 0.9 to 2.1 (br, m, 9H); 0.85 (s, 9H) IR (CCl_4): 8 bands in the range 2907 to 900 cm^{-1}	[10]
16*	$O{=}S{=}N(CH_2)_2N{=}S{=}O$	like No. 1	bright yellow liquid; b.p. 100°C/25 Torr, resinification on storage in sealed vessels	[11]
17*	$O{=}S{=}N(CH_2)_3N{=}S{=}O$	like No. 1	gold-yellow liquid; b.p. 117°C/26 Torr, penetrating odor	[11]
18*	$O{=}S{=}N(CH_2)_4N{=}S{=}O$	like No. 1 in benzene, 40% yield	b.p. 60°C/0.002 Torr	[12]
19	$O{=}S{=}N(CH_2)_2Cl$	NH + $SOCl_2$ $(N(C_2H_5)_3$ as base in CCl_4, 48% yield [13] $HOCH_2CH_2NH_2 + SOCl_2$ pyridine as base in $CHCl_3$, 22% yield [14], 40% yield [15] 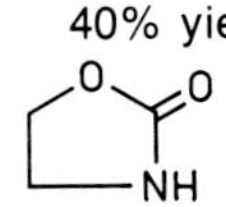 + $SOCl_2$ in toluene at 95°C [16]	colorless liquid; b.p. 36 to 37°C/6 Torr [15]; 50 to 53°C/6 Torr [14]; 62 to 64°C/33 Torr [13] ^1H NMR ($CHCl_3$): δ (in ppm) = 3.70 (t); 4.85 (t) ≙ A_2M_2 pattern [13] IR: 1260 s, 1123 s, 581 m cm^{-1} [13] UV (CH_3OH): λ_{max} = 230 nm (log ε = 3.27); (c-C_6H_{12}): λ_{max} = 233 nm (log ε = 3.55) [5] the compound fumes on exposure to humid atmosphere and decomposes rapidly upon contact with H_2O [13]	[5, 13 to 16]

20	$O{=}S{=}N(CH_2)_3Cl$	$HO(CH_2)_3NH_2 + SOCl_2$ pyridine as base in $CHCl_3$, 48% [14], 60% [15] yield	b.p. 58 to 60°C/8 Torr [15]; 61 to 63°C/4 Torr [14] UV (CH_3OH): $\lambda_{max} = 231$ nm ($\log \varepsilon = 3.49$); ($c\text{-}C_6H_{12}$): $\lambda_{max} = 231$ nm ($\log \varepsilon = 3.82$) [5]	[5, 14, 15]
21	$O{=}S{=}NCH_2C(H)ClCH_3$	$H_3CCH(OH)CH_2NH_2 + SOCl_2$ pyridine as base 36% yield [14]  NH + $SOCl_2$, ($N(C_2H_5)_3$) CH_3 the reaction yields a mixture of No. 21 and $O{=}S{=}NCH(CH_3)CH_2Cl$ [13]	b.p. 44 to 46°C/2 Torr	[13, 14]
22	$O{=}S{=}NC(CH_3)_2CH_2Cl$	$HOCH_2C(CH_3)_2NH_2 + SOCl_2$ pyridine as base 25% yield [14]	b.p. 42 to 43°C/2 Torr	[14]
23	$O{=}S{=}NCH_2C(CH_3)_2Cl$	CH_3 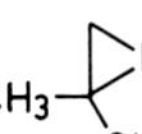 NH + $SOCl_2$, ($N(C_2H_5)_3$) CH_3 in $CHCl_3$	b.p. 69 to 70.5°C/42 Torr ^{1}H NMR: δ (in ppm) = 4.12 (s), 1.62 (s)	[13]
24	$O{=}S{=}NCCl(CN)CCl_3$	$Cl_3CC(CN){=}NSi(CH_3)_3 + SOCl_2$ 4 h reflux, 62% yield	b.p. 94 to 95°C/0.05 Torr $n_D^{20} = 1.5375$ IR (film): $\nu = 2240, 1250, 1180$ cm^{-1}	[17, 18]
25*	$O{=}S{=}NC(CN)(CH_3)_2$	like No. 1 82% yield (from the hydrochloride) $(H_3C)_2C(CN)N{=}SCl_2 + SOCl_2$ 65% yield $(H_3C)_2C(CN)N{=}SCl_2 + CH_3COOH$ 70 to 80% yield [19]	b.p. 60°C/12 Torr $n_D^{20} = 1.3742$ $d_4^{20} = 1.4538$ g/cm^3 [19] $\mu = 2.90$ D (temperature-dependent, t = 20°C) [20]	[19, 20]
26*	$O{=}S{=}NC(CN)(CH_3)C_2H_5$	probably like No. 25	$\mu = 2.91$ D (temperature-dependent, t = 20°C)	[20]

Table 38 (continued)

No.	compound O=S=NR	synthesis	properties, spectra	Ref.
27	$O{=}S{=}NCH(CN)CH(CH_3)_2$	like No. 1 81% yield (from the hydrochloride) $(CH_3)_2CHCH(CN)NCl_2 + SOCl_2$ 69% yield $(CH_3)_2CHCH(CN)N{=}SCl_2 +$ CH_3COOH in benzene 70 to 80% yield	b.p. 73 to 73.5°C/12 Torr $n_D^{20} = 1.3489$ $d_4^{22} = 1.4598$ g/cm³	[19]
28*	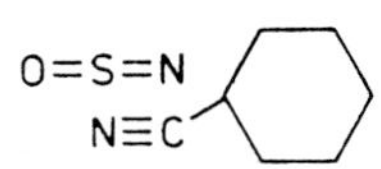	like No. 1 64% yield 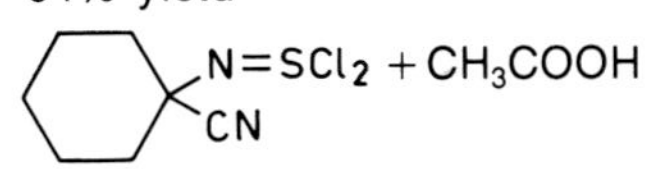 70 to 80% yield [19]	b.p. 110 to 111°C/12 Torr $n_D^{20} = 1.2974$ $d_4^{20} = 1.4982$ g/cm³ [19] $\mu = 3.46$ D (cis) [20]	[19, 20]
29*	$O{=}S{=}NC(CH_3)_2C(O)OCH_3$	$H_3CO(O)CC(CH_3)_2N{=}SCl_2 +$ $SOCl_2$ 90% yield	colorless, mobile liquid; b.p. 66 to 67°C/13 Torr $n_D^{24} = 1.4452$ $d_4^{24} = 1.1075$ g/cm³	[21]
30	$O{=}S{=}NCH(CH_3)(CH_2)_2CH{=}CH_2$	like No. 1	colorless liquid; b.p. 156 to 158°C $d_4^{15} = 0.9986$ g/cm³	[22]
31	$O{=}S{=}NC(Cl){=}C(CN)_2$	like No. 1 [23] $KC(CN)_3 + SOCl_2$ [23] $C(CN)_4 + SOCl_2$ 40 h at 20 to 25°C [24]	red liquid; b.p. 96°C/2 Torr the compound is readily hydrolyzed yielding $(CN)_2C{=}CClNH_2$	[23, 24]
32*	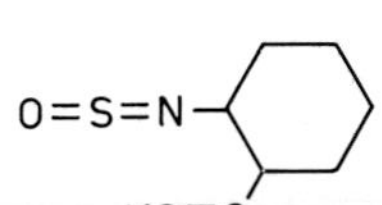	like No. 1 65% yield	b.p. 203°C	[12]

33	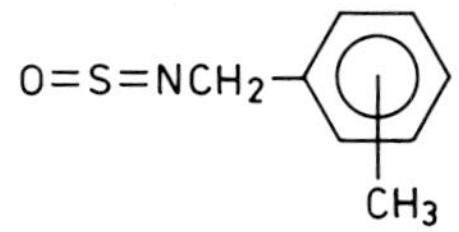	like No. 1	bright brownish liquids; heating at 130 to 140°C yields the corresponding aldehydes in low yields	[25]
34	$O{=}S{=}N(CH_2)_2C_6H_5$	like No. 1	yellow liquid; b.p. 170 to 173°C/25 Torr	[25]
35	$O{=}S{=}NCH(CH_3)C_6H_5$	like No. 1 a distillation is not possible	brownish yellow liquid; heating in a sealed glass tube at 200°C yields acetophenone	[25]
36	$O{=}S{=}N(CH_2)_3C_6H_5$	like No. 1	brown-yellow liquid distillation leads to decomposition	[22]
37	$O{=}S{=}NCH(C_6H_5)_2$	like No. 1	bright yellow liquid; b.p. 88°C/35 Torr heating at 150 to 200°C yields benzophenone	[25]
38*	$O{=}S{=}NCCl(CF_3)C_6H_5$	$C_6H_5(F_3C)C{=}NH + SOCl_2$ $N(C_2H_5)_3$ as base in benzene, 20°C, 2d, 72% yield	colorless liquid; b.p. 47 to 48°C/0.08 Torr. $n_D^{20} = 1.5011$ IR spectrum: $\nu_s(NSO) = 1190$, $\nu_{as}(NSO) = 1260\ cm^{-1}$ readily dissolving in CH_2Cl_2, $C_2H_4Cl_2$, $CH_3C(O)CH_3$, C_6H_6, and $CHCl_3$ the substance is easily hydrolyzed by water	[26]
39	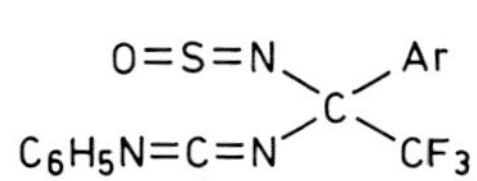 Ar = C_6H_5 (and p-$CH_3C_6H_4$, p-ClC_6H_4)	Ar = C_6H_5: No. 38 + $(H_3C)_3Si{-}N{=}C{=}N{-}C_6H_5$ mole ratio 1:1, 24 h at 20 to 25°C, 52% yield	Ar = C_6H_5: b.p. 140°C/0.1 Torr $n_D^{20} = 1.5620$ IR (CCl_4): $\nu(NCN) \approx 2160$, $\nu_{as}(NSO) \approx 1260$, $\nu_s(NSO) \approx 1190\ cm^{-1}$	[27]

Table 38 (continued)

No.	compound O=S=NR	synthesis	properties, spectra	Ref.
40	$O{=}S{=}NC(C_6H_5)_2N{=}C(C_6H_5)_2$	$(C_6H_5)_2C{=}NLi + SOCl_2$ in hexane–ether (ratio 2:1); 87% yield	yellow crystals; m.p. 181 to 183°C IR (Nujol): 32 bands in the range 3050 to 427 cm^{-1}, $\nu_s(NSO) = 1130$, $\nu_{as}(NSO) = 1273$ cm^{-1} MS: m/e = 408 M^+ and 13 peaks with assignments	[28]
41	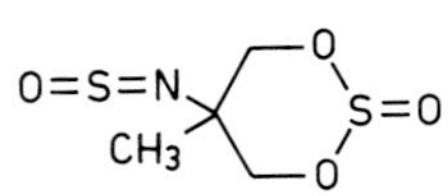	$CH_3C(NH_2)(CH_2OH)_2 + SOCl_2$ 25% yield	b.p. 65 to 70°C/2 Torr	[14]
42	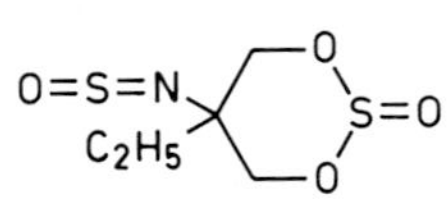	like No. 41 39% yield	b.p. 82 to 85°C/2 Torr	[14]

* Further information:

$O{=}S{=}N(CH_2)_9CH_3$ (Table **38**, No. **10**). The compound reacts with $C_4H_9(C_2H_5)CHCH_2OP(O)$-$HO^-NH_4^+$ to yield $CH_3(CH_2)_9N(H)S(O)P(OH)(O)OCH_2CH(C_2H_5)C_4H_9$ [7].

$O{=}S{=}N(CH_2)_2N{=}S{=}O$ (Table **38**, No. **16**). The compound reacts with $(CH_3)_2B{-}SCH_3$ to yield the corresponding diamine [12]:

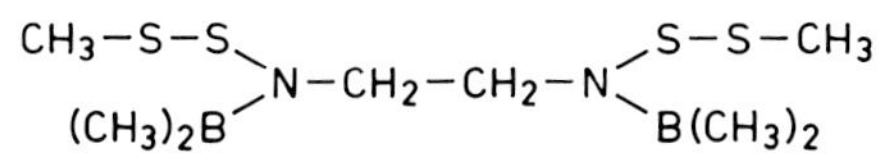

$O{=}S{=}N(CH_2)_nN{=}S{=}O$, n = 2, 3, 4 (Table **38**, Nos. **16** to **18**). The compounds react with CH_3B-$(SCH_3)_2$ to yield five-, six-, or seven-membered ring compounds [12]:

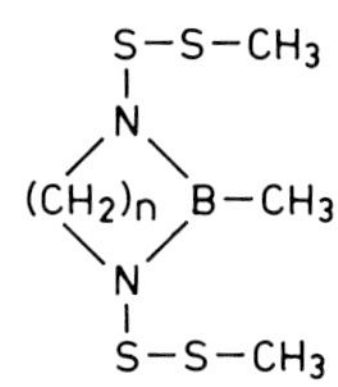

$O{=}S{=}NC(CN)(CH_3)_2$ (Table **38**, No. **25**). The density of the liquid compound between 20 and 120°C varied from D = 1.030 to 0.898 g/cm³, the dipole moment from 2.90 to 3.24 D, the dielectric constant from 16.14 to 12.51. In the temperature range from 20 to 100°C the electrical conductivity varied from 1.5 to $7.5 \times 10^{-6}\ \Omega^{-1} \cdot cm^{-1}$. In the solid phase at t = −190°C is D = 1.275 g/cm³, atom polarization $P_A = 14.3$ cm³, deformation polarization $P_{def} = 53.1$ cm³, molar refraction $R_D = 38.8$ cm³, and deformation limit of dielectric constant $\varepsilon_\infty = 3.01$ [20].

The compound reacts with PCl_5 in CCl_4 at 70 to 75°C for 2 h to give a mixture of $1/2((CH_3)_2(CN)CN{=}PCl_3)_2$ (isolated on cooling by filtration, 19% yield) and $(CH_3)_2(CN)CN{=}SCl_2$ (isolated by distillation of the filtrate, 75% yield) along with $SOCl_2$ and $POCl_3$. The reaction with PBr_5 in CCl_4 at 60 to 70°C for 6 to 8 h and at 80 to 90°C for 30 h afforded $(CH_3)_2(CN)CN{=}PBr_3$ and an unidentified yellow liquid after working up by distillation [21].

$O{=}S{=}NC(CN)(CH_3)C_2H_5$ (Table **38**, No. **26**). The density of the liquid compound between 20 and 100°C varied from D = 0.998 to 0.932 g/cm³, the dipole moment from 2.91 to 3.05 D, the electrical conductivity from 1.9 to $8.2 \times 10^{-6}\ \Omega^{-1} \cdot cm^{-1}$. In the solid phase at t = −190°C is D = 1.18 g/cm³, $\varepsilon_\infty = 2.94$, $P_A = 16.7$ cm³, $P_{def} = 61.7$ cm³, $R_D = 45.0$ cm³ [20].

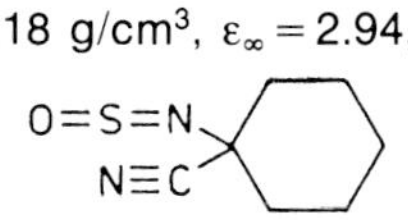

(Table **38**, No. **28**). The density of the liquid compound between 20 and 100°C varied from D = 1.006 to 0.984 g/cm³, the electrical conductivity from 0.45 to $2.8 \times 10^{-6}\ \Omega^{-1} \cdot cm^{-1}$. The dipole moment $\mu \approx 3.46$ D (temperature-independent). In the solid phase at t = −190°C is $\varepsilon_\infty = 2.87$, D = 1.20 g/cm³, $P_A = 12.2$ cm³, $P_{def} = 69.3$ cm³, $R_D = 57.1$ cm³ [20].

$O{=}S{=}NC(CH_3)_2C(O)OCH_3$ (Table **38**, No. **29**). The compound reacts with PCl_5 at 50°C in CCl_4 to produce $CH_3OC(O)C(CH_3)_2N{=}SCl_2$ [21].

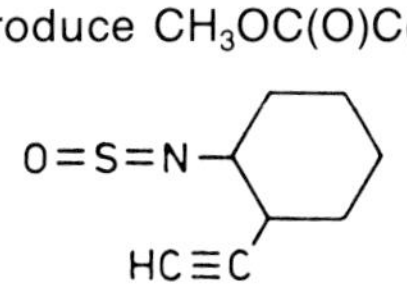

(Table **38**, No. **32**). The compound reacts with $(CH_3)_2B{-}SCH_3$ to give [12]:

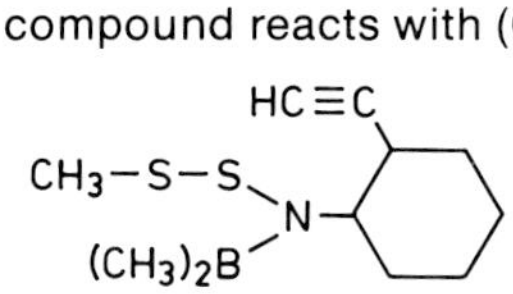

O=S=NCCl(CF$_3$)C$_6$H$_5$ (Table **38**, No. **38**). The compound reacts with N-trimethylsilylmorpholine to form the derivative I [26]:

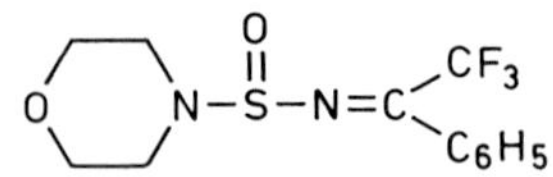

I

References:

[1] Michaelis, A.: Storbeck, O. (Liebigs Ann. Chem. **274** [1893] 187/96, 191).
[2] Grunwell, J. R.; Kochan, A. (J. Org. Chem. **38** [1973] 1610/1).
[3] Leandri, G.; Mangini, A. (Spectrochim. Acta **1959** 421/31; Ric. Sci. **27** [1957] 1920/5).
[4] Schell, F. M.; Carter, J. P.; Wiaux-Zamar, C. (J. Am. Chem. Soc. **100** [1978] 2894/6).
[5] Smith, W. T., Jr.; Grinninger, L. D. (J. Org. Chem. **26** [1961] 2133/4).
[6] Clemens, D. H.; Bell, A. J.; O'Brien, J. L. (Tetrahedron Letters **20** [1965] 1491/5).
[7] Laskorin, B. N.; Yakshin, V. V.; Bulgakova, V. B. (Zh. Obshch. Khim. **46** [1976] 2477/81; J. Gen. Chem. [USSR] **46** [1976] 2372/6).
[8] Birum, G. H.; Heininger, S. A. (U.S. 2893910 [1957/59]; C.A. **1960** 1434).
[9] Sasaki, T. (Japan. Kokai Tokkyo Koho 79-44654 [1977/79] from C.A. **91** [1979] No. 174913).
[10] Beak, P.; Adams, J. T.; Barron, J. A. (J. Am. Chem. Soc. **96** [1974] 2494/510, 2506).

[11] Michaelis, A.; Graentz, P. (Ber. Deut. Chem. Ges. **30** [1897] 1009/14).
[12] Meller, A.; Maringgele, W.; Fetzer, H. (Chem. Ber. **113** [1980] 1950/61).
[13] Tomalia, D. A. (Tetrahedron Letters **1967** 2559/62; U.S. 3449396 [1966/69]; C.A. **71** [1969] No. 80684).
[14] Smith, W. T., Jr.; Chen, W.-J.; Linfield, M. (J. Med. Chem. **8** [1965] 718/9).
[15] Smith, W. T., Jr.; Grasley, M. (Chim. Ther. **3** [1968] 266/9).
[16] Poindexter, G. S. (U.S. 4505859 [1981/85]; C.A. **103** [1985] No. 6044).
[17] Lazukina, L. A.; Kukhar, V. P. (Synthesis **1979** 747/8).
[18] Lazukina, L. A.; Kukhar, V. P. (Zh. Obshch. Khim. **53** [1983] 2239/43; J. Gen. Chem. [USSR] **53** [1983] 2018/21).
[19] Markovskii, L. N.; Fedyuk, G. S.; Levchenko, E. S. (Zh. Org. Khim. **6** [1970] 2426/31; J. Org. Chem. [USSR] **6** [1970] 2438/42).
[20] Egorov, Yu. P.; Borovikov, Yu. Ya.; Fedyuk, G. S.; Levchenko, E. S. (Zh. Obshch. Khim. **43** [1973] 2470/6; J. Gen. Chem. [USSR] **43** [1973] 2456/61).

[21] Markovskii, L. N.; Fedyuk, G. S.; Levchenko, E. S. (Zh. Org. Khim. **8** [1972] 2057/62; J. Org. Chem. [USSR] **8** [1972] 2104/8).
[22] Michaelis, A.; Jacobi, W. (Ber. Deut. Chem. Ges. **26** [1893] 2158/62).
[23] Zefirov, N. S.; Chapovskaya, N. K.; D'yachkova, L. Ya.; Trach, S. S. (Zh. Org. Khim. **11** [1975] 1981; J. Org. Chem. [USSR] **11** [1975] 2003/4).
[24] Makhon'kov, D. I.; Zefirov, N. S. (Zh. Org. Khim. **15** [1979] 2441/5; J. Org. Chem. [USSR] **15** [1979] 2207/10).
[25] Michaelis, A.; Linow, E. (Ber. Deut. Chem. Ges. **26** [1893] 2162/71).
[26] Markovskii, L. N.; Shermolovich, Yu. G.; Gorbatenko, V. I.; Shevchenko, V. I. (Zh. Org. Khim. **11** [1975] 751/6; J. Org. Chem. [USSR] **11** [1975] 745/9).
[27] Shermolovich, Yu. G.; Gorbatenko, V. I. (Zh. Org. Khim. **12** [1976] 1129/30; J. Org. Chem. [USSR] **12** [1976] 1139).
[28] Banister, A. J.; Durrant, J. A.; Padley, J. S.; Wade, K. (J. Inorg. Nucl. Chem. **41** [1979] 1415/9).

2.6.22.2 N-Sulfinyl-arenamines, O=S=NR, R = aryl

In this section, representative examples of N-sulfinyl-arenamines, which are intensively studied, are described.

2.6.22.2.1 N-Sulfinyl-benzenamine, N-Sulfinyl-aniline, $O{=}S{=}NC_6H_5$

2.6.22.2.1.1 Formation and Preparation

The standard method of preparation $O{=}S{=}NC_6H_5$ is the sulfinylation of $C_6H_5NH_2$ with $SOCl_2$ in dry benzene [1 to 7] or toluene [8]. The reaction is complete after refluxing the mixture for 2 to 6 h until a clear solution is obtained. Purification is by distillation. The yields are in the range 83 [6] to 95% [4]. The reaction was described in the year 1878, however, the title compound was not identified [9]. Using the hydrochloride of aniline as the starting material is somewhat more advantageous, because the reaction sets in after heating [2, 10, 11, 12]. The yields for this procedure are in the range 74 [12] to 96.2% [11]. The related reaction of $C_6H_5N(Si(CH_3)_3)_2$ with $SOCl_2$ at −78°C gave the title compound in 90% yield [13].

The compound was obtained by treatment of $C_6H_5NH_2$ with N,N′-sulfinyl-bisimidazole and N-(chlorosulfinyl)imidazole in CH_2Cl_2 at 20°C for 0.5 h in 73 and 89% yields, respectively. The sulfinylation agents were prepared in situ from imidazole and $SOCl_2$ [14]. It can also be prepared in 90% yield from the reaction of $C_6H_5NH_2$ with $SO(OC_4H_9)_2$ [15]. The transsulfinylation reaction of $C_6H_5NH_2$ with $O{=}S{=}NC_6H_4NO_2$-4 (or $O{=}S{=}NC_6H_4NO_2$-2) in anhydrous benzene produced $O{=}S{=}NC_6H_5$ in an 68% (80%) yield [16].

Several preparation methods of $O{=}S{=}NC_6H_5$ are based on reactions with SO_2 [17 to 22]. Dissolving $(C_6H_5)_3P{=}NC_6H_5$ in liquid SO_2 led to $O{=}S{=}NC_6H_5$ which was separated from the by-product $(C_6H_5)_3P{=}O$ by treating the residue with petroleum ether or ether after evaporation of excess SO_2 [17]. $O{=}S{=}NC_6H_5$ was formed analogously when dry SO_2 gas was bubbled through a solution of $(C_6H_5)_3As{=}NC_6H_5$ [18]. The $Fe(CO)_5$-catalyzed reaction between $C_6H_5N{=}C{=}O$ and SO_2 in an autoclave at 130°C (overnight) gave $O{=}S{=}NC_6H_5$ in a yield >90% [19]. Aniline was treated with n-C_4H_9Li and SO_2 in THF at room temperature giving $C_6H_5NHS(O)OLi$, which was allowed to react again with n-C_4H_9Li and subsequently with SO_2 to yield the title compound (46% yield) [20]. A similar, however experimentally less complex and more productive procedure, started from the monosilylated aniline. A solution of the anion, obtained by lithiating with butyllithium in hexane–THF at 0°C was added to an excess of SO_2 in THF at −78°C. The resulting mixture was stirred for 1 h at room temperature; yield 74%. A modified procedure was described in the same paper by treating a solution of $C_6H_5NHSi(CH_3)_3$ in ether containing one equivalent of pyridine with a large excess of gaseous SO_2; 69% yield [21]. The reactions of SO_2 with n-$C_4H_9SnOC_2H_5$ and n-$C_4H_9SnOSnC_4H_9$ yielded corresponding sulfites $(n\text{-}C_4H_9)_3SnOS(O)OC_2H_5$ which reacted with the isocyanate $C_6H_5N{=}C{=}O$ to form the carbamate $(n\text{-}C_4H_9)_3SnN(C_6H_5)C(O)OC_2H_5$. Subsequent reaction with SO_2 led to the formation of $O{=}S{=}NC_6H_5$. The reactions of $(n\text{-}C_4H_9)_3SnN(C_6H_5)SO(C_6H_5)NSn(C_4H_9\text{-}n)_3$ (which was obtained by treatment of $C_6H_5N{=}S{=}NC_6H_5$ with $(n\text{-}C_4H_9)_3SnOSn(C_4H_9\text{-}n)_3$) with protic agents such as alcohols or acetic acid, isocyanates, SO_2, and CCl_3CHO yielded $O{=}S{=}NC_6H_5$, together with by-products. Heating the same educts at ~95 to 100°C also gave the title compound. It should be mentioned that $O{=}S{=}NC_6H_5$ was only detected by NMR spectroscopic investigation and was not isolated in most cases [22].

$C_6H_5SO_2N{=}S{=}NC_6H_5$ and $(CH_3)_2SO$ react together when heated for 1 h at 120 to 125°C to form $O{=}S{=}NC_6H_5$ and $C_6H_5SO_2N{=}S(CH_3)_2$ [23]. Heating equimolar amounts of $C_6H_5N{=}C{=}S$ and $O{=}S{=}NAr$ (Ar = 2-$CH_3C_6H_4$, 4-$CH_3C_6H_4$, 4-$CH_3OC_6H_4$, 2-ClC_6H_4, 4-$CH_3C_6H_4SO_2$) together in an atmosphere of N_2 at temperatures in the range 170 to ~200°C for 8 to 9 h yielded a mixture of

$O{=}S{=}NC_6H_5$ (yields 8.2 to 22.6%), $ArN{=}C{=}S$, various carbodiimides, and sulfur. The compounds were identified by comparison to chromatography (glpc) and IR spectrum of authentic samples [24].

The title compound was also prepared by exchange reactions of $C_6H_5N{=}C{=}NC_6H_5$ with $O{=}S{=}NSO_2C_6H_4CH_3$-4 (yield 48%) and similarly with $O{=}S{=}NSO_2CH_3$ (yield 62%) in benzene at ambient temperature (cycloaddition products could not be isolated). The reaction of $C_6H_5N{=}C{=}NC_6H_5$ with $O{=}S{=}NC(O)C_6H_5$ led to an oily mixture of 1,2- and 1,4-cycloadducts which decomposed on distilling in vacuum to give $O{=}S{=}NC_6H_5$ (47% yield) (and $C_6H_5N{=}C{=}O$, C_6H_5CN, $C_6H_5C(O)NHC_6H_5$) [25].

Thiirane 1-oxide $H_2\overline{C{-}S(O){-}C}H_2$ reacts with $C_6H_5N_3$ in an inert solvent at a temperature of 80 to 140°C to give $O{=}S{=}NC_6H_5$ (10 to 30% yield) [26]. Analogously, the title compound was obtained by reacting equimolar amounts of $H(C_6H_5)\overline{C{-}S(O){-}C}(C_6H_5)H$ and $C_6H_5N_3$ in boiling CH_2Cl_2 for ca. 3 h (yield 22%) [27].

$O{=}S{=}NC_6H_5$ is a yellow liquid with a penetrating, aromatic odor [2].

References:

[1] Michaelis, A.; Herz, R. (Ber. Deut. Chem. Ges. **23** [1890] 3480/2).
[2] Michaelis, A. (Ber. Deut. Chem. Ges. **24** [1891] 745/57).
[3] Kresze, G.; Smalla, H. (Chem. Ber. **92** [1959] 1042/8).
[4] Kresze, G.; Maschke, A.; Albrecht, R.; Bederke, K.; Patzschke, H. P.; Smalla, H.; Trede, A. (Angew. Chem. **74** [1962] 135/44, 135; Angew. Chem. Intern. Ed. Engl. **1** [1962] 89/98).
[5] Rajagopalan, P.; Advani, B. G.; Talaty, C. N. (Org. Syn. Collective Vol. **5** [1973] 504/8).
[6] Kataev, E. G.; Plemenkov, V. V.; Pluzhnov, V. K. (Zh. Org. Khim. **3** [1967] 336/41; J. Org. Chem. [USSR] **3** [1967] 319/23).
[7] Birum, G. H.; Heininger, S. A. (U.S. 2893910 [1957/59]; C.A. **1960** 1434).
[8] Pesin, V. G.; Khaletsky, A. M. (Zh. Obshch. Khim. **27** [1957] 2599/604; J. Gen. Chem. [USSR] **27** [1957] 2656/60).
[9] Böttinger, C. (Ber. Deut. Chem. Ges. **11** [1878] 1407/8).
[10] Wichterle, O.; Roček, J. (Collection Czech. Chem. Commun. **19** [1954] 282/96; Chem. Listy **47** [1953] 1768/80).

[11] Klamann, D.; Sass, C.; Zelenka, M. (Chem. Ber. **92** [1959] 1910/3).
[12] Al-Ajely, M. S.; Al-Rawi, J. M. A.; Elvidge, J. A. (J. Chem. Soc. Perkin Trans. **1** [1982] 1575/8).
[13] Abel, E. W.; Armitage, D. A. (J. Chem. Soc. **1964** 3122/6).
[14] Kim, Y. H.; Shin, J. M. (Tetrahedron Letters **26** [1985] 3821/4).
[15] Carré, P.; Libermann, D. (Bull. Soc. Chim. France **6** [1939] 579/80).
[16] Carpanelli, C.; Leandri, G. (Ann. Chim. [Rome] **50** [1960] 147/55).
[17] Staudinger, H.; Hauser, E. (Helv. Chim. Acta **4** [1921] 861/86, 867, 874).
[18] Froeyen, P. (Acta Chem. Scand. **27** [1973] 141/5).
[19] Drapier, J.; Hubert, A. J.; Teyssie, P. (Synthesis **1975** 649).
[20] Sukai, S.; Fujinami, T.; Komizo, K. (J. Org. Chem. **40** [1975] 3291/2).

[21] Porskamp, P. A. T. W.; Zwanenburg, B. (Synthesis **1981** 368/9).
[22] Davies, A. G.; Kennedy, J. D. (J. Chem. Soc. C **1968** 2630/40).
[23] Levchenko, E. S.; Ugarov, B. N. (Zh. Org. Khim. **4** [1968] 1413/20; J. Org. Chem. [USSR] **4** [1968] 1361/7).
[24] Minami, T.; Agawa, T. (Tetrahedron Letters **1968** 2651/3).
[25] Minami, T.; Fukuda, M.; Abe, M.; Agawa, T. (Bull. Chem. Soc. Japan **46** [1973] 2156/9).

[26] Bonini, B. F.; Maccagnani, G.; Mazzanti, G.; Franceschini, M. (Chim. Ind. [Milan] **58** [1976] 805).
[27] Bonini, B. F.; Maccagnani, G.; Mazzanti, G. (Tetrahedron Letters **1977** 1185/6).

2.6.22.2.1.2 Molecular Properties. Spectra

Geometrical Structure

The exact geometrical structure of $O{=}S{=}NC_6H_5$ is still unknown. A complete structure analysis has not yet been performed. However, dipole moment [1 to 7], microwave [5, 6], and NMR studies [4, 8] were employed to analyze the geometrical conformation of the molecule. The most widely accepted opinion is that $O{=}S{=}NC_6H_5$ preferentially adopts a planar cis structure with respect to the S=N bond [9]. The planarity of the molecule was shown by a microwave spectral study by variation of the inertial defect [6]. The first statements about the conformation of $O{=}S{=}NC_6H_5$ were based upon dipole moment studies [1, 2, 3]. Some workers assumed an equilibrium mixture of *cis*- and *trans*-$O{=}S{=}NC_6H_5$ in solution (cis: 56%) [1]. This result was deduced from the experimentally measured μ value which was compared to theoretical dipole moments of *cis*- and *trans*-$O{=}S{=}NC_6H_5$. The workers were also influenced in their opinion through earlier kinetic studies of the alcoholysis of $O{=}S{=}NC_6H_5$ which were interpreted as indicating the presence of two rotamers [10, 11]. The possible existence of two rotational isomers was also concluded from an IR study in a mixed solvent of C_2Cl_4–CCl_4 (1:2) at temperatures down to −40°C in the range 1325 to 1125 cm^{-1}, which showed band doubling of the O=S=N group [12]. However, reinvestigation of the reaction kinetic [13], temperature-dependent ^1H and ^{13}C NMR studies (range 40 to −60°C [8]), and IR spectroscopic investigations of the compound as solid and dissolved in different solvents [4] did not confirm the presence of rotamers. Other groups concluded from their dipole moment studies [2, 3, 7] and, in addition, from the value of the Kerr constant [2], that the title compound may exist only in the trans form. In contrast, several more recent papers assume a planar syn conformation for $O{=}S{=}NC_6H_5$ [4, 5, 6, 8]. This was first shown by reinvestigation of the dipole moment of $O{=}S{=}NC_6H_5$ [4, 5] and of several mono- and polysubstituted polysulfinylanilines [4]. The same result was obtained through comparisons of the ^1H NMR spectrum with that of E-α-sulfinylbenzylchloride, E-$O{=}S{=}C(Cl)C_6H_5$, [4] and both ^1H and ^{13}C NMR spectra with those of cis aryl groups in sulfur diimides [8], which showed good agreement in their chemical shifts. Further evidence for a cis conformation came from microwave spectral studies [5, 6]. The observed LRMW spectrum (low-resolution microwave) of $O{=}S{=}NC_6H_5$ (gas) was compared with calculated spectra of four molecular models of $O{=}S{=}NC_6H_5$. The best agreement was achieved in the value of $B+C$ ($=1575\pm5$ MHz) for the syn model [5]. Other workers agree that experimental rotational constants could be reproduced only by a cis structure (assuming a regular phenyl ring as in phenol, CN as in $O{=}S{=}NCH_3$, r(SN) and r(SO) as in O=S=NH, ∢CNS = 130° and ∢NSO = 123°). The barrier to internal rotation was determined to be 2.3 kcal/mol [6]. The above results were confirmed by an X-ray structure analysis of a transition metal complex of $O{=}S{=}NC_6H_4CH_3$-4 which showed a cis conformation of the ligand [8], see p. 279.

References:

[1] Jannelli, L.; Lamanna, U.; Lumbroso, H. (Bull. Soc. Chim. France **1966** 3626/30).
[2] Bonham, J. S.; Cheng, C. L.; Le Fèvre, R. J. W.; Ritchie, G. L. D. (Australian J. Chem. **26** [1973] 421/4).
[3] Egorov, Yu. P.; Borovikov, Yu. Ya.; Fedyuk, G. S.; Levchenko, E. S. (Zh. Obshch. Khim. **43** [1973] 2470/6; J. Gen. Chem. [USSR] **43** [1973] 2456/61).

[4] van Woerden, H. F.; Bijl-Vlieger, S. H. (Recl. Trav. Chim. **93** [1974] 85/90).
[5] Bak, B.; Svanholt, H.; Larsen, C. (J. Mol. Struct. **36** [1977] 55/9).
[6] Caminati, W.; Mirri, A. M.; Maccagnani, G. (J. Mol. Spectrosc. **66** [1977] 368/74).
[7] Roy, S. K. (Advan. Mol. Relaxation Interact. Processes **12** [1978] 229/37).
[8] Meij, R.; Oskam, A.; Stufkens, D. J. (J. Mol. Struct. **51** [1979] 37/49, 39).
[9] Bussas, R.; Kresze, G.; Münsterer, H.; Schwöbel, A. (Sulfur Rept. **2** [1983] 215/377, 246).
[10] Jannelli, L.; Senatore, L.; Carpanelli, C. (Ann. Chim. [Rome] **53** [1963] 1150/60).

[11] Jannelli, L.; Senatore, L. (Ann. Chim. [Rome] **53** [1963] 1356/69; **55** [1965] 980/90).
[12] Glass, W. K.; Pullin, A. D. E. (Trans. Faraday Soc. **57** [1961] 546/56).
[13] Collins, N. C.; Glass, W. K. (J. Chem. Soc. B **1971** 2156/9).

Nuclear Magnetic Resonance

The ^{1}H and ^{13}C NMR spectra recorded by several groups are compiled in the following tables.

^{1}H NMR spectra of $O{=}S{=}NC_6H_5$ (δ in ppm, relative to internal TMS).

solvent	δ(H2)	δ(H3)	δ(H4)	δ(H5)	δ(H6)	Ref.
CCl_4	7.76	7.22	7.22	7.22	7.69	[1]
$CDCl_3$	7.78		7.2 to 7.4		7.78	[2]
$CDCl_3$	7.70	7.2	7.2	7.2	7.70	[3]
C_7D_8	7.63	6.97	6.97	6.97	7.63	[3]

^{13}C NMR spectra of $O{=}S{=}NC_6H_5$ (δ in ppm, relative to TMS).

solvent	δ(C1)	δ(C2,6)	δ(C3,5)	δ(C4)	Ref.
$CDCl_3$	142.65	127.06	129.14	130.37	[4]
$CDCl_3$	142.7	127.1	129.1	130.4	[5]
$CDCl_3$	142.66	127.07	129.14	130.36	[3]

^{14}N NMR (neat/satd. NH_4NO_3 in H_2O): $\delta = 294$ ppm [6] (referenced to CH_3NO_2: $\delta = -65.5$ ppm); (neat/liquid NH_3 via external NH_4Cl): $\delta = 316$ ppm [7] (referenced to CH_3NO_2: $\delta = -64.2$ ppm); (neat/external CH_3NO_2): $\delta = -62.1$ ppm; (1.7 M in ether/external CH_3NO_2): $\delta = -63.5$ ppm; (2.0 M in acetone/external CH_3NO_2): $\delta = -63.2$ ppm [8].

^{15}N NMR (neat/external $H^{15}NO_3$): $\delta = -57.6$ ppm (referenced to CH_3NO_2: $\delta = -62$ ppm) [9]; (CCl_4/external NH_3): $\delta = 319.4$ ppm (referenced to CH_3NO_2: $\delta = -60.8$ ppm) [10, 11]; (CCl_4/external CH_3NO_2): $\delta = -60.8$ ppm [12].

^{17}O NMR (CCl_4/natural content of the isotope/external D_2O, 80°C): $\delta = 410 \pm 3$ ppm [10 to 13].

^{33}S NMR (neat/2 M Cs_2SO_4 via external sulfolane): $\delta = 261$ ppm [7].

References:

[1] Mirek, J.; Czerwiec, Z. (Zesz. Nauk Univ. Jagiel. Proc. Chem. No. 18 [1973] 183/6).
[2] van Woerden, H. F.; Bijl-Vlieger, S. H. (Recl. Trav. Chim. **93** [1974] 85/90).

[3] Meij, R.; Oskam, A.; Stufkens, D. J. (J. Mol. Struct. **51** [1979] 37/49).
[4] Brands, G.; Golloch, A. (Z. Naturforsch. **38b** [1983] 326/31).
[5] Kresze, G.; Berger, M.; Claus, P. K.; Rieder, W. (Org. Magn. Resonance **8** [1976] 170/2).
[6] Armitage, D. A.; Mason, J.; Vinter, J. G. (Inorg. Chem. **17** [1978] 776/7).
[7] Belton, P. S.; Woollins, J. D. (Magn. Resonance Chem. **24** [1986] 1080/2).
[8] Sicińska, W.; Stefaniak, L.; Witanowski, M.; Webb, G. A. (Org. Magn. Resonance **15** [1981] 241/3).
[9] Yavari, I.; Staral, J. S.; Roberts, J. D. (Org. Magn. Resonance **12** [1979] 340/2).
[10] Dolenka, G. N.; Zibarev, A. V.; Krupoder, S. A.; Furin, G. S.; Yakobson, G. G. (Izv. Sibirsk. Otd. Akad. Nauk SSSR Ser. Khim. Nauk **1980** No. 6, pp. 74/7).

[11] Furin, G. G.; Fedotov, M. A.; Yakobson, G. G.; Zibarev, A. V. (J. Fluorine Chem. **28** [1985] 273/90).
[12] Zibarev, A. V.; Furin, G. G.; Korobeinicheva, I. K.; Fedotov, M. A.; Yakobson, G. G. (Izv. Akad. Nauk SSSR Ser. Khim. **1983** 2259/65; Bull. Acad. Sci. USSR Div. Chem. Sci. **32** [1983] 2039/44).
[13] Rezvukhin, A. I.; Furin, G. G.; Yakobson, G. G. (Izv. Akad. Nauk SSSR Ser. Khim. **1981** 2512/8; Bull. Acad. Sci. USSR Div. Chem. Sci. **30** [1981] 2078/83).

Dipole Moment

In an early measurement of the dipole moment μ a value of 2.06 D was obtained [1]. Later several measurements were carried out in benzene solution at temperatures of 20 to 25°C, yielding μ values of 1.90 D [2, 3], 2.00 D [4], 1.95 D [5], 1.91 D [6], and in CCl_4 at 25°C 1.93 ± 0.05 D [7]. A special method (dielectric absorption of microwaves) gave a value of 1.90 D in benzene solution at 25°C, in good agreement with the other reported data [8]. Dipole moments of the pure liquid phase are 1.95 D (at 19.9°C) [3], obtained by the Onsager method and 1.60 D [9]. Microwave analysis with the aid of the Stark effect gave a gas-phase μ value of 2.30 ± 0.06 D (at 25°C) [10]. A smaller value of 2.07 D was obtained by extrapolation of dilute solution data [5]. The dipole moment components along the molecular inertial axes a and b were determined: $\mu_a = 1.81$ D [5], 2.20 ± 0.06 D [10] and $\mu_B = -1.00$ D [5], 0.664 ± 0.005 D [10]. The moment of the S=N bond was calculated to be 2.0 D [5] or 1.18 D [9] and that of the S=O bond 1.21 D [9]. The dipole moments of the isomers *cis*- and theoretical *trans*-O=S=NC_6H_5 were calculated to be 1.0 D and 2.8 D, respectively [4]. The authors stated that both isomers may exist in solution, yielding the experimental dipole moment value of ~2.0 D [4]. However, the dipole moment is almost temperature-independent (range 30 to 70°C), indicating that only one conformation of the compound is present in solution [8]. For the recent result that O=S=NC_6H_5 exists in the cis conformation, see p. 115, μ_{tot} is almost parallel to the N=S bond and points from N(+) to S(−) [5].

References:

[1] Bergmann, E.; Tschudnowsky, M. (Z. Physik Chem. B **17** [1932] 100/6).
[2] Jensen, K. A.; Hofman Bang, N. (Liebigs Ann. Chem. **548** [1941] 95/105).
[3] Kresze, G.; Smalla, H. (Chem. Ber. **92** [1959] 1042/8).
[4] Jannelli, L.; Lamanna, U.; Lumbroso, H. (Bull. Soc. Chim. France **1966** 3626/30).
[5] Bak, B.; Svanholt, H.; Larsen, C. (J. Mol. Struct. **36** [1977] 55/9).
[6] Waclawek, W.; Puchala, C.; Czerwiec, Z. (Bull. Acad. Polon. Sci. Ser. Sci. Chim. **26** No. 4 [1978] 337/41).
[7] Bonham, J. S.; Cheng, C. L.; Le Fèvre, R. J. W.; Ritchie, G. L. D. (Australian J. Chem. **26** [1973] 421/4).
[8] Roy, S. K. (Advan. Mol. Relaxation Interact. Processes **12** [1978] 229/37, 233).

[9] Egorov, Yu. P.; Borovikov, Yu. Ya.; Fedyuk, G. S.; Levchenko, E. S. (Zh. Obshch. Khim. **43** [1973] 2470/6; J. Gen. Chem. [USSR] **43** [1973] 2456/61).
[10] Caminati, W.; Mirri, A. M.; Maccagnani, G. (J. Mol. Spectrosc. **66** [1977] 368/74).

Electronic Structure. Electronic Spectra (X-PES, UPS, UV)

The ionization energy E_i of the N1s electron was determined from X-ray photoelectron spectra (X-PES), obtained by use of AlK_α radiation, to be 400.2 eV [2]. The sulfur K_α line (inner-shell electron transition $2p \rightarrow 1s$) was measured by X-ray emission spectroscopy to be 2308.01 eV (the mean number of valence electrons is 5.07 ± 0.08) [1]. The sulfur K_β X-ray emission [4] and fluorescence [5, 12] spectra were measured by two groups: 5.058 ± 0.001 Å; 5.038 ± 0.001 Å, 5.0288 ± 0.0008 Å, 5.018 ± 0.002 Å [13] (converted to eV: 2451.08, 2461.00, 2465.51, 2471.80 eV) and 2466.2, 2469.8, 2471.5 eV [12]. The experimental data are in good agreement with theoretical results obtained by an MNDO calculation [5]. $O{=}S{=}NC_6H_5$ can be identified within the N-sulfinyl-arenamines with the help of the SK_β band form and SK_α line chemical shift [21].

UV photoelectron spectra (UPS) of gaseous $O{=}S{=}NC_6H_5$ were recorded in the energy range 8 to 20 eV by use of HeI and HeII radiation [3 to 6]. Other workers measured only the first two E_i's using UPS [2], see Table 39. In the former studies, the assignments of experimental ionization energies are based on CNDO/S LCAO MO [3, 4] and MNDO [5] calculations. (According to Koopmans' theorem the ionization potentials are equal to the one-electron energies of the corresponding MO taken with reverse sign.) In contrast to [2, 3, 5] the order of orbitals b_1 and a_2 was reversed (from F-substitution experiments) [6]. Bonding parameters (Wiberg indices) indicate that the SO bond is a σ bond. The title compound may be represented by the formula: $^{\ominus}|\underline{\overline{O}}{-}S^{\oplus}{=}NC_6H_5$ [4] in agreement with [3]. The net atomic charges (Mulliken populations) were calculated by CNDO/S [3, 4] and CNDO/2 [7]: at N 0.04 [4], −0.17 [3], −0.066 [7]; S 0.12 [4], 0.45 [3], 0.256 [7]; O −0.5 [4], −0.49 [3], −0.301 [7], and by modified CNDO/S: N −0.06, S −0.02, O −0.29 [3]. The electronic structure was investigated using the HMO method for studying cycloadditions [8, 9], nucleophilic and electrophilic attack [10] in $O{=}S{=}NC_6H_5$. Optimized Hückel frontier orbitals [8, 9] and frontier electron densities [10] are presented. Valence and π-electron densities calculated with the CNDO/2 method are shown in [11]. A comparison between the distributions of the energies of the Hückel orbitals and those of "modified" CNDO/S orbitals was made in [9]. It must be mentioned that all these theoretical calculations are not based on the real structure of $O{=}S{=}NC_6H_5$.

The UV spectrum of $O{=}S{=}NC_6H_5$ was measured (λ in nm, ε in $L \cdot mol^{-1} \cdot cm^{-1}$) by several groups [4, 14 to 19] in different solvents. Two strong absorptions were observed: a long wavelength π–π* transition at 321 (log ε = 3.97) in cyclohexane [14], 314 (log ε = 4.00) [15], 315 [18], and 321 (log ε ≈ 4.0) [4, 19] in n-heptane, 314 (log ε = 4.00) in ether [16], 313 to 314 ($\varepsilon = 9.01 \times 10^3$) in 1,4-dioxane [17], and an n–π* transition at 232 to 237 (log ε = 3.8) in cyclohexane [14], 231 (log ε = 3.75) [15] and 230 (log ε ≈ 4.0) [19] in n-heptane, 231 (log ε = 3.78) in ether [16], and 232 to 233 ($\varepsilon = 5.08 \times 10^3$) in 1,4-dioxane [17]. No changes in the UV spectrum were observed in heptane solutions containing 10^{-4} M $O{=}S{=}NC_6H_5$ and 10^{-2} M $N(C_2H_5)_3$ or in dichloroethane solutions containing 10^{-4} M $O{=}S{=}NC_6H_5$ and 2×10^{-3} M TCNE (tetracyanoethylene). The authors concluded that $O{=}S{=}NC_6H_5$ is unlikely to function generally as a ligand since the electron donating and accepting capacities are of the same order as benzene [19].

The title compound was studied by the position annihilation method. Angular correlation and position lifetime parameters were measured. A long-life component τ_2 was not observed [20].

Table 39

Ionization Energies E_i in eV of $O{=}S{=}NC_6H_5$*).

method	b_1	a_2	σ_1	σ_2	π_3	σ_3	σ_4	
UPS [3]	9.12	9.65	10.62	11.09	11.42	12.37	12.71	—
UPS [4, 5]	9.06	9.52	10.38 10.87	11.18	—	12.26	12.85	13.75
UPS [6]	9.52	9.06	10.38 10.87	11.18	—	12.26	12.85	13.75
UPS [2]	9.12	9.65	—	—	—	—	—	—
mod. CNDO/S [3]	10.39	10.71	12.24	13.04	13.40	—	—	—
CNDO/S [4]	9.37	10.19	10.57	11.39	—	13.14	13.45	—
MNDO [5]	9.0	9.82	—	—	—	—	—	—

*) Assignments from [2, 3] (b_1 and a_2 also from [5] and [6]).

References:

[1] Takahashi, Y.; Yabe, K.; Sato, T. (Bull. Chem. Soc. Japan **42** [1969] 2707/8).
[2] Colonna, F. P.; Distefano, G.; Jones, D.; Modelli, A.; Galasso, V. (Z. Naturforsch. **36a** [1981] 385/90).
[3] Louwen, J. N.; van Dam, H.; Stufkens, D. J.; Oskam, A.; Jaffe, H. H. (J. Electron Spectrosc. Relat. Phenom. **26** [1982] 235/46).
[4] Furin, G. G.; Fedotov, M. A.; Yakobson, G. G.; Zibarev, A. V. (J. Fluorine Chem. **28** [1985] 273/90, 279).
[5] Zibarev, A. V.; Furin, G. G.; Konchenko, S. N.; Voityuk, A. A. (J. Fluorine Chem. **35** [1987] 627/41).
[6] Zibarev, A. V.; Sultanov, A. Sh.; Furin, G. G.; Furlei, I. I. (Izv. Akad. Nauk SSSR Ser. Khim. **35** [1986] 474/6; Bull. Acad. Sci. USSR Div. Chem. Sci. **35** [1986] 439/41).
[7] Glass, W. K.; King, I. J.; Shiels, A. (Inorg. Chim. Acta **25** [1977] 157/8).
[8] Hanson, P.; Stockburn, W. A. (J. Chem. Soc. Perkin Trans. II **1985** 589/95).
[9] Hanson, P.; Wren, S. A. C. (J. Chem. Soc. Perkin Trans. II **1987** 197/203).
[10] Plemenkov, V. V.; Kataev, E. G. (Str. Mol. Kvantovaya Khim. **1970** 146/50; C.A. **74** [1971] No. 87103).

[11] Fabian, J.; Mayer, R. (Phosphorus Sulfur **3** [1977] 179/84).
[12] Dolenko, G. N.; Zibarev, A. V.; Krupoder, S. A.; Furin, G. G.; Yakobson, G. G. (Izv. Sibirsk. Otd. Akad. Nauk SSSR Ser. Khim. Nauk **1980** No. 6, pp. 74/7).
[13] Takahashi, Y.; Yabe, K. (Bull. Chem. Soc. Japan **42** [1969] 3064/72, 3068).
[14] Leandri, G.; Mangini, A. (Spectrochim. Acta **15** [1959] 421/31, 426; Ric. Sci. **27** [1957] 1920/5; Boll. Sci. Fac. Chim. Ind. Bologna **17** [1959] 15/8).
[15] Kresze, G.; Smalla, H. (Chem. Ber. **92** [1959] 1042/8).
[16] Smith, W. T., Jr.; Trimnell, D.; Grinninger, L. D. (J. Org. Chem. **24** [1959] 664/6).
[17] Stephenson, C. V.; Cobum, W. C., Jr.; Wilcox, W. S. (Spectrochim. Acta **17** [1961] 933/46, 935).
[18] Zibarev, A. V.; Furin, G. G.; Korobeinicheva, I. K.; Fedotov, M. A.; Yakobson, G. G. (Izv. Akad. Nauk SSSR Ser. Khim. **1983** 2259/65; Bull. Acad. Sci. USSR Div. Chem. Sci. **32** [1983] 2039/44).
[19] Collins, N. C.; Glass, W. K. (Spectrochim. Acta A **30** [1974] 1335/9).

[20] Filipecki, J.; Czerwiec, Z.; Dworakowski, J.; Koziol, J.; Glass, J. C. (Chem. Scr. **25** [1985] 157/9).

[21] Treiger, B. A.; Mazalov, L. N.; Dolenko, G. N.; Guzhavina, T. I. (Comput. Enhanced Spectrosc. **3** [1986] 153/63).

Microwave Spectrum

A microwave spectrum was recorded in the frequency range 8000 to 40000 MHz. RF-MWDR (radiofrequency-microwave double resonance) technique was employed to identify individual a- and b-type transitions [1]. The measured transition frequencies and differences between the observed and calculated values for the ground and three first excited torsional states are presented in two tables. The rotational (A, B, C) and centrifugal distortion constants D_J and D_{JK} obtained by least-squares fitting of the rotational spectra, the inertia defects $\Delta = I_{cc} - I_{bb} - I_{aa}$ and the differences $\Delta_{v_{t+1}} - \Delta_{v_t}$ are given in Table 40.

Table 40
Molecular Constants of $O{=}S{=}NC_6H_5$ for the Ground and Three First Excited Torsional States, v_t [1].

	$v_t = 0$	$v_t = 1$	$v_t = 2$	$v_t = 3$
A in MHz	4026.669(6)	4013.42(2)	4000.0(6)	3987.6(8)
B in MHz	860.636(1)	861.359(4)	862.106(13)	862.873(17)
C in MHz	709.531(1)	711.246(4)	713.013(17)	714.813(22)
D_J in Hz	37.1(1.2)	29.5(4.9)	35(18)	45(21)
D_{JK} in Hz	−160(20)	−236(51)	—	—
Δ in u·Å^2	−0.4518(10)	−2.0906(40)	−3.765(20)	−5.426(30)
$\Delta_{v_{t+1}} - \Delta_{v_t}$	−1.639(4)	−1.674(20)	−1.661(30)	

LRMW (low-resolution microwave) spectra [1, 2] show the presence of an a-type band separated by $\Delta\nu \simeq B + C$. The sum of B + C was found to be 1575 ± 5 MHz [2] in good agreement with 1570.2 MHz from Table 40 [1]. The experimental rotational constants could only be reproduced by a cis structure with respect to the central S–N bond [1, 2]. The torsional frequency was estimated as $\nu_t = 41.1 \pm 1\ cm^{-1}$ [1].

References:

[1] Caminati, W.; Mirri, A. M.; Maccagnani, G. (J. Mol. Spectrosc. **66** [1977] 368/74).
[2] Bak, B.; Svanholt, H.; Larsen, C. (J. Mol. Struct. **36** [1977] 55/9).

IR and Raman Spectra

IR spectra of $O{=}S{=}NC_6H_5$ were recorded in the range 4000 to 700 cm^{-1}, depicted in [1], from 4000 to 365 cm^{-1} in CCl_4 (in n-hexane or cyclohexane from 850 to 650 cm^{-1}) and in pure liquid depicted in [2], from 4000 to 650 cm^{-1} (NaCl) and 700 to 285 cm^{-1} (CsBr) [3], from 1350 to 1010 cm^{-1} in liquid [4], and from 1576 to 300 cm^{-1} also for the ^{15}N-substituted species in pure liquid depicted in [5]. The attempt to record a vapor spectrum of $O{=}S{=}NC_6H_5$ was not very successful. Only the contours of the stronger bands were obtained [2]. Raman spectra were measured from pure liquids in the ranges 3100 to ~150 cm^{-1} using the 5461 Å mercury line [3], from 1585 to ~146 cm^{-1}, also for the ^{15}N-substituted species [5] and with the 6328 Å He–Ne laser line [7]. In [2, 3, 5] the full spectra are listed and tentatively assigned. The most recent complete IR and Raman data from [5] are given in Table 41. The assignments were ascertained by observed band shifts through ^{15}N substitution and complexation of $O{=}S{=}NC_6H_5$. Some discrepancies exist to the assignments of the earlier works [5]. Symmetry group of the compound is C_s. The most important result is that hardly any coupling exists between ν(NS) (1284 cm^{-1}) and ν(SO) (1155 cm^{-1}) [5]. In earlier works [3, 6] the 638 cm^{-1} bond was ascribed to a δ(N=S=O) absorption, now it is attributed to a $\delta(C_6H_5{-}N{=}S)$, and the band at 748 cm^{-1} is δ(N=S=O) [5]. The absorption at 368 cm^{-1} is assigned to a ϱ(N=S=O) [5] in contrast to [3]. The vibrational frequencies of the N=S=O group above 1000 cm^{-1} in the Raman spectrum were reinvestigated by [7]. The absorptions at 1296, 1282, and 1154 (of $O{=}S{=}^{14}NC_6H_5$) were assigned in close agreement to [5] to ν(C–N), ν(N=S), and ν(S=O), respectively [7]. In a more recent publication: ν(N=S) = 1292 cm^{-1}, ν(S=O) = 1164 cm^{-1}, $\nu_8(C_6H_5)$ = 1594 cm^{-1}. Intensities were also measured [8]. The absorption at 1031 cm^{-1}, which was attributed to a δ(C–H) vibration [5], is now tentatively ascribed to a δ(N=S=O) vibration since it was observed in the spectra of perfluorinated analogs of $O{=}S{=}NC_6H_5$ and undergoes an important shift on substitution of ^{14}N by ^{15}N [7].

Table 41

IR and Raman[a)] Spectra of $O{=}S{=}NC_6H_5$ and of the ^{15}N-Enriched Compound (frequencies in cm^{-1})[b)] [5].

IR			Raman			assignment
^{14}N	^{15}N	Δ[c)]	^{14}N	^{15}N	Δ[c)]	
			1585 m	1585 m		ring stretch
1576 m	1576 m		1576 sh			ring stretch
1486 s	1484 s	2	1483 m	1483 m		ring stretch
1454 m	1454 m					ring stretch
1299 m			1299 s	1292 m	7	$\nu(C_6H_5{-}N)$
1284 s	1272 s	12	1284 s	1272 s	12	ν(N=S)
1181 m	1178 m	3				δ(C–H)
1161 sh 1155 s	1160 sh 1154 s		1158	1154 s		δ(C–H) ν(S=O)
1100 w	1100 w					δ(C–H)
1074 m	1070 m	4				δ(C–H)
1035 m	1031 w	4	1034 m	1030 m	4	δ(C–H)
1020 m	1016 m	4	1020 w	1015 w	5	δ(C–H)
1000 w	1000 w		999 m	1000 m		ring deformation
920 m	920 m					δ(C–H)
767 s	765 s	2				δ(C–H)
748 w	740 w	8	747 m	739 m	8	δ(N=S=O)
683 s	682 s					δ(C–H)
638 w	631 w	7	638 m	630 m	8	δ(C–N=S)
606 w	604 w	2	606 w	606 w		ring deformation
533 w	530 w	3				ring deformation
436 m	434 m	2	437 m	438 m		ring deformation

Table 41 (continued)

IR			Raman			assignment
^{14}N	^{15}N	$\Delta^{c)}$	^{14}N	^{15}N	$\Delta^{c)}$	
368 w	362 w	6				ϱ(N=S=O)
300 w	300 w		300 w	300 w		skeletal deformation
			166 w			
			146 w			

a) All bands in Raman spectra are polarized. – b) Accuracy of measurements about $1\ cm^{-1}$. – c) Δ = Lowering of frequency on substitution of ^{15}N.

The integrated infrared intensity of the ν_8 (ν_{16} in Herzberg's nomenclature) ring band of $O{=}S{=}NC_6H_5$ was calculated to be $523\ L \cdot mol^{-1} \cdot cm^{-2}$ [9], see also [8]. The extinction coefficient ε of the ν_{13} band (Herzberg's nomenclature) ($1484\ cm^{-1}$) was estimated $\varepsilon^{1/2} = 21.2$ [10].

References:

[1] Kresze, G.; Maschke, A. (Chem. Ber. **94** [1960] 450/6).
[2] Glass, W. K.; Pullin, A. D. E. (Trans. Faraday Soc. **57** [1961] 546/56).
[3] Stephenson, C. V.; Coburn, W. C., Jr.; Wilcox, W. S. (Spectrochim. Acta **17** [1961] 933/46).
[4] van Woerden, H. F.; Bijl-Vlieger, S. H. (Recl. Trav. Chim. **93** [1974] 85/90).
[5] Meij, R.; Oskam, A.; Stufkens, D. J. (J. Mol. Struct. **51** [1979] 37/49).
[6] Kresze, G.; Uhlig, U.; Ropte, E.; Schrader, B. (Z. Anal. Chem. **197** [1963] 283/94, 292).
[7] Zibarev, A. V.; Furin, G. G.; Korobeinicheva, I. K.; Fedotov, M. A.; Yakobson, G. G. (Izv. Akad. Nauk SSSR Ser. Khim. **1983** 2259/65; Bull. Acad. Sci. USSR Div. Chem. Sci. **32** [1983] 2039/44).
[8] Korobeinicheva, I. K.; Fugaeva, O. M.; Furin, G. G. (Izv. Akad. Nauk SSSR Ser. Khim. **1987** 1766/9; Bull. Acad. Sci. USSR Div. Chem. Sci. **36** [1987] 1634/7).
[9] Brownlee, R. T. C.; Hutchinson, R. E. J.; Katritzky, A. R.; Tidwell, T. T.; Topsom, R. D. (J. Am. Chem. Soc. **90** [1968] 1757/67).
[10] Brownlee, R. T. C.; English, P. J. Q.; Katritzky, A. R.; Topsom, R. D. (J. Phys. Chem. **73** [1969] 557/64, 559).

2.6.22.2.1.3 Other Physical Properties

Density. Viscosity

The density of the liquid was determined to be D = 1.236 (t = 15°C) [1], 1.2329 (t = 19.9°C) [2], and $1.231\ g/cm^3$ (t = 25°C) [3]. In the solid phase, t = −190°C is D = $1.37\ g/cm^3$ [3]. The relative vapor density is 5.13 (method of Meyer) and calculated 4.81 [1].

The viscosity was determined with an Oswald viscosimeter to be 0.58 cP at 30°C [4].

References:

[1] Michaelis, A. (Liebigs Ann. Chem. **274** [1893] 173/266, 201).
[2] Kresze, G.; Smalla, H. (Chem. Ber. **92** [1959] 1042/8).
[3] Egorov, Yu. P.; Borovikov, Yu. Ya.; Fedyuk, G. S.; Levchenko, E. S. (Zh. Obshch. Khim. **43** [1973] 2470/6; J. Gen. Chem. [USSR] **43** [1973] 2456/61).
[4] Roy, S. K. (Advan. Mol. Relaxation Interact. Processes **12** [1978] 229/37).

Boiling Point

The boiling point measured by many workers is, e.g., 198 to 200°C/760 Torr [1], 88 to 95°C/17 to 20 Torr [2].

References:

[1] Michaelis, A. (Ber. Deut. Chem. Ges. **24** [1891] 745/57).
[2] Rajagopalan, P.; Advani, B. G.; Talaty, C. N. (Org. Syn. Collective Vol. **5** [1973] 504/8).

Refraction. Kerr Constant. Polarization. Dielectric Properties

The refraction index $n_D^{19.5} = 1.6259$ [1], $n_D^{21} = 1.6253$ [2], $n_D^{25} = 1.6247$ was measured [3]. The molar Kerr constant, K_m, was determined in CCl_4 at 25°C to be $(500 \pm 10) \times 10^{-12}$ e.s.u./mol [4].

The polarization $P_{2\infty}$ (extrapolated to infinite dilution) was determined in benzene solution to be 121.8 cm³ [5] and in CCl_4 to 122 ± 3 cm³ [4]. The molar refraction R_D was determined to 39.9 cm³ in benzene [5] and to 44 ± 3 cm³ in CCl_4 at 25°C [4]. Measurements were also carried out on the solid phase of the compound at −190°C: $R_D = 49.0$ cm³; atom polarization $P_A =$ 2 cm³; deformation polarization $P_{def} = 50.7$ cm³ [6].

The dielectric constant ε_∞ (deformation limit) was measured to 2.41 at 30°C [7] and 2.92 at −190°C [6].

Dielectric permittivities ε' and dielectric losses ε'' were estimated in benzene solution using different microwave frequencies at 30, 50, and 70°C. These data were analyzed in terms of molecular and intramolecular relaxation processes. Static dielectric constants, ε_0, were measured at 1 MHz using a dielectrometer. Activation energies for dielectric relaxation ΔH_τ and for viscous flow ΔH_η and relaxation times are given [7].

References:

[1] Kresze, G.; Smalla, H. (Chem. Ber. **92** [1959] 1042/8).
[2] Rajagopalan, P.; Advani, B. G.; Talaty, C. N. (Org. Syn. Collective Vol. **5** [1973] 504/8).
[3] Abel, E. W.; Armitage, D. A. (J. Chem. Soc. **1964** 3122/6).
[4] Bonham, J. S.; Cheng, C. L.; Le Fèvre, R. J. W.; Ritchie, G. L. D. (Australian J. Chem. **26** [1973] 421/4).
[5] Jannelli, L.; Lamanna, U.; Lumbroso, H. (Bull. Soc. Chim. France **1966** 3626/30).
[6] Egorov, Yu. P.; Borovikov, Yu. Ya.; Fedyuk, G. S.; Levchenko, E. S. (Zh. Obshch. Khim. **43** [1973] 2470/6; J. Gen. Chem. [USSR] **43** [1973] 2456/61).
[7] Roy, S. K. (Advan. Mol. Relaxation Interact. Processes **12** [1978] 229/37, 233).

2.6.22.2.1.4 Chromatography

$R_f = 0.26$ in benzene (10 other solvent systems with corresponding R_f value are presented) [1]. Gas-chromatographic separation from other N-sulfinyl-arenamines is possible (silicone rubber SE on cellite C-22) [2, 3].

References:

[1] Czerwiec, Z. (Chem. Anal. [Warsaw] **21** [1976] 751/3).
[2] Czerwiec, Z.; Mirek, J. (Zesz. Nauk Univ. Jagiel. Proc. Chem. Nc. 14 [1969] 103/6).
[3] Czerwiec, Z.; Markowski, J. (Chem. Anal. [Warsaw] **20** [1975] 213/5).

2.6.22.2.1.5 Hammett Constant of the O=S=N Group

Several numerical σ values were obtained by IR [1 to 4] and NMR [4, 5, 6] methods. A σ_R° value of ±0.09 was derived from the intensity of the ν_8 IR phenyl ring vibration (ν_{16} in Herzberg's nomenclature) of $O{=}S{=}NC_6H_5$ [1, 2]. In reasonable agreement other IR studies gave values of ~0.1 to 0.2 (σ_P°) [3] and 0.13 (σ_P) [4] (measured from $O{=}S{=}NC_6H_4NO_2$-4 and $O{=}S{=}NC_6H_4CH_3$-4, respectively). Close to these results, a ^{19}F NMR study gave also a value of 0.13 [5] (measured from N-sulfinyl-(3 or 4)-fluoro-anilines). However, divergent values of 0.3 (σ^+) [6] and 0.44 (σ_P) [4] were obtained by ^{13}C and ^1H NMR spectroscopy, respectively (measured from $O{=}S{=}NC_6H_5$ and $O{=}S{=}NC_6H_4CH_3$-4, respectively).

References:

[1] Brownlee, R. T. C.; Hutchinson, R. E. J.; Katritzky, A. R.; Tidwell, T. T.; Topsom, R. D. (J. Am. Chem. Soc. **90** [1968] 1757/67).
[2] Butt, G.; Davis, M.; Tau Pang, Y.; Topsom, R. D.; Katritzky, A. R. (J. Chem. Soc. Perkin Trans. II **1974** 260/3).
[3] Egorov, Yu. P.; Borovikov, Yu. Ya.; Fedyuk, G. S.; Levchenko, E. S. (Zh. Obshch. Khim. **43** [1973] 2470/6; J. Gen. Chem. [USSR] **43** [1973] 2456/61).
[4] Collins, N. C.; Glass, W. K. (Spectrochim. Acta A **30** [1974] 1335/9).
[5] Taft, R. W.; Sheppard, W. A. (private communication in [1, p. 1759]).
[6] Kresze, G.; Berger, M.; Claus, P. K.; Rieder, W. (Org. Magn. Resonance **8** [1976] 170/2).

2.6.22.2.1.6 Polarography

$O{=}S{=}NC_6H_5$ is polarographically active and provides a three-step wave. A suitable solvent is DMF (or DMF in C_6H_6). The halfwave potential (towards the internal mercury electrode) is −0.584 V [1] or −0.701 V vs. the calomel electrode [2].

References:

[1] Czerwiec, Z.; Bogaczeck, J. (Z. Anal. Chem. **293** [1978] 415).
[2] Walter, D.; Pfützenreuter, C. (Z. Chem. [Leipzig] **17** [1977] 426/7).

2.6.22.2.1.7 Chemical Reactions

Thermal Analysis. Pyrolysis

$O{=}S{=}NC_6H_5$ was thermally analyzed in static air in the temperature range 20 to 300°C by TG (thermogravimetry), DTG (derivative thermogravimetry), and DTA (differential thermal analysis). The measured curves are shown in a figure. On heating, the compound undergoes only evaporation without degradation [1].

High-temperature gas-phase pyrolysis at 600 to 800°C (10^{-3} to 10^{-2} Torr) gave SO_2, H_2S, and probably CO and/or N_2 as condensable gases and aniline (10%) besides the starting material (60%). Heating at 1000°C (30 min, 0.05 to 0.1 Torr) led to the formation of nearly the same species as observed in the mass spectrum: 1-cyanocyclopentadiene (4%), benzonitrile (5.1%), benzene (0.05%), thiophene (0.3%), 2- and 3-cyanothiophene (3% each), 2-cyanothiacyclohexa-2,4-diene (2%), and traces of aniline. The products were isolated and analyzed by chromatography (GLC) [2]. The formation of 1-cyano-cyclopentadiene (and some of the aniline) is ascribed to the generation of phenylnitrene by SO loss. SO (in the ground state $^3\Sigma$) was identified by microwave analysis. SO_2 was also observed in the temperature range 650 to 980°C (ca. 8% partial pressure at 980°C) detected by a rotational line and by its mass spectrum [3].

References:

[1] Czerwiec, Z.; Pielichowski, J. (J. Therm. Anal. **7** [1975] 465/8).
[2] Wentrup, C. (Tetrahedron **27** [1971] 1027/32).
[3] Saito, S.; Wentrup, C. (Helv. Chim. Acta **54** [1971] 273/5).

Ionization and Fragmentation by Electron Impact

Electron-impact mass spectra of $O{=}S{=}NC_6H_5$ were measured [1 to 4] (taken at 75 eV [2] and 70 eV [3, 4]), and molecular ions were recorded. MS (100°C, 70 eV): m/e (rel. int. in %) = M^+ (100); M^+-16 (1); M^+-17 (1); M^+-CO (19); M^+-CHO (10); M^+-33 (1); M^+-SO (19); $M^+-CO{-}HCN$ (16); $C_4H_5N^+$ (15); M^+-CNOS (7); $M^+-CHNOS$ (11); M^+-CH_2NOS (14); $(M-CO)^{2+}$ (2); 52 (5); 51 (11); 50 (5) [4].

The compositions of the major fragment ions were confirmed by high-resolution measurements [2, 4]. The major electron-impact fragmentation pathway is the loss of carbon monoxide from the molecular ion. This is a low-energy process which even occurs at nominal 10 eV [1, 2, 5]. Two different mechanistic pathways were suggested for the loss of CO: via loss of C-1 carbon atom [2] and via C-2 carbon atom [1]. The latter was confirmed by mass spectroscopic study of $O{=}S{=}NC_6H_5$-^{13}C-1 [5]. Another important pathway is the loss of SO by S–N cleavage from the molecular ion. Subsequent loss of CN leads to the cyclopentadienyl cation [2]. DADI (direct analysis of daughter ions) spectrum of the molecular ion was also recorded: m/z = 139 (mother); 111.2 (daughter, $C_5H_5NS^+$) [3]. The CID (collision induced dissociation) spectrum of $C_6H_5N^{+\bullet}$ radical, obtained by electron impact (70 eV) on $O{=}S{=}NC_6H_5$ was recorded. Two competing mechanisms for the formation of $C_6H_5N^{+\bullet}$ are suggested: via loss of SO from $O{=}S{=}NC_6H_5$ (direct formation) or formation via intermediate cyclization [6].

References:

[1] Job, B. E. (Chem. Commun. **1967** 44/5).
[2] Bowie, J. H.; Larsson, F. C. V.; Schroll, G.; Lawesson, S.-O.; Cooks, R. G. (Tetrahedron **23** [1967] 3743/52).
[3] Brands, G.; Golloch, A. (Phosphorus Sulfur **13** [1982] 221/34, 224).
[4] Wentrup, C. (Tetrahedron **27** [1971] 1027/32).
[5] Siegel, A. S. (Org. Mass. Spectrom. **3** [1970] 875/7).
[6] Maquesbian, A.; van Haverbeke, Y.; Flammang, R.; Menu, A.; Wentrup, C. (Org. Mass Spectrom. **13** [1978] 518/26).

Reactions with Metals, Bases, and Other Catalysts

Treatment of $O{=}S{=}NC_6H_5$ with Na suspended in toluene [1, 2] or strongly basic anions, such as NH_2^-, $C_2H_5O^-$, and t-$C_4H_9O^-$ in benzene or ether [3, 4] led to the formation of $C_6H_5N{=}S{=}N$-C_6H_5 (yield ~70%). In earlier communications an azo-thio-benzene structure was assumed for the product [1, 2]. BipyNiCOD was also used as a catalyst. However, the reaction in refluxing benzene or without solvent is slow and small yields in the range 12 to 20% are achieved [5]. When a mixture of $O{=}S{=}NC_6H_5$ and finely divided copper (mole ratio 1:1) is heated at 130°C for 4 h $C_6H_5N{=}NC_6H_5$ is formed in 30% yield. At lower temperature (110°C) and in the presence of toluene the reaction is slower (8 h) and the yield is small (3.3%) [6].

References:

[1] Leandri, G.; Rebora, P. (Experimentia **13** [1957] 71; Gazz. Chim. Ital. **87** [1957] 503/9).
[2] Cramer, R. (J. Org. Chem. **26** [1961] 3476/8).
[3] Hörhold, H. H.; Floßmann, K.-D. (Z. Chem. [Leipzig] **7** [1967] 345/6).
[4] Hörhold, H. H.; Beck, J. (J. Prakt. Chem. **311** [1969] 621/9).
[5] Walther, D.; Dinjus, E.; Wolf, H. (Z. Chem. [Leipzig] **19** [1979] 381/2).
[6] Minami, T.; Agawa, T. (Tetrahedron Letters **1968** 4109/10).

Hydrolysis

Slow decomposition was observed on hydrolysis using pure H_2O or diluted acids. The decomposition was faster on shaking. Decomposition takes place immediately if $O{=}S{=}NC_6H_5$ is treated with diluted alkali [1]. The decomposition products of hydrolysis are $C_6H_5NH_2$ and SO_2 [1, 2]. The kinetic of the hydrolysis was studied at 30°C (20 and 40°C). The uncatalyzed reaction is slow and has an induction period. The formed aniline autocatalyzes the reaction. The catalytic effect is much stronger when pyridine is added as a catalyst. In this case the rate equation is of third-order overall and first-order with respect to $O{=}S{=}NC_6H_5$, H_2O, and pyridine, respectively. The pyridine and H_2O form an H-bonded complex, increasing the nucleophilic properties of H_2O. The rate constant in CH_3CN at 30°C is 2.96×10^3 $mol^2\cdot L^2\cdot sec^{-1}$, $\Delta H^{\neq}=5.7$ kcal/mol, $\Delta S^{\neq}=-51$ e.u. The reaction is made faster by adding a carboxylic acid to the reaction mixture. A decrease of the reaction rate by a factor of almost two was observed by using D_2O (isotopic effect 1.73). The reaction was also slower if deuteroacetic acid was added instead of acetic acid ($k(CH_3COOH)/k(CH_3COOD)=2.23$). The reaction mechanism is discussed [2].

References:

[1] Michaelis, A. (Ber. Deut. Chem. Ges. **24** [1891] 745/57, 747).
[2] Mironova, D. F.; Konoplya, O. Ya. (Ukr. Khim. Zh. **42** [1976] 374/8; Soviet Progr. Chem. **42** No. 4 [1976] 35/8).

Reactions with Halogens and Inorganic Compounds

Passing dry Cl_2 gas through a solution of $O{=}S{=}NC_6H_5$ in petroleum ether for hours led to the formation of 2,4,6-trichloroaniline and $SOCl_2$ [2]. In a similar way, $O{=}S{=}NC_6H_5$ reacts with Br_2 in ether or ligroine to produce 2,4,6-tribromoaniline and $SOBr_2$ [1].

Dry HCl reacts with $O{=}S{=}NC_6H_5$ to form the unstable intermediate $O{=}S{=}NC_6H_5\cdot2HCl$ which converts to $C_6H_5NH_2$ and $SOCl_2$ when treated further with HCl (reversion of the formation of $O{=}S{=}NC_6H_5$) [1].

The reaction with PCl_5 in dry CCl_4 presumably led to the formation of $C_6H_5N{=}SCl_2$ and $POCl_3$. $C_6H_5N{=}SCl_2$, however, could not be isolated and seems only to be stable in solution [2].

Treatment of $O{=}S{=}NC_6H_5$ with S_2Cl_2 in dry benzene at 60 to 70°C for 6 h yields I, the same product as in the case of aniline (Herz reaction) [3].

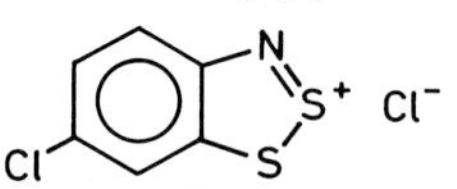

I

$O{=}S{=}NC_6H_5$ reacts with $COCl_2$ in refluxing benzene (85 min) with pyridine or N,N-dimethylformamide as a catalyst to yield $C_6H_5N{=}C{=}O$ (60% yield) and $SOCl_2$. In o-dichlorobenzene at 180°C no reaction was observed [4].

O=S=NC_6H_5 reacts with N_2O_4 (−10°C, in ether) and NOCl (−70°C, ether–CCl_4) to produce $C_6H_5N_2^+NO_3^-$ and $C_6H_5N_2^+Cl^-$, respectively (note: $C_6H_5N_2^+NO_3^-$ tends to decompose violently upon slight shock!) [13].

When O=S=NC_6H_5 and $P(C_6H_5)_3$ in 1:1 mole ratio are heated without a solvent at 175°C for 14 h equal amounts of $(C_6H_5)_3P$=S and unchanged O=S=NC_6H_5 result. The same reaction, when carried out at or below 125°C, yields in addition $(C_6H_5)_3P$=O. $P(OC_2H_5)_3$, $P(OC_6H_5)_3$, and $P(SC_6H_5)_3$ are reported to not react with O=S=NC_6H_5 [7]. Other authors observed that O=S=NC_6H_5 does not react at room temperature, but with boiling $P(OC_2H_5)_3$ (155°C) in a strongly exothermic reaction with formation of $(C_2H_5O)_3P$=O and $(C_2H_5O)_3P$=S in a 1:1 mole ratio, and $(C_2H_5O)_3P$=NC_6H_5 in 72% yield [8]. With $P(N(CH_3)_2)_3$, O=S=NC_6H_5 reacts at temperatures below 0°C (0 to −80°C) to yield all three types of products stated above in equal proportions [9]. O=S=NC_6H_5 behaves indifferently towards $(C_6H_5)_3P$=O, $(C_6H_5)_3P$=S, and $(C_6H_5)_3P$=NC_6H_5. With $(C_2H_5)_3P$=NC_2H_5 the reaction is unclear and yields only small amounts of $(C_2H_5)_3P$=S and $C_6H_5NH_2$ [7]. The title compound reacts with O-alkyl esters of dithiophosphoric acid derivatives as shown below [10]:

$$O{=}S{=}NC_6H_5 + 3(RO)_2P(S)SH \rightarrow ((RO)_2P(S)S{-})_2 + C_6H_5NH_3^+(RO)_2PS_2^- + [SO]$$

$$R = CH_3,\ i\text{-}C_3H_7$$

The silylated derivatives give a more complex mixture of products (room temperature, no solvent):

$$O{=}S{=}NC_6H_5 + 4(RO)_2P(S)S{-}Si(CH_3)_3 \rightarrow$$
$$((RO)_2P(S)S{-})_2 + ((RO)_2P(S)S)_2S + ((CH_3)_3Si)_2O + C_6H_5N(Si(CH_3)_3)_2$$

$$R = C_2H_5,\ i\text{-}C_3H_7$$

A reaction mechanism is proposed. The initial step is assumed to be the addition of the S=O bond to the S–Si bond [11].

The title compound reacts with $LiN(Si(CH_3)_3)_2$ in a single-stage reaction to afford C_6H_5-N=S=$NSi(CH_3)_3$ (in hexane at room temperature) [16]:

$$O{=}S{=}NC_6H_5 + LiN(Si(CH_3)_3)_2 \rightarrow C_6H_5N{=}S{=}NSi(CH_3)_3 + 1/n\ (LiOSi(CH_3)_3)_n$$

With alkyl-trimethylsilylamines, mole ratio 1:1, in ice-cooled petroleum ether air-sensitive compounds are formed in 70% yield [17]:

$$O{=}S{=}NC_6H_5 + LiN(R){-}Si(CH_3)_3 \rightarrow LiN(C_6H_5){-}S(O){-}N(R){-}Si(CH_3)_3$$

$$R = CH_3,\ C_2H_5$$

The title compound reacts with CS_2 in a sealed tube to form the corresponding isothiocyanate (and sulfur) in a quantitative yield [12].

It reacts quantitatively at room temperature with $(CH_3)_2B$–SCH_3 in CH_2Cl_2 to yield C_6H_5N-$(B(CH_3)_2)$–S–S–CH_3 [14]. With 3,5-dimethyl-1,2,4-trithia-3,5-diborolane (II) the formation of several ring compounds occurs with III as the major product [15].

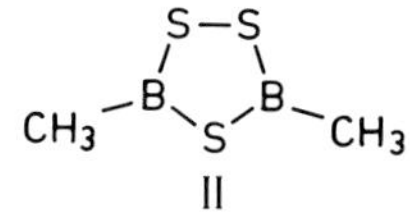

II

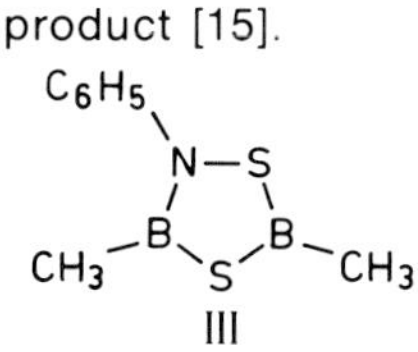

III

The reactions of O=S=NC_6H_5 with $TiCl_4$ (in pentane, 24 h), $TiCl_3(OCH_3)$ (in CH_2Cl_2, 24 h), and $ZrCl_4$ (in CCl_4, 24 h) gave 1:1 complexes (see p. 270). In the case of $TiCl_4$ and excess O=S=NC_6H_5 a 1:2 adduct was also isolated. Attempts to synthesize complexes with $NbCl_5$ and $TaCl_5$ were not successful [5]. $VOCl_3$ reacts with two moles of O=S=NC_6H_5 in benzene to form the adduct $VOCl_3 \cdot 2O$=S=NC_6H_5 which readily liberates SO_2 at room temperature [6].

References:

[1] Michaelis, A.; Herz, R. (Ber. Deut. Chem. Ges. **24** [1891] 745/57, 747/8).
[2] Michaelis, A. (Liebigs Ann. Chem. **274** [1893] 173/266, 201/4).
[3] Maior, O.; Arsenescu, N. (Analele Univ. Bucaresti Ser. Stiint Nat. **14** [1965] 25/33).
[4] Ulrich, H.; Tucker, B.; Sayigh, A. A. R. (J. Org. Chem. **34** [1969] 3200/1).
[5] Wilkins, J. D. (J. Inorg. Nucl. Chem. **38** [1976] 673/5).
[6] Slawisch, A. (Naturwissenschaften **56** [1969] 369; Z. Anorg. Allgem. Chem. **374** [1970] 291/6).
[7] Senning, A. (Acta Chem. Scand. **19** [1965] 1755/9).
[8] Wieczorkowski, J. (Bull. Acad. Polon. Sci. Ser. Sci. Chim. **13** [1965] 155/61, 157).
[9] Pudovik, M. A.; Morozova, N. P.; Pudovik, A. N. (Zh. Obshch. Khim. **48** [1978] 495/9; J. Gen. Chem. [USSR] **48** [1978] 446/9).
[10] Kutyrev, G. A.; Mavrin, V. Yu.; Cherkasov, R. A.; Pudovik, A. N. (Izv. Akad. Nauk SSSR Ser. Khim. **1980** 1402/4; C.A. **93** [1980] No. 167773).

[11] Kutyrev, G. A.; Lygin, A. V.; Cherkasov, R. A.; Pudovik, A. N. (Zh. Obshch. Khim. **53** [1983] 1009/13; J. Gen. Chem. [USSR] **53** [1983] 891/4).
[12] Minami, T.; Agawa, T. (Tetrahedron Letters **1968** 2651/3).
[13] Kobayashi, M.; Honda, K. (Bull. Chem. Soc. Japan **39** [1966] 1778/82).
[14] Meller, A.; Maringgele, W.; Fetzer, H. (Chem. Ber. **113** [1980] 1950/61, 1951/2).
[15] Meller, A.; Habben, C. (Monatsh. Chem. **113** [1982] 139/53, 140/3).
[16] Ruppert, I.; Bastian, V.; Appel, R. (Chem. Ber. **108** [1975] 2329/39, 2330, 2336).
[17] Scherer, O. J.; Hornig, P. (Angew. Chem. **78** [1966] 776/7; Angew. Chem. Intern. Ed. Engl. **5** [1966] 729/30).

2.6.22.2.2 Monosubstituted N-Sulfinyl-benzenamines

2.6.22.2.2.1 N-Sulfinyl-4-fluoro-aniline, $O{=}S{=}NC_6H_4F\text{-}4$

The compound was obtained by the standard method of Michaelis as a yellow oil with b.p. 107°C/16 Torr [1].

The SK_β X-ray fluorescent spectrum shows peaks at 2471.9, 2469.2, and 2466.5 eV [2] ($\Delta SK_\alpha = 0.46 \pm 0.02$ [2]).

HeI photoelectron spectrum (in eV): 9.12 (b_1 [3], a_2 [4]); 9.96 (a_2 [3], b_1 [4]); 10.66; 11.26; 12.61; 13.63; 14.49; 15.32 [3, 4]; MNDO calculation: 9.17 (b_1); 10.23 (a_2) [3].

^{15}N NMR (CCl_4/external liquid NH_3): $\delta = 300.2$ ppm [5, 6] (referenced to CH_3NO_2: $\delta = -80$ ppm).

^{19}F NMR (CCl_4/internal C_6F_6): $\delta = 60.5$ ppm [5] (referenced to $CFCl_3$: $\delta = -102.4$ ppm).

^{17}O NMR (CCl_4/external D_2O, t = 20°C): $\delta = 412$ ppm [5, 6].

Raman (neat): ν (in cm^{-1}) = 1300 ν(C–N), 1280 ν(N=S), 1162 ν(S=O), 1025 δ(NSO) [5]. Raman spectrum of the compound enriched with the ^{15}N isotope: 1288 ν(C–N), 1270 ν(N=S), 1161 ν(S=O), 1014 δ(NSO) [5]. The intensities of some Raman frequencies were determined (in acetone): ν (in cm^{-1}) = 1602 (I_{rel} = 1.25) ν_8(C=C, arom.); 1290 (I_{rel} = 1.60) ν(N=S); 1162, 1156 (I_{rel} = 3.25) ν(S=O) [7].

The integrated intensity of the ν_8 ring vibration (1598, 1587 cm^{-1}) in the IR spectrum is 3495 $L \cdot mol^{-1} \cdot cm^{-2}$ [1].

UV (n-heptane): $\lambda_{max} = 319$ nm [5, 6].

The dipole moment in the gas phase is 1.43 D [8].

The compound reacts with $(CH_3)_2BSCH_3$ to yield 4-$FC_6H_4N(B(CH_3)_2)SSCH_3$ [9]. On refluxing it with 3,5-dimethyl-1,2,4-trithia-3,5-diborolane (I) in CCl_4 during 24 h two boraxines (II, III) are formed [10, 11]:

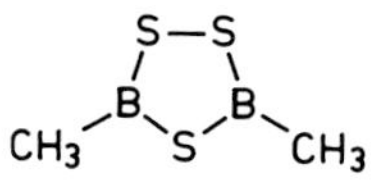

I

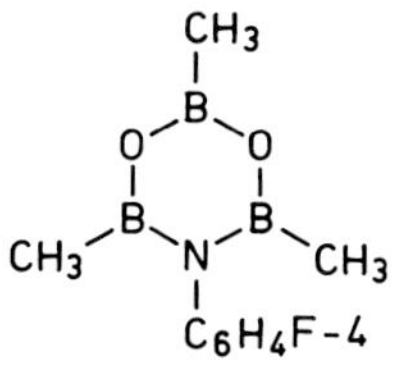

39%

II

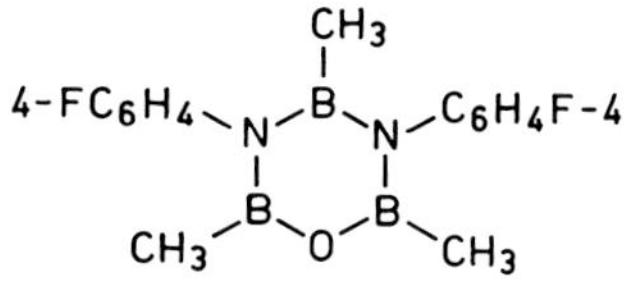

45%

III

References:

[1] Butt, G.; Davis, M.; Pang, Yuk Tau; Topsom, R. D.; Katritzky, A. R. (J. Chem. Soc. Perkin Trans. II **1974** 260/3).

[2] Dolenko, G. N.; Zibarev, A. V.; Krupoder, S. A.; Furin, G. G.; Yakobson, G. G. (Izv. Sibirsk. Otd. Akad. Nauk SSSR Ser. Khim. Nauk **1980** No. 6, pp. 74/7).

[3] Zibarev, A. V.; Furin, G. G.; Konchenko, S. N.; Voityuk, A. A. (J. Fluorine Chem. **35** [1987] 627/41 631/2).

[4] Zibarev, A. V.; Sultanov, A. Sh.; Furin, G. G.; Furlei, I. I. (Izv. Akad. Nauk SSSR Ser. Khim. **1986** 474/6; Bull. Acad. Sci. USSR Div. Chem. Sci. **35** [1986] 439/41).

[5] Zibarev, A. V.; Furin, G. G.; Korobeinicheva, I. K.; Fedotov, M. A.; Yakobson, G. G. (Izv. Akad. Nauk SSSR Ser. Khim. **1983** 2259/65; Bull. Acad. Sci. USSR Div. Chem. Sci. **32** [1983] 2039/44).

[6] Furin, G. G.; Fedotov, M. A.; Yakobson, G. G.; Zibarev, A. V. (J. Fluorine Chem. **28** [1985] 273/90, 276).

[7] Korobeinicheva, I. K.; Fugaeva, O. M.; Furin, G. G. (Izv. Akad. Nauk SSSR Ser. Khim. **36** [1987] 1766/9; Bull. Acad. Sci. [USSR] Div. Chem. Sci. **36** [1987] 1634/7).

[8] Bak, B.; Svanholt, H.; Larsen, C. (J. Mol. Struct. **36** [1977] 55/9).

[9] Meller, A.; Maringgele, W.; Fetzer, H. (Chem. Ber. **113** [1980] 1950/61, 1951/2).

[10] Meller, A.; Habben, C.; Noltemeyer, M.; Sheldrick, G. M. (Z. Naturforsch. **37 b** [1982] 1504/6).

[11] Meller, A.; Habben, C. (Monatsh. Chem. **113** [1982] 139/53).

2.6.22.2.2.2 N-Sulfinyl-(2, 3, or 4)-chloro-aniline

N-Sulfinyl-2-chloro-aniline, $O{=}S{=}NC_6H_4Cl$-2

The compound was synthesized by the reaction of 2-$ClC_6H_4NH_2$ with $SOCl_2$ in benzene. The yellow oil boils at 207°C/46 Torr [1], 130°C/18 Torr [2], and solidifies at −8°C [1].

^{1}H NMR (CCl_4/TMS): δ (in ppm) = 6.9 to 7.42 m (H3, H4, H5); 8.06 m (H6) [3].

UV (n-hexane): λ_{max} (in nm) = 238 (log ε = 3.84, flat maximum); 240 to 243 (log ε = 3.81, flat region); 317 (log ε = 3.90) [4].

Mass spectrum: The EI mass spectrum (recorded at 75 eV) is shown in a figure. Molecular peaks are observed at m/e = 173 and 175. Exact mass measurements of the fragments are given [5]. The negative-ion spectrum was also studied [6].

The dipole moment was determined in benzene at 25°C to be 1.97 D. A calculation with the Thomson vector scheme gave 2.90 D (free rotation about the N–C_{ar} bond between the N=S=O group and the phenyl ring is assumed). The difference $\Delta\mu$ between experimental and calculated value was interpreted as a measure of a hindrance of the rotations of the N=S=O group [7].

The compound was studied, together with other substituted N-sulfinyl-anilines, by TG, DTA, and DTG [8], gas chromatography [9, 10], thin-layer chromatography (R_f = 0.70 in benzene) [11].

It easily decomposes on treatment with H_2O [1]. It reacts with copper (mole ratio 1:1) to form the corresponding azo compound (110°C, 8 h in toluene, 31.9% yield; 140°C, 5 h in xylene, 39.8% yield) [19].

N-Sulfinyl-3-chloro-aniline, O=S=NC_6H_4Cl-3

The compound was obtained by the reaction of 3-$ClC_6H_4NH_2$ with $SOCl_2$ in benzene [12, 13, 14] (yield 98.3% [12]). The yellow substance melts at 40°C [13], 22°C [14, 15], 33°C (DTA) [8] and boils at 233°C [13], 126 to 127°C/16 Torr [14], 107°C/12 Torr [15]. The density at −190°C is 1.50 and at 60°C (liquid) 1.388 g/cm³ [16]. It is easily soluble in benzene, alcohol, ether, and petroleum ether [13].

^1H NMR spectrum (CCl_4/TMS): δ (in ppm) = 7.65 (s, H2), 7.21 (d, H4), 7.19 (t, H5), 7.62 (d, H6); ^3J (H4, H5) = 3 Hz, ^3J (H5, H6) = 6 Hz [3].

IR spectrum (organic solvents): ν_s(NSO) = 1169 cm^{-1} [15]; ν = 1313, 1302, 1291 cm^{-1} (without assignment) [15]. The integrated intensity of the ν_8 ring vibration (1583, 1565 cm^{-1}) is 2292 $L \cdot mol^{-1} \cdot cm^{-2}$ [17].

UV spectrum (heptane): λ_{max} (in nm) = 236 (log ε = 3.84), 312 (log ε = 4.00) (in the original paper given as $\bar{\nu}_{max}$ (in cm^{-1}) = 42370, 32000) [14].

The positive-ion mass spectrum (at 75 eV) exhibits the molecular peaks at m/e = 173 (63%) and 175 (24%), and 34 fragment peaks (of intensity greater than 2% of the base peak) [5]. The negative-ion mass spectrum (recorded at 70 eV) is shown in a figure [6].

Dielectric properties were determined at −190°C: deformation limit of the dielectric constant ε_∞ = 2.98; deformation polarization P_{def} = 59.5 cm³, refraction R_D = 55.3 cm³; atomic polarization P_A = 4 cm³ [16]; in benzene solution at 25°C: $P_{2\infty}$ = 115.5 cm³; R_D = 44.9 cm³ [18].

The dipole moment of the compound in benzene solution at 20°C is 1.81 D (1.75 D) [14], at 25°C it is 1.82 D [7], 1.86 D [18] and in dioxane at 20°C 1.72 D [14]. The dipole moment of the liquid phase is 1.95 D (at 25°C) from Onsager's formula [14] or 1.59 D (at 20°C) [16]. It was also calculated by the Guggenheim method to be 1.93 D [14] and from the Thomson vector scheme to be 1.98 D [7]. A calculation of the dipole moments of hypothetical cis and trans isomer gave μ_{cis} = 1.6 D and μ_{trans} = 2.5 D [18]. It was proposed from sterical considerations that the compound consists of 46% cis and 54% trans isomer, however, without further explanations [16].

The title compound was studied, together with other substituted N-sulfinyl-anilines, by TG, DTA, and DTG [8], gas chromatography [9, 10], thin-layer chromatography (R_f = 0.50 in benzene) [11].

The compound converts to 3-ClC_6H_4–N=S=N–C_6H_4Cl-3 when treated with t-$C_4H_9O^-K^+$ in benzene at 80°C for 5 h (yield 50%) [20].

N-Sulfinyl-4-chloro-aniline, $O{=}S{=}NC_6H_4Cl$-4

The title compound was obtained by the reaction of 4-$ClC_6H_4NH_2$ or 4-$ClC_6H_4NH_3Cl$ with $SOCl_2$ in benzene [13, 21] (yield 61%) [21]. It can also be prepared by the reaction of 4-$HOC_6H_4NH_2$ and $SOCl_2$ (without solvent), 71% yield [20]. The transsulfinylation reaction of 4-$ClC_6H_4NH_2$ with $O{=}S{=}NC_6H_4NO_2$-4 or $O{=}S{=}NC_6H_4NO_2$-2 in dry benzene yielded $O{=}S{=}NC_6H_4Cl$-4 in 83 and 72% yield, respectively [22]. In analogy to $O{=}S{=}NC_6H_4CH_3$-4, the compound was synthesized by sulfinylation with SO_2 gas of the corresponding lithiated amine (45% yield) [23] or N-trimethylsilylamine in the presence of a base (65 and 73% yield in two procedures) [24]. Recently, two new in situ sulfinylation reactions were performed. Treatment of 4-$ClC_6H_4NH_2$ in CH_2Cl_2 at 20°C for 20 min with N, N'-sulfinylbisimidazole (prepared from imidazole and $SOCl_2$) yielded the title compound in 69% yield. A higher yield of ~87% was obtained by using N-(chlorosulfinyl)-imidazole as sulfinylation reagent (prepared from N, N'-sulfinylbisimidazole and $SOCl_2$ or by treatment of imidazole with excess amount of $SOCl_2$) in CH_2Cl_2 at 20°C for 0.4 h [25]. It was also produced by the reaction of 4-$ClC_6H_4N_3$ with $H(C_6H_5)C{-}S(O){-}C(C_6H_5)H$ in boiling CH_2Cl_2 for 3 h (besides *trans*-stilbene) [26].

The compound forms weak yellow crystals which melt at 36°C [13], 35 to 36°C [20] and boil at 237°C [13], 110°C/17 Torr [14], 100 to 105°C/12 Torr [23], 114°C/16 Torr [9], 102 to 106°C/12 Torr [24], 60 to 63°C/1 Torr [25]. It is easily soluble in benzene, alcohol, ether, and petroleum ether [1]. The density D = 1.398 g/cm^3 at 20°C; in the liquid phase D = 1.347 g/cm^3 at 60°C [16].

^{1}H NMR (neat/(($CH_3)_3Si)_2O$ (= HMDS)): δ (in ppm) = 7.16 (d, H3, H5), 7.60 (d, H2, H6); ^3J(H2, H3) = ^3J(H5, H6) = 8 Hz [3]; (solvent ?/TMS): δ (in ppm) = 7.27 and 7.77 (C_6H_4, ca. AB_q, J = 9 Hz) [21]; (CCl_4/TMS): δ (in ppm) = 7.40, 7.55, 7.87, 8.02 for the ring protons [28].

^{13}C NMR ($CDCl_3$/TMS): δ (in ppm) = 141.2 (C1), 128.4 (C2), 129.5 (C3), 136.2 (C4) [29].

IR (organic solvent): ν (in cm^{-1}) = 1300, 1277 ν_{as}(NSO) and ν(CN); 1162 ν_s(NSO) [15]; (neat?): ν (in cm^{-1}) = 1292, 1166 ν(NSO) [23]; (neat or KBr): ν (in cm^{-1}) = 1291, 1170 ν(NSO) [24]. The integrated intensity of the ν_8 (C=C, arom.) vibration (1585 and 1564 cm^{-1}) was determined to be 1389 L·mol^{-1}·cm^{-2}. The electronic intramolecular interaction was discussed [17].

The HeI and HeII photoelectron spectra of $O{=}S{=}NC_6H_4Cl$-4 were recorded and assigned on the basis of an CNDO LCAO MO calculation [27].

Ionization energies E_i in eV:

E_i (obs)	9.10	9.99	10.85	11.29	—	11.65	12.71	13.37
E_i (calc)	9.62	10.37	11.46	12.70	11.98	—	—	—
assignment	"b_1"	"a_2"	σ_1	σ_2	π_3	Cl	σ_3	σ_4

The HeI spectrum is shown in a figure [27]. The charge densities ϱ (in electron units) were calculated by CNDO/2 (a positive sign indicates electron deficiency): $\varrho_C = 0.207$, $\varrho_N = -0.051$, $\varrho_S = 0.270$, $\varrho_O = -0.315$ [2].

UV (cyclohexane): λ_{max} (in nm) = 237 (log ε = 3.83, with an inflexion at 223 nm (log ε = 3.64)), 328 (log ε = 4.09) [4].

Dielectric measurements were performed at 20°C: deformation limit of the dielectric constant $\varepsilon_\infty = 4.28$; deformation polarization $P_{def} = 101$ cm^3; refraction $R_D = 55.3$ cm^3; atomic polarization $P_A = 46$ cm^3 [16]; in benzene solution $P_{2\infty} = 78.6$ cm^3 and $R_D = 44.9$ cm^3 [18]. The dipole moment was determined in benzene solution at 25°C to 1.29 D [18], 1.30 D [7].

The positive-ion mass spectrum was recorded at 75 eV. Two molecular peaks at m/e = 173 (rel. int. 54%) and 175 (20%) are observed together with 38 fragment peaks (all greater than 2%

of the base peak) [5]. The two M^+ peaks m/e = 173 and 175 are also cited in [21]. The negative-ion mass spectrum was recorded at 70 eV. It is very simple and the molecular peak is the base peak [6].

The compound was studied, together with other substituted N-sulfinyl-anilines, by TG, DTA, and DTG [8], gas chromatography [9, 10], thin-layer chromatography (R_f = 0.32 in benzene) [11].

The compound is slowly hydrolyzed by cold H_2O, rapidly by hot H_2O or alkali to yield $4\text{-}ClC_6H_4NH_2$ [13]. When treated with $C_2H_5O^-Na^+$ or $NaNH_2$ in benzene at 80°C or with NaH in DMF or $t\text{-}C_4H_9O^-K^+$ in THF at 20°C it converts to $4\text{-}ClC_6H_4\text{–}N{=}S{=}N\text{–}C_6H_4Cl\text{-}4$ (yield 60, 74, 64, and 64%, respectively) [20]. With nitrosyl chloride it produces in CCl_4–ether at −70°C the corresponding diazonium salt $4\text{-}ClC_6H_4N_2^+Cl^-$ and SO_2 [30]. Treatment with S_2Cl_2 at about 60 to 70°C on a water bath for 14 h with occasional shaking gives the "Herz compound" 6-chloro-1,2,3-benzodithiazol-2-ium chloride [31]:

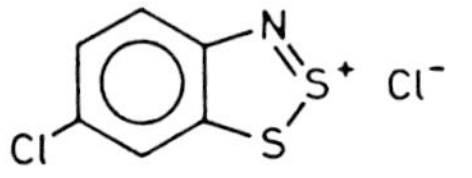

Reaction of the title compound with copper powder in toluene at 110°C for 8 h yielded the corresponding azo compound (yield 39.4%). The reaction of $O{=}S{=}NC_6H_5$, $O{=}S{=}NC_6H_4Cl\text{-}4$, and copper (mole ratio 1:1:2) without solvent at 130°C for 8 h, gave a mixture of compounds: $C_6H_5\text{–}N{=}N\text{–}C_6H_5$ (trace), $4\text{-}ClC_6H_4\text{–}N{=}N\text{–}C_6H_5$ (22% yield), and $4\text{-}ClC_6H_4\text{–}N{=}N\text{–}C_6H_4Cl\text{-}4$ (9% yield) [19].

References:

[1] Michaelis, A.; Hümme, C. (J. Liebigs. Ann. Chem. **274** [1893] 173/266, 217/9).
[2] Glass, W. K.; King, I. J.; Shiels, A. (Inorg. Chim. Acta **25** [1977] 157/8).
[3] Mirek, J.; Czerwiec, Z. (Zesz. Nauk Uniw. Jagiel. Proc. Chem. No. 18 [1973] 183/6; C.A. **79** [1973] No. 145570).
[4] Leandri, G.; Mangini, A. (Spectrochim. Acta **1959** 421/31; Ric. Sci. **27** [1957] 1920/5).
[5] Bowie, J. H.; Larsson, F. C. V.; Schroll, G.; Lawesson, S. O.; Cooks, R. G. (Tetrahedron **23** [1967] 3743/52).
[6] Bowie, J. H.; Duus, F.; Lawesson, S. O.; Larsson, F. C. V.; Madsen, J. Oe. (Australian J. Chem. **22** [1969] 153/9).
[7] Waclawek, W.; Puchala, C.; Czerwiec, Z. (Bull. Acad. Polon. Sci. Ser. Sci. Chim. **26** [1978] 337/41).
[8] Czerwiec, Z.; Pielichowski, J. (J. Therm. Anal. **7** [1975] 465/8).
[9] Czerwiec, Z.; Mirek, J. (Zesz. Nauk. Uniw. Jagiel. Proc. Chem. No. 14 [1969] 103/6; C.A. **72** [1970] No. 28195).
[10] Czerwiec, Z.; Markovski, J. (Chem. Anal. [Warsaw] **20** [1975] 213/5).

[11] Czerwiec, Z. (Chem. Anal. [Warsaw] **21** [1976] 751/3).
[12] Wichterle, O.; Roček, J. (Chem. Listy **47** [1953] 1768/80; Collection Czech. Chem. Commun. **19** [1954] 282/96).
[13] Michaelis, A.; Hümme, C. (Chem. Ber. **24** [1891] 745/57, 754).
[14] Kresze, G.; Smalla, H. (Chem. Ber. **92** [1959] 1042/8).
[15] Kresze, G.; Maschke, A. (Chem. Ber. **94** [1961] 450/6).
[16] Egorov, Yu. P.; Borovikov, Yu. Ya.; Fedyuk, G. S.; Levchenko, E. S. (Zh. Obshch. Khim. **43** [1973] 2470/6; J. Gen. Chem. [USSR] **43** [1973] 2456/61).
[17] Butt, G.; Davis, M.; Pang, Yuk Tau; Topsom, R. D.; Katritzky, A. R. (J. Chem. Soc. Perkin Trans. II **1974** 260/3).
[18] Jannelli, L.; Lamanna, U.; Lumbroso, H. (Bull. Soc. Chim. France **1966** 3626/30).

[19] Minami, T.; Agawa, T. (Tetrahedron Letters **1968** 4109/10).
[20] Hörhold, H. H.; Beck, J. (J. Prakt. Chem. **311** [1969] 621/9).

[21] Al-Ajely, M. S.; Al-Rawi, J. M. A.; Elvidge, J. A. (J. Chem. Soc. Perkin Trans. I **1982** 1575/8).
[22] Carpanelli, C.; Leandri, G. (Ann. Chim. [Rome] **50** [1960] 147/55).
[23] Sakai, S.; Fujinami, T.; Komizo, K. (J. Org. Chem. **40** [1975] 3291/2).
[24] Porskamp, P. A. T. W.; Zwanenburg, B. (Synthesis **1981** 368/9).
[25] Kim, Y. H.; Shin, J. M. (Tetrahedron Letters **1985** 3821/4).
[26] Bonini, B. F.; Maccagnani, G.; Mazzanti, G. (Tetrahedron Letters **1977** 1185/6).
[27] Louwen, J. N.; van Dam, H.; Stufkens, D. J.; Oskam, A.; Jaffé, H. H. (J. Electron. Spectrosc. Relat. Phenom. **26** [1982] 235/46).
[28] Yamada, F.; Nishiyama, T.; Emi, T. (Bull. Chem. Soc. Japan **45** [1972] 271/4).
[29] Kresze, G.; Berger, M.; Claus, P. K.; Rieder, W. (Org. Magn. Resonance **8** [1976] 170/2).
[30] Kobayashi, M.; Honda, K. (Bull. Chem. Soc. Japan **39** [1966] 1778/82).

[31] Arsenescu, N. (Rev. Roumaine Chim. **12** [1967] 427/8).

2.6.22.2.2.3 N-Sulfinyl-4-bromo-aniline, $O{=}S{=}NC_6H_4Br$-4

The title compound was prepared by heating a solution of $H_2NC_6H_4Br$-4 in benzene with $SOCl_2$ [1]. It is unstable and reverts to $H_2NC_6H_4Br$-4 rapidly at ambient temperature [2].

X-ray diffraction data show that the compound crystallizes with a monoclinic cell, space group $P2_1/n–C^5_{2h}$ (No. 14) with a = 7.752(1), b = 6.119(1), c = 15.876(1) Å, β = 90.10(2)°, V = 753.03 Å^3, Z = 4; R = 0.0602, R_W = 0.0542. The compound is fully planar and has Z structure, see **Fig. 13** [2].

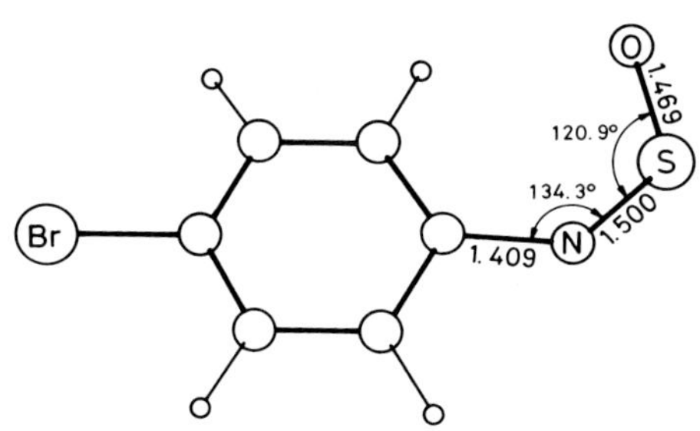

Fig. 13. Molecular structure of $O{=}S{=}NC_6H_4Br$-4
(selected bond lengths (in Å) and bond angles (in °)) [2].

The substance forms yellow crystals which melt at 60 to 61 [1], or 65 to 66°C [3]. The density was determined to be 2.18 at 20°C and 1.97 g/cm^3 at 81°C (liquid phase) [4].

The HeI and HeII photoelectron spectra were recorded and assigned on the basis of a CNDO/S LCAO MO calculation [5].

Ionization energies in eV:

E_i (obs)	8.91	9.90	10.57	10.82	—	10.96	12.71	12.92
E_i (calc)	9.85	10.36	11.66	12.78	12.17	—	—	—
assignment	"b_1"	"a_2"	σ_1	σ_2	π_3	Br	σ_3	σ_4

The dipole moment was determined in CCl_4 to be 1.53 [3], in the liquid phase at 81°C to 0.45 D [4].

^{1}H NMR (CCl_4/TMS): δ (in ppm) = 7.78 (H2), 7.46 (H3, H5), and 7.67 (H6) [3].

IR (CCl_4): 10 absorptions in the range 1350 to 1000 cm^{-1} [3]; other workers measured a band at 1166 cm^{-1}, ν_s(NSO) [6]. The integrated intensity of the ν_8 (ring) IR absorption (1589, 1575, 1561 cm^{-1}) was estimated to be 1609 $L \cdot mol^{-1} \cdot cm^{-2}$ [7].

UV (ethyl ether): λ_{max} (log ε) = 239 nm (3.83), 325 nm (4.02) [8]; (cyclohexane): λ_{max} (log ε) = 239 nm (3.81; flat maximum with an inflexion at 227 nm, log ε = 3.72), 243 nm (3.78; flat region), 330 nm (4.13) [9].

Dielectric properties were determined at 20°C (solid phase): deformation limit of the dielectric constant $\varepsilon_\infty = 3.44$; deformation polarization $P_{def} = 62.0$ cm^3; refraction $R_D = 59.0$ cm^3, atom polarization $P_A = 3$ cm^3; at 81°C (liquid phase): dielectric constant $\varepsilon = 3.46$ [4].

References:

[1] Michaelis, A.; Hümme, C. (Liebigs Ann. Chem. **274** [1893] 173/266, 219/20).
[2] Butler, R. N.; Duffy, J. P.; McArdle, P.; Cunningham, D.; O'Halloran, G. A. (J. Chem. Soc. Chem. Commun. **1989** 1210/1).
[3] van Woerden, H. F.; Bijl-Vlieger, S. H. (Recl. Trav. Chim. **93** [1974] 85/90).
[4] Egorov, Yu.-P.; Borovikov, Yu. Ya.; Fedyuk, G. S.; Levchenko, E. S. (Zh. Obshch. Khim. **43** [1973] 2470/6; J. Gen. Chem. [USSR] **43** [1973] 2456/61).
[5] Louwen, J. N.; van Dam, H.; Stufkens, D. J.; Oskam, A. (J. Electron. Spectrosc. Relat. Phenom. **26** [1982] 235/46).
[6] Kresze, K.; Maschke, A. (Chem. Ber. **94** [1961] 450/6).
[7] Butt, G.; Davis, M.; Pang, Yuk Tau; Topsom, R. D.; Katritzky, A. R. (J. Chem. Soc. Perkin Trans. II **1974** 260/3).
[8] Smith, W. T., Jr.; Trimnell, D.; Grinninger, L. D. (J. Org. Chem. **24** [1959] 664/6).
[9] Leandri, G.; Mangini, A. (Spectrochim. Acta **15** [1959] 421/31; Ric. Sci. **27** [1957] 1920/5).

2.6.22.2.2.4 2-Sulfinyl-amino-phenol, $O{=}S{=}NC_6H_4OH$-2

The compound was obtained by the reaction of $H_2NC_6H_4OH$-2 with $O{=}S{=}NSO_2C_6H_4CH_3$-4 (mole ratio 1:1) in dry ether (30 min stirring). After working up, the substance was recrystallized from petroleum ether; orange-yellow prisms of melting point 39 to 40°C (84% yield).

^{1}H NMR spectrum ($CDCl_3$/TMS): δ (in ppm) = 8.25 (Ar-H6), 7 (m, Ar-H3,4,5), 6.75 (OH).

IR spectrum (CCl_4): ν (in cm^{-1}) = 3505 ν(OH); 1250 ν_{as}(N=S=O) and ν(C–N); 1165, 1145 ν_s(N=S=O) and ν(C–O).

Mass spectrum: m/e = 155 M^+; 138 M^+ − OH; 106 M^+ − SOH; 78 M^+ − SOH − CO.

Reference:

Wucherpfennig, W. (Chem. Ber. **101** [1968] 371/2).

2.6.22.2.2.5 N-Sulfinyl-(2,3, or 4)-methoxy-aniline

N-Sulfinyl-2-methoxy-aniline, O=S=NC$_6$H$_4$OCH$_3$-2

The compound was synthesized by the reaction of 2-methoxy-aniline [1, 2, 3] or the corresponding hydrochloride [4] with $SOCl_2$ (yield 70% [1], 83% [4]). It was also prepared by sulfinylation of 2-$CH_3OC_6H_4NHLi$ (prepared in situ) with SO_2 gas (the intermediate 2-$CH_3OC_6H_4N$-S(O)OLi was lithiated again with butyllithium and subsequently allowed to react with a second mole of SO_2). After refluxing, the title compound was isolated by distillation (yield 60%) [5].

The red-brown liquid boils at 203°C/65 Torr [3] (distillation at atmospheric pressure lead to decomposition [3]), 94°C/3 Torr [2], 95°C/1.1 Torr [1], 75 to 80°C/0.2 Torr [5], and 244°C [4]. Melting point is 30°C (from hexane) [4].

UV spectrum (n-hexane): λ_{max} (in nm) = 240 to 245 (log ε = 3.76, flat region), 312 (log ε = 3.71), 364 (log ε = 3.77); the spectrum is shown in a figure [2]; (dioxane): λ_{max} (in nm) = 243 (log ε = 3.94), 313 (log ε = 3.84), 370 (log ε = 3.90) [1].

^1H NMR spectrum (solvent?/TMS): δ (in ppm) = 4.2 (s, CH_3); 7.2 to 8.5 (m, C_6H_4) [4].

IR spectrum (CCl_4): ν(NSO) = 1298, 1289, 1162 cm^{-1}; ν(C=C, arom.) = 1590, 1489 cm^{-1} [1]; (solvent?): ν(NSO) = 1286, 1156 cm^{-1} [5].

Mass spectrum: The EI mass spectrum was recorded at 75 eV; it is shown in a figure (m/e = 169 M^+). Exact mass measurement gave the following result for the fragments: m/e (rel. int. in %) = 66 $C_5H_5^+$ (50) and $C_4H_4N^+$ (50), 78 $C_5H_4N^+$, 93 $C_6H_7N^+$ (70), and $C_6H_5O^+$ (30), 98 $C_4H_4NS^+$, 106 $C_6H_4NO^+$, 110 $C_5H_4NS^+$, 120 $C_7H_6NO^+$, 121 $C_7H_7NO^+$, 138 $C_6H_4NOS^+$, 139 $C_6H_5NOS^+$ [6]. The very simple negative-ion mass spectrum was recorded at 70 eV; the molecular ion is the base peak [7]. The +E spectrum was also measured. Rearrangement $M^{+\bullet} \rightarrow$ [M − HSO$^\bullet$]$^+$ was observed [8].

The substance decomposes when refluxed in H_2O for a long time. It is easily decomposed by NaOH yielding 2-methoxy-aniline [3]. Treatment with t-$C_4H_9O^-K^+$ in benzene at 80°C for 5 to 6 h yields the corresponding sulfur diimide (yield 52%) [9]. 6-Chloro-4-methoxy-1,2,3-benzodithiazol-2-ium chloride (I) is obtained from the reaction with S_2Cl_2 [10].

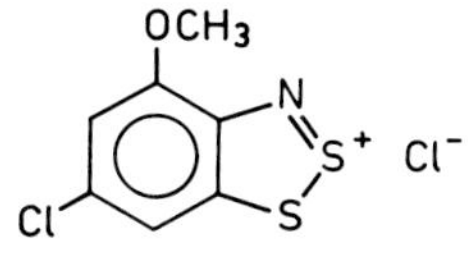

I

N-Sulfinyl-3-methoxy-aniline, O=S=NC$_6$H$_4$OCH$_3$-3

The title compound was obtained by the method of Michaelis [2, 11] by reacting $SOCl_2$ with 3-$CH_3OC_6H_4NH_2$. It was also prepared by sulfinylation with SO_2 gas starting from 3-$CH_3OC_6H_4$-NHLi (prepared in situ) in THF (yield 50%) [5] and also from 3-$CH_3OC_6H_4NH$–Si$(CH_3)_3$ in THF–hexane using n-butyllithium and pyridine as a base, yield 64 and 70%, respectively [12].

The substance boils at 98°C/2 Torr [13], 133 to 134°C/17 Torr [2], 138 to 139°C/25 Torr [14], 66 to 72°C/0.2 Torr [5], and 69 to 71°C/0.2 Torr [12].

IR: ν (in cm^{-1}) = 1324; 1303 and 1293 ν_{as}(NSO) and ν(CN); 1160 ν_s(NSO) (organic solvents, not specified) [13]; 1292, 1154 ν(NSO) (medium not given) [5]; 1288, 1148 ν(NSO) (neat or KBr) [12]. The integrated intensity of the ν_8(C=C, arom.) vibration (1595, 1589, 1575 cm^{-1}) is 5608 L·mol^{-1}·cm^{-2} [14].

UV (n-hexane): λ_{max} (in nm) = 239 (log ε = 3.68, inflexion), 247 (log ε = 3.80), 315 (log ε = 3.92), 352 (log ε = 3.65); the spectrum is shown in a figure [2].

The EI mass spectrum was recorded at 75 eV. It is shown in a figure (m/e = 169 M^+) [6]. The negative-ion mass spectrum was recorded at 70 eV and is also shown in a figure. It is very simple; the molecular ion is the base peak [7].

N-Sulfinyl-4-methoxy-aniline, $O{=}S{=}NC_6H_4OCH_3$-4

The title compound was obtained by the general method of Michaelis by the reaction of 4-$CH_3OC_6H_4NH_2$ [1, 15, 16, 17] (yield 93.6% [15], 77% [1]) or 4-$CH_3OC_6H_4NH_3^+Cl^-$ (yield 75%) [4] in benzene with $SOCl_2$. The transsulfinylation reaction of 4-$CH_3OC_6H_4NH_2$ in dry benzene with $O{=}S{=}NC_6H_4NO_2$-4 or $O{=}S{=}NC_6H_4NO_2$-2 yielded $O{=}S{=}NC_6H_4OCH_3$-4 in 83 and 88% yield, respectively [18]. It was also produced using N-(chlorosulfinyl)imidazole (prepared by a direct treatment of N, N′-sulfinylbisimidazole with $SOCl_2$ or by the treatment of imidazole with excess amount of $SOCl_2$) as the sulfinylation agent. A solution of N-(chlorosulfinyl)imidazole was added to 4-$CH_3OC_6H_4NH_2$ in CH_2Cl_2 and the mixture was stirred for 0.5 h at 20°C (yield 86%, purified product) [19]. The reaction of N-lithio-4-methoxy-aniline (prepared from 4-methoxy-aniline and butyllithium) with SO_2 in dry THF gave 4-$CH_3OC_6H_4$NHS(O)OLi which was again lithiated with butyllithium and subsequently treated with a second mole of SO_2 gas, yielding the title compound after refluxing (yield 47%) [5]. It was also synthesized in 45% yield by the reaction of 4-$CH_3OC_6H_4$–N_3 with $HC(C_6H_5)$–S(O)–$C(C_6H_5)H$ in refluxing CH_2Cl_2 for 3 h (besides *trans*-stilbene) [20].

The substance melts at 26 [16], 24 to 25 [19], 24 [1, 15], 34°C (from CCl_4) [4] and boils at 134°C/12 Torr [16], 80 to 84°C/0.5 Torr [5], 80 to 84°C/1 Torr [19], 132°C/15 Torr [15], 144°C/15 Torr [2]. The density D at −190°C is 1.26, at 40°C = 1.167 [21], and at 25°C = 1.3818 g/cm^3 [16]. For the refraction index at 25°C a value of n_D = 1.6300 is given [16].

^{1}H NMR (CCl_4/TMS): δ (in ppm) = 7.83 (H2); 6.81 (H3, H5) [24]; 8.20, 7.20 (C_6H_4, ca. AB_q, J = 8 Hz); 4.20 (s, OCH_3) [4]; ($CDCl_3$/TMS) at −20°C: 7.93 (H2); 6.89 (H3, H5); 3.85 (OCH_3); J = 8 Hz [25].

^{13}C NMR ($CDCl_3$/TMS): δ (in ppm) = 137.4 (C1); 129.5 (C2); 114.2 (C3); 161.0 (C4); 55.4 ($\underline{C}H_3O$) [26].

^{15}N NMR (30°C, neat or 3 M in DMSO/external 1.0 M $H^{15}NO_3$ (98% enriched) in D_2O): δ (in ppm) = −58.2 or −58.8 [11] (referenced to CH_3NO_2: δ = −64.4 or −65.0, respectively).

IR: ν (in cm^{-1}) = 1311, 1296 ν_{as}(NSO) and ν(CN); 1148 ν_s(NSO) (organic solvents, not specified) [13]; 1308, 1296, 1283 ν(NSO); 1601, 1504 ν(C=C, arom.) (in CCl_4) [1]; 1305, 1157 ν(NSO) (medium not given) [5]. The integrated intensity of the ν_8(C=C, arom.) vibration (1596, 1569 cm^{-1}) was determined to be 8172 L·mol^{-1}·cm^{-2}. Electronic interactions were discussed [14].

The SK_β X-ray spectrum shows peaks at 2471.3, 2469.1, and 2466.0 eV (ΔSK_α = 0.38 ± 0.02 eV) [22]. For an attempt to identify the compound by this spectrum, see [23].

UV (cyclohexane): λ_{max} (in nm) = 247 (log ε = 3.86, with an inflexion at 228, log ε = 3.57), 254 (log ε = 3.78, shoulder), 356 (log ε = 4.22) [2]; (dioxane): λ_{max} (in nm) = 247 (log ε = 3.94), 356 (log ε = 4.22) [1]; (heptane): $\bar{\nu}_{max}$ (in cm^{-1}) = 40400 (log ε = 3.88), 27970 (log ε = 4.20) [16]; (converted to nm: λ_{max} = 248 and 358 nm).

Dielectric properties of the compound were measured at −190°C: deformation limit of dielectric constant ε_∞ = 3.22; deformation polarization P_{def} = 76.5 cm^3; refraction R_D = 53.8 cm^3; atomic polarization P_A = 22.7 cm^3. The electrical conductivity was determined at 20, 40, and 60°C to be 1.3, 3.0, and 5.1 × 10^{-6} Ω^{-1}·cm^{-1}, respectively [21]. The dipole moment μ in benzene at 20°C is 2.90 D (calculated by the Guggenheim method: 1.73 D) [16]. The dipole moment without solvent is 2.29 D in the temperature range 20 to 60°C [21].

The El mass spectrum was recorded at 75 eV; it is shown in a figure; m/e = 169 M^+; the spectrum is different from those of $O{=}S{=}NC_6H_4OCH_3$-2 and $O{=}S{=}NC_6H_4OCH_3$-3. Exact mass measurements of the fragments were performed: m/e (rel. int. in %) = 64 C_5H_4 (50), C_4H_2N (30), and SO_2 (20); 66 C_4H_4N (50) and C_5H_6 (50); 69 C_3HS; 78 C_5H_4N [6]. The negative-ion mass spectrum was recorded at 70 eV; it is very simple. The molecular ion is the base peak [7]. The +E spectrum was also studied; the rearrangement ion $[M-CH_3-2CO]^+$ has a very small abundance [8].

The reaction with copper leads to the formation of the corresponding azo compound 4-$CH_3OC_6H_4$–N=N–$C_6H_4OCH_3$-4 (at 140°C for 8 h, without solvent 8.3%, in xylene 11.6% yield) [27]. NOCl reacts with $O{=}S{=}NC_6H_4OCH_3$-4 in CCl_4–ether at −70°C to produce the corresponding diazonium salt 4-$CH_3OC_6H_4N_2^+Cl^-$ and SO_2 [28]. S_2Cl_2 reacts with the title compound to yield 6-methoxy-1,2,3-benzodithiazol-2-ium chloride (I) [10].

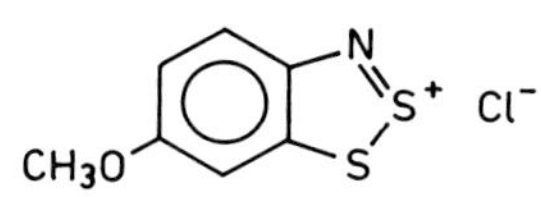

I

Treatment of the title compound with $C_2H_5O^-Na^+$ in di-isopropyl ether at 29°C for 1 h yields the corresponding sulfur diimide (yield 75%) [9, 29].

References:

[1] Uher, M.; Jendrichovsky, J.; Novomestska, D. (Zb. Pr. Chemickotechnol. Fak. SVST **1972/74** 93/6; C.A. **82** [1974] No. 97771).
[2] Leandri, G.; Mangini, A. (Spectrochim. Acta **1959** 421/31; Ric. Sci. **27** [1957] 1920/5).
[3] Michaelis, A.; Haegele, E. (Liebigs. Ann. Chem. **274** [1893] 243/7).
[4] Al-Ajely, M. S.; Al-Rawi, J. M. A.; Elvidge, J. A. (J. Chem. Soc. Perkin Trans. I **1982** 1575/8).
[5] Sakai, S.; Fujinami, T.; Komizo, K. (J. Org. Chem. **40** [1975] 3291/2).
[6] Bowie, J. H.; Larsson, F. C. V.; Schroll, G.; Lawesson, S.-O.; Cooks, R. G. (Tetrahedron **23** [1967] 3743/52).
[7] Bowie, J. H.; Duus, F.; Lawesson, S.-O.; Larsson, F. C. V.; Madsen, J. Oe. (Australian J. Chem. **22** [1969] 153/9).
[8] Bowie, J. H.; White, P. Y.; Wilson, J. G.; Larsson, F. C. V.; Lawesson, S.-O.; Madsen, J. Ø.; Nolde, C.; Schroll, G. (Org. Mass. Spectrom. **12** [1977] 191/5).
[9] Hörhold, H.-H.; Beck, J. (J. Prakt. Chem. [4] **311** [1969] 621/9).
[10] Maior, O.; Arsenescu, N. (Analele Univ. Bucuresti Ser. Stiint. Nat. Chim. **16** [1967] 101/6; C.A. **71** [1969] No. 3331).

[11] Yavari, I.; Staral, J. S.; Roberts, J. D. (Org. Magn. Resonance **12** [1979] 340/2).
[12] Porskamp, P. A. T. W.; Zwanenburg, B. (Synthesis **1981** 368/9).
[13] Kresze, G.; Maschke, A. (Chem. Ber. **94** [1961] 450/6).
[14] Butt, G.; Davis, M.; Pang, Yuk Tau; Topsom, R. D.; Katritzky, A. R. (J. Chem. Soc. Perkin Trans. II **1974** 260/3).
[15] Wichterle, O.; Roček, J. (Chem. Listy **47** [1953] 1768/80; Collection Czech. Chem. Commun. **19** [1954] 282/96).
[16] Kresze, G.; Smalla, H. (Chem. Ber. **92** [1959] 1042/8).
[17] Sugihara, H.; Sasaki, M.; Kawamatsu, Y.; Miromoto, H. (Liebigs Ann. Chem. **763** [1972] 121/7).

[18] Carpanelli, C.; Leandri, G. (Ann. Chim. [Rome] **50** [1960] 147/55).
[19] Kim, Y. H.; Shin, J. M. (Tetrahedron Letters **26** [1985] 3821/4).
[20] Bonini, B. F.; Maccagnani, G.; Mazzanti, G. (Tetrahedron Letters **1977** 1185/6).

[21] Egorov, Yu. P.; Borovikov, Yu. Ya.; Fedyuk, G. S.; Levchenko, E. S. (Zh. Obshch. Khim. **43** [1973] 2470/6; J. Gen. Chem. [USSR] **43** [1973] 2456/61).
[22] Dolenko, G. N.; Zibarev, A. V.; Krupoder, S. A.; Furin, G. G.; Yakobson, G. G. (Izv. Sibirsk. Otd. Akad. Nauk SSSR Ser. Khim. Nauk **1980** No. 6, pp. 74/7).
[23] Treiger, B. A.; Mazalov, L. N.; Dolenko, G. N.; Guzhavina, T. I. (Comput. Enhanced Spectrosc. **3** [1986] 153/63).
[24] van Woerden, H. F.; Bijl-Vlieger, S. H. (Recl. Trav. Chim. **93** [1974] 85/90).
[25] Kuyper, J.; Vrieze, K. (J. Organometal. Chem. **86** [1975] 127/38).
[26] Kresze, G.; Berger, M.; Claus, P. K.; Rieder, W. (Org. Magn. Resonance **8** [1976] 170/2).
[27] Minami, T.; Agawa, T. (Tetrahedron Letters **38** [1968] 4109/10).
[28] Kobayashi, M.; Honda, K. (Bull. Chem. Soc. Japan **39** [1966] 1778/82).
[29] Hörhold, H.-H.; Floßmann, K. D. (Z. Chem. [Leipzig] **7** [1967] 345/6).

2.6.22.2.2.6 N-Sulfinyl-1,(2, 3, or 4)-phenylenediamine and N′-Sulfinyl-N,N-dimethyl-1,4-phenylenediamine

N-Sulfinyl-1,2-phenylenediamine, $O{=}S{=}NC_6H_4NH_2$-2

The title compound formed on treatment of o-phenylenediamine with N-sulfinyl-p-toluenesulfonamide (mole ratio 1:1) in dry benzene (stirring for 3 h, under cooling). The mixture was filtered and concentrated by evaporation below 25°C (bath temperature) in vacuum. The substance is stable for some time only in cooled solution. The red oily residue shows the following ^{1}H NMR and IR data:

^{1}H NMR ($CDCl_3$/TMS): δ (in ppm) = 8.32 (Ar-H6); 6.7 to 7.1 (m, Ar-H3,4,5); 3.95 (NH_2).

IR (CCl_4): ν (in cm^{-1}) = 3520, 3420 $\nu(NH_2)$; 1615 $\delta(NH_2)$; 1265 ν_{as}(NSO) and ν(CN); 1140 ν_s(NSO).

On standing at room temperature the compound becomes hot, decolorizes and liberates SO_2. Work-up yields p-toluenesulfonamide (96%), o-phenylenediamine (42%), and 2,1,3-benzothiadiazole (49%) [1].

N-Sulfinyl-1,3-phenylenediamine, $O{=}S{=}NC_6H_4NH_2$-3

The compound formed on treatment of 3-$H_2N{-}C_6H_4{-}NH_2$ with 4-$O{=}S{=}NC_6H_4NO_2$ (mole ratio 1:1) in anhydrous benzene; yield 18%. The red oil boils at 109 to 110°C [2].

N-Sulfinyl-1,4-phenylenediamine, $O{=}S{=}NC_6H_4NH_2$-4

The compound was obtained during the synthesis of $O{=}S{=}NC_6H_4N{=}S{=}O$-4 from hydrochloric 1,4-phenylenediamine and $SOCl_2$. The workers observed that before this compound crystallized in yellow prisms, a red-colored substance separated which was identified to be the title compound based on elemental analysis [3].

The substance was also prepared by transsulfinylation of 4-$H_2N{-}C_6H_4{-}NH_2$ with $O{=}S{=}N$-$C_6H_4NO_2$-4 (mole ratio 1:1) in anhydrous benzene; yield 76 % [2]. The red crystals melt at 67°C [3] or 67 to 68°C [2] and boil at 165 to 167°C/13 Torr [2]. They decompose gradually on treatment with H_2O and immediately with diluted acids [3].

References:

[1] Wucherpfennig, W. (Chem. Ber. **101** [1968] 371/2).
[2] Carpanelli, C.; Leandri, G. (Ann. Chim. [Rome] **50** [1960] 147/55).
[3] Michaelis, A.; Petou, K. (Ber. Deut. Chem. Ges. **31** [1898] 984/97, 995).

N'-Sulfinyl-N,N-dimethyl-1,4-phenylenediamine, $O{=}S{=}NC_6H_4N(CH_3)_2$-4

The title compound was obtained by the reaction of 4-$(CH_3)_2NC_6H_4NH_2$ with $SOCl_2$ in benzene. After refluxing the mixture until the reaction was finished the dark red-brown solution was filtered and most of the solvent evaporated. The compound crystallized on standing [1]. In one paper, the use of triethylamine as neutralizing agent was recommended [2]. The compound was prepared by transsulfinylation of 4-$(CH_3)_2NC_6H_4NH_2$ with $O{=}S{=}NC_6H_5$ in ether [3] or with $O{=}S{=}NSO_2C_6H_4CH_3$-4 in benzene [4, 5]. It formed also along with 4,4'-bis(dimethylamino)-azobenzene (4-$((CH_3)_2NC_6H_4N)_2$) and 4,4'-bis(dimethylamino)azoxybenzene (4-$((CH_3)_2NC_6H_4$-$N(O){=}NC_6H_4N(CH_3)_2$)-4') by refluxing a solution of p-dimethylaminonitrosobenzene (4-$(CH_3)_2N$-$C_6H_4N{=}O$) and di-n-propylsulfoxylate ($S(OC_3H_7\text{-}n)_2$) in benzene under nitrogen atmosphere (yield 45%) [6]. Purification was performed by distillation (b.p. 110 to 115°C/0.05 Torr [5], 150 to 151°C/4 Torr [2]; 115°C/2 Torr [4]), by recrystallization from a mixture of benzene and petroleum ether [1], from CCl_4 [5], from ether [2], from hexane [3], or by chromatography on silica gel (using benzene as eluent) [6].

The compound forms red crystals [1, 2, 6] or red-orange plates [3], which melt at 72°C [1, 4, 5, 6], 68 to 70°C [2], or 70 to 73°C [3]. The density at 25°C is 1.19 g/cm³ [7].

^{1}H NMR spectrum (δ in ppm, TMS):

solvent	δ(H2, H6)	δ(H3, H5)	δ($N(CH_3)_2$)	Ref.
$CDCl_3$	7.88	6.60	3.01	[5]
CCl_4	A_2B_2 pattern with components centered at 7.73 and 6.44		2.93	[3]
CCl_4	7.78	6.52	3.06(s)	[6]

IR spectrum (organic solvents): ν (in cm^{-1}) = 1309, 1292, 1272 ν_{as}(NSO) or ν(CN); 1137 ν_s(NSO) [8]; (Nujol): 1140 cm^{-1} ν(SO) [6]. The integrated intensity of the ν_8(C=C, arom.) vibration (1599 cm^{-1}) is 22831 $L \cdot mol^{-1} \cdot cm^{-2}$ [9].

UV spectrum (n-hexane): λ_{max} (in nm) = 256 (log ε = 3.92), 319 (log ε = 3.21), 331 (log ε = 3.21), 408 (log ε = 4.45). The spectrum is shown in a figure [2]; (c-C_6H_{12}): λ_{max} (in nm) = 415 (ε = 29000) [3].

The dipole moment at 25°C is 4.3 D. The dielectric constant of the liquid at 25°C is ε = 34.2 [7].

The compound gradually decomposes by treatment with cold H_2O but immediately by treatment with acids or alkali. On standing in the air it gradually converts to 4-$(H_3C)_2NC_6H_4NHSO_2H$ [1]. On treatment with t-$C_4H_9O^-K^+$ in DMF at 20°C for 12 h the compound forms the corresponding sulfur diimide [10]. It reacts with P_4S_{10} in benzene under N_2 for several hours to give N'-thiosulfinyl-N,N-dimethyl-1,4-phenylenediamine ($S{=}S{=}NC_6H_4N(CH_3)_2$-4) [10]. With $(CH_3)_2$-$BSCH_3$ the formation of 4-$(CH_3)_2NC_6H_4N(B(CH_3)_2)S{-}SCH_3$ occurs [11].

References:

[1] Francke, A. (Ber. Deut. Chem. Ges. **31** [1898] 2179/82).
[2] Leandri, G.; Mangini, A. (Spectrochim. Acta **1959** 421/31; Ric. Sci. **27** [1957] 1920/5).

[3] Middleton, W. J. (J. Am. Chem. Soc. **88** [1966] 3842/4).
[4] Kresze, G.; Maschke, A.; Albrecht, R.; Bederke, K.; Patzschke, H. P.; Smalla, H.; Trede, A. (Angew. Chem. **74** [1962] 135/44; Angew. Chem. Intern. Ed. Engl. **1** [1962] 89/98).
[5] van Woerden, H. F.; Bijl-Vlieger, S. H. (Recl. Trav. Chim. **93** [1974] 85/90).
[6] Kagami, H.; Motoki, S. (J. Org. Chem. **43** [1978] 1267/8).
[7] Egorov, Yu. P.; Borovikov, Yu. Ya.; Fedyuk, G. S.; Levchenko, E. S. (Zh. Obshch. Khim. **43** [1973] 2470/6; J. Gen. Chem. [USSR] **43** [1973] 2456/61).
[8] Kresze, G.; Maschke, A. (Chem. Ber. **94** [1961] 450/6).
[9] Butt, G.; Davis, M.; Pang, Yuk Tau; Topsom, R. D.; Katritzky, A. R. (J. Chem. Soc. Perkin Trans. II **1974** 260/3).
[10] Barton, D. R. H.; Robson, M. J. (J. Chem. Soc. Perkin Trans. I **1974** 1245/7).

[11] Meller, A.; Maringgele, W.; Fetzer, H. (Chem. Ber. **113** [1980] 1950/61, 1951/2).

2.6.22.2.2.7 N-Sulfinyl-(2, 3, or 4)-nitro-aniline

N-Sulfinyl-2-nitro-aniline, $O{=}S{=}NC_6H_4NO_2$-2

The title compound was obtained by the reaction of $2\text{-}NO_2C_6H_4NH_2$ [1, 2] or $2\text{-}NO_2C_6H_4NH_3^+Cl^-$ [3] with $SOCl_2$ (yield 75% [3]). For purification, the compound was distilled in vacuum (b.p. 155 to 156°C/14 Torr [1], 120°C/540 Torr [3]) or recrystallized from petroleum ether. It forms yellow [2] or chrome-yellow [1] prisms. Two different melting points were given: 52 [2] or 28°C [1]. The substance crystallizes very slowly at room temperature but faster on cooling in carbon dioxide snow [1]. It is soluble in petroleum ether and very soluble in benzene [1, 2].

UV spectrum ($c\text{-}C_6H_{12}$): λ_{max} (in nm) = 270 (log ε = 3.81, inflexion), 311 (log ε = 3.65, broad band); the spectrum is shown in a figure [4].

Mass spectrum: m/e = 184 M^+ [3].

The dipole moment in benzene solution at 25°C is 3.71 D [1], in CCl_4 at 25°C it is 3.74 D [5]. The compound was used as transsulfinylation reagent for the preparation of several N-sulfinylanilines [6].

N-Sulfinyl-3-nitro-aniline, $O{=}S{=}NC_6H_4NO_2$-3

The title compound was obtained by the reaction of $3\text{-}NO_2C_6H_4NH_2$ with $SOCl_2$ in benzene [1, 4, 8, 9, 10]. The compound forms yellow, flat prisms, which melt at 63.5°C [8, 9], 64 to 64.5°C [4, 10] and boil at 114 to 115°C/3 Torr [10]. The density at 20°C (solid phase) is 1.56 and at 70°C (liquid phase) 1.407 g/cm³ [11].

UV spectrum ($c\text{-}C_6H_{12}$): λ_{max} (in nm) = 259 to 260 (log ε = 4.03), 308 (log ε = 3.98) [4]; (heptane): $\bar{\nu}_{max}$ (in cm^{-1}) = 39060 (log ε = 4.00), 32680 (log ε = 3.92) [9] (converted to nm: λ_{max} = 256 and 306 nm).

IR spectrum: ν_s(NSO) = 1178 cm^{-1}; ν = 1275, 1303, 1316 cm^{-1} (without assignment) [12].

Dielectric properties were measured at 20°C (solid phase): deformation limit of the dielectric constant ε_∞ = 3.94, deformation polarization P_{def} = 87.4 cm³, refraction R_D = 56.8 cm³, atomic polarization P_A = 30.6 cm³ [11]. In benzene solution at 25°C $P_{2\infty}$ = 314.6 cm³ and R_D = 46 cm³ [13]. The dipole moment μ in benzene solution at 25°C is 3.47 D [1], 3.62 D [13] and at 20°C 3.50 D [9]. The μ value of the liquid phase (at 70°C) is 2.81 D [11]. The dipole moment was also calculated by the Guggenheim method to be 3.66 D [9] and 3.75 D (from the Thomson vector scheme, assuming free rotation about the N–C_{ar} bond) [14]. The value for the cis and hypothetical trans conformers were also calculated to be 4.0 D and 3.3 D, respectively [13].

The compound decomposes to $H_2NC_6H_4NO_2$-3 by treatment with hot H_2O or alkali [8]. Treatment with t-$C_4H_9O^-K^+$ in DMF at 20°C for 4 h leads to the formation of 2-$NO_2C_6H_4N$=S=N-$C_6H_4NO_2$-2 (yield 61.2%) [15]. With 4-$CH_3C_6H_4SO_2H$ (at −15°C) the corresponding sulfinylsulfone is formed [16]: O=S=$NC_6H_4NO_2$-3 + 2RSO_2H → $H_2NC_6H_4NO_2$-3 + RS(O)S(O)$_2$R + SO_2 (R = 4-$CH_3C_6H_4$).

N-Sulfinyl-4-nitro-aniline, O=S=$NC_6H_4NO_2$-4

The compound was obtained by the reaction of 4-$NO_2C_6H_4NH_2$ [8, 9, 11, 17, 18] or 4-NO_2-$C_6H_4NH_3^+Cl^-$ [3] with $SOCl_2$ in benzene (yields 80 [18] and 87% [3, 17]). Treatment of 4-$NO_2C_6H_4$-NH_2 with the sulfinylation reagent N-(chlorosulfinyl)-imidazole (prepared from the reaction of N,N′-sulfinylbisimidazole with $SOCl_2$ or by reaction of imidazole with excess amount of $SOCl_2$) in CH_2Cl_2 at 20°C for 0.5 h also led to the formation of the title compound (yield 80%, after purification) [19]. Another method of preparation is the reaction of 4-$NO_2C_6H_4N$=SCl_2 with acetic acid in boiling benzene for ~4 h; yield 64% [20].

It forms fine yellow or yellow-red crystals, which melt at 70 [8, 18], 72 [3], 72 to 73 [4], 71.5 to 72 [9], 68 to 70 [17], 70 to 72 [20], or 67 to 69°C [19]. Heating above the melting point leads to decomposition [8]. The density D_m at 20°C is 1.565, at 75°C (liquid) 1.38 [11], and from an X-ray diffraction study D_x = 1.50 (calc. 1.515) g/cm³ [21].

The molecular structure of the compound in the 1:1 complex with 4-$NO_2C_6H_4NH_2$ was analyzed by X-ray diffraction [21]. The crystals belong to the monoclinic space group $P2_1/n-C_{2h}^5$ (No. 14) with a = 10.672(6), b = 18.229(9), c = 7.358(5) Å, and β = 99.25(3)°, V = 1412.8 Å³; Z = 4; R = 0.077. In a first approximation the two components of the complex are planar. In analogy to other structurally examined "OSN" compounds (O=S=NH, O=S=NCl, O=S=NCH_3, (O=S=N)$_2$S) and in complexes of O=S=NAr the S=N bond is cis-configured. Bond lengths and bond angles of both components are given in **Fig. 14** [21].

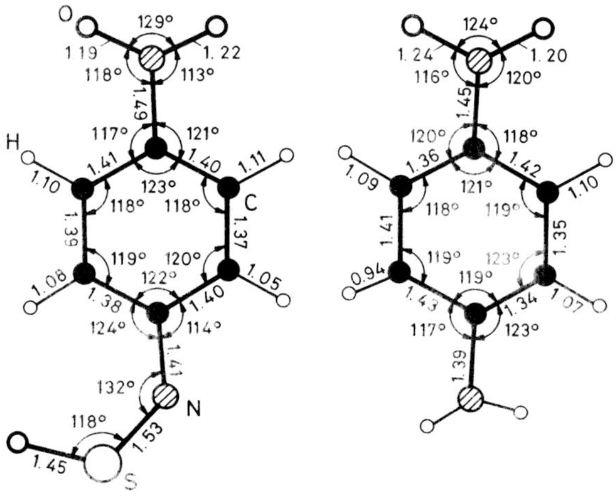

1 2

Fig. 14. Molecular structure of O=S=$NC_6H_4NO_2$-4 (1) and 4-$NO_2C_6H_4NH_2$ (2) in their 1:1 complex. Bond lengths in Å, bond angles in °. Mean standard deviations of bond lenghts 0.01 Å, of bond angles 1°. Positions of H atoms not refined [21].

^1H NMR ($CDCl_3$/TMS): δ (in ppm) = 7.95 (H2, H6), 8.25 (H3, H5) [5]; (solvent?/TMS): δ (in ppm) = 8.16; 8.54 (2d, $^3J(H2,H3) = {}^3J(H5,H6) = 90$ Hz) [25]; (solvent?/TMS): δ (in ppm) = 7.92, 8.20 (C_6H_4, ca. AB_q, J = 9 Hz) [3].

^{13}C NMR ($CDCl_3$/TMS): δ (in ppm) = 146.9 (C1), 127.9 (C2), 125.5 (C3), 148.0 (C4) [26].

^{15}N NMR (3 M in DMSO, t = 30°C, external standard 1.0 M solution of 98% enriched $H^{15}NO_3$ in D_2O): δ (in ppm) = −64.7 (NSO), −6.7 (NO_2) [27]; (referenced to CH_3NO_2): δ (in ppm) = −70.9 (NSO), −12.9 (NO_2).

IR (CCl_4): ν (in cm^{-1}) = 1340 (s), 1305 (w), 1295 (m), 1285 (s), 1170 (s), 1105 (m), 1010 (w) (without assignment) [5]; (CCl_4): ν (in cm^{-1}) = 1301, 1290, 1179 ν(NSO); 1603, 1485 ν(C=C); 1528 $\nu_{as}(NO_2)$; 1346 $\nu_s(NO_2)$ [17]; (Nujol): ν (in cm^{-1}) = 1293, 1175 ν(NSO) [25]; (neat): ν (in cm^{-1}) = 1303 ν_{as}(NSO) or ν(CN); 1293, 1175 ν(NSO) [12]; (CH_2ClCH_2Cl): 1525 cm^{-1} $\nu_{as}(NO_2)$ [11]. The integrated intensity A of the ν_8(C=C, arom.) vibration (at 1621, 1601, 1588 cm^{-1}) is 2501 $L \cdot mol^{-1} \cdot cm^{-2}$. The intramolecular interaction was discussed [28].

The sulfur Kβ X-ray fluorescence spectrum shows peaks at 2466.3, 2469.5, and 2471.2 eV ($\Delta SK_\alpha = 0.39 \pm 0.01$ eV) [22]. The compound may be identified by its valence band spectrum [23]. The charge densities ϱ were calculated in electron units by the CNDO/2 method (based on a hypothetical trans conformer): $\varrho_C = 0.156$, $\varrho_N = -0.284$, $\varrho_S = 0.313$, and $\varrho_O = -0.128$ (a positive sign indicates electron deficiency) [24].

UV (heptane): λ_{max} (in nm) ≈ 270 sh, 314 (log ε = 4.12) (in the original paper in cm^{-1}): $\bar{\nu}_{max} \approx 37000$ (sh, 31850) [9]; (solvent?): λ_{max} (in nm) = 226 (log ε = 3.89), 355 (log ε = 4.19) [17]; ($c\text{-}C_6H_{12}$): λ_{max} (in nm) = 320 (log ε = 4.14) [4]. The spectrum is shown in a figure [4].

Dielectric properties of the compound were measured at 20°C (solid phase): deformation limit of the dielectric constant $\varepsilon_\infty = 3.58$; deformation polarization $P_{def} = 77.2$ cm^3; refraction $R_D = 56.8$ cm^3; atomic polarization $P_A = 20.4$ cm^3 [11]. In benzene solution at 25°C $P_{2\infty} = 263.9$ cm^3 and $R_D = 46.4$ cm^3 [13]. The conductivity ϰ of the compound in liquid phase (at 75°C) is $3 \times 10^{-6} \cdot \Omega^{-1} \cdot cm^{-1}$ [11]. The dipole moment μ in benzene solution at 20°C is 3.06 D [1], 3.20 D [9], at 25°C 3.26 D [13], in CCl_4 solution at 25°C 3.16 D [5]. The dipole moment in the liquid phase (at 75°C) is 2.61 [11, 29]. It was also calculated by the Guggenheim method to be 2.71 D [9] and 2.90 D (from the Thomson vector scheme, assuming free rotation about the $N\text{–}C_{ar}$ bond) [14]. The calculated μ values for the cis and hypothetical trans conformers are 3.7 and 1.7 D, respectively [13]. The angle of the direction of the dipole moment of the NSO group to the C–N bond was estimated to be Θ(NSO) = 40° [11].

Mass spectrum: m/e = 184 M^+ [3].

The compound easily decomposes on treatment with hot H_2O and by alkali [8]. Treatment of the substance in DMF with $K^+t\text{-}C_4H_9O^-$ at 20°C for 3 h produces the corresponding $4\text{-}NO_2C_6H_4\text{-}N{=}S{=}NC_6H_4NO_2\text{-}4$ (yield 67%) [15]. The reaction with NOCl in CCl_4–ether at −70°C yields p-nitrobenzenediazonium chloride $4\text{-}NO_2C_6H_4N_2^+Cl^-$ [30]. Treatment with S_2Cl_2 at about 60 to 70°C on a water bath for 14 h with occasional shaking gives the "Herz compound" 6-nitro-1,2,3-benzodithiazol-2-ium chloride (I) [31]. The title compound was used as a transsulfinylation reagent for the preparation of several N-sulfinyl-anilines [6]. The reaction with $(CH_3)_2BSCH_3$ yields $4\text{-}NO_2C_6H_4N(B(CH_3)_2)S\text{–}S\text{–}CH_3$ [7].

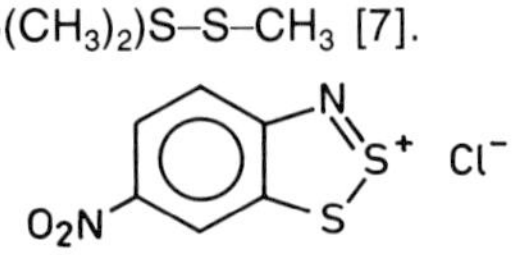

I

References:

[1] Jensen, K. A.; Hofman-Bang, N. (Liebigs Ann. Chem. **548** [1941] 95/105, 105).
[2] Michaelis, A.; Hümme, C. (Liebigs Ann. Chem. **274** [1893] 217/27, 225/6).
[3] Al-Ajely, M. S.; Al-Rawi, J. M. A.; Elvidge, J. A. (J. Chem. Soc. Perkin Trans. I **1982** 1575/8).
[4] Leandri, G.; Mangini, A. (Spectrochim. Acta **1959** 421/31; Ric. Sci. **27** [1957] 1920/5).
[5] van Woerden, H. F.; Bijl-Vlieger, S. H. (Recl. Trav. Chim. **93** [1974] 85/90).
[6] Carpanelli, C.; Leandri, G. (Ann. Chim. [Rome] **50** [1960] 147/55).
[7] Meller, A.; Maringgele, W.; Fetzer, H. (Chem. Ber. **113** [1980] 1950/61, 1951/2).
[8] Michaelis, A.; Hümme, C. (Ber. Deut. Chem. Ges. **24** [1891] 745/57; 755/6).
[9] Kresze, G.; Smalla, H. (Chem. Ber. **92** [1959] 1042/8).
[10] Jannelli, L.; Senatore, L. (Ann. Chim. [Rome] **53** [1963] 1356/69).

[11] Egorov, Yu. P.; Borovikov, Yu. Ya.; Fedyuk, G. S.; Levchenko, E. S. (Zh. Obshch. Khim. **43** [1973] 2470/6; J. Gen. Chem. [USSR] **43** [1973] 2456/61).
[12] Kresze, G.; Maschke, A. (Chem. Ber. **94** [1961] 450/6).
[13] Jannelli, L.; Lamanna, U.; Lumbroso, H. (Bull. Soc. Chim. France **1966** 3626/30).
[14] Waclawek, W.; Puchala, C.; Czerwiec, Z. (Bull. Acad. Polon. Sci. Ser. Sci. Chim. **26** [1978] 337/41).
[15] Hörhold, H.-H.; Beck, J. (J. Prakt. Chem. [4] **311** [1969] 621/9).
[16] Bederke, K. (Dipl.-Arbeit T.H. Berlin 1961 from Bussas, R.; Kresze, G.; Münsterer, H.; Schwöbel, A., Sulfur Rept. **2** [1983] 215/377, 259).
[17] Uher, M.; Jendrichovský, J.; Novomestská, D. (Zb. Pr. Chemickotechnol. Fak. SVST **1972/74** 93/6; C.A. **82** [1974] No. 97771).
[18] Sugihara, H.; Sasaki, M.; Kawamatsu, Y.; Miromoto, H. (Liebigs Ann. Chem. **1972** 121/7).
[19] Kim, Y. H.; Shin, J. M. (Tetrahedron Letters **26** [1985] 3821/4).
[20] Markovskii, L. N.; Fedyuk, G. S.; Levchenko, E. S. (Zh. Org. Khim. **6** [1970] 2426/31; J. Org. Chem. [USSR] **6** [1970] 2438/42).

[21] Dederer, B.; Gieren, A. (Naturwissenschaften **66** [1979] 470/1).
[22] Dolenko, G. N.; Zibarev, A. V.; Krupoder, S. A.; Furin, G. G.; Yakobson, G. G. (Izv. Sibirsk. Otd. Akad. Nauk SSR Ser. Khim. Nauk **1980** No. 6, pp. 74/7).
[23] Treiger, B. A.; Mazalov, L. N.; Dolenko, G. N.; Guzhavina, T. I. (Comput. Enhanced Spectrosc. **3** [1986] 153/63).
[24] Glass, W. K.; King, I. J.; Shiels, A. (Inorg. Chim. Acta **25** [1977] 157/8).
[25] Davies, A. G.; Kennedy, J. D. (J. Chem. Soc. C **1968** 2630/40).
[26] Kresze, G.; Berger, M.; Claus, P. K.; Rieder, W. (Org. Magn. Resonance **8** [1976] 170/2).
[27] Yavari, I.; Staral, J. S.; Roberts, J. D. (Org. Magn. Resonance **12** [1979] 340/2).
[28] Butt, G.; Davis, M.; Pang, Yuk Tau; Topsom, R. D.; Katritzky, A. R. (J. Chem. Soc. Perkin Trans. II **1974** 260/3).
[29] Borovikov, Yu. Ya.; Egorov, Yu. P. (Teor. Eksperim. Khim. **7** [1971] 663/9; Theor. Exptl. Chem. [USSR] **7** [1971] 539/44).
[30] Kobayashi, M.; Honda, K. (Bull. Chem. Soc. Japan **39** [1966] 1778/82).

[31] Arsenescu, N. (Revue Roumaine Chim. **12** [1967] 427/8).

2.6.22.2.2.8 N-Sulfinyl-(2, 3, or 4)-methyl-benzenamine

N-Sulfinyl-2-methyl-benzenamine, $O{=}S{=}NC_6H_4CH_3$-2

The compound was obtained by the reaction of 2-$CH_3C_6H_4NH_2$ with $SOCl_2$ in benzene [1 to 5] (yield 85% [2], 96% [5]). It was also prepared analogously to $O{=}S{=}NC_6H_4CH_3$-4 by a sulfinylation reaction of 2-$CH_3C_6H_4NH_2$ with SO_2 gas (after the compound was lithiated) (in 59% yield) [6].

The bright yellow liquid ($n_D^{25}=1.6169$ [3], $n_D^{20}=1.6225$ [7]) cannot be distilled without decomposition [1]. The boiling point is 184°C/100 Torr [1], 117°C/18 Torr [9], 101.5°C/16 Torr [7], 105°C/16 Torr [3], 99.5°C/15 Torr [5], 83 to 85°C/13 Torr [6], 60°C/1.6 Torr [2].

The density $d_4^{22}=1.182$ g/cm³ [3].

^{1}H NMR spectrum (δ in ppm, relative to TMS):

solvent	δ(H3)	δ(H4)	δ(H5)	δ(H6)	δ(CH_3)	Ref.
CCl_4	—	7.15 to 7.35(m)	—	8.50	2.35(s)	[3]
CCl_4	6.90(d?)[a)]	6.88(t?)[a)]	6.88(t?)[a)]	8.12(m)	2.13(s)	[10]
$CDCl_3$	—	7.17(m)	—	8.22(m)	2.30(s)	[11]

a) J = 5 Hz

^{13}C NMR ($CDCl_3$/TMS): δ (in ppm) = 141.59 (C–N); 134.89 (C2); 127.86 (C6); 129.33 (C3); 126.58 (C5); 139.26 (C4); 18.20 (CH_3) [11].

^{14}N NMR (2.0 M in $(C_2H_5)_2O$/external CH_3NO_2): δ = −66.7 ppm [12].

IR: ν (in cm^{-1}) = 1300 sh, 1285 s, 1205 m, 1170 vs, 1120 w, 1050 m, 1030 m [3]; ν(N=S) = 1284 [11], 1283 [2], 1285 [6]; ν(S=O) = 1164 [11], 1169 [2], 1167 [6]; δ(N=S=O) = 720 vw, sh [11]; ϱ(N=S=O) = 372 sh [11]; δ(CNS) = 640 m [11]; ν(C=C, arom.) = 1596, 1483 [2]; ν(C_6H_4–N) = 1294 sh, s [11], 1291 [2]. State of measured specimens: neat liquid [3, 6 (?), 11], in CCl_4 [2].

Raman (neat liquid): ν (in cm^{-1}) = 1296 s ν(N=S); 1162 s ν(S=O); 724 m δ(N=S=O); 638 m δ(CNS); 1296 s ν(C_6H_4–N) (all bands are polarized) [11].

The HeI and HeII photoelectron spectra of $O{=}S{=}NC_6H_4CH_3$-2 were recorded and assigned using a CNDO LCAO MO calculation [8].

Ionization energies E_i in eV:

E_i (obs)	8.92	9.36	10.58	10.88	11.25	12.05	12.40
E_i (calc)	9.22	9.85	11.10	12.20	11.50	—	—
assignment	"b_1"	"a_2"	σ_1	σ_2	π_3	σ_3	σ_4

UV (n-hexane): λ_{max} (in nm) = 235 to 237 (log ε = 3.73, flat region); 248 (log ε = 3.79); 325 (log ε = 3.97) [9]; (dioxane): λ_{max} (in nm) = 237 (log ε = 3.76); 328 (log ε = 3.95) [2].

Dielectric measurements were performed in CCl_4 at 25°C: $P_{2\infty} = 139 \pm 1$ cm³; $R_D = 48 \pm 2$ [4] or $R_D^{22} = 45.35$ cm³ [3]. The dipole moment in CCl_4 at 25°C is 2.08 ± 0.03 D [4] or 2.20 D [3]. The molar Kerr constant was found to be $K_m = (530 \pm 20) \times 10^{-12}$ e.s.u./mol [4].

The title compound was studied, together with other substituted N-sulfinyl-anilines, by TG, DTA, and DTG [16], gas chromatography [6, 7, 17], thin-layer chromatography ($R_f = 0.25$ in benzene) [18], and polarography (half-wave potential in DMF against the internal Hg electrode $E_{1/2} = -0.739$ V) [19].

The positive-ion mass spectrum (75 eV) is shown in a figure (m/e = 153 M^+ (35%)) [13]. The molecule peak and 7 fragment peaks were also observed in a mass spectrum recorded at 70 eV [3]. The negative-ion mass spectrum (recorded at 70 eV) is very simple and the molecular ion is the base peak [14]. The +E spectrum of the molecular anion (at 70 eV) is shown in a figure. It exhibits abundant peaks originating from the rearrangements $[M]^{+\cdot} \rightarrow [M-HO^{\cdot}]^+$, $[M]^{+\cdot} \rightarrow [M-SH^{\cdot}]^+$, and $[M]^{+\cdot} \rightarrow [M-HSO^{\cdot}]^+$ [15].

The compound evaporates without degradation at 220 to 228°C [16]. $O{=}S{=}NC_6H_4CH_3$-2 reacts with S_2Cl_2 to yield I, the same product as in the case of the Herz reaction with H_2N-$C_6H_4CH_3$-2 [20].

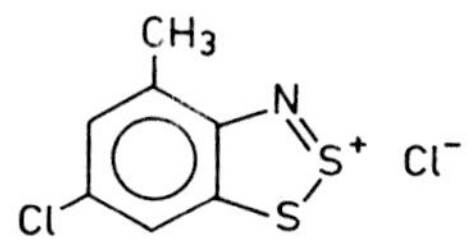

I

It converts to 2-$CH_3C_6H_4N{=}S{=}NC_6H_4CH_3$-2 when treated with t-$C_4H_9O^-K^+$ in benzene at 80°C for 5 h (yield 44%) [21]. The reaction with copper (mole ratio 1:1) at 130°C for 8 h leads to traces of the azo compound 2-$CH_3C_6H_4N{=}NC_6H_4CH_3$-2 [22].

References:

[1] Michaelis, A.; Herz, R. (Ber. Deut. Chem. Ges. **24** [1891] 745/54).

[2] Uher, M.; Jendrichovsky, J.; Novomestská, D. (Zb. Pr. Chemickotechnol. Fak. SVST **1972/74** 93/6; C.A. **82** [1974] No. 97771).

[3] van Woerden, H. F.; Bijl-Vlieger, S. H. (Recl. Trav. Chim. **93** [1974] 85/90).

[4] Bonham, J. S.; Cheng, C. L.; Le Fèvre, R. J. W.; Ritchie, G. L. D. (Australian J. Chem. **26** [1973] 421/4).

[5] Wichterle, O.; Roček, J. (Chem. Listy **47** [1953] 1768/80; Collection Czech. Chem. Commun. **19** [1954] 282/97).

[6] Sakai, S.; Fujinami, T.; Komizo, K. (J. Org. Chem. **40** [1975] 3291/2).

[7] Czerwiec, Z.; Mirek, J. (Zesz. Nauk. Univ. Jagiel. Proc. Chem. No. 14 [1969] 103/6; C.A. **72** [1970] No. 28195).

[8] Louwen, J. N.; van Dam, H.; Stufkens, D. J.; Oskam, A. (J. Electron Spectrosc. Relat. Phenom. **26** [1982] 235/46).

[9] Leandri, G.; Mangini, A. (Spectrochim. Acta **15** [1959] 421/31, Ric. Sci. **27** [1957] 1920/5).

[10] Mirek, J.; Czerwiec, Z. (Zesz. Nauk. Univ. Jagiel. Proc. Chem. No. 18 [1973] 183/6; C.A. **79** [1973] No. 145570).

[11] Meij, R.; Oskam, A.; Stufkens, D. J. (J. Mol. Struct. **51** [1979] 37/49).

[12] Sicińska, W.; Stefaniak, L.; Witanowski, M.; Webb, G. A. (Org. Magn. Resonance **15** [1981] 241/3).

[13] Bowie, J. H.; Larsson, F. C. V.; Schroll, G.; Lawesson, S.-O.; Cooks, R. G. (Tetrahedron **23** [1967] 3743/52).

[14] Bowie, J. H.; Duus, F.; Lawesson, S.-O.; Larsson, F.C.V.; Madsen, J. Ø. (Australian J. Chem. **22** [1969] 153/9).

[15] Bowie, J. H.; White, P. Y.; Wilson, J. C.; Larsson, F. C. V.; Lawesson, S.-O.; Madsen, J. Ø.; Nolde, C.; Schroll, G. (Org. Mass Spectrom. **12** [1977] 191/5).

[16] Czerwiec, Z.; Pielichowski, J. (J. Therm. Anal. **7** [1975] 465/8).

[17] Czerwiec, Z.; Markowski, J. (Chem. Anal. [Warsaw] **20** [1975] 213/5).

[18] Czerwiec, Z. (Chem. Anal. [Warsaw] **21** [1976] 751/3).

[19] Czerwiec, Z.; Bogaczek, J. (Z. Anal. Chem. **293** [1978] 415).

[20] Maior, O.; Arsenescu, N. (An. Univ. Bucuresti Ser. Stiint. Nat. **14** [1965] 25/33).

[21] Hörhold, H.-H.; Beck, J. (J. Prakt. Chem. [4] **311** [1969] 621/9).
[22] Minami, T.; Agawa, T. (Tetrahedron Letters **1968** 4109/10).

N-Sulfinyl-3-methyl-benzenamine, $O{=}S{=}NC_6H_4CH_3$-3

The title compound was prepared by the reaction of 3-$CH_3C_6H_4NH_2$ with $SOCl_2$ in benzene [1, 2, 3] (yield 83% [3]).

The yellow liquid (n_D^{20} = 1.1641 [4], 1.6132 [2]) boils at 220°C [1], 58°C/2 Torr [5], 70°C/2 Torr [3], 103°C/15 Torr [4], 105°C/16 Torr [2]. The density is d_4^{20} = 1.1908 g/cm³ [2].

^{1}H NMR (CCl_4/TMS): δ (in ppm) = 7.39 (s, H2); 6.93 (d, J = 8 Hz, H4); 6.98 (t, J = 8 Hz, H5); 7.45 (d, J = 8 Hz, H6); 2.08 (CH_3) [6].

^{14}N NMR (2.0 M in $(C_2H_5)_2O$/external CH_3NO_2): δ = −63.0 ppm [7].

IR (CS_2): ν (in cm^{-1}) = 1295, 1281 ν_{as}(NSO) and ν(CN); 1156 ν_s(NSO) [5]; (CCl_4): ν (in cm^{-1}) = 1601, 1484 ν(C=C, arom.); 1294, 1276, 1158 ν(N=S=O) [3] (for the assignment of ν = 1294 cm^{-1} compare [8]); (KBr or dioxane): 632 δ(N=S=O), ε_{max} = 28 L·mol^{-1}·cm^{-1} [9] (this assignment was not confirmed by [8]). The integrated intensity of the ν_8(C=C, arom.) vibration (1603, 1594, 1573 cm^{-1}) is 1227 L·mol^{-1}·cm^{-2} [19].

UV (in heptane): $\bar{\nu}_{max}$ = 42190 cm^{-1} (log ε = 3.75) and 31350 cm^{-1} (log ε = 4.00) [2] (converted to nm: λ_{max} = 237 and 319 nm); (in dioxane): λ_{max} = 239 nm (log ε = 4.14), 321 nm (log ε = 4.28) [3].

The dipole moment μ in benzene at 20°C is 2.14 D [2] and at 25°C 2.16 D [12]. The dipole moment of the pure liquid is 2.16 D at 20°C (from Onsager's formula) [2]. Calculated values are 2.08 D (from the Thomson vector scheme) [12] and 2.09 D (from the Guggenheim method) [2]. Calculated values for the cis and hypothetical trans isomer are 1.1 D and 3.0_5 D, respectively [13].

The compound was studied, together with other substituted N-sulfinyl-anilines, by TG, DTA, and DTG [14], gas chromatography [4, 15], thin-layer chromatography (R_f = 0.23 in benzene) [16], and polarography (half-wave potential against the internal mercury electrode $E_{1/2}$ = −0.572 V in DMF) [17]. The compound evaporates without degradation at 223 to 225°C [14].

Mass spectrum (75 eV positive-ion spectrum): m/e = 153 M^+ (100%), and 38 fragment peaks [10]. The negative-ion mass spectrum (recorded at 70 eV) is shown in [11]. It is very simple. The molecular ion is the base peak [11].

$O{=}S{=}NC_6H_4CH_3$-3 converts to 3-$CH_3C_6H_4N{=}S{=}NC_6H_4CH_3$-3 when treated with t-$C_4H_9O^-K^+$ in benzene at 80°C for 3 h (62% yield) [18].

References:

[1] Michaelis, A.; Hümme, C. (Liebigs. Ann. Chem. **274** [1893] 217/27, 226).
[2] Kresze, G.; Smalla, H. (Chem. Ber. **92** [1959] 1042/8).
[3] Uher, M.; Jendrichovský, J.; Novomestská, D. (Zb. Pr. Chemickotechnol. Fak. SVST **1972/74** 93/6; C.A. **82** [1974] No. 97771).
[4] Czerwiec, Z.; Mirek, J. (Zesz. Nauk. Univ. Jagiel. Proc. Chem. No. 14 [1969] 103/6; C.A. **72** [1970] No. 28195).
[5] Kresze, G.; Maschke, A. (Chem. Ber. **94** [1961] 450/6).
[6] Mirek, J.; Czerwiec, Z. (Zesz. Nauk. Univ. Jagiel. Proc. Chem. No. 18 [1973] 183/6; C.A. **79** [1973] No. 145570).

[7] Sicińska, W.; Stefaniak, L.; Witanowski, M.; Webb, G. A. (Org. Magn. Resonance **15** [1981] 241/3).
[8] Meij, R.; Oskam, A.; Stufkens, D. J. (J. Mol. Struct. **51** [1979] 37/49).
[9] Kresze, G.; Uhlich, U.; Ropte, E.; Schrader, B. (Z. Anal. Chem. **197** [1963] 283/94, 292).
[10] Bowie, J. H.; Larsson, F. C. V.; Schroll, G.; Lawesson, S.-O.; Cooks, R. G. (Tetrahedron **23** [1967] 3743/52).

[11] Bowie, J. H.; Duus, F.; Lawesson, S.-O.; Larsson, F. C. V.; Madsen, J. Ø. (Australian J. Chem. **22** [1969] 153/9).
[12] Waclawek, W.; Puchala, C.; Czerwiec, Z. (Bull. Acad. Polon. Sci. Ser. Sci. Chim. **26** [1978] 337/41).
[13] Jannelli, L.; Lamanna, U.; Lumbroso, H. L. (Bull. Soc. Chim. France **1966** 3626/30).
[14] Czerwiec, Z.; Pielichowski, J. (J. Therm. Anal. **7** [1975] 465/8).
[15] Czerwiec, Z.; Markowski, J. (Chem. Anal. [Warsaw] **20** [1975] 213/5).
[16] Czerwiec, Z. (Chem. Anal. [Warsaw] **21** [1976] 751/3).
[17] Czerwiec, Z.; Bogaczek, J. (Z. Anal. Chem. **293** [1978] 415).
[18] Hörhold, H.-H.; Beck, J. (J. Prakt. Chem. **311** [1969] 621/9).
[19] Butt, G.; Davis, M.; Pang, Yuk Tau; Topsom, R. D.; Katritzky, A. R. (J. Chem. Soc. Perkin Trans. II **1974** 260/3).

N-Sulfinyl-4-methyl-benzenamine, $O{=}S{=}NC_6H_4CH_3$-4

Preparation

The title compound was obtained by reaction of 4-$CH_3C_6H_4NH_2$ with $SOCl_2$ in dry benzene [1 to 7] (mole ratio ~1:1 [6] or ~1:2 [2]). The yields are in the range 70 [4] to 93% [6]. It was prepared by the reaction of 4-$CH_3C_6H_4NH_3^+Cl^-$ with $SOCl_2$ in dry benzene (yield 56%) [8].

The transsulfinylation reaction of 4-$CH_3C_6H_4NH_2$ with $O{=}S{=}NC_6H_4NO_2$-4 or $O{=}S{=}NC_6H_4NO_2$-2 (mole ratio 1:1) in dry benzene led to the formation of $O{=}S{=}NC_6H_4CH_3$-4 in 85 and 78% yield, respectively [9]. Treatment of 4-$CH_3C_6H_4NH_2$ with the sulfinylation reagents N,N'-sulfinylbisimidazole (prepared from imidazole and $SOCl_2$) in CH_2Cl_2 at 20°C for 25 min or with N-(chlorosulfinyl)imidazole (prepared from N,N'-sulfinylbisimidazole and $SOCl_2$ or by treatment of imidazole with excess amount of $SOCl_2$) in CH_2Cl_2 at 20°C for 0.5 h yielded the title compound in 72 and 86% (purified) yields, respectively [10].

Another novel synthetic method for sulfinylation was also applied: 4-$CH_3C_6H_4NH_2$ was treated in dry THF with butyllithium (in petroleum ether) under a dry nitrogen atmosphere at room temperature and subsequently with SO_2 gas. The resulting lithium-N-tolylaminosulfinate was again lithiated with butyllithium and treated with SO_2 gas (~2 mol each of butyllithium and SO_2 per ~1 mol of p-toluidine were used). The title compound was isolated by distillation in 40% yield [11].

N-Trimethylsilylamines can be sulfinated also by the use of SO_2. Two procedures were employed: a) The silylated p-toluidine was treated with butyllithium in THF at 0°C. The resulting solution is then added to a solution of excess SO_2 in THF at −78°C. The title compound was isolated by distillation in 90% yield. b) Gaseous SO_2 was introduced at room temperature to a solution of N-trimethylsilyl-p-toluidine in ether containing one equivalent of pyridine. After the reaction was complete, $Si(CH_3)_3Cl$ was added to the mixture. Distillation gave the title compound in 76% yield [12].

$O{=}S{=}NC_6H_4CH_3$-4 was also synthesized by the reaction of $H(C_6H_5)C{-}S(O){-}C(C_6H_5)H$ with 4-$CH_3C_6H_4N_3$ in boiling CH_2Cl_2 for ~3 h (yield 24%) [13].

References:

[1] Michaelis, A.; Herz, R. (Ber. Deut. Chem. Ges. **23** [1890] 3480/2).
[2] Kresze, G.; Smalla, H. (Chem. Ber. **92** [1962] 1042/8).
[3] Leandri, G.; Mangini, A. (Spectrochim. Acta **15** [1959] 421/31, Ric. Sci. **27** [1957] 1920/5).
[4] Tsuge, O.; Mataka, S.; Tashiro, M.; Mashiba, F. (Bull. Chem. Soc. Japan **40** [1967] 2709/11).
[5] Davies, A. G.; Kennedy, J. D. (J. Chem. Soc. C **1968** 2630/40).
[6] Bowlus, S. B.; Katzenellenbogen, J. A. (Syn. Commun. **4** [1974] 137/41).
[7] Uher, M.; Jendrichovský, J.; Novomestská, D. (Zb. Pr. Chemickotechnol. Fak. SVST **1972/74** 93/6; C.A. **82** [1974] No. 97771).
[8] Al-Ajely, M. S.; Al-Rawi, J. M. A.; Elvidge, J. A. (J. Chem. Soc. Perkin Trans. I **1982** 1575/8).
[9] Carpanelli, C.; Leandri, G. (Ann. Chim. [Rome] **50** [1960] 147/55).
[10] Kim, Y. H.; Shin, J. M. (Tetrahedron Letters **26** [1985] 3821/4).

[11] Sakai, S.; Fujinami, T.; Komizo, K. (J. Org. Chem. **40** [1975] 3291/2).
[12] Porskamp, P. A. T. W.; Zwanenburg, B. (Synthesis **1981** 368/9).
[13] Bonini, B. F.; Maccagnani, G.; Mazzanti, G. (Tetrahedron Letters **1977** 1185/6).

Molecular Properties. Spectra

Electronic Structure

The HeI and HeII photoelectron spectra of $O{=}S{=}NC_6H_4CH_3$-4 were measured and assigned on the basis of a CNDO/S LCAO MO calculation [1].

Ionization energy E_i (in eV) and assignment [1]:

E_i (obs)	8.84	9.53	10.52	11.05	11.26	12.15	12.54
E_i (calc)	9.11	10.09	11.00	12.28	11.38	—	—
assignment	"b_1"	"a_2"	σ_1	σ_2	π_3	σ_3	σ_4

The sulfur K_β X-ray fluorescence spectrum shows peaks at 2471.5, 2469.4, and 2466.2 eV ($\Delta SK_\alpha = 0.43 \pm 0.01$ eV) [2]. Identification of the compound by means of this spectrum was tried [3].

The UV spectrum was recorded in several solvents: (heptane): $\lambda_{max} = 42190$ cm^{-1} (log ε = 3.80), 30580 cm^{-1} (log ε = 4.08) [4] (converted to nm: λ_{max} = 237 and 327 nm); (c-C_6H_{12}): λ_{max} = 243 nm (log ε = 3.78); 329 nm (log ε = 4.08) [5]; (dioxane): λ_{max} = 240 nm (log ε = 3.91); 332 nm (log ε = 4.13) [6]; λ_{max} = 329 nm ($\pi \to \pi^*$) [7].

References:

[1] Louwen, J. N.; van Dam, H.; Stufkens, D. J.; Oskam, A. (J. Electron Spectrosc. Relat. Phenon. **26** [1982] 235/46).
[2] Dolenko, G. N.; Zibarev, A. V.; Krupoder, S. A.; Furin, G. G.; Yakobson, G. G. (Izv. Sibirsk. Otd. Akad. Nauk SSSR Ser. Khim. Nauk **1980** No. 6, pp. 74/7).
[3] Treiger, B. A.; Mazalov, L. N.; Dolenko, G. N.; Guzhavina, T. I. (Comput. Enhanced Spectrosc. **3** [1986] 153/63).
[4] Kresze, G.; Smalla, H. (Chem. Ber. **92** [1959] 1042/8).
[5] Leandri, G.; Mangini, A. (Spectrochim. Acta **15** [1959] 421/31, Ric. Sci. **27** [1957] 1920/5).
[6] Uher, M.; Jendrichovský, J.; Novomestská, D. (Zb. Pr. Chemickotechnol. Fak. SVST **1972/74** 93/6; C.A. **82** [1974] No. 97771).
[7] Collins, N. C.; Glass, W. K. (Spectrochim. Acta A **30** [1974] 1335/9).

Nuclear Magnetic Resonance

^1H NMR and ^{13}C NMR spectral data of $O{=}S{=}NC_6H_4CH_3$-4 are listed in the following tables.

^1H NMR spectrum: δ in ppm, relative to internal TMS.

solvent	δ(H2, H6)	δ(H3, H5)	$\delta(CH_3)$	Ref.
$CDCl_3$	7.81	7.21	2.39	[1][a)]
$CDCl_3$	7.68	7.03	2.32(s)	[2]
C_7D_8	7.63	6.77	2.00(s)	[2]
CCl_4	7.48	6.83	2.11(s)	[3]
?	7.90	7.26	2.33	[4][b)]
CCl_4	7.84/7.97	7.22/7.35	2.62	[5][b)]
?	7.36 center of an AA′BB′ pattern		2.29	[6]
?	7.75	7.02	2.8(s)	[7]

[a)] t = −20°C, J = 8.5 Hz [1]. – [b)] Without assignment in the paper.

^{13}C NMR spectrum ($CDCl_3$): δ in ppm, relative to internal TMS.

δ(C–NSO)	δ(C2, C6)	δ(C3, C5)	δ(C4)	$\delta(CH_3)$	Ref.
141.2	127.3	129.7	140.8	21.6	[8]
140.4	126.79	129.33	140.79	21.7	[2]

The ^{14}N NMR spectrum (2.0 M in $(C_2H_5)_2O$/external CH_3NO_2) exhibits a singlet with δ = −62.5 ppm [9].

References:

[1] Kuyper, J.; Vrieze, K. (J. Organometal. Chem. **86** [1975] 127/38, 132).
[2] Meij, R.; Oskam, A.; Stufkens, D. J. (J. Mol. Struct. **51** [1979] 37/49).
[3] Mirek, J.; Czerwiec, Z. (Zesz. Nauk. Univ. Jagiel. Proc. Chem. No. 18 [1973] 183/6; C.A. **79** [1973] No. 145570).
[4] Davies, A. G.; Kennedy, J. D. (J. Chem. Soc. C **1968** 2630/40).
[5] Yamada, F.; Nishiyama, T.; Emi, T. (Bull. Chem. Soc. Japan **45** [1972] 271/4).
[6] Bowlus, S. B.; Katzenellenbogen, J. A. (Syn. Commun. **4** [1974] 137/41).
[7] Al-Ajely, M. S.; Al-Rawi, J. M. A.; Elvidge, J. A. (J. Chem. Soc. Perkin Trans. I **1982** 1575/8).
[8] Kresze, G.; Berger, M.; Claus, P. K.; Rieder, W. (Org. Magn. Resonance **8** [1976] 170/2).
[9] Sicińska, W.; Stefaniak, L.; Witanowski, M.; Webb, G. A. (Org. Magn. Resonance **15** [1981] 241/3).

IR and Raman Spectra

The following IR frequencies of $O{=}S{=}NC_6H_4CH_3$-4 were measured: ν(N=S) = 1286 [1], 1285 [2], 1284 [7], 1282 [3], 1280 cm^{-1} [4]; ν(S=O) = 1161 [2], 1157 [4], 1156 [1, 7], 1150 cm^{-1} [3]; ϱ(N=S=O) = 366 cm^{-1} [7]; $\nu(C_6H_4{-}N)$ = 1300 cm^{-1} [1, 2, 7]; ν(C=C, arom.) = 1605, 1501 cm^{-1} [2]. State of specimen: neat [3, 7], solutions in CCl_4 [2] and CS_2 [1]; not given in [4].

Different frequency values were assigned to the N=S=O bending vibration: δ(N=S=O) = 638 cm^{-1} (ε = 2.0 $L\cdot mol^{-1}\cdot cm^{-1}$; in KBr) [5]; 806 cm^{-1} (sh, vw, pure liquid) [7].

In [7] a band at 638 cm^{-1} in the IR spectrum (of pure liquid) was not observed, however, a frequency of this value in the Raman spectrum was assigned to a δ(C–N=S) deformation vibration. The integrated intensity of the ν_8(C=C, arom.) vibration (at 1678, 1602, and 1564 cm^{-1}) is 700 $L \cdot mol^{-1} \cdot cm^{-2}$. The intramolecular interaction was quantitatively studied [6].

Raman spectrum (pure liquid): ν(N=S) = 1284 cm^{-1}, ν(S=O) = 1156 cm^{-1}, δ(N=S=O) = 806 cm^{-1} [7].

References:

[1] Kresze, G.; Maschke, A. (Chem. Ber. **94** [1961] 450/6).
[2] Uher, M; Jendrichovský, J.; Novomestská, D. (Zb. Pr. Chemickotechnol. Fak. SVST **1972/74** 93/6; C.A. **82** [1974] No. 97771).
[3] Porskamp, P. A. T. W.; Zwanenburg, B. (Synthesis **1981** 368/9).
[4] Sakai, S.; Fujinami, T.; Komizo, K. (J. Org. Chem. **40** [1975] 3291/2).
[5] Kresze, G.; Uhlich, U.; Ropte, E.; Schrader, B. (Z. Anal. Chem. **197** [1963] 283/94, 291/2).
[6] Butt, G.; Davis, M.; Pang, Yuk Tau; Topsom, R. D.; Katritzky, A. R. (J. Chem. Soc. Perkin Trans. II **1974** 260/3).
[7] Meij, R.; Oskam, A.; Stufkens, D. J. (J. Mol. Struct. **51** [1979] 37/49).

Melting Point. Boiling Point. Density

$O=S=NC_6H_4CH_3$-4 forms yellow crystals which melt at 7°C [1]. The boiling point of the substance was measured by several groups: e.g., 224°C [1], 50 to 53°C/1 Torr [2].

The density D (in g/cm^3) = 1.1685 at 15°C [3], 1.847 [4] and 1.190 [5] at 20°C, and 1.315 at −190°C [5].

References:

[1] Michaelis, A.; Herz, R. (Ber. Deut. Chem. Ges. **23** [1890] 3480/2).
[2] Kim, Y. H.; Shin, J. M. (Tetrahedron Letters **26** [1985] 3821/4).
[3] Michaelis, A.; Hümme, C. (Liebigs Ann. Chem. **274** [1893] 217/27, 226).
[4] Kresze, G.; Smalla, H. (Chem. Ber. **92** [1962] 1042/8).
[5] Egorov, Yu. P.; Borovikov, Yu. Ya.; Fedyuk, G. S.; Levchenko, E. S. (Zh. Obshch. Khim. **43** [1973] 2470/6; J. Gen. Chem. [USSR] **43** [1973] 2456/61).

Dielectric Constant. Polarization. Dipole Moment. Refraction. Kerr Constant

Dielectric properties of $O=S=NC_6H_4CH_3$-4 were measured in the solid phase at −190°C: deformation limit of the dielectric constant $\varepsilon_\infty = 3.00$; deformation polarization $P_{def} = 60.4$ cm^3; refraction $R_D = 54.5$ cm^3, and atom polarization $P_A = 6$ cm^3 [1]. Measurements in CCl_4 solution at 25°C gave $P_{2\infty} = 164 \pm 2$ cm^3 and $R_D = 49 \pm 2$ cm^3 [2].

The dipole moment of the compound was determined from dielectric measurements: in the liquid phase at 20°C $\mu = 2.39$ D (from Onsager's formula) [3] or $\mu = 1.92$ D [1]. In C_6H_6 solution at 20°C $\mu = 2.40$ D [3], at 25°C $\mu = 2.38$ D [4], and in CCl_4 at 25°C $\mu = 2.36 \pm 0.03$ D [2]. The dipole moment was also derived by the Guggenheim method $\mu = 2.23$ D [3] and from the Thomson vector scheme $\mu = 2.20$ D [4]. A calculation of the dipole moment of the cis and (hypothetical) trans isomer gave 1.3 and 3.2 D, respectively [5].

Refractive index $n_D = 1.6241$ at 20°C [3]. The molar Kerr constant in CCl_4 at 25°C was determined to be $K_m = (920 \pm 30) \times 10^{-12}$ e.s.u./mol [2].

References:

[1] Egorov, Yu. P.; Borovikov, Yu. Ya.; Fedyuk, G. S.; Levchenko, E. S. (Zh. Obshch. Khim. **43** [1973] 2470/6; J. Gen. Chem. [USSR] **43** [1973] 2456/61).
[2] Bonham, J. S.; Cheng, C. L.; Le Fèvre, R. J. W.; Ritchie, G. L. D. (Australian J. Chem. **26** [1973] 421/4).
[3] Kresze, G.; Smalla, H. (Chem. Ber. **92** [1959] 1042/8).
[4] Waclawek, W.; Puchala, C.; Czerwiec, Z. (Bull. Acad. Polon. Sci. Ser. Sci. Chim. **26** [1978] 337/41).
[5] Jannelli, L.; Lamanna, U.; Lumbroso, H. (Bull. Soc. Chim. France **1966** 3626/30).

Chromatography

$O{=}S{=}NC_6H_4CH_3$-4 was studied together with several other N-sulfinyl-anilines by gas chromatography [1, 2, 3] and thin-layer chromatography ($R_f = 0.18$ in benzene) [4].

References:

[1] Czerwiec, Z.; Mirek, J. (Zesz. Nauk Univ. Jagiel. Proc. Chem. No. 14 [1969] 103/6; C.A. **72** [1970] No. 28195).
[2] Czerwiec, Z.; Markovski, J. (Chem. Anal. [Warsaw] **20** [1975] 213/5).
[3] Czerwiec, Z.; Budahegyi, M.; Takács, J. (Magy. Kem. Folyoirat **87** [1981] 279/83).
[4] Czerwiec, Z. (Chem. Anal. [Warsaw] **21** [1976] 751/3).

Polarography

The half-wave potential in DMF is $E_{1/2} = -0.674$ V (against the internal mercury electrode).

Reference:

Czerwiec, Z.; Bogaczek, J. (Z. Anal. Chem. **293** [1978] 415).

Thermal Analysis

$O{=}S{=}NC_6H_4CH_3$-4 was thermally analyzed by TG (thermogravimetry), DTA (differential thermal analysis) and DTG (derivative thermogravimetry) in static air between 20 to 300°C. It was thermally stable in that range. The temperature corresponding to maximum rate of mass loss by evaporation in DTG curve $t^{DTG} = 215$°C. The end-temperature of conversion $t^{end} = 217$°C.

Reference:

Czerwiec, Z.; Pielichowski, J. (J. Therm. Anal. **7** [1975] 465/8).

Ionization and Fragmentation by Electron Impact

The EI mass spectrum (75 eV) is shown in **Fig. 15**, p. 152. The compositions of the major fragment ions were determined by exact mass measurement. m/e (rel. int. in %) = 138 $C_6H_4NOS^+$; 136 $C_7H_6NS^+$; 125 $C_6H_7NS^+$; 124 $C_6H_6NS^+$; 110 $C_5H_4NS^+$; 105 $C_7H_7N^+$; 104 $C_7H_6N^+$; 98 $C_5H_6S^+$; 92 $C_6H_6N^+$; 91 $C_7H_7^+$ (40), $C_6H_5N^+$ (60) [1].

The +E mass spectrum derived from the molecular anion of $O{=}S{=}NC_6H_4CH_3$-4 was recorded and is shown in a figure [2, 3]. The rearrangement peaks $[M]^{+\bullet} \rightarrow [M-CO]^{+\bullet}$ and $[M-CH_3^\bullet]^+ \rightarrow [(M-CH_3^\bullet)-CO]^+$ are of small abundance in comparison with other peaks in the spectrum [2, 3]. The negative-ion mass spectrum is very simple with the molecular ion always being the base peak [3].

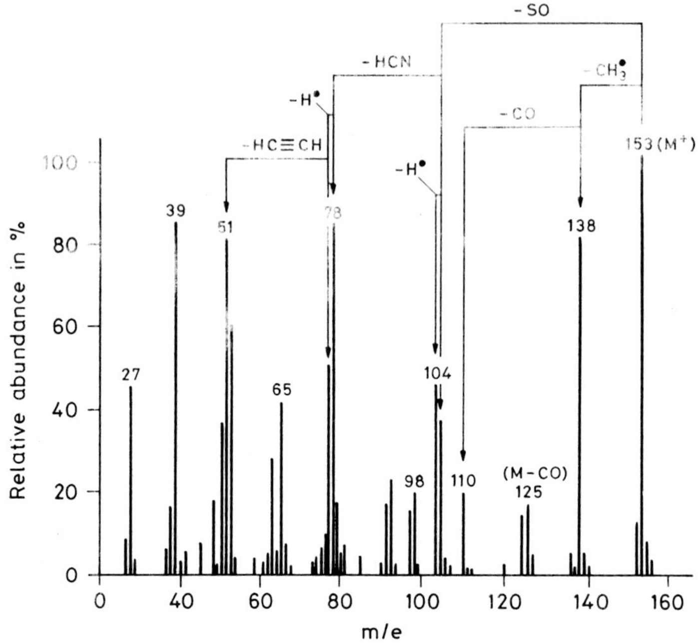

Fig. 15. EI Mass spectrum (75 eV) of $O{=}S{=}NC_6H_4CH_3$-4 [1].

References:

[1] Bowie, J. H.; Larsson, F. C. V.; Schroll, G.; Lawesson, S.-O.; Cooks, R. G. (Tetrahedron **23** [1967] 3743/52).

[2] Bowie, J. H.; White, P. Y.; Wilson, J. C.; Larsson, F. C. V.; Lawesson, S.-O.; Madsen, J. Ø.; Nolde, C.; Schroll, G. (Org. Mass Spectrom. **12** [1977] 191/5).

[3] Bowie, J. H.; Duus, F.; Lawesson, S.-O.; Larsson, F. C. V., Madsen, J. Ø. (Australian J. Chem. **22** [1969] 153/9).

Chemical Reactions

The hydrolysis of $O{=}S{=}NC_6H_4CH_3$-4 with H_2O in THF and CH_3CN at 30°C (using pyridine as a catalyst) is analogous to that of $O{=}S{=}NC_6H_5$, see p. 126. However, the title compound is hydrolyzed more slowly [1]. The compound converts to 4-$CH_3C_6H_4N{=}S{=}NC_6H_4CH_3$-4 when treated with NaH in DMF for 3 h at room temperature (yield 76.7%) [2] or with t-$C_4H_9O^-K^+$ in benzene or $C_2H_5O^-Na^+$ in ether (85% yield) [3]. The reaction with finely divided copper or copper powder (mole ratio 1:1) leads to the formation of the corresponding azo compound (130°C for 7 h, without solvent 52.8% yield; 110°C for 8 h in toluene 2.9% yield) [4]. With 4-$CH_3C_6H_4SO_2H$ at −15°C, it forms bis-(p-tolyl)sulfinylsulfone: $O{=}S{=}NR + 2RSO_2H \rightarrow RNH_2 + RS(O)S(O)_2R + SO_2$ ($R = 4\text{-}CH_3C_6H_4$) [5].

References:

[1] Mironova, D. F.; Konoplya, O. Ya. (Ukr. Khim. Zh. **42** [1976] 374/8; Soviet Progr. Chem. **42** No. 4 [1976] 35/8).
[2] Hörhold, H.-H.; Beck, J. (J. Prakt. Chem. **311** [1969] 621/9).
[3] Hörhold, H.-H.; Floßmann, K. D. (Z. Chem. [Leipzig] **7** [1967] 345/6).
[4] Minami, T.; Agawa, T. (Tetrahedron Letters **1968** 4109/10).
[5] Bederke, K. (Dipl.-Arbeit T.H. Berlin 1961 from Bussas, R.; Kresze, G.; Münsterer, H.; Schwöbel, A., Sulfur Rept. **2** [1983] 215/377, 259).

2.6.22.2.2.9 N-Sulfinyl-3-trifluoromethyl-benzenamine, $O{=}S{=}NC_6H_4CF_3$-3

The title compound formed by the reaction of 3-$CF_3C_6H_4NH_2$ with $SOCl_2$ (without solvent). The reaction mixture was heated at 100°C until a clear solution was obtained. The substance was isolated and purified by distillation; b.p. 57°C/0.1 Torr; yield 91%.

^1H NMR (solvent?/TMS): $\delta = 7.47$, 7.82 ppm.

^{19}F NMR (solvent?/$CFCl_3$): $\delta = -63.59$ ppm.

IR spectrum: 21 bands in the range 3095 to 673 cm^{-1}.

Mass spectrum: m/e (rel. int. in%) = 207 M^+ (100), 191 $M^+ - O$ (5), 188 $M^+ - F$ (15), 161 $CF_3C_6H_4NH_2^+$ (21) [1].

The compound reacts with $(CH_3)_2BSCH_3$ to yield 3-$CF_3C_6H_4N(B(CH_3)_2)SSCH_3$ [2].

References:

[1] Roesky, H. W.; Ambrosius, K. (Z. Naturforsch. **33b** [1978] 759/62).
[2] Meller, A.; Maringgele, W.; Fetzer, H. (Chem. Ber. **113** [1980] 1950/61, 1951/2).

2.6.22.2.3 Multisubstituted N-Sulfinyl-benzenamines

2.6.22.2.3.1 N-Sulfinyl-2,6-difluoro-benzenamine, $O{=}S{=}NC_6H_3F_2$-2,6

The compound was prepared by the reaction of 2,6-difluoroaniline with $SOCl_2$ [1].

^{15}N NMR (neat, CCl_4 and CH_2Cl_2/external liquid NH_3): $\delta = 293.8$ ppm [1 to 4] (referenced to CH_3NO_2: $\delta = -86.4$ ppm).

^{17}O NMR (t = 20°C, CCl_4/external D_2O): $\delta = 437$ ppm [2, 4].

^{19}F NMR (CCl_4/internal C_6F_6): $\delta = +53.9$ ppm (F2, F6) [2] (referenced to $CFCl_3$: $\delta = -109.0$ ppm).

UV (n-heptane): $\lambda_{max} = 346$ nm [2, 4].

References:

[1] Rezvukhin, A. I.; Furin, G. G.; Yakobson, G. G. (Izv. Akad. Nauk SSSR Ser. Khim. **1981** 2512/8; Bull. Acad. Sci. USSR Div. Chem. Sci. **1981** 2078/83).
[2] Zibarev, A. V.; Furin, G. G.; Korobeinicheva, I. K.; Fedotov, M. A.; Yakobson, G. G. (Izv. Akad. Nauk SSSR Ser. Khim. **1983** 2259/65; Bull. Acad. Sci. USSR Div. Chem. Sci. **1983** 2039/44).

[3] Furin, G. G.; Rezvukhin, A. I.; Fedotov, M. A.; Yakobson, G. G. (J. Fluorine Chem. **22** [1983] 231/52, 236, 245).
[4] Furin, G. G.; Fedotov, M. A.; Yakobson, G. G.; Zibarev, A. V. (J. Fluorine Chem. **28** [1985] 273/90, 276).

2.6.22.2.3.2 N-Sulfinyl-2,3,5,6-tetrafluoro-benzenamine, $O{=}S{=}NC_6HF_4$-2,3,5,6

The preparation method is not given.

The HeI photoelectron spectrum shows the following ionization energies E_i (in eV): 9.71 (a_2 and b_1); 11.77; 13.04; 14.03; 14.93 [1, 2]; MNDO calculations give the MO energies $-\varepsilon$ (in eV) = 10.02 (a_2); 10.66 (b_1) [2].

The ν_8(C=C, arom.), ν(N=S), and ν(S=O) Raman frequencies were measured in acetone: 1644, 1252, and 1146 cm^{-1}, respectively. Their intensities (in relative units) are: 0.40, 0.60, and 1.40 [3].

References:

[1] Zibarev, A. V.; Sultanov, A. Sh.; Furin, G. G.; Furlei, I. I. (Izv. Akad. Nauk SSSR Ser. Khim. **1986** 474/6; Bull. Acad. Sci. USSR Div. Chem. Sci. **1986** 439/41).
[2] Zibarev, A. V.; Furin, G. G.; Konchenko, S. N.; Voityuk, A. A. (J. Fluorine Chem. **35** [1987] 627/41, 631/2).
[3] Korobeinicheva, I. K.; Fugaeva, O. M.; Furin, G. G. (Izv. Akad. Nauk SSSR Ser. Khim. **1987** 1766/9; Bull. Acad. Sci. USSR **1987** 1634/7).

2.6.22.2.3.3 N-Sulfinyl-pentafluoro-benzenamine, $O{=}S{=}NC_6F_5$

The compound is prepared by dropping $SOCl_2$ in excess into $C_6F_5NH_2$ (mole ratio 1.5:1) and refluxing for 10 h. After distillation $O{=}S{=}NC_6F_5$ is obtained in 87% yield (with respect to $C_6F_5NH_2$) as a water-sensitive, viscous, yellow liquid [1].

M.p. 18°C, b.p. 175.5 ± 2°C. Enthalpy and entropy of vaporization $\Delta H_v = 11.5$ kcal/mol, $\Delta S_v = 25.5$ $cal \cdot mol^{-1} \cdot K^{-1}$. Vapor pressure equation: log (p/Torr) = 8.447 − 2510/T [1].

^{19}F NMR (neat/$CFCl_3$): δ (in ppm) = −141.7 (ortho-F), −156.7 (para-F), −164.0 (meta-F); ^3J(p-F, m-F) = 19.3 Hz, ^4J(p-F, o-F) = 1.2 Hz [1].

IR (capillary film): 12 bands between 1535 and 510 cm^{-1} are given. Those at ν = 1255 and 1070 cm^{-1} are tentatively assigned to ν(NSO).

MS: m/e (rel. int. in %) = 229 M^+ (100); 213 $C_6F_5NS^+$ (3); 210 $C_6F_4NSO^+$ (4); 198 $C_5F_4NSO^+$ (43); 181 $C_6F_5N^+$ (17); 131 $C_5F_3N^+$ (88); 48 SO^+ (41); 46 SN^+ (40) [1].

The HeI photoelectron spectrum shows the following ionization potentials, partially assigned by aid of MNDO calculations: E_i (in eV) = 9.97 (a_2 and b_1), 11.78, 12.29, 13.19, 13.98. MNDO-calculated energies for the highest occupied molecular orbitals: $-\varepsilon = 10.35$ eV (a_2) and 10.84 eV (b_1) [2].

$O{=}S{=}NC_6F_5$ reacts quantitatively at room temperature with $(CH_3)_2BSCH_3$ in CH_2Cl_2. The SN group is inserted into the B–S bond [3]:

$O{=}S{=}NC_6F_5 + 3(CH_3)_2BSCH_3 \rightarrow C_6F_5N(S{-}SCH_3)B(CH_3)_2 + (CH_3)_2BOB(CH_3)_2 + CH_3SSCH_3$

$O{=}S{=}NC_6F_5$ splits the sulfane bridge of the 1,2,4-trithia-3,5-diborolane ring (in CCl_4, reflux) [4]:

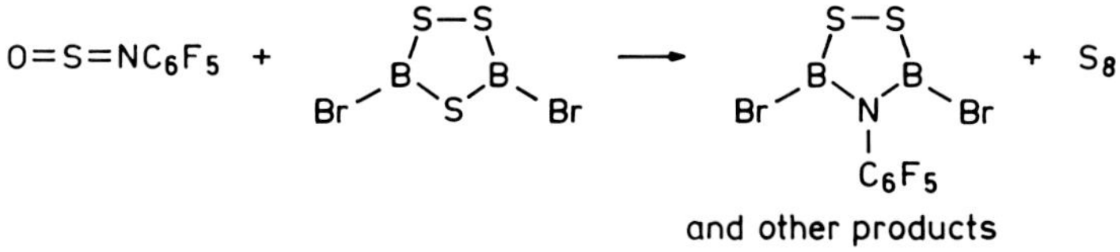

References:

[1] Glemser, O.; von Halasz, S. P.; Biermann, U. (Inorg. Nucl. Chem. Letters **4** [1968] 591/6).
[2] Zibarev, A. V.; Furin, G. G.; Konchenko, S. N. (J. Fluorine Chem. **35** [1987] 627/41, 631/2).
[3] Meller, A.; Maringgele, W.; Fetzer, H. (Chem. Ber. **113** [1980] 1950/61, 1951/2).
[4] Meller, A.; Habben, C. (Monatsh. Chem. **113** [1982] 139/53, 140, 148).

2.6.22.2.3.4 N-Sulfinyl-2,3,5,6-tetrafluoro-4-trifluoromethyl-benzenamine, $O{=}S{=}NC_6F_4CF_3$-4

The compound was obtained by the method of Michaelis; b.p. 65 to 66°C/2 Torr, 88% yield [1].

The SK_β X-ray fluorescence spectrum shows peaks at 2471.8 and 2466.6 eV ($\Delta SK_\alpha = 0.45 \pm 0.01$ eV) [1, 4].

^{13}C NMR (solvent?/TMS): δ (in ppm) = 122.9 (C1), 139.2 (C2,6), 144.6 (C3,5), 108.2 (C4) [8].

^{15}N NMR (neat, CCl_4 and CH_2Cl_2/liquid NH_3): δ = 292.0 ppm [1, 5, 6, 7] (referenced to CH_3NO_2: δ = −88.2 ppm).

^{17}O NMR (t = 80°C, CH_2Cl_2/external H_2O): $\delta = 441 \pm 3$ ppm [5].

^{19}F NMR (neat and CCl_4/internal C_6F_6): δ (in ppm) = +22.7 (F2,3,5,6), +107.0 (CF_3) (referenced to $CFCl_3$: δ = −140.2 (F2,3,5,6), −55.9 (CF_3) ppm) [1, 2].

Raman (neat): ν (in cm^{-1}) = 1343 ν(C–N), 1273 ν(N=S), 1129 ν(S=O), 1102 (NSO)? [2]. The ν_8(C=C, arom.), ν(N=S), and ν(S=O) frequencies were measured in acetone to be 1660, 1283, and 1139 cm^{-1}, respectively. The intensities of these bands (in relative units) are 0.40, 0.30, and 0.50, respectively [3].

UV (n-heptane): $\lambda_{max} = 338$ nm [2, 7].

References:

[1] Dolenko, G. N.; Zibarev, A. V.; Krupoder, S. A.; Furin, G. G.; Yakobson, G. G. (Izv. Sibirsk. Otd. Akad. Nauk SSSR Ser. Khim. Nauk **1980** No. 6, pp. 74/7).
[2] Zibarev, A. V.; Furin, G. G.; Korobeinicheva, I. K.; Fedotov, M. A.; Yakobson, G. G. (Izv. Akad. Nauk SSSR Ser. Khim. **1983** 2259/65; Bull. Acad. Sci. USSR Div. Chem. Sci. **1983** 2039/44).
[3] Korobeinicheva, I. K.; Fugaeva, O. M.; Furin, G. G. (Izv. Akad. Nauk SSSR Ser. Khim. **1987** 1766/9; Bull. Acad. Sci. USSR Div. Chem. Sci. **1987** 1634/7).
[4] Treiger, B. A.; Mazalov, L. N.; Dolenko, G. N.; Guzhavina, T. I. (Comput. Enhanced Spectrosc. **3** [1986] 153/63).
[5] Furin, G. G.; Rezvukhin, A. I.; Fedotov, M. A.; Yakobson, G. G. (J. Fluorine Chem. **22** [1983] 231/52, 236, 245).
[6] Rezvukhin, A. I.; Furin, G. G.; Yakobson, G. G. (Izv. Akad. Nauk SSSR Ser. Khim. **1981** 2512/8; Bull. Acad. Sci. USSR Div. Chem. Sci. **1981** 2078/83).

[7] Furin, G. G.; Fedotov, M. A.; Yakobson, G. G.; Zibarev, A. V. (J. Fluorine Chem. **28** [1985] 273/90, 276).
[8] Rezvukhin, A. I.; Furin, G. G.; Kalabin, G. A.; Krupodev, S. A.; Storozhenko, V. G.; Polonov, V. M.; Kushnarev, D. F.; Yakobson, G. G. (Izv. Sibirsk. Otd. Akad. Nauk SSSR Ser. Khim. Nauk **1982** No. 5, pp. 94/102; C.A. **98** [1983] No. 16213).

2.6.22.2.3.5 Other N-Sulfinyl-fluoro-benzenamines

For spectral data and further information of other N-sulfinyl-fluoro-benzenamines, see [1, 2, 3].

References:

[1] Zibarev, A. V.; Furin, G. G.; Korobeinicheva, I. K.; Fedotov, M. A.; Yakobson, G. G. (Izv. Akad. Nauk SSSR Ser. Khim. **1983** 2259/65; Bull. Acad. Sci. USSR Div. Chem. Sci. **1983** 2039/44).
[2] Furin, G. G.; Fedotov, M. A.; Zibarev, A. V. (J. Fluorine Chem. **28** [1985] 273/90).
[3] Korobeinicheva, I. K.; Fugaeva, O. M.; Furin, G. G. (J. Fluorine Chem. **46** [1990] 179/209, 196/8).

2.6.22.2.3.6 N-Sulfinyl-pentachloro-benzenamine, $O{=}S{=}NC_6Cl_5$

The compound was obtained by the general method of Michaelis, 72% yield, m.p. 122 to 123°C. The SK_β X-ray fluorescence spectrum (in eV) shows peaks at 2472.1, 2469.6, and 2466.3 ($\Delta SK_\alpha = 0.41 \pm 0.01$ eV) [1]. Identification of the compound by means of its SK_β spectrum was tried [2].

References:

[1] Dolenko, G. N.; Zibarev, A. V.; Krupoder, S. A.; Furin, G. G.; Yakobson, G. G. (Izv. Sibirsk. Otd. Akad. Nauk SSSR Ser. Khim. Nauk **1980** No. 6, pp. 74/7).
[2] Treiger, B. A.; Mazalov, L. N.; Dolenko, G. N.; Guzhavina, T. I. (Comput. Enhanced Spectrosc. **3** [1986] 153/63).

2.6.22.2.3.7 N-Sulfinyl-2,6-dimethyl-benzenamine, $O{=}S{=}NC_6H_3(CH_3)_2$-2,6

The compound was obtained by the standard method of Michaelis by the reaction of 2,6-dimethylaniline with $SOCl_2$ in benzene [1, 2, 3] (yield 84% [1]). Other methods of preparation are the sulfinylation of the corresponding lithioaniline [4] or 2,6-dimethyl-N-trimethylsilylaniline [5] with SO_2 gas (yields 52 [4] and 85% [5]).

The substance boils at 102°C/11 Torr [1], 108 to 110°C/9 Torr [2], or 97 to 100°C/15 Torr [4, 5].

Refractive index $n_D^{25} = 1.5778$ [1].

^{1}H NMR (CCl_4/TMS): δ (in ppm) = 6.9 (arom. H3,4,5), 2.13 (CH_3) [6].

IR: ν (in cm^{-1}) = 1280, 1178 ν(NSO) [4]; (neat or KBr): 1282, 1171 ν(NSO) [5].

MS (75 eV): m/e = 167 M^+. The spectrum is very simple and no rearrangement ions are observed. The fragmentation proceeds via $M^+ - HSO - N_2 - HC{\equiv}CH$ [1].

The dipole moment in CCl_4 and benzene at 25°C is 1.52 ± 0.03 D [2] and 1.53 D [7], respectively. It was also calculated to 1.66 D by the Thomson vector scheme, assuming that free rotation occurs about the N–C_{ar} bond [7]. The molar Kerr constant at 25°C in CCl_4 is K_m = (101 ± 2) × 10^{-12} e.s.u./mol. The polarization $P_{2\infty}$ = 99 ± 2 cm^3, the refraction R_D = 49.1 ± 0.2 cm^3 [2].

The compound was thermally analyzed in the range 20 to 300°C by simultaneous thermogravimetry, derivative thermogravimetry, and differential thermal analysis: the temperature corresponding to maximum rate of mass loss in DTG curve t^{DTG} = 213°C; the end-temperature of conversion t^{end} = 217°C [8]. The compound, together with several isomers, was also studied by gas-chromatographic analysis [9] and by thin-layer chromatography [10].

References:

[1] Bowie, J. H.; Larsson, F. C. V.; Schroll, G.; Lawesson, S.-O.; Cooks, R. G. (Tetrahedron **23** [1967] 3743/52).
[2] Bonham, J. S.; Cheng, C. L.; Le Fèvre, R. J. W.; Ritchie, G. L. D. (Australian J. Chem. **26** [1973] 421/4).
[3] Singerman, G. M. (J. Heterocycl. Chem. **12** [1975] 877/82).
[4] Sakai, S.; Fujinami, T.; Komizo, K. (J. Org. Chem. **40** [1975] 3291/2).
[5] Porskamp, P. A. T. W.; Zwanenburg, B. (Synthesis **1981** 368/9).
[6] van Woerden, H. F.; Bijl-Vlieger, S. H. (Recl. Trav. Chim. **93** [1974] 85/90).
[7] Waclawek, W.; Puchala, C.; Czerwiec, Z. (Bull. Acad. Polon. Sci. Ser. Sci. Chim. **26** [1978] 337/41).
[8] Czerwiec, Z. (J. Therm. Anal. **10** [1976] 441/5).
[9] Czerwiec, Z.; Markowski, J. (J. Chromatog. **124** [1976] 76/8).
[10] Czerwiec, Z.; Maleta, E. (J. Chromatog. **132** [1977] 168/71).

2.6.22.2.3.8 N-Sulfinyl-2,4,6-trimethyl-benzenamine, $O{=}S{=}NC_6H_2(CH_3)_3$-2,4,6

The compound was prepared by the reaction of 2,4,6-trimethyl-aniline with $SOCl_2$ [1, 2, 3] or by transsulfinylation with $O{=}S{=}NC_6H_4NO_2$-4 (or $O{=}S{=}NC_6H_4NO_2$-2), yield 45% [4]. Another method is the sulfinylation of the corresponding N-trimethyl-silylaniline with excess SO_2 in THF, yield 85% [5].

The orange-red oil boils at 241°C (slight decomposition) [1], 244 to 246°C [2], 86 to 88°C/2.5 Torr [5], 113 to 114°C/8 Torr (or 241°C/760 Torr) [3], 239 to 241°C/760 Torr (decomposition), 85 to 89°C/3 Torr [4]. It solidifies at −11°C. The density is 1.121 g/cm^3 at 14°C [1].

^{1}H NMR ($CDCl_3$/TMS): δ (in ppm) = 6.78 (arom. H), 2.17 (s, CH_3); (C_7D_8/TMS): δ = 6.60 (arom. H), 2.18 and 2.02 (s, CH_3) [6]; (CCl_4/TMS): δ = 6.82 (arom. H), 2.20 and 2.25 (CH_3) [3]; ($CDCl_3$/TMS at −20°C): δ = 6.93 (arom. H), 2.23 and 2.28 (intensity 1:2, CH_3) [12].

^{13}C NMR ($CDCl_3$/TMS): δ (in ppm) = 138.03 (C–N), 129.78 (C2, C6), 128.71 (C3, C5), 137.06 (C4), 19.53 and 20.84 (CH_3) [6].

IR (neat): ν (in cm^{-1}) = 1314 ν(C_6H_2–N), 1276 ν(N=S), 1152 ν(S=O) [6]; (neat): ν (in cm^{-1}) = 1275, 1185 ν(N=S=O) [5]; (solvent?): ν (in cm^{-1}) = 1305 m, 1270 vs, 1180 s, 1140 s, 1030 m [3].

Raman (neat): ν (in cm^{-1}) = 1312 ν(C_6H_2–N), 1278 ν(S=N), 1150 ν(S=O) [6].

Ionization energies E_i (in eV) were determined using HeI and HeII photoelectron spectroscopy: 8.46 ("b_1"); 8.94 ("a_2"); 10.67 (σ_1); 10.86 (σ_2); 11.13 (π_3); 13.00 (σ_3); 12.00 (σ_4). The electronic structure is different from other investigated substituted N-sulfinylanilines, probably reflecting a different conformation of that compound [10].

Dielectric properties were determined in CCl_4 at 25°C: polarization $P_{2\infty} = 119 \pm 1$ cm³; refraction $R_D = 55.5 \pm 0.5$ cm³; $\mu = 1.73 \pm 0.02$ D; molar Kerr constant $K_m = (284 \pm 6) \times 10^{-12}$ e.s.u./mol [2]; in benzene at 25°C: $P_{2\infty} = 131.6$ cm³, $R_D = 54.7$ cm³, and $\mu = 1.94$ D [7]. $\mu = 1.91$ D calculated on assuming free rotation about the N–C_{ar} bond [11].

The compound is relative stable towards H_2O and dilute H_2SO_4. It can be steam distilled without appreciable hydrolysis [1]. The reaction with KOH–$CHCl_3$ in benzene with an addition of dicyclohexyl-18-crown-6 leads to the formation of the corresponding isocyanide [8, 9].

References:

[1] Michaelis, A.; Siebert, H. (Liebigs Ann. Chem. **274** [1893] 230/42, 240).
[2] Bonham, J. S.; Cheng, C. L.; Le Fèvre, R. J. W.; Ritchie, G. L. D. (Australian J. Chem. **26** [1973] 421/4).
[3] van Woerden, H. F.; Bijl-Vlieger, S. H. (Recl. Trav. Chim. **93** [1974] 85/90).
[4] Carpanelli, C.; Leandri, G. (Ann. Chim. [Rome] **50** [1960] 147/55).
[5] Porskamp, P. A. T. W.; Zwanenburg, B. (Synthesis **1981** 368/9).
[6] Meij, R.; Oskam, A.; Stufkens, D. J. (J. Mol. Struct. **51** [1979] 37/49).
[7] Jannelli, L.; Lamanna, U.; Lumbroso, H. (Bull. Soc. Chim. France **1966** 3626/30).
[8] Domschke, G.; Beckert, R.; Mayer, R. (Synthesis **1977** 275/6).
[9] Mayer, R.; Beckert, R.; Domschke, G. (Ger. 128531 [1976/77]; C.A. **89** [1977] No. 6127).
[10] Louwen, J. N.; van Dam, H.; Stufkens, D. J.; Oskam, A. (J. Electron Spectrosc. Relat. Phenom. **26** [1982] 235/46, 239, 244).
[11] Waclawek, W.; Puchala, C.; Czerwiec, Z. (Bull. Acad. Polon. Sci. Ser. Sci. Chim. **26** [1978] 337/41).
[12] Kuyper, J.; Vrieze, K. (J. Organometal. Chem. **86** [1975] 127/38).

2.6.22.2.3.9 N-Sulfinyl-2,3,5,6-tetramethyl-4-nitro-benzenamine, $O{=}S{=}NC_6(CH_3)_4$-2,3,5,6-NO_2-4

The title compound was obtained by the reaction of 4-NO_2-2,3,5,6-$(CH_3)_4C_6NH_2 \cdot HCl$ with $SOCl_2$ at 60 to 70°C for 1.5 h; "quantitative yield"; yellow needles (from petroleum ether), m.p. 81 to 82°C [1]. Density at 20°C D = 1.175, at 86.3°C 1.105 g/cm³ [2].

The dipole moment at 86.3°C in the liquid phase or in C_6H_5Cl or $C_6H_4Cl_2$-1,2 solution is 2.20 D. The electrical conductivity at 86.3°C is $2.4 \times 10^{-6}\ \Omega^{-1} \cdot cm^{-1}$. Dielectric properties were determined at 20°C (solid phase): deformation limit of the dielectric constant $\varepsilon_\infty = 3.06$, deformation polarization $P_{def} = 109$ cm³, refraction $R_D = 80$ cm³, and atomic polarization $P_A = 29$ cm³ [2].

The substance reacts with PCl_5 (mole ratio 1:1) in CCl_4 at 50°C for 4 h to give $Cl_2S{=}N$-$C_6(CH_3)_4$-2,3,5,6-NO_2-4 in quantitative yield and $POCl_3$ [1].

References:

[1] Markovskii, L. N.; Fedyuk, G. S.; Levchenko, E. S. (Zh. Org. Khim. **8** [1972] 2057/62; J. Org. Chem. [USSR] **8** [1972] 2104/8).
[2] Egorov, Yu. P.; Borovikov, Yu. Ya.; Fedyuk, G. S.; Levchenko, E. S. (Zh. Obshch. Khim. **43** [1973] 2470/6; J. Gen. Chem. [USSR] **43** [1973] 2456/61).

2.6.22.2.3.10 N-Sulfinyl-2,4-di-*tert*-butyl-6-methyl-benzenamine, $O{=}S{=}NC_6H_2(C_4H_9\text{-t})_2\text{-}2,4\text{-}CH_3\text{-}6$

The title compound was prepared by reactions analogous to those used for the synthesis of N-sulfinyl-2,4,6-tri-*tert*-butyl-benzenamine [1, 2, 3] (see below). In addition, it was synthesized by the reaction of the anion of N-trimethylsilyl-6-methyl-2,4-di-*tert*-butylaniline ($2,4\text{-t-}C_4H_9\text{-}6\text{-}CH_3{-}C_6H_2NHSi(CH_3)_3$) with SO_2 in THF at −78°C and then at room temperature (yield 80%) [4].

The compound forms orange crystals which melt at 55.5 to 56.5°C [2], 53 to 55°C [1, 4].

^1H NMR (CCl_4/TMS): δ (in ppm) = 1.31 (s, 9H), 1.43 (s, 9H), 2.25 (s, 3H), and 7.20 (AB_q, $\Delta\delta$ = 0.13, 2H; ^4J = 2 Hz) [2].

IR (KBr): ν(NSO) = 1272 cm^{-1} [1, 2, 4] and 1181 [1, 4] or 1182 [2] cm^{-1}.

UV (hexane): λ_{max} (in nm) = 226 (ε = 10100), 255 (ε = 5390), 282sh (ε = 2700), 392 (ε = 2590) [2].

Mass spectrum: m/e (rel. int. in %) = 265 M^+ (23), 250 (77), 209 (42), 208 (81), 202 (32), 57 (100) [2].

The dipole moment was determined from the dielectric properties in dilute benzene solution at 25°C to be 2.05 D [5].

The title compound reacts with $ArP(Li)Si(CH_3)_2C_4H_9\text{-t}$ (Ar = $2,4,6\text{-}(t\text{-}C_4H_9)_3C_6H_2$) at −78°C in THF to yield the corresponding imino phosphine-P-sulfide ArP(S)=NR (17%), diphosphene Ar–P=P–Ar (6%), thiadiphosphirane S–P(Ar)–PAr (8%), and the sulfur diimide RN=S=NR (R = $C_6H_2(C_4H_9\text{-t})_2\text{-}2,4\text{-}CH_3\text{-}6$) (62%) [6].

References:

[1] Porskamp, P. A. T. W.; Zwanenburg, B. (Synthesis **1981** 368/9).
[2] Inagaki, Y.; Okazaki, R.; Inamoto, N. (Bull. Chem. Soc. Japan **52** [1979] 1998/2001).
[3] Inagaki, Y.; Okazaki, R.; Inamoto, N. (Bull. Chem. Soc. Japan **52** [1979] 3615/8).
[4] Inagaki, Y.; Okazaki, R.; Inamoto, N. (Chem. Letters **1978** 1095/8).
[5] Inagaki, Y.; Okazaki, R.; Inamoto, N.; Shimozawa, T. (Chem. Letters **1978** 1217/8).
[6] Yoshifuji, M.; Shibayama, K.; Toyota, K.; Inamoto, N.; Nagase, S. (Chem. Letters **1985** 237/40).

2.6.22.2.3.11 N-Sulfinyl-2,4,6-tri-*tert*-butyl-benzenamine, $O{=}S{=}NC_6H_2(C_4H_9\text{-t})_3\text{-}2,4,6$

The title compound can be prepared by the standard method of Michaelis, adding $SOCl_2$ in hexane dropwise to an ice-cold mixture of 2,4,6-tri-*tert*-butylaniline ($2,4,6\text{-}(t\text{-}C_4H_9)_3C_6H_2NH_2$) and $(C_2H_5)_3N$ in hexane (83% yield) [1]. The compound forms in several other reactions along with other products and in a smaller yield. It (IV) was formed by the reaction of 2,4,6-tri-*tert*-butyl-7,8-dithia-9-azabicyclo[4.3.0]nona-2,4,9-triene (I) with m-chloroperbenzoic acid (MCPBA) in CH_2Cl_2 at 22°C (yield 5.8%) along to 2,4,r-6-tri-*tert*-butyl-7,8-dithia-9-azabicyclo[4.3.0]nona-2,4,9-triene t-7-oxide (II) and 4,6-di-*tert*-butyl-3H-1,2,3-benzodithiazole 2-oxide (III).

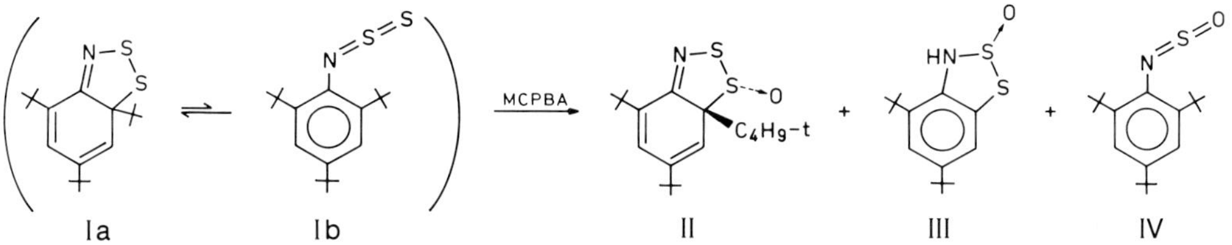

Ia Ib II III IV

The thermolysis of II (refluxing benzene, 3.5 h under N_2) also gave the title compound along with other compounds (yield 24.4%) [2, 3]. $O{=}S{=}NC_6H_2(C_4H_9\text{-}t)_3$-2,4,6 formed in the reaction of $H_2NC_6H_2(C_4H_9\text{-}t)_3$-2,4,6 with SCl_2 in ether (11.9%) along with 2,4,6-$(t\text{-}C_4H_9)_3C_6H_2N{=}S{=}NC_6H_2$-$(C_4H_9\text{-}t)_3$-2,4,6 (12.6%) and I (23.7%) [1]. Treatment of I with an equimolar amount of $P(C_6H_5)_3$ in ether gave the title compound in 11% yield together with 2,4,6-$(t\text{-}C_4H_9)_3C_6H_2N{=}S{=}NC_6H_2$-$(C_4H_9\text{-}t)_3$-2,4,6 (34.4%) and $H_2NC_6H_2(C_4H_9\text{-}t)_3$-2,4,6 and an unidentified compound. The yield of the title compound increased to 38.6% when the reaction was performed in the presence of oxygen [4].

The title compound forms orange crystals which melt at 120.5 to 122°C [1]. X-ray analysis shows that it crystallizes in the monoclinic space group $P2_1/n\text{-}C^5_{2h}$ (No. 14) with a = 15.871(2), b = 11.584(2), c = 10.062(2) Å and β = 90.12(1)°; Z = 4, V = 1849.8(5) Å^3. R = 0.051 for 2491 observed reflexions. The calculated density is D_x = 1.104 g/cm^3. Atomic and thermal parameters are given in the paper. The main feature of the molecular structure is the cis arrangement of the NSO group. This result is in agreement with spectroscopically determined structures of other N-sulfinylamines such as O=S=NH, $O{=}S{=}NCH_3$, and $O{=}S{=}NC_6H_5$. The SO and SN bond lengths are 1.436 and 1.494 Å, respectively [5]. In an earlier paper 1.441 and 1.479 Å, respectively were reported [6]. The OSN and SNC angles are 121.9° and 131.7°, respectively. The remaining bond lengths and bond angles are also presented [5].

^{1}H NMR spectrum (CCl_4): δ (in ppm) = 1.27 (s, 9H); 1.38 (s, 18H), and 7.28 (s, 2H) [1].

IR spectrum (KBr): 1295, 1180 cm^{-1} ν(N=S=O) [1].

UV spectrum (hexane): λ_{max} (in nm) (ε in L·mol^{-1}·cm^{-1}) = 230.5 (12400), 410 (350) [1].

Mass spectrum: m/e (rel. int. in %) = 307 M$^+$ (13), 292 (65), 251 (100), 245 (90) [1].

References:

[1] Inagaki, Y.; Okazaki, R.; Inamoto, N. (Bull. Chem. Soc. Japan **52** [1979] 1998/2001).
[2] Inagaki, Y.; Okazaki, R.; Inamoto, N. (Bull. Chem. Soc. Japan **52** [1979] 3615/8).
[3] Inagaki, Y.; Okazaki, R.; Inamoto, N. (Chem. Letters **1978** 1095/8).
[4] Inagaki, Y.; Hosogai, T.; Okazaki, R.; Inamoto, N. (Bull. Chem. Soc. Japan **53** [1980] 205/9).
[5] Iwasaki, F. (Acta Cryst. B **36** [1980] 1700/3).
[6] Iwasaki, K. (in: Inagaki, Y.; Okazaki, R.; Inamoto, N., Chem. Letters **1978** 1217/8).

2.6.22.2.3.12 N,N'-Bis(sulfinyl)-2,2'-disulfanediyl-bis(3,5-di-*tert*-butyl-benzenamine), $(O{=}S{=}NC_6H_2(C_4H_9\text{-}t)_2\text{-}3,5\text{-}S\text{-}2\text{-})_2$

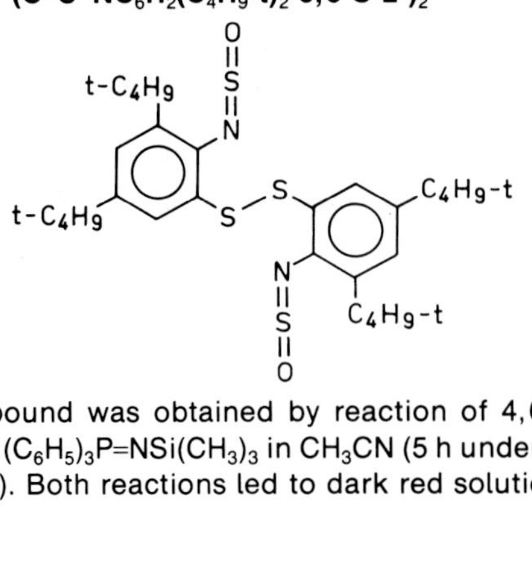

The compound was obtained by reaction of 4,6-di-*tert*-butyl-1,2,3-benzodithiazol-2-ium chloride with $(C_6H_5)_3P{=}NSi(CH_3)_3$ in CH_3CN (5 h under reflux) or with $Na[N(Si(CH_3)_3)_2]$ in CH_2Cl_2 (24 h at 20°C). Both reactions led to dark red solutions of an intermediate compound which gave the title compound on treatment with atmospheric oxygen (24 h at 20°C). The yields are 38 and 31%, respectively. The compound forms yellow crystals which melt at 134°C.

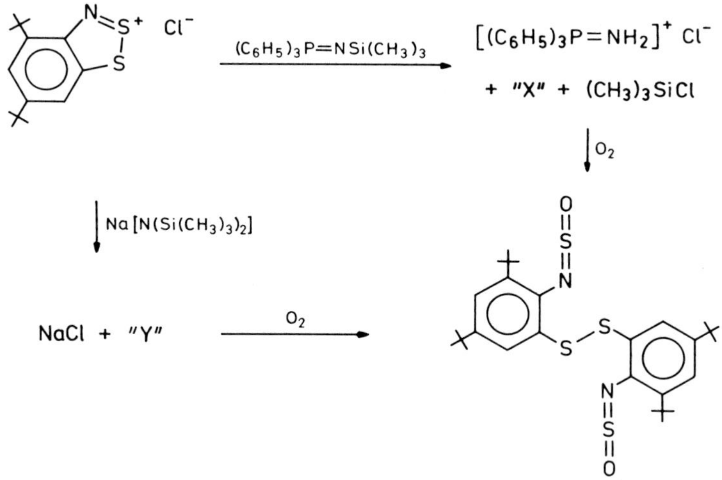

^{1}H NMR: (CH_2Cl_2/TMS): δ (in ppm) = 1.22 (s, $C(CH_3)_3$), 1.36 (s, $C(CH_3)_3$), 7.38 (d, arom. H, J = 2 Hz), 7.53 (d, arom. H, J = 2 Hz).

IR (Nujol): ν (in cm^{-1}) = 1590 ν(C=C, arom.), 1289 ν_{as}(NSO), 1184 ν_s(NSO).

Mass spectrum (70 eV, 130°C): m/e (rel. int. in %) = 564 M^+ (16), 516 $M^+ - SO$ (1), 282 $M^+/2$ (56), 266 $M^+/2 - O$ (100).

The structure of the compound was established by X-ray diffraction. The compound forms triclinic crystals, space group $P\bar{1}-C_i^1$ (No. 2) with a = 1043.6(4), b = 1088.8(7), c = 1451.5(7) pm, α = 76.69(5)°, β = 80.67(3)°, γ = 77.91(4)°; Z = 2, V = 1.5581 nm^3 and D_x = 1.204 g/cm^3. R = 0.078 from 2162 reflections, R_w = 0.080. The molecular structure is shown in **Fig. 16**. The molecule has a disulfide structure with normal geometry: S(1')–S(1) = 204.6(3) pm, dihedral angle

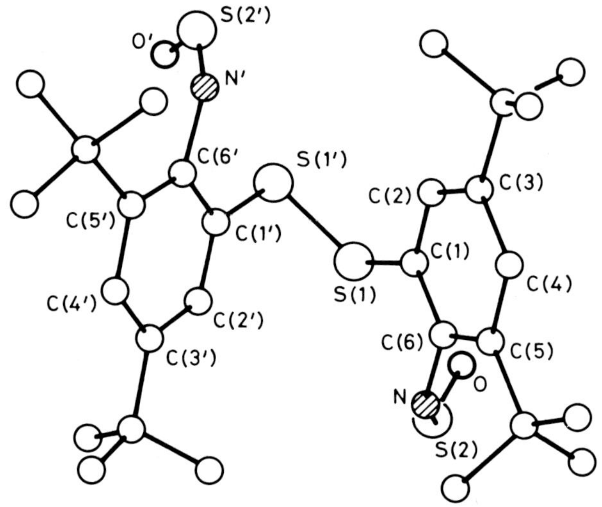

Fig. 16. Molecular structure of $(O{=}S{=}NC_6H_2(C_4H_9\text{-}t)_2\text{-}3,5\text{-}S\text{-}2\text{-})_2$.

C(1)–S(1)–S(1′)–C(1′) = −68.2(4)°. The S=N bonds are cis configured. Selected bond lengths and bond angles are given in Table 42.

Table 42
Selected Bond Lengths and Bond Angles of $(O{=}S{=}NC_6H_2(C_4H_9\text{-}t)_2\text{-}3,5\text{-}S\text{-}2\text{-})_2$.

bond length	in pm	bond angle	in °
S(1)–C(1)	178.4(8)	C(1)–S(1)–S(1′)	104.0(3)
S(2)–N	150.6(7)	S(2)–N–C(6)	129.3(6)
N–C(6)	143.0(9)	S(1)–C(1)–C(6)	119.8(5)
S(2′)–N′	150.1(9)	N–C(6)–C(1)	120.3(7)
N′–C(6′)	141.9(10)	N′–S(2′)–O′	119.9(4)
S(1)–S(1′)	204.6(3)	N′–C(6′)–C(5′)	120.2(6)
S(2)–O	143.2(8)	N–S(2)–O	120.6(4)
S(1′)–C(1′)	176.3(8)	S(1)–C(1)–C(2)	120.8(7)
S(2′)–O′	144.8(8)	N–C(6)–C(5)	118.4(8)
		S(1)–S(1′)–C(1′)	103.8(3)
		S(2′)–N′–C(6′)	130.2(6)
		S(1′)–C(1′)–C(6′)	121.0(6)
		N′–C(6′)–C(1′)	118.1(7)

Reference:

Roesky, H. W.; Weber, K.-L.; Noltemeyer, M.; Sheldrick, G. M. (Z. Naturforsch. **39b** [1984] 163/6).

2.6.22.2.4 N,N′-Bis(sulfinyl)-1,(2, 3, or 4)-phenylenediamine

N,N′-Bis(sulfinyl)-1,2-phenylenediamine, $O{=}S{=}NC_6H_4N{=}S{=}O\text{-}2$

In contrast to earlier communications [1, 2] which reported that o-phenylenediamine reacts with $SOCl_2$ via a cyclization to 2,1,3-benzothiadiazole, it was found that under anhydrous conditions the title compound formed [3]. It was also obtained by transsulfinylation of o-phenylenediamine with N-sulfinyl-p-toluenesulfonamide ($O{=}S{=}N{-}SO_2C_6H_4CH_3$-4) in dry $CHCl_3$ (mole ratio 1:2), yield 71% [4]. The compound forms yellow rods which melt at 56 to 57°C [3] (53 to 54°C [4]).

^{1}H NMR ($CDCl_3$/TMS): A_2B_2 system $\delta(H_A) = 8.01$, $\delta(H_B) = 7.30$ ppm [4].

IR (CCl_4): ν_{as}(NSO) and ν(C–N) = 1300 cm^{-1}; ν_s(NSO) = 1170 cm^{-1} [4], 1166 [3].

UV (c-C_6H_{12}): λ_{max} (in nm) = 237 (log ε = 3.90), 314 (log ε = 4.09), 350 to 360sh (log ε = 3.90) [3].

MS: m/e = 200 M^+, 154 M^+ − SN, 152* M^+ − SO, 137* 154 − OH, 136* M^+ − SO_2, 104* 152 − SO, 77* 104 − HCN (* with metastable peak) [4].

The title compound can be distilled without decomposition [3, 4]; b.p. 72 to 74°C/0.001 Torr [3].

At t = 20°C the deformation limit of the dielectric constant $\varepsilon_\infty = 5.93$, deformation polarization $P_{def} = 160\ cm^3$, the refraction $R_D = 67.8\ cm^3$. The density is 1.49 g/cm³ (at 66.5°C, 1.36 g/cm³) [5]. The dipole moment in the liquid phase at 66.5°C is 1.07 D [5, 6].

The compound is very sensitive to moisture. The color changes to deep red and a smell of SO_2 and 2,1,3-benzothiadiazole can be noticed [3]. The hydrolysis of the substance was intensively studied under three different conditions: a) stirring in H_2O at 70°C, b) steam distillation, and c) moisture at 20 to 25°C. The isolated products are o-phenylenediamine and 2,1,3-benzothiadiazole in a) 78 and 21%, b) 36 and 64%, and c) 27 and 73%, respectively [3].

N,N′-Bis(sulfinyl)-1,3-phenylenediamine, O=S=NC$_6$H$_4$N=S=O-3

The compound was obtained by reacting m-phenylenediamine or hydrochloric m-phenylenediamine with $SOCl_2$ in dry benzene [1].

It forms yellow needles which melt at 44°C [1] (44 to 45°C [7]). The mass spectrum was recorded at 75 eV: m/e = 200 M^+ (relative intensity 20%) and 30 further unassigned peaks [8]. It is soluble in petroleum ether, ether, and absolute alcohol. Slow decomposition takes place by treatment with cold H_2O, and more rapidly with hot H_2O or alkali [1]. A red-brown polymer with sulfur diimide units is formed by the condensation using t-$C_4H_9O^-K^+$ in benzene, toluene, or THF [7].

N,N′-Bis(sulfinyl)-1,4-phenylenediamine, O=S=NC$_6$H$_4$N=S=O-4

The title compound was obtained by the reaction of hydrochloric p-phenylenediamine [1, 9] or p-phenylenediamine [7, 10, 11] in benzene (or without solvent [11]), yield 69% [9], 90 to 95% [7], 65% [10], 92% [11].

The substance forms yellow needles which melt at 115 to 116°C [1], 113 to 115°C [9], 115°C [7], 111 to 112°C [12], 114 to 116°C [13], 116°C [14].

The solid state structure of the compound is unknown. In solution a mixture of two planar trans forms [13] and equal amounts of a cis and a trans isomer [12] have been assumed.

^1H NMR ($CDCl_3$/TMS): δ (in ppm) = 7.90(s) [11].

^{13}C NMR ($CDCl_3$/TMS): δ (in ppm) = 127.0(s) and 148.0 (s, C1, C4) [11].

IR: ν(N=S=O) bands at 1161, 1278, and 1301 cm^{-1}; ν(C=C) at 1488 cm^{-1} [9]. The integrated intensity of the ν_8(C=C, arom.) vibration (1585 cm^{-1}) is 506 $L \cdot mol^{-1} \cdot cm^{-2}$ [12].

UV: λ_{max} (in nm) = 240 (log ε = 4.02), 370 (log ε = 4.39) [9].

The density at 20, 25, and 119°C is 1.53, 1.515, and 1.18 g/cm³, respectively [5].

At t = 20°C the deformation limit of the dielectric constant $\varepsilon_\infty = 3.51$, the deformation polarization $P_{def} = 83.2\ cm^3$, the refraction $R_D = 67.8\ cm^3$ [5, 13]. This R_D value is in agreement with values obtained from solutions in CCl_4 (at 25°C) and in benzene (at 25°C) to be 67 ± 2 cm³ [13] and 58.7 [14], respectively. The polarization is 116 ± 3 [13] and 112.1 cm³ [14]. The dipole moment in benzene at 25°C is 1.60 D [14], in CCl_4 at 25°C 1.49 ± 0.06 D [13], and (neat liquid) 1.50 D at 25°C [5, 6]. The molar Kerr constant was estimated in CCl_4 solution at 25°C to be $K_m = (310 \pm 20) \times 10^{-12}$ e.s.u./mol [13].

The R_f value in $CH_3CO_2C_2H_5$–C_6H_{14} (1:1) is 0.61 [11].

O=S=NC$_6$H$_4$N=S=O-4 is easily soluble in hot benzene but not as soluble in ether, ligroin, and absolute alcohol [1], CCl_4 [10]. The compound is not soluble in n-hexane, cyclohexane, pentane [11]. It is sensitive to moisture [12] and light [11], and is rapidly hydrolyzed by hot H_2O yielding p-phenylenediamine and SO_2 [1]. O=S=NC$_6$H$_4$N=S=O-4 reacts with PCl_5 (mole ratio 1:2) in CCl_4 at 60 to 65°C to produce $Cl_2S{=}NC_6H_4N{=}SCl_2$-4 in 69% yield [15]. A black polymer containing the sulfur diimide unit is formed by the condensation reaction with naphthalene sodium in THF or t-$C_4H_9O^-K^+$ in THF, benzene or toluene under inert-gas atmosphere [7]. The reaction with benzidine (mole ratio 1:1) in NMP (N-methylpyrrolidone) or DMF at 110°C (oil bath temperature) yielded after ~6 h a black solid which was purified by reprecipitation from DMF–ether. The substance was identified as a polymer having an –NH–SO–NH– functionality [11]. An analogous black polymer was formed by the reaction with $H_2N{-}C_6H_4{-}NH_2$-4 at 150°C in DMF (conductivity σ at 80 kHz is $9.5\times10^{-6}\ \Omega^{-1}\cdot cm^{-1}$) [16]. Attempts to prepare a corresponding bis(sulfur diimide) with $LiN(Si(CH_3)_3)_2$ gave only poor yields [17]:

$$O{=}S{=}N{-}C_6H_4{-}N{=}S{=}O\text{-}4 + 2\,LiNR'_2 \xrightarrow[-2\,LiCl]{+2\,R'Cl} R'N{=}S{=}N{-}C_6H_4{-}N{=}S{=}NR'\text{-}4 + 2\,R'_2O$$

$R' = Si(CH_3)_3$

References:

[1] Michaelis, A.; Buntrock, A. (Liebigs Ann. Chem. **274** [1893] 259/66).
[2] Dolars, A. (Houben-Weyl, Methoden der Org. Chem. 4th Ed. **11/2** [1958] 738/44, 740).
[3] Beecken, H. (Chem. Ber. **100** [1967] 2170/7).
[4] Wucherpfennig, W. (Chem. Ber. **101** [1968] 371/2).
[5] Egorov, Yu. P.; Borovikov, Yu. Ya.; Fedyuk, G. S.; Levchenko, E. S. (Zh. Obshch. Khim. **43** [1973] 2470/6; J. Gen. Chem. [USSR] **43** [1973] 2456/61).
[6] Borovikov, Yu. Ya.; Egorov, Yu. P. (Teor. Eksperim. Khim. **7** [1971] 663/9; Theor. Exptl. Chem. **7** [1971] 539/44).
[7] Hörhold, H.-H.; Märtin, R.; Wolf, H. (Ger. [East] 75778 [1969/70]; C.A. **75** [1971] No. 21453).
[8] Bowie, J. H.; Larsson, F. C. V.; Schroll, G.; Lawesson, S.-O.; Cooks, R. G. (Tetrahedron **23** [1967] 3743/52).
[9] Uher, M.; Jendrichovsky, J.; Novomestska, D. (Zb. Pr. Chemickotechnol. Fak. SVST **1972/74** 93/6; C.A. **82** [1974] No. 97771).
[10] Wichterle, O.; Roček, J. (Chem. Listy **47** [1953] 1768/80; Collection Czech. Chem. Commun. **19** [1954] 282/96).

[11] Ghosh, M. (Polym. Bull. [Berlin] **19** [1988] 1/6).
[12] Butt, G.; Davis, M.; Pang, Yuk Tau; Topsom, R. D.; Katritzky, A. R. (J. Chem. Soc. Perkin Trans. II **1974** 260/3).
[13] Bonham, J. S.; Cheng, C. L.; Le Fèvre, R. J. W.; Ritchie, G. L. D. (Australian J. Chem. **26** [1973] 421/4).
[14] Jensen, K. A.; Hofman-Bang, N. (Liebigs Ann. Chem. **548** [1941] 95/105).
[15] Markovskii, L. N.; Fedyuk, G. S.; Levchenko, E. S. (Zh. Org. Khim. **8** [1972] 2057/62; J. Org. Chem. [USSR] **8** [1972] 2104/8).
[16] Ghander, A. M.; El-Sonbati, A. Z.; Diab, M. A. (J. Polym. Sci. Polym. Chem. Ed. **27** [1989] 447/52).
[17] Scherer, O. J.; Wolmershäuser, G.; Jotter, R. (Z. Naturforsch. **37b** [1982] 432/6).

2.6.22.2.5 N-Sulfinyl-(1, or 2)-naphthalenamine

N-Sulfinyl-1-naphthalenamine, 1-$O{=}S{=}NC_{10}H_7$

The title compound was prepared by the reaction of 1-naphthylamine and $SOCl_2$ in dry benzene [1, 2] (yield 60% [2]). The compound forms yellow or red-yellow needles which melt at 33°C [1, 2] or 37°C [3] and boil at 226°C/100 Torr [1] or 150 to 152°C/1 Torr [2].

^1H NMR (CCl_4/TMS): δ (in ppm) = 8.30 (d, H2, J = 7 Hz), 6.63 to 7.15 (m, H3 to H6), 7.90 (m, H7) [5].

UV (n-hexane, 0.4 g/L): λ_{max} (in nm) = 390 (log ε = 3.96) inflexion, 381 (log ε = 3.99), 270 (log ε = 4.06) [3].

A thermal analysis was made using TG, DTG, and DTA. The measured curves are shown in a figure. Degradation was observed by heating the substance in air at 296°C (DTA) [4]. The substance is easily soluble in benzene, ligroin, ether, and absolute alcohol. It is stable against H_2O [1].

The compound converts to the corresponding N,N'-bis(1-naphthyl) sulfur diimide when treated with t-$C_4H_9O^-K^+$ in benzene at 80°C for 1 h (50% yield) [6, 7].

N-Sulfinyl-2-naphthalenamine, 2-O=S=N$C_{10}H_7$

The title compound was obtained analogously to 1-O=S=N$C_{10}H_7$ [1, 2, 8] (yield 90.5% [8], 45% [2]). The compound forms yellow needles which melt at 53°C [1, 2], 55 to 58°C [8] and boil at 123°C/0.8 Torr [8] or 142 to 143°C/1 Torr [2].

^1H NMR spectrum (neat/(($CH_3)_3Si)_2O$): δ (in ppm) = 8.27 (s, H1), 7.27 to 7.70 (m, H3 to H8) [5].

It was thermally analyzed by TG, DTG, and DTA. Decomposition was observed at 247°C in the DTA curve [4]. The relations of solubility of the compound are the same as for 1-O=S=N-$C_{10}H_7$. However, the compound is not so stable against H_2O. Heating with H_2O yields a dark yellow oil which converts to 2-naphthylamine and SO_2 [1]. S_2Cl_2 reacts with the title compound to yield naphtho[2,1-d]-1,2,3-dithiazol-2-ium chloride [9].

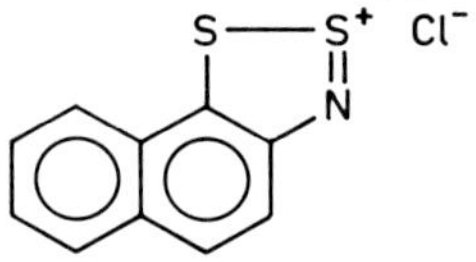

The compound reacts with t-$C_4H_9O^-K^+$ in benzene at 80°C for 1 h to produce N,N'-bis(2-naphthyl) sulfur diimide (yield 85%) [6, 7].

References:

[1] Michaelis, A. (Liebigs Ann. Chem. **274** [1893] 200/66, 253/9).
[2] Kataev, E. G.; Plemenkov, V. V.; Pluzhnov, V. K. (Zh. Org. Khim. **3** [1967] 336/41; J. Org. Chem. [USSR] **3** [1967] 319/23).
[3] Leandri, G.; Mangini, A. (Spectrochim. Acta **15** [1959] 421/31, Ric. Sci. **27** [1957] 1920/5).
[4] Czerwiec, Z.; Pielrichowski, J. (J. Therm. Anal. **7** [1975] 465/8).
[5] Mirek, J.; Czerwiec, Z. (Zesz. Nauk. Uniw. Jagiel. Proc. Chem. No. 18 [1973] 183/6; C.A. **79** [1973] No. 145570).
[6] Hörhold, H.-H.; Beck, J. (J. Prakt. Chem. **311** [1969] 621/9).
[7] Hörhold, H.-H.; Floßmann, K.-D. (Z. Chem. [Leipzig] **7** [1967] 345/6).
[8] Wichterle, O.; Roček, J. (Chem. Listy **47** [1953] 1768/80; Collection Czech. Chem. Commun. **19** [1954] 282/96).
[9] Maior, O.; Arsenescu, N. (Analele Univ. Bucuresti Ser. Stiint. Nat. Chim. **16** [1967] 101/6; C.A. **71** [1969] No. 3331).

2.6.22.2.6 N,N′-Bis(sulfinyl)-1,(2, or 5)-naphthalenediamine

N,N′-Bis(sulfinyl)-1,2-naphthalenediamine, 1-O=S=NC$_{10}$H$_6$N=S=O-2

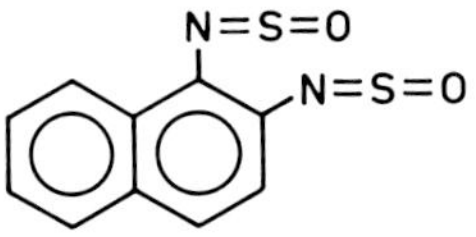

The compound can be prepared in 63% yield by the reaction of freshly distilled 1,2-naphthylene diamine with $SOCl_2$ in dry benzene. The compound forms yellow-orange prisms with melting point of 103 to 104°C. Boiling point is 129 to 130°C/0.001 Torr.

UV (c-C_6H_{12}): λ_{max} (in nm) = 223 (log ε = 4.66), 283 (log ε = 4.17), 341 (log ε = 4.24), 415 (log ε = 3.89) [1].

N,N′-Bis(sulfinyl)-1,5-naphthalenediamine, 1-O=S=NC$_{10}$H$_6$N=S=O-5

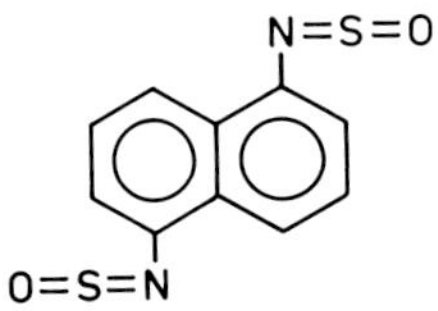

The title compound was prepared by the standard method of Michaelis by reacting 1,5-naphthylene diamine with $SOCl_2$ in benzene. Green crystals were obtained which melt at 187 to 190°C with decomposition.

^{1}H NMR (CH_2Cl_2/TMS): δ (in ppm) = 8.72 (H2), 7.62 (H3), and 8.51 (H4).

IR (KBr): ν (in cm^{-1}) = 1325 m, 1285 m, 1220 m, 1200 m, 1160 m, 1080 m.

The dipole moment was determined at 25°C in benzene to be 0.96 D [2].

References:

[1] Beecken, H. (Chem. Ber. **100** [1967] 2170/7).
[2] van Woerden, H. F.; Bijl-Vlieger, S. H. (Recl. Trav. Chim. **93** [1974] 85/90).

2.6.22.3 N-Sulfinylamino-Substituted Heterocycles

2.6.22.3.1 5-Sulfinyl-amino-2-amino-3,4-thiophenedicarbonitrile

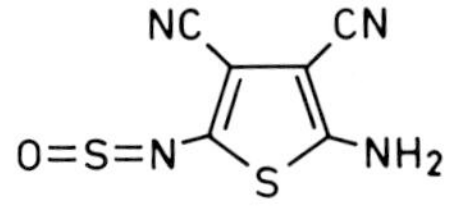

The compound was formed by partial hydrolysis on exposure of 2,5-bis(sulfinylamino)-3,4-thiophenedicarbonitrile to atmospheric moisture. It was also obtained by refluxing 2,5-bis(sulfinylamino)-3,4-thiophenedicarbonitrile in toluene (in an open flask) for 1 h (yield 36%). The substance forms shiny red crystals.

IR (KBr): ν (in cm^{-1}) = 3380 s, 3280 s, 3175 s, 2225 s, 1625 s, 1510 s, 1465 s, 1440 w, 1318 s, 1139 s, 1020 m, 940 m.

The reaction with $COCl_2$ produced complex mixtures. IR spectroscopic study showed that the NSO function was destroyed.

Reference:

Wudl, F.; Zellers, E. T.; Nalewajek, D. (J. Org. Chem. **45** [1980] 3211/5).

2.6.22.3.2 2,5-Bis(sulfinyl-amino)-3,4-thiophenedicarbonitrile

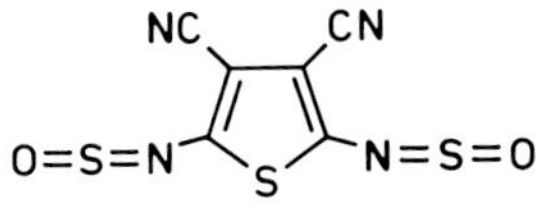

The compound was obtained by the reaction of 2,5-diamino-3,4-thiophenedicarbonitrile with $SOCl_2$ (without solvent). It was purified by recrystallization from $SOCl_2$ (yield 45%) to give gold needles which turn red on prolonged exposure to the atmosphere.

^{13}C NMR (in THF-d_8/TMS): δ (in ppm) = 149.6, 111.1, and 107.8.

IR (KBr): ν (in cm^{-1}) = 2230 s, 1440 m, 1379 m, 1298 w, 1270 s, 1231 w, 1205 w, 1165 s.

The compound is very moisture-sensitive. Exposure to atmospheric moisture led to the formation of 5-sulfinyl-amino-2-amino-3,4-thiophenedicarbonitrile. Treatment with moist DMF produced the complete hydrolysis product 2,5-diamino-3,4-thiophenedicarbonitrile. On stirring with dry pyridine under N_2 in a dry flask at room temperature for 15 min and subsequent standing overnight, a black, amorphous solid is formed (room temperature conductivity (pressed pellet) $\sigma_{RT} = 1/700\ \Omega^{-1}\cdot cm^{-1}$). Polymerization attempt with $AlCl_3$ also yielded indifferent, black, intractable materials which could not be separated from Al salts. The reaction with TTF (tetrathiafulvalene) in CH_3CN produced shiny black platelets (room temperature conductivity $\sigma_{RT} = 10^{-3}\ \Omega^{-1}\cdot cm^{-1}$). The calculated formula is $C_{27}H_{15}N_{10}O_5S_{12}$.

Reference:

Wudl, F.; Zellers, E. T.; Nalewajek, D. (J. Org. Chem. **45** [1980] 3211/5).

2.6.22.3.3 5-Sulfinyl-amino-2-amino-3,4-selenophenedicarbonitrile

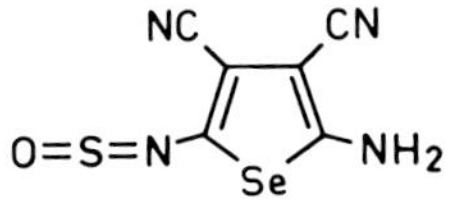

The preparation of this compound is similar to that of the analogous sulfur compound; see p. 166. However, a correct elemental analysis could not be obtained.

Reference:

Wudl, F.; Zellers, E. T.; Nalewajek, D. (J. Org. Chem. **45** [1980] 3211/5).

2.6.22.3.4 2,5-Bis(sulfinyl-amino)-3,4-selenophenedicarbonitrile

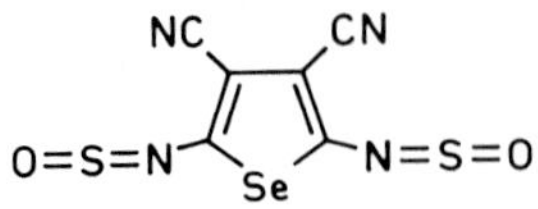

The compound was obtained by adding 2,5-diamino-3,4-selenophenedicarbonitrile to a vigorously stirred mixture of dry CH_2Cl_2, pyridine, and freshly distilled $SOCl_2$. The crude product, which was obtained after 24 h by filtration and subsequently washing with CH_2Cl_2, was again treated with a mixture of CH_2Cl_2, pyridine, and $SOCl_2$. After 1 h stirring golden yellow crystals were obtained (69% yield).

IR (KBr): ν (in cm^{-1}) = 2230 m, 1500 m, 1457 m, 1413 m, 1375 w, 1250 m, 1150 s, 1015 w, 910 w, 750 w.

The substance is very moisture-sensitive. Exposure to atmospheric moisture led to the formation of 5-sulfinyl-amino-2-amino-3,4-selenophenedicarbonitrile. Treatment with moist DMF produced the complete hydrolysis product 2,5-diamino-3,4-selenophenedicarbonitrile. With pyridine, a similar reaction takes place as that with the thiophene compound yielding a black solid with g = 2.0025 G in the ESR spectrum and the room temperature conductivity σ_{RT} = 1/3000 $\Omega^{-1} \cdot cm^{-1}$.

Reference:

Wudl, F.; Zellers, E. T.; Nalewajek, D. (J. Org. Chem. **45** [1980] 3211/5).

2.6.22.3.5 Other N-Sulfinylamino-Substituted Heterocycles

In Table 43, several other N-sulfinylamino-substituted heterocycles have been compiled. The compounds were prepared by two methods:

Method I: Reaction of the corresponding amine (Het-NH_2, Het = heterocyclyl) with $SOCl_2$ (standard method of Michaelis).

Method II: Transsulfinylation of the corresponding amine (Het-NH_2, Het = heterocyclyl) with $O{=}S{=}NSO_2C_6H_5$ (Method IIa) or with $O{=}S{=}NC_6H_5$ (Method IIb).

Table 43
Preparation and Properties of N-Sulfinylamino-Substituted Heterocycles.
Further information on compounds marked with an asterisk is found on pp. 170/1.

No.	compound	method of preparation (yield in %) properties	Ref.
1	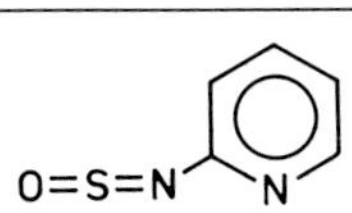	IIa, was handled in solution [1] or isolated in pure state, but decomposed within hours [2]	[1, 2]
2	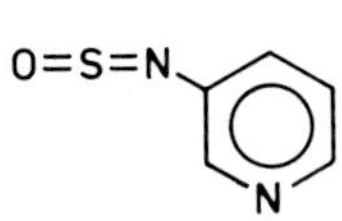	IIa (92), in dry benzene, yellow crystals m.p. 24 to 27°C, b.p. 73 to 73.5°C	[1]

Table 43 (continued)

No.	compound	method of preparation (yield in %) properties	Ref.
3	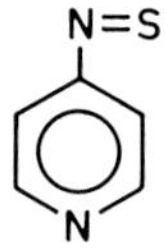	prone to hydrolysis and to deterioration if neat, even when dry	[2]
4*	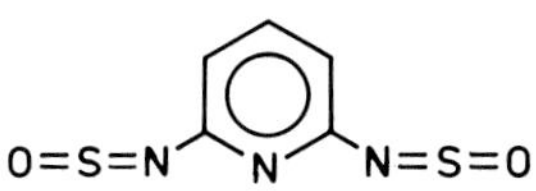	I, refluxing for 5 to 6 h at 80°C dark brown crystals m.p. 103°C IR: 1295 and 1260 cm^{-1} ν(NSO)	[8]
5*	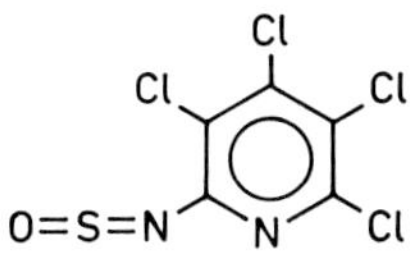	I (92), 3 h refluxing orange crystals m.p. 58°C readily hydrolyzed by air moisture	[3]
6*	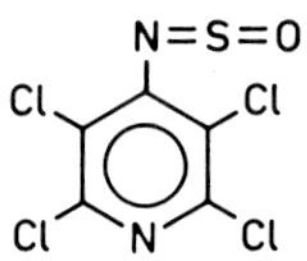	I (68), 15 h refluxing yellow crystals m.p. 70 to 72°C readily hydrolyzed by air moisture	[3]
7	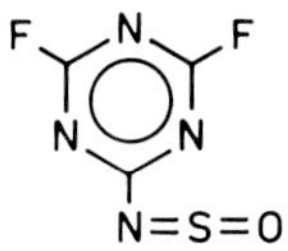	I (5), 4 d refluxing weak yellow, crystalline solid m.p. 51 to 52°C ^{19}F NMR ($CH_3CN/CFCl_3$): $\delta = -35.7$ ppm IR (Nujol/KBr): 14 bands in the range 1605 to 480 cm^{-1} vigorous reaction with H_2O under SO_2 cleavage	[4]
8	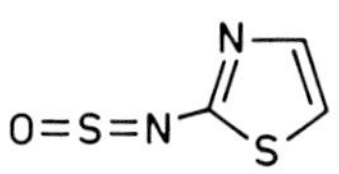	IIa orange-red substance extremely sensitive to moisture	[1]
9	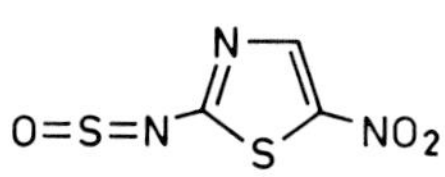	I (83), in benzene m.p. 84°C	[5]
10	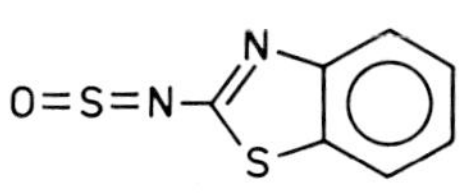	I, in benzene, 6 h refluxing	[1]
11	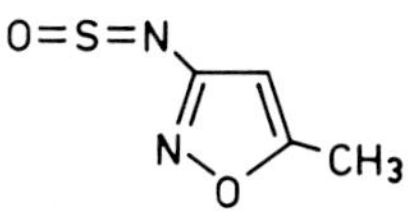	I (90), in benzene oil ^{13}C NMR ($CDCl_3$/TMS): $\delta = 159.8$ ppm (C–N=S=O)	[5]
12	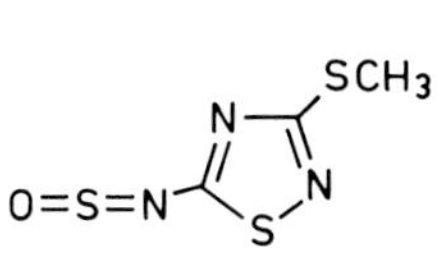	I (78), in benzene red prisms m.p. 74 to 76°C (purified by high-vaccum sublimation)	[1]

Table 43 (continued)

No.	compound	method of preparation (yield in %) properties	Ref.
13	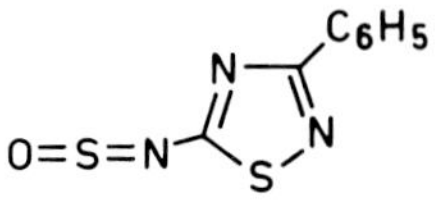	I (85), in benzene red crystals [1], oil [5] ^{13}C NMR ($CDCl_3$/TMS): δ = 163.6 ppm (C–N=S=O) [5]	[1, 5]
14	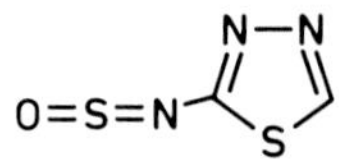	I (89), in benzene m.p. 105 to 107°C	[5]
15	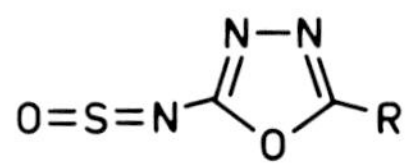 R = C_6H_5; 4-$CH_3C_6H_4$	I (R = C_6H_5: 86; R = 4-$CH_3C_6H_4$: 84), in benzene R = C_6H_5: m.p. = 99°C R = 4-$CH_3C_6H_4$: m.p. = 132 to 134°C	[5]
16*	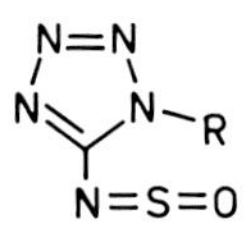 R = CH_3, $CH_2C_6H_5$	I (R = CH_3: 78; R = $CH_2C_6H_5$: 90), in dry benzene or p-xylene [5] unstable, moisture-sensitive compounds [5, 9] R = CH_3: m.p. = 42.4°C ^{13}C NMR ($CDCl_3$/TMS): δ = 150.3 ppm (C–N=S=O) [5] R = $CH_2C_6H_5$, oil ^{13}C NMR ($CDCl_3$/TMS): δ = 149.8 ppm (C–N=S=O) [5, 9], 50.5 ppm (N–$\underline{C}H_2$–C_6H_5) [9]	[5, 6, 9]
17	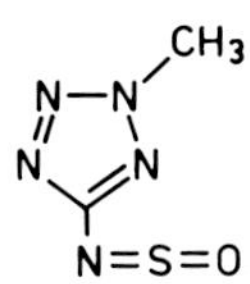	I (89), in benzene oil ^{13}C NMR ($CDCl_3$/TMS): δ = 160.1 ppm (C–N=S=O)	[5]
18	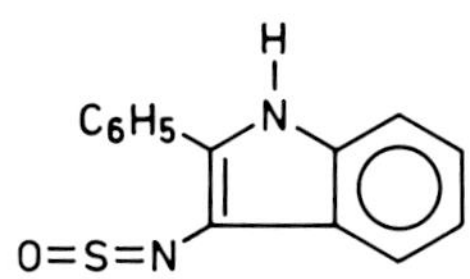	IIb (90), in benzene, 12 h at ambient temperature m.p. 228 to 229°C IR: 1139 ν(NSO), 3240 cm^{-1} ν(N–H)	[7]

* Further information:

Table **43**, No. **4**: Reaction with 2,6-diaminopyridine in DMF at 180 to 200°C for 10 h (mole ratio 1:1) yields dark black poly(bis-2,6-diaminopyridinesulfoxide). The electrical conductivity of the polymer at various frequencies was measured [8].

Table **43**, Nos. **5** and **6**: The compounds are converted by reaction with PCl_5 in CCl_4 to iminothionylchlorides. They produce trichlorophosphato tetrachloropyridines when treated with ground PCl_5 at ~150°C for ~30 min. The compounds react with dimethylformamide and aromatic aldehydes to give formamidines and benzylidene derivatives. 2-(2,2,2-trichloroethylidene)amino-3,4,5,6-tetrachloropyridine is obtained by the reaction with chloral [3].

Table **43**, No. **16**: The compounds were characterized as Diels-Alder adducts with 2,3-dimethylbuta-1,3-diene. Heating the compound (R = CH_3) with $SOCl_2$ in p-xylene yields the azide

S–N=C(N_3)–N=C(H), which is also obtained by direct heating Het-NH_2 with excess $SOCl_2$ in dry p-xylene [6].

References:

[1] Beecken, H. (Chem. Ber. **100** [1967] 2159/63).
[2] Hanson, P.; Wren, S. A. C. (J. Chem. Soc. Perkin Trans. II **1987** 197/203).
[3] Kukhar, V. P.; Matsnev, V. V.; Pavlenko, N. G.; Pavlenko, A. F. (Khim. Geterotsikl. Soedin. **1978** 1360/2; Chem. Heterocycl. Compounds [USSR] **1978** 1104/6).
[4] Roesky, H. W.; Giere, H. H. (Chem. Ber. **102** [1969] 2330/5).
[5] Butler, R. N.; O'Halloran, G. A. (Chem. Ind. [London] **1986** 750/1).
[6] Butler, R. N.; O'Donoghue, D. A.; O'Halloran, G. A. (J. Chem. Soc. Chem. Commun. **1986** 800/1).
[7] Bruni, P.; Poloni, M. (Gazz. Chim. Ital. **100** [1970] 684/8).
[8] El-Sonbati, A. Z.; Ahmed, M. A.; Reicha, F. M.; Diab, M. A. (Eur. Polym. J. **25** [1989] 125/7).
[9] Butler, R. N.; O'Halloran, G. A.; O'Donoghue, D. A. (J. Chem. Res. S **1988** 188/9).

2.6.22.4 N-Sulfinyl-carboxamides, O=S=NC(O)R. N-Sulfinyl-carbamic Acid Esters, O=S=NC(O)OR; R = alkyl, aryl

Preparation of O=S=NC(O)R and O=S=NC(O)OR (R = alkyl, aryl) was first reported in the sixties [1 to 10]. Until then the standard method was the reaction of $SOCl_2$ with equimolar quantities of an appropriate nitrogen compound:

$$R^1_2NC(O)R^2 + SOCl_2 \rightarrow O{=}S{=}NC(O)R^2 + 2\,R^1Cl$$

$$R^1N{=}CR^2(OR^3) + SOCl_2 \rightarrow O{=}S{=}NC(O)R^2 + R^1Cl + R^3Cl$$

Some compounds were also prepared by thermolysis of $R^2C(OR^3)$=N–S(O)Cl and $R^2C(O)N{=}S{=}N$-$C(O)R^2$. R^1 = H, Cl, $Si(CH_3)_3$, R^2 = alkyl, aryl, O-alkyl, O-aryl, and R^3 = alkyl.

The compounds are purified by vacuum distillation. The compounds, conditions of preparation, yields, and boiling points (melting point) are compiled in Tables 44 and 45, pp. 172/3. Besides the desired products, appreciable amounts of the nitriles RC≡N [12, 13] or diimides RC(O)N=S=NC(O)R [11] may form. Nearly all the compounds are colorless or yellow oily liquids which may be distilled in vacuum (see Tables 44 and 45). Only O=S=NC(O)$C_6H_4NO_2$-4 [14] and O=S=NC(O)OC_6H_5 [16, 20] form colorless crystals. For O=S=NC(O)OCH_3 and O=S=NC(O)OC_2H_5 an unpleasant pungent odor is reported [4].

Data for density, refractive indices, and molecular refraction, dielectric constants, and dipole moments are compiled in Table 46, p. 174; vibrational spectra in Table 47, p. 175.

Table 44
Preparation, Yields, and Boiling Points (Melting Point) of N-Sulfinyl-carboxamides, O=S=NC(O)R.

compound	preparation	yield in %	b.p. in °C/ p in Torr	Ref.
O=S=NC(O)CH_3	$((CH_3)_3Si)_2NC(O)CH_3 + SOCl_2$; −60°C to room temperature; 1 h	70 to 74	36 to 37/11	[8, 9]
O=S=NC(O)C_6H_5	$((CH_3)_3Si)_2NC(O)C_6H_5 + SOCl_2$; in ether; room temperature	91	68 to 70/0.1	[9]
	$C_6H_5C(OC_2H_5)$=NH + $SOCl_2$, mole ratio 2:1; in CCl_4; 20°C for 1 h, and then at 120 to 140°C for 1.5 h	33	77 to 79/0.3	[10]
	$C_6H_5C(O)NH_2 + SOCl_2 + C_5H_5N$; in C_6H_6; room temperature, overnight; >10°C for 12 h	62	76/0.16	[11]
	$C_6H_5C(O)NH_2 + SOCl_2 + C_5H_5N$; in C_6H_6; −3 to −5°C	65 to 69	61 to 63/0.08 135 to 140/15 to 17	[12]
	$C_6H_5C(O)NCl_2 + SOCl_2$; boiled in CCl_4 until Cl_2 evolution ceased	74		[13]
	$(CH_3)_3SiNHC(O)C_6H_5 + SOCl_2 + C_5H_5N$; in C_6H_6; ice-water cooling	46	75/0.08	[13]
	$C_6H_5C(OCH_3)$=NSOCl; thermolysis at 130 to 140°C	74	62 to 64/0.07	[13]
O=S=NC(O)C_6H_4Cl-2	2-$ClC_6H_4C(O)NCl_2 + SOCl_2$; boiled in CCl_4 until Cl_2 evolution ceased	87	83 to 85/0.06	[13]
O=S=NC(O)C_6H_4Cl-4	4-$ClC_6H_4C(O)NH_2 + SOCl_2 + C_5H_5N$; in C_6H_6; 10 to 14°C	62 to 65	91 to 95/0.08	[12]
O=S=NC(O)C_6H_4Br-4	4-$BrC_6H_4C(O)NH_2 + SOCl_2 + C_5H_5N$; in C_6H_6; 10 to 14°C	50 to 60	101 to 104/0.08	[12]
	4-$BrC_6H_4C(OR')$=NH + $SOCl_2$; in C_6H_6; 70 to 75°C; ~3 h; R′ = CH_3, C_2H_5	77	77 to 78/0.05	[14]
O=S=NC(O)$C_6H_4NO_2$-4	4-$NO_2C_6H_4C(O)NCl_2 + SOCl_2$; in C_6H_6; 70 to 75°C; ~3 h	92	m.p. 98 to 99	[14]
	4-$NO_2C_6H_4C(OR')$=NH + $SOCl_2$; in C_6H_6; 70 to 75°C; ~3 h; R′ = CH_3, C_2H_5	75	m.p. 98 to 99	[14]
O=S=NC(O)$C_6H_4OCH_3$-4	4-$CH_3OC_6H_4C(O)NH_2 + SOCl_2 + C_5H_5N$; in C_6H_6; 14 to 16°C	48 to 50	109 to 110/0.05	[12]
O=S=NC(O)$C_6H_4CH_3$-4	4-$CH_3C_6H_4C(O)NH_2 + SOCl_2 + C_5H_5N$; in C_6H_6; 60 to 70°C	50 to 52	84 to 85/0.07	[12]
O=S=NC(O)C(O)N=S=O	$(((CH_3)_3Si)_2NC(O))_2 + 2SOCl_2$; in ether; −50 to −15°C	~100		[9]

Table 45
Preparation, Yields, and Boiling Points of N-Sulfinyl-carbamic Acid Esters, O=S=NC(O)OR.

compound	preparation	yield in %	b.p in °C/ p in Torr	Ref.
$O{=}S{=}NC(O)OCH_3$	$CH_3OC(O)NCl_2 + SOCl_2$; 75 to 95°C until Cl_2 evolution ceased	75 to 80	53 to 54/18	[1, 4]
	$(CH_3O)_2C{=}NCl + SOCl_2$; 0°C until Cl_2 evolution ceased	41	44/12	[6, 7]
	$CH_3OC(O)N{=}S{=}NC(O)OCH_3$; thermolysis at 180 to 210°C	36		[15]
$O{=}S{=}NC(O)OC_2H_5$	$C_2H_5OC(O)NH_2 + SOCl_2 + C_5H_5N$;			
	in benzene; room temperature	20	≤30/0.1	[2]
	in ether; room temperature	73	27 to 28/0.8	[18]
	in ether; ice cooling	40	60 to 61/3.33	[15, 16]
	$C_2H_5OC(O)NCl_2 + SOCl_2$; 75 to 95°C until Cl_2 evolution ceased	75 to 80	60 to 61/25	[1, 4]
	$(C_2H_5O)_2C{=}NCl + SOCl_2$; 0°C until Cl_2 evolution ceased	50	48/12	[6, 7]
	$C_2H_5OC(O)N{=}S{=}NC(O)OC_2H_5$; thermolysis at 150°C for 40 min and then at 180 to 210°C	49		[15]
$O{=}S{=}NC(O)OC_3H_7$-n	$C_3H_7OC(O)NH_2 + SOCl_2 + C_5H_5N$; in ether; ice cooling for 1 h	51	73 to 74/18 (2.4 kPa)	[15, 16]
$O{=}S{=}NC(O)OC_4H_9$-n	n-$C_4H_9OC(O)NH_2 + SOCl_2 + C_5H_5N$; in ether; ice cooling for 1 h	43	86 to 88/18 (2.4 kPa)	[15, 16]
	n-$C_4H_9OC(O)NH_2 + SOCl_2 + C_5H_5N$; in ether; ice cooling for 2 h	56	75 to 78/12	[16, 17]
$O{=}S{=}NC(O)OC_4H_9$-t	t-$C_4H_9OC(O)NH_2 + SOCl_2 + C_5H_5N$; in ether; ice cooling for 2 h	71	28/3	[16]
$O{=}S{=}NC(O)OCH_2CH_2Cl$	$(ClCH_2CH_2O)_2C{=}NCl + SOCl_2$; 0°C, later 95 to 100°C	14	78/2	[7]
$O{=}S{=}NC(O)OCH_2CCl_3$	$Cl_3CCH_2OC(O)NH_2 + SOCl_2 + C_5H_5N$; in ether; ice cooling for 2 h	91	63/0.4	[16]
$O{=}S{=}NC(O)OC_6H_5$	$C_6H_5OC(O)NH_2 + SOCl_2 + C_5H_5N$; in ether; ice cooling for 2 h	66	51 to 52/0.2	[16]

Table 46

Densities, d (in g/cm³), Refractive Indices, n, Dielectric Constants, ε, and Dipole Moments, μ (in D), of O=S=NC(O)R.
Densities, d (in g/cm³), Refractive Indices, n, and Molecular Refraction, MR_D (in cm³), of O=S=NC(O)OR.

O=S=NC(O)R

R	C_6H_5	4-ClC_6H_4	4-BrC_6H_4	4-$NO_2C_6H_4$	4-$CH_3OC_6H_4$	4-$CH_3C_6H_4$	Ref.
d_4^{20}	1.2326	1.4503	1.4675	—	1.3221	1.3119	[12]
	1.2298	—	—	—	—	—	[10]
d_4^{-190}	1.37	1.52	—	1.46	—	—	[22]
d_4^t (liq.)	1.29 (20°)	1.42 (20°)	—	1.34 (96°)	—	—	[22]
n_D^{20}	1.5775	1.6050	1.6225	—	1.5995	1.6123	[12]
	1.5735	—	—	—	—	—	[10]
ε (liq.)	26.7 (20°)	10.4 (20°)	—	8.5 (96°)	—	—	[22]
ε (−190°)	1.37	2.91	—	3.46	—	—	[22]
μ	3.5	2.28	—	—	—	—	[22]

O=S=NC(O)OR

OR	OCH_3	OC_2H_5	Ref.	OR = OC_4H_9-n	Ref.
d_4^{20}	1.3690	1.2620	[1, 4]	1.3200	[17]
n_D^{20}	1.4525	1.4431	[1, 4]	1.4530	[17]
MR_D	23.85	28.39	[4]	—	

Table 47

Vibrational Spectra (frequencies in cm^{-1}) of N-Sulfinyl-carboxamides O=S=NC(O)R and N-Sulfinyl-carbamic Acid Esters O=S=NC(O)OR.

compound	state	spectrum	ν(C=O)	ν_{as}(N=S=O)	ν_s(N=S=O)	δ(N=S=O)	Ref.
$O{=}S{=}NC(O)CH_3$ *)	in CCl_4	IR	1740	1245	1090	585	[8, 9]
$O{=}S{=}NC(O)C_2H_5$	in Nujol	IR	—	1240, 1265	1126	—	[19]
$O{=}S{=}NC(O)C_6H_5$ *)	film	IR	1700	1240	1135	520	[9]
	?	IR	1690	1250	1135	—	[12]
	in CCl_4	IR	1715	1250	1100	—	[22]
$O{=}S{=}NC(O)C_6H_4Cl$-4	?	IR	1690	1250	1130	—	[12]
$O{=}S{=}NC(O)C_6H_4Br$-4	?	IR	1690	1250	1070	—	[12]
$O{=}S{=}NC(O)C_6H_4NO_2$-4	in CCl_4	IR	1719	—	1113	—	[22]
$O{=}S{=}NC(O)C_6H_4OCH_3$-4	?	IR	1690	1250	1140	—	[12]
$O{=}S{=}NC(O)C_6H_4CH_3$-4	?	IR	1690	1250	1140	—	[12]
O=S=NC(O)C(O)N=S=O	film	IR	1750	1260	1170	565	[9]
$O{=}S{=}NC(O)OCH_3$	gas	IR	1790	1250	1135	—	[7]
	liquid	Ra	1741	1230	1122	—	[7]
$O{=}S{=}NC(O)OC_2H_5$	gas	IR	1790	1250	1135	—	[7]
	liquid	Ra	1737	1228	1115	—	[7]
$O{=}S{=}NC(O)OC_2H_4Cl$	liquid	IR	1760	1240	1140	—	[7]
$O{=}S{=}NC(O)OC_6H_5$	?	IR	1750	1260	1100	—	[20]

*) Other tentatively assigned bands in the original paper [9].

^1H NMR (CCl_4/TMS):

O=S=NC(O)CH_3: $\delta = 2.33$ ppm (s), $^1J(^{13}C,H) = 131.0$ Hz [8, 9].
O=S=NC(O)C_6H_5: $\delta = 7.7$ ppm (m) [9].
O=S=NC(O)OC_2H_5: δ (in ppm) = 1.37 (t, CH_3); 4.35 (q, OCH_2) [15].
O=S=NC(O)OC_3H_7-n: δ (in ppm) = 1.00 (t, CH_3); 1.48 to 2.06 (m, CH_2); 4.25 (t, OCH_2) [15].
O=S=NC(O)OC_4H_9-n: δ (in ppm) = 0.82 to 1.86 (m); 4.30 (t, OCH_2) [15].

The mass spectrum of O=S=NC(O)OC_2H_5 shows the molecular ion at m/e = 135 (1%) [15].

There exist no structural data for any of the compounds of this type. The O=S=N group normally exists in a stable cis conformation. The ν(C=O) vibration frequency of O=S=NC(O)C_6H_5 points to a small mesomeric interaction of C=O and O=S=N groups. Dipole moments of S=O and S=N bonds are plotted against the electronegativity of R in O=S=NR, and are shown to vary nearly linearly for μ(S=O), while μ(S=N) remains approximately constant. Estimated bond moments for O=S=NC(O)C_6H_5 are μ(S=N) = 0.2 D, and μ(S=O) = 1.2 D [22].

O=S=NC(O)C_6H_5 [10], O=S=NC(O)OCH_3, and O=S=NC(O)OC_2H_5 [4] are soluble in nonpolar organic solvents. O=S=NC(O)CH_3 is not soluble in petroleum ether, but is readily soluble in ether, CCl_4, $CHCl_3$, and CH_2Cl_2 [8].

In the absence of moisture most N-sulfinyl-carboxamides O=S=NC(O)R are stable. O=S=N-C(O)CH_3 may be preserved for weeks in a refrigerator without decomposition. But, when exposed to daylight at room temperature, slow red-brown coloring occurs [8, 9]. O=S=NC(O)C_6H_5 is stable at room temperature in a dry atmosphere [12], and even when stored for months in daylight [9]. Compounds O=S=NC(O)C_6H_4R-4 (R = Cl, Br, CH_3O, CH_3) are completely stable in a dry atmosphere [12]. O=S=NC(O)C(O)N=S=O, however, begins to polymerize with SO_2 evolution above −15°C. At room temperature vigorous SO_2 evolution and polymerization occurs [9]. N-Sulfinyl-carbamic acid esters, O=S=NC(O)OR (R = CH_3, C_2H_5), at room temperature slowly change into ROC(O)NHC(O)NH_2, and O=S=NC(O)OCH_2CCl_3 slowly transforms into Cl_3CCH_2OC(O)NHC(O)-NHC(O)OCH_2CCl_3 [21].

All compounds (except O=S=NC(O)C(O)N=S=O) may be distilled in vaccum (see Tables 44 and 45, pp. 172/3). When O=S=NC(O)CH_3 is distilled under normal pressure at 120 to 150°C SO_2 evolves, and brown-black resins remain [8]. Thermal instability of O=S=NC(O)C_6H_5, reported by [5] and [11], was disproved by [9] and [12]. They found that the compound may be distilled without decomposition even above 100°C [9], and decomposition into C_6H_5CN and SO_2 is observed at ~250°C [12]. The relatively high boiling points of O=S=NC(O)C_6H_4R-4 (R = Cl, Br, CH_3O, CH_3) show their thermal stability [12]. N-Sulfinyl-carbamic acid esters, O=S=NC(O)OR (R = C_2H_5, n-C_3H_7, n-C_4H_9), when heated at about 150°C, evolve SO_2 and sulfur diimides form [2, 15]: O=S=NC(O)OR → ROC(O)N=S=NC(O)OR + SO_2.

All O=S=NC(O)R and O=S=NC(O)OR compounds show extraordinary sensitivity to H_2O. When exposed to moist air the N-sulfinyl-carboxamides are hydrolyzed according to: O=S=NC(O)R + H_2O → RC(O)NH_2 + SO_2 [6 to 12]. O=S=NC(O)C(O)N=S=O + 2H_2O → H_2NC(O)C(O)NH_2 + 2SO_2 [9]. Slow hydrolysis of O=S=NC(O)C_2H_5 in air, however, gives a 1:1 complex of C_2H_5OC(O)NH-SNHC(O)OC_2H_5 and C_2H_5OC(O)$NHSO_2NHC$(O)OC_2H_5 [2]. The N-sulfinyl-carbamic acid esters O=S=NC(O)OR (R = CH_3, C_2H_5) are easily hydrolyzed to give carbamic acid esters and SO_2: O=S=NC(O)OR + H_2O → ROC(O)NH_2 + SO_2 [1, 2, 12].

Traces of bases transform O=S=NC(O)CH_3 and O=S=NC(O)C_6H_5 to polymers by splitting off SO_2 [8, 9]. In the presence of pyridine O=S=NC(O)C_6H_4X-4 (X = H, Cl, Br) are converted into the corresponding sulfur diimides: 2 O=S=NC(O)C_6H_4X-4 $\xrightarrow{\text{pyridine}}$ 4-XC_6H_4C(O)N=S=NC(O)C_6H_4X-4 [12].

Dry HCl cleaves O=S=NC(O)C_6H_5 into C_6H_5C(O)NH_2 and $SOCl_2$ [12].

With $NO^+SbF_6^-$ the oxocarbonium compound $C_6H_5CO^+SbF_6^-$ is formed, providing an effective acylating agent. Thus, in the presence of toluene benzoylation takes place: $O{=}S{=}N{-}C(O)C_6H_5 + C_6H_5CH_3 + NO^+SbF_6^- \rightarrow CH_3C_6H_4C(O)C_6H_5 + N_2 + SO_2 + HSbF_6$. When the reaction is carried out with $NO^+BF_4^-$ at room temperature in the absence of aromatics, the acyl fluoride C_6H_5COF is obtained through the decomposition of the relatively unstable $C_6H_5CO^+BF_4^-$ [3].

Compounds O=S=NC(O)Ar (Ar = aryl) failed to react with diaryl selenones Ar_2SeO_2, even when heated to 150°C [27, 28]. $(CH_3)_2BSCH_3$ reacts with $O{=}S{=}NC(O)C_6H_5$ to form the coordinatively dimerized compound I [23]:

$2\,O{=}S{=}NC(O)C_6H_5 + 2\,(CH_3)_2BSCH_3 \rightarrow$ 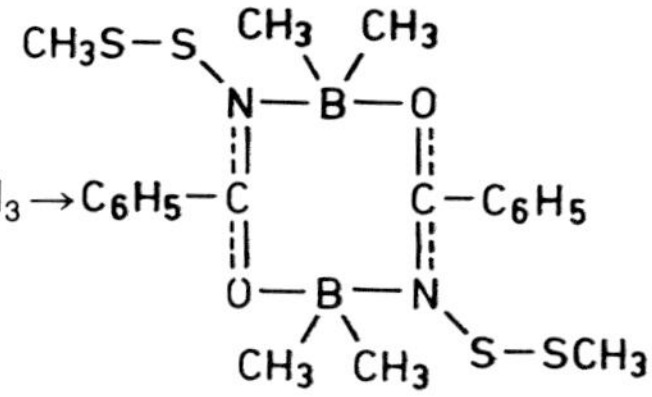

I

N-Sulfinyl-acetamide and bis-trimethylsilyl-acetamide at low temperatures (−70 to −35°C, in ether) give a nearly quantitative yield of crystalline 1:1 adduct, which decomposes already above −5°C [9]:

$O{=}S{=}NC(O)CH_3 + CH_3C(O)N(Si(CH_3)_3)_2 \rightarrow$ 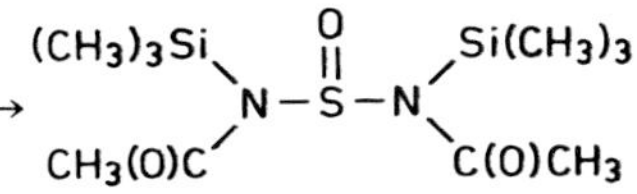

$O{=}S{=}NC(O)CH_3$ and $P(OC_2H_5)_3$ in ether at −60 to −70°C yield $(C_2H_5O)_3P{=}O$ and $(C_2H_5O)_3P{=}S$ in the ratio 4:1 [24]. With $P(N(CH_3)_2)_3$ $O{=}S{=}NC(O)CH_3$ reacts at temperatures at 0 to −80°C to yield three products, $((CH_3)_2N)_3P{=}O$, $((CH_3)_2N)_3P{=}S$, $((CH_3)_2N)_3P{=}NC(O)CH_3$, in equal portions [24]. $O{=}S{=}NC(O)CH_3$ oxidizes 1,3,2-oxazaphospholidines in dry CH_2Cl_2, $CHCl_3$, or ether at −80°C.

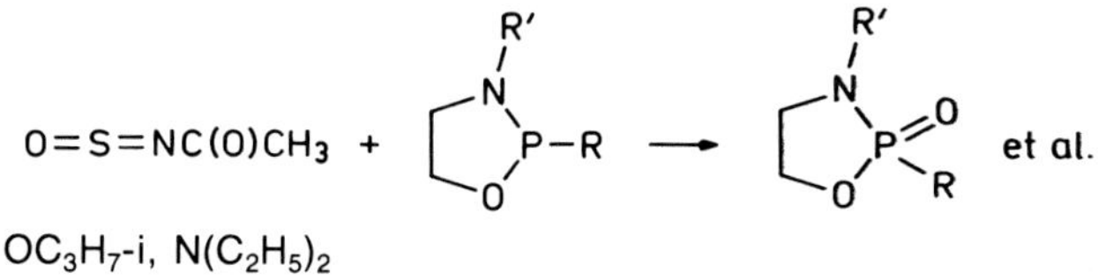

R = Cl, OC_2H_5, OC_3H_7-i, $N(C_2H_5)_2$
R′ = H, CH_3, C_6H_5

Similarly, the 1,3,2-dioxaphospholane is oxidized:

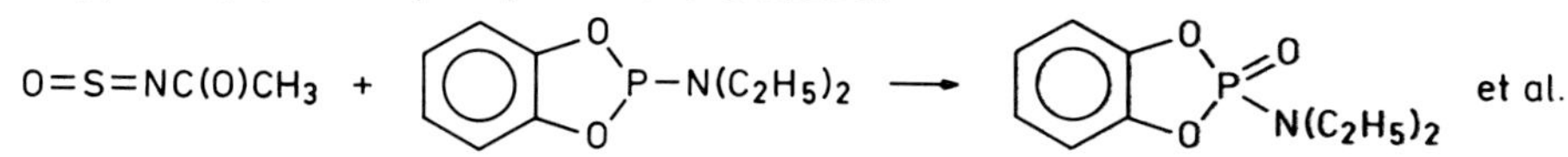

The mechanism is discussed, and the primary step is supposed to be the electrophilic attack of the S atom on P(III) with subsequent bond-breaking [24]. With O-alkyl esters of dithiophosphoric acid derivatives the following reaction occurs [25]:

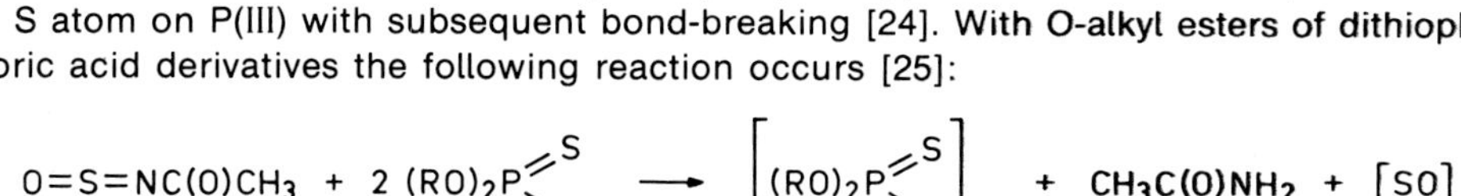

R = CH_3, i-C_3H_7

The silylated derivatives give a more complex mixture of products (room temperature, no solvent):

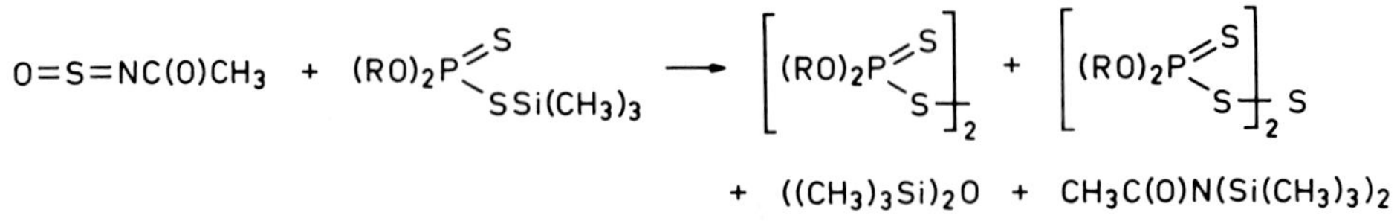

$R = C_2H_5$, $i\text{-}C_3H_7$, $(RO)_2 = OCH(CH_3)CH(CH_3)O$

A reaction mechanism is proposed. The initial step is assumed to be the addition of the S=O bond to the S–Si bond [26].

References:

[1] Institute of Organic Chemistry, Academy of Sciences, Ukr. SSR (Fr. 1417457 [1964/65] 1/5; from C.A. **64** [1966] 9599).

[2] Hancock, J.; Markert, A. R. (Tetrahedron Letters **1966** 6157/61).

[3] Olah, G. A.; Friedman, N.; Bollinger, J. M.; Lukas, J. (J. Am. Chem. Soc. **88** [1966] 5328/30).

[4] Levchenko, E. S.; Bal'on, Ya. G.; Kirsanov, A. V. (Zh. Org. Khim. **3** [1967] 1883/9; J. Org. Chem. [USSR] **3** [1967] 1838/42).

[5] Wucherpfennig, W. (Tetrahedron Letters **1967** 3235/7).

[6] Hörhold, H.-H. (Angew. Chem. **79** [1967] 312/3; Angew. Chem. Intern. Ed. Engl. **6** [1967] 357/8).

[7] Hörhold, H.-H.; Eibisch, H. (Chem. Ber. **101** [1968] 3567/78, 3568/9, 3572/3).

[8] Scherer, O. J.; Schmitt, R. (Z. Naturforsch. **22b** [1967] 224).

[9] Scherer, O. J.; Schmitt, R. (Chem. Ber. **101** [1968] 3302/12, 3303, 3305, 3309/10).

[10] Samarai, L. I.; Belaya, V. P.; Derkach, G. I. (Khim. Geterotsikl. Soedin. **1968** No. 4, pp. 368/9; Chem. Heterocycl. Compounds [USSR] **1968** No. 4, pp. 271/2).

[11] Tsuge, O.; Mataka, S. (Bull. Chem. Soc. Japan **44** [1971] 2836/9).

[12] Levchenko, E. S.; Dorokhova, E. M. (Zh. Org. Khim. **8** [1972] 2526/31; J. Org. Chem. [USSR] **8** [1972] 2573/7).

[13] Dorokhova, E. M.; Levchenko, E. S.; Pel'kis, N. P. (Zh. Org. Khim. **11** [1975] 762/6; J. Org. Chem. [USSR] **11** [1975] 755/8).

[14] Borovikova, G. S.; Levchenko, E. S.; Borovik, E. I. (Zh. Org. Khim. **15** [1979] 2313/8; J. Org. Chem. [USSR] **15** [1979] 2095/9).

[15] Niclas, H. J.; Habisch, D.; Martin, D. (Tetrahedron **35** [1979] 2353/7).

[16] Wald, L.; Wucherpfennig, W. (Liebigs Ann. Chem. **746** [1971] 28/31).

[17] Sluysarenko, E. I.; Pel'kis, N. P.; Levchenko, E. S. (Zh. Org. Khim. **14** [1978] 1092/4; J. Org. Chem. [USSR] **14** [1978] 1016/8).

[18] Solyom, S.; Szilagyi, K.; Toldy, L. (Liebigs Ann. Chem. **1983** 1001/19, 1012).

[19] Lutskii, A. E.; Shevchenko, Z. A.; Kochergina, L. A.; Levchenko, E. S.; Uzarov, B. N.; Men'shova, I. I.; Bal'on, Ya. G. (Zh. Obshch. Khim. **37** [1967] 2095/101; J. Gen. Chem. [USSR] **37** [1967] 1986/91).

[20] Ichimura, K.; Ichikawa, S.; Imamura, K. (Bull. Chem. Soc. Japan **49** [1976] 1157/8).

[21] Bussas, R.; Kresze, G. (Liebigs Ann. Chem. **1982** 545/63, 548).

[22] Borovikov, Yu. Ya.; Borovikova, G. S.; Koval, N. V.; Borovik, E. I. (Ukr. Khim. Zh. **52** [1986] 875/9; Soviet Progr. Chem. **52** No. 8 [1986] 96/9).

[23] Meller, A.; Maringgele, W.; Fetzer, H. (Chem. Ber. **113** [1980] 1950/61, 1952).

[24] Pudovik, M. A.; Morozova, N. P.; Pudovik, A. N. (Zh. Obshch. Khim. **48** [1978] 495/9; J. Gen. Chem. [USSR] **48** [1978] 446/9).

[25] Kutyrev, G. A.; Mavrin, V. Yu.; Cherkasov, R. A.; Pudovik, A. N. (Izv. Akad. Nauk SSSR Ser. Khim. **1980** 1402/4; C.A. **93** [1980] No. 167773).
[26] Kutyrev, G. A.; Lygin, A. V.; Cherkasov, R. A.; Pudovik, A. N. (Zh. Obshch. Khim. **53** [1983] 1009/13; J. Gen. Chem. [USSR] **53** [1983] 891/4).
[27] Derkach, N. Ya.; Pasmurtseva, N. A.; Levchenko, E. S. (Zh. Obshch. Khim. **42** [1972] 1652/3; J. Gen. Chem. [USSR] **42** [1972] 1646).
[28] Derkach, N. Ya.; Lyapina, T. V.; Pasmurtseva, N. A. (Zh. Org. Khim. **10** [1974] 807/10; J. Org. Chem. [USSR] **10** [1974] 810/2).

2.6.22.5 N-Sulfinyl-cyanamide, O=S=NC≡N

The compound was obtained in 80% yield when $SOCl_2$ was slowly dropped with stirring into an equimolar amount of $((CH_3)_3SiN{=})_2C$ at −70°C, and then allowed to warm to 0°C:

$$(CH_3)_3SiN{=}C{=}NSi(CH_3)_3 + SOCl_2 \rightarrow O{=}S{=}NC{\equiv}N + 2\,(CH_3)_3SiCl.$$

$(CH_3)_3SiCl$ was distilled off in an oil-pump vacuum and the residue fractionated. But caution! Rapid polymerization with detonation may occur during preparation.

O=S=NC≡N is a yellow liquid, m.p. −37°C, b.p. 23 to 24°C/4 Torr.

IR: ν (in cm^{-1}) = 2190 s, ν(C≡N); 1450 m, ν(NC); 1260 s, ν_{as}(N=S=O); 1170 s, ν_s(N=S=O); 530 s, δ(N=S=O).

O=S=NC≡N is readily soluble in $CHCl_3$, CH_2Cl_2, ether, and benzene, moderately soluble in CCl_4, but insoluble in petroleum ether. It can be stored for some days in a refrigerator without decomposition. When exposed to daylight at room temperature, polymerization with evolution of SO_2 occurs within some hours. The pure compound decomposes at ~100°C. Traces of water or alcohol lead to explosive decomposition. Treating with N,N-bis-trimethylsilyl or N,N-bis-trimethylstannyl methylamine in ether at −70°C, and then at room temperature, yields the corresponding carbodiimides: $O{=}S{=}NC{\equiv}N + CH_3N(M(CH_3)_3)_2 \rightarrow (CH_3)_3MN{=}C{=}NM(CH_3)_3 + SN$ polymers, M = Si, Sn.

Reference:

Scherer, O. J.; Schmitt, R. (Angew. Chem. **79** [1967] 691; Angew. Chem. Intern. Ed. Engl. **6** [1967] 701/2; Chem. Ber. **101** [1968] 3302/12, 3305/6, 3308, 3310/1).

2.6.22.6 N-Sulfinyl-carboxamidines, O=S=NC(R)=NR′, R = C_6H_5, R′ = $C(O)C_6H_5$; R = CCl_3, R′ = C_6H_5 or $C_6H_4CH_3$-4

2.6.22.6.1 N-(Sulfinyl-amino-benzylidene)-benzamide, O=S=NC(C_6H_5)=NC(O)C_6H_5

The compound was obtained in 84% yield, when a solution of $SOCl_2$ in benzene was added dropwise to $C_6H_5C({=}NC(O)C_6H_5)NHCl$ and pyridine in benzene (mole ratio 1:1:1) at 0 to 5°C, and then stirred at 20°C for 1 h. The thick oil crystallized on cooling to give light yellow prisms; m.p. 81 to 82°C. The compound was identified by its reaction with isoprene.

It decomposes with elimination of SO_2, when heated in C_6H_6 or CCl_4:

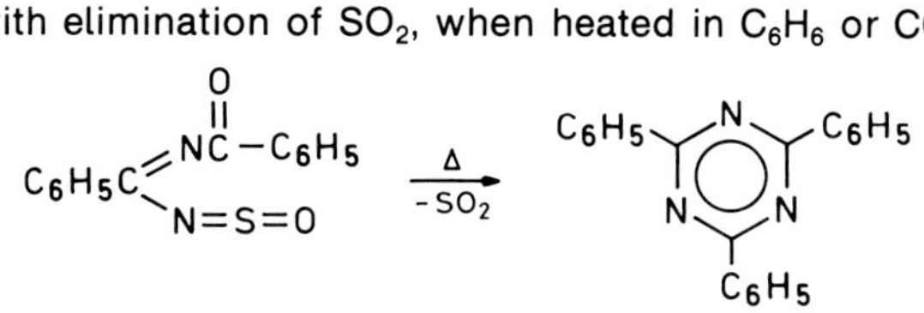

When the compound is heated at 80°C for 2 h in the presence of SCl_2 in an inert solvent, the 1,2,4-thiadiazole I forms.

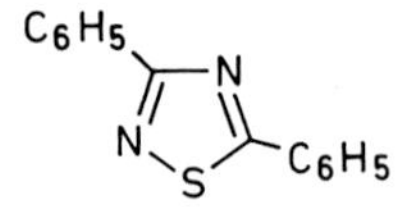

I

The compound hydrolyzes to give $C_6H_5C(=NC(O)C_6H_5)NH_2$ [1].

2.6.22.6.2 N-Sulfinyl-N′-aryl-trichloroacetamidines, $O{=}S{=}NC(CCl_3){=}NC_6H_5$, $O{=}S{=}NC(CCl_3){=}NC_6H_4CH_3$-4

The compounds were obtained in about 95% yield by the reaction of $SOCl_2$ with the corresponding amidine, mole ratio 1:3, in CCl_4 at 50°C for 2 to 3 h. They can be recrystallized from petroleum ether.

$$3\,Cl_3CC(=NR)NH_2 + SOCl_2 \rightarrow O{=}S{=}NC(CCl_3){=}NR + 2\,Cl_3CC(=NR)NH_2 \cdot HCl$$

$R = C_6H_5$, 4-$CH_3C_6H_4$

The compounds form orange-colored needles. M.p. (in a sealed capillary) 40 to 42°C ($R = C_6H_5$), and 70 to 72°C ($R = 4\text{-}CH_3C_6H_4$). IR: $\nu(N{=}S{=}O) = 1260\ cm^{-1}$ (both).

They are readily hydrolyzed by atmospheric moisture. On being stored in a closed vessel at room temperature the compound with $R = C_6H_5$ is converted, after 2 to 4 d, into a heterocyclic compound [2]:

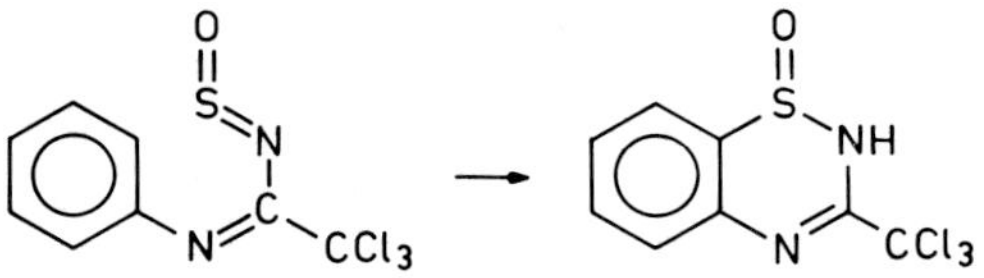

References:

[1] Borikova, G. S.; Levchenko, E. S.; Borovik, E. I.; Darmokhval, E. A. (Zh. Org. Khim. **20** [1984] 190/6; J. Org. Chem. [USSR] **20** [1984] 171/6).

[2] Samarai, L. I.; Belaya, V. P.; Derkach, G. I. (Khim. Geterotsikl. Soedin. **1968** No. 4, pp. 368/9; Chem. Heterocycl. Compounds [USSR] **1968** No. 4, pp. 271/2).

2.6.22.7 N,N′-Bis(sulfinyl)-carbonic Diamide, N,N′-Bis(sulfinyl)-urea, $O{=}S{=}NC(O)N{=}S{=}O$

The compound was obtained, when $SOCl_2$ was condensed onto $Hg(NCO)_2$ (mole ratio ~1:1), and the reaction mixture slowly allowed to warm to room temperature; yield 62%. It formed also when excess $ClC(O)N{=}C{=}O$ was condensed onto $(O{=}S{=}N)_2Hg$, and the mixture was slowly allowed to warm; yield 35%, based on $(O{=}S{=}N)_2Hg$. The slightly yellow liquid was distilled at low temperature in oil-pump vacuum. B.p. 18°C/0.13 mbar (~0.1 Torr).

IR (KBr, film): ν (in cm^{-1}) = 2240 vs, $\nu_{as}(NCO)$; 2179 w; 1719 s, br, $\nu(C{=}O)$; 1390 m, $\nu_s(NCO)$; 1245 s, $\nu(SO)$; 1203 s, $\nu(NS)$; 809 m; 755 w; 643 w. 15 Raman lines are given in the paper.

MS (70 eV): m/e (rel. int. in %) = 132 (8.3) M^+; 90 (62.7) $SONCO^+$; 70 (100) $OCNCO^+$; 64 (28.8) SO_2^+; 59 (22); 48 (47.5) SO^+; 46 (22) SN^+; 43 (15.3) $HNCO^+$; 42 (15.3) NCO^+.

The compound is extremely sensitive towards hydrolysis. It polymerizes at room temperature within 1 d to form a red-brown solid.

Reference:

Krannich, H. J.; Sundermeyer, W. (Z. Anorg. Allgem. Chem. **450** [1979] 45/9).

2.6.22.8 N-Sulfinyl-perhalogenoorganylamines, O=S=NR$_f$, R$_f$ = C_nF_{2n+1} (n = 1, 2, 3), CF_2Cl, $C(CF_3)_2Cl$, $C(O)CF_3$, $C(CF_3)_2N{=}C(CF_3)_2$

2.6.22.8.1 N-Sulfinyl-trifluoromethanamine, O=S=NCF$_3$

The compound was prepared by distilling $CF_3N{=}SCl_2$ into a Pyrex bulb containing Ag_2O in nitrobenzene at −196°C (mole ratio ~1:11), and rapidly warming to 7°C. Separation of the products by trap-to-trap distillation afforded pure $O{=}S{=}NCF_3$ in 80% yield [1]. The same procedure in perfluoromethyldecalin at t < 0°C gave 77% yield [2]. Preparation from $CF_3N{=}SCl_2$ and HgO without solvent is reported without giving any details [3]. Formation of $O{=}S{=}NCF_3$ was also observed in the reaction of CF_3NCl_2 with $SOCl_2$. On UV irradiation of a CH_2Cl_2 solution of $Hg(CF_3)_2$ and O=S=NCl, formation of $O{=}S{=}NCF_3$ was observed [2].

The compound could not be made to crystallize, although it was separated by pumping into a trap at −126°C [1]. Boiling point, enthalpy and entropy of vaporization, vapor pressure equation, and ^{19}F NMR data are compiled in Table 48, p. 182. Vapor density and effusion rate show the compound to be a monomer in the vapor phase. IR (gas): 9 bands are listed between 1430 and 480 cm^{-1}.

MS (100 eV): m/e (rel. int. in %) = 131 (37.3) CF_3NSO^+; 112 (6.5) CF_2NSO^+; 69 (87.3) CF_3^+; 50 (5.0) CF_2^+; 48 (46.1) SO^+ or $^{34}SN^+$; 46 (100) SN^+; 32 (8.2) S^+; 31 (6.4) CF^+ [1].

When $O{=}S{=}NCF_3$ is irradiated with UV-light, or treated with pyridine, $CF_3N{=}S{=}NCF_3$ is formed [2]. $O{=}S{=}NCF_3$ reacts with excess ClF at −78°C for 3 h to give CF_3NCl_2 (60% yield) and SOF_2 [3]. In the reaction with XeF_2 in perfluoromethyldecalin at room temperature for 8 h, S^{IV} is oxidized to S^{VI} with formation of $CF_3N{=}SOF_2$ and small amounts of $CF_3N{=}SF_2{=}NCF_3$ [2]. Exposure to HCl at low temperature affords $CF_3NH_2 \cdot HCl$ [2]. Reaction with SO_3 at 20 to 45°C yields cyclic compounds I [8, 9] and II [8]:

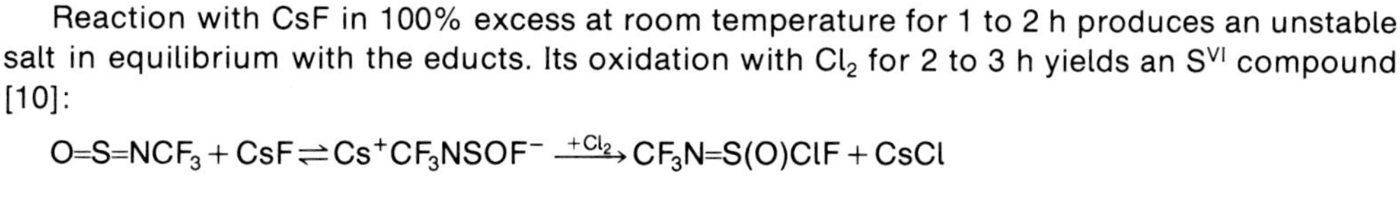

I (60.7%) II (22.2%)

Reaction with CsF in 100% excess at room temperature for 1 to 2 h produces an unstable salt in equilibrium with the educts. Its oxidation with Cl_2 for 2 to 3 h yields an S^{VI} compound [10]:

$$O{=}S{=}NCF_3 + CsF \rightleftharpoons Cs^+CF_3NSOF^- \xrightarrow{+Cl_2} CF_3N{=}S(O)ClF + CsCl$$

Table 48
Properties of N-Sulfinyl-perhalogenoorganylamines.

compound	m.p. in °C	b.p. in °C	ΔH_v in kcal/mol	ΔS_v in $cal \cdot mol^{-1} \cdot K^{-1}$	log p (p in Torr)	^{19}F NMR (standard $CFCl_3$) δ in ppm	Ref.
$O{=}S{=}NCF_3$	—	18.0	6.47	22.20	7.33 − 1413/T	−53.6 (CF_3)	[1]
$O{=}S{=}NC_2F_5$	−84.5 ± 1.5	45.1	6.94	21.80	7.643 − 1516/T	−86.7 (CF_3); −90.8 (CF_2)	[1]
$O{=}S{=}NC_3F_7$-n	—	51.8	7.1	21.8	7.65 − 1550/T	−81.3 (t, CF_3); −86.4 (q, α-CF_2); −128.5 (q, β-CF_2); $^3J(CF_3, \alpha\text{-}CF_2) = 8.6$ Hz	[4]
$O{=}S{=}NC_3F_7$-i	−108 ± 2	60 to 61 (743 Torr)	8.08	24.1	8.156 − 1766/T	−81.2 (CF_3); −146.0 (CF); $^3J(CF_3, CF) = 4.6$ Hz	[5]
$O{=}S{=}NCCl(CF_3)_2$	—	90.2 (calc.)	8.94	24.6	8.26 − 1954/T	−76.3 (s, CF_3)	[7]
$O{=}S{=}NC(CF_3)_2N{=}C(CF_3)_2$	—	102.3	9.4	25.2	8.38 − 2068/T	−65.5 (br); −74 (m)*)	[12]
$O{=}S{=}NC(O)CF_3$	−54 ± 1	75 (749 Torr)	8.46	24.2	8.179 − 1849/T	−78.0 (s, CF_3)	[5]

*) At room temperature; at −60°C the signal at −65.5 ppm resolves to $\delta = -60.5$ ppm (m, complex) and −70.5 ppm (q); $J_{AB} = 6.8$ Hz; the resonance at $\delta = -74$ ppm simplifies to a quartet, $J_{AC} = 2.8$ Hz; at 90°C $\delta = -66.1$ (m) and −74 ppm (m).

2.6.22.8.2 N-Sulfinyl-pentafluoroethanamine, $O{=}S{=}NC_2F_5$

The compound was prepared in the same way as $O{=}S{=}NCF_3$ (see above) from $C_2F_5N{=}SCl_2$ and Ag_2O in nitrobenzene [1] or perfluoromethyldecalin [2] in 65.7% yield [1]. It was obtained in 70% yield, when $C_2F_5N{=}SCl_2$ was condensed onto excess yellow HgO at −196°C, and the solution warmed gradually (5 h) to 25°C. Purification is conducted by trap-to-trap distillation [4]. A 1:1 mixture of $O{=}S{=}NC_2F_5$ and $(CF_3C(O))_2O$ formed by oxygenation of $C_2F_5NSCl_2$ with $AgOC(O)CF_3$ (mole ratio 1:2.5) at 25°C for 2 h [4].

The vapor density shows the compound to be monomeric in the gas phase. Melting point, boiling point, enthalpy and entropy of vaporization, vapor pressure equation, and ^{19}F NMR data are compiled in Table 48.

IR (gas): 13 bands are listed between 1330 and 440 cm^{-1}.

MS (100 eV): m/e (rel. int. in %) = 112 (100.0) CF_2NSO^+; 69 (85.4) CF_3^+; 50 (13.0) CF_2^+; 48 (63.7) SO^+ or $^{34}SN^+$; 46 (52.8) SN^+; 32 (12.0) S^+; 31 (18.0) CF^+ [1].

Thermolysis of $O{=}S{=}NC_2F_5$ at 164°C for 10 h yields only small quantities of CF_3CN, SOF_2, and SO_2F_2 (14% decomposition), and mainly unreacted $O{=}S{=}NC_2F_5$ remains. However, thermolysis in the presence of CsF at 140°C for 11 h yields, almost quantitatively, CF_3CN and SOF_2 in equal amounts [4]. Reaction with CsF (100% excess) at room temperature produces an unstable salt in equilibrium with the starting educts. It can be oxidized with Cl_2 at low temperatures to give a stable S^{VI} compound [10]:

$$O{=}S{=}NC_2F_5 + CsF + Cl_2 \rightleftharpoons Cs^+C_2F_5NSOF^- + Cl_2 \rightarrow C_2F_5N{=}S(O)ClF + CsCl.$$

When $((CH_3)_2N)_3S^+(CH_3)_3SiF_2^-$ ("TAS fluoride") is used instead of CsF a stable salt, $((CH_3)_2N)_3S^+$-$C_2F_5NSOF^-$, could be isolated in 54% yield (in CH_3CN–ether, −55°C) [6]. Reaction with Cl_2 in the presence of CsF at higher temperature results in oxidative chlorofluorination or fluorination [4]:

$$O{=}S{=}NC_2F_5 + Cl_2 \xrightarrow[74°C/33\ h]{CsF} \underset{41\%}{C_2F_5N{=}S(O)Cl_2} + \underset{46\%}{C_2F_5N{=}S(O)ClF}$$

$$O{=}S{=}NC_2F_5 + Cl_2 \xrightarrow[52°C/26\ h]{CsF} \underset{29\%}{C_2F_5N{=}S(O)ClF} + \underset{trace}{C_2F_5N{=}S(O)F_2}$$

XeF_2 oxidizes S^{IV} to S^{VI} with formation of $C_2F_5NS(O)F_2$ and small amounts of $C_2F_5N{=}SF_2{=}NC_2F_5$. Exposure to HCl at low temperatures affords $C_2F_5NH_2 \cdot HCl$ [2].

With SO_3 at room temperature ring closure as well as rearrangement occurs:

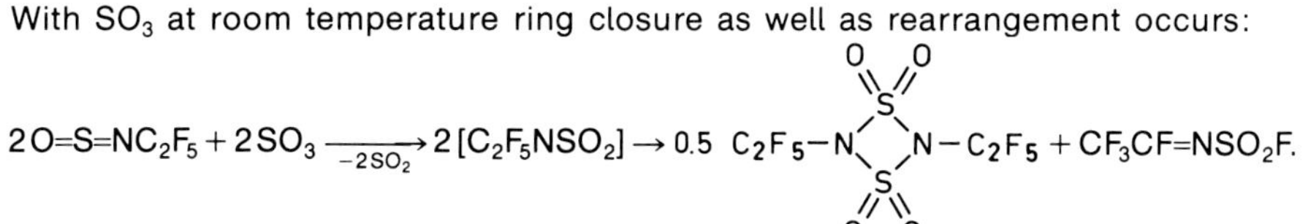

The mass spectrum of the products indicates the formation of a trimeric SN compound too, but isolation has not succeeded to date [8]. Reaction with $ClOSO_2F$ at low temperatures (−78 to −40°C) gives only degraded products: $O{=}S{=}NC_2F_5 + 2ClOSO_2F \rightarrow C_2F_5NCl_2 + SO(OSO_2F)_2$. With $CH_3C(O)CH_3$ at 25°C only tarry materials are obtained. When $O{=}S{=}NC_2F_5$ is heated with $CF_3C(O)CH_3$ or $CF_3C(O)CF_3$ at 144°C, or with $CF_3C(O)OH$ at 84°C, no reaction occurs [4].

2.6.22.8.3 N-Sulfinyl-heptafluoro-1-propanamine, $O{=}S{=}NC_3F_7$-n

The transparent colorless liquid was prepared from n-$C_3F_7N{=}SCl_2$ and HgO in the same manner as $O{=}S{=}NC_2F_5$ (see above). Purification by trap-to-trap (−64°C) distillation yields 88.1% $O{=}S{=}NC_3F_7$-n as a transparent colorless liquid [4]. Boiling point, enthalpy and entropy of vaporization, vapor pressure equation, and ^{19}F NMR data are compiled in Table 48, p. 182.

IR (gas): 10 bands between 1330 and 918 cm^{-1} are listed in the paper.

MS: m/e = 212 M^+-F, 169 $C_3F_7^+$, 112 $C_3F_4^+$ [4].

Reaction of $O{=}S{=}NC_3F_7$-n with CsF (100% excess) at room temperature yields unstable salts, as described above for $O{=}S{=}NCF_3$ [10]. Reaction with Cl_2 in the presence of CsF results in chlorofluorination and fluorination [4]:

$$O{=}S{=}NC_3F_7\text{-n} + Cl_2 \xrightarrow[53°C/23\,h]{CsF} \underset{16\%}{\text{n-}C_3F_7N{=}S(O)ClF} + \underset{trace}{\text{n-}C_3F_7N{=}S(O)F_2}$$

2.6.22.8.4 N-Sulfinyl-heptafluoro-2-propanamine, $O{=}S{=}NC_3F_7$-i

The pale yellow liquid was obtained when $(CF_3)_2C{=}NH$ and excess SOF_2 were condensed onto CsF (mole ratio ~1:2:1) in an autoclave, and then heated at 100°C for 24 h. Fractional distillation gives $O{=}S{=}NC_3F_7$-i in 23% yield, based on $(CF_3)_2C{=}NH$. Melting point, boiling point, enthalpy and entropy of vaporization, vapor pressure equation, and ^{19}F NMR data are compiled in Table 48, p. 182.

IR (gas): ν (in cm^{-1}) = 1332 m, $\nu_{as}(N{=}S{=}O)$; 1293 vs, 1259 vs, and 1192 s, $\nu(CF_3)$; 1218 s, $\nu_s(N{=}S{=}O)$; 1119 s, ν(CF); 994 vs, 929 w, ν(CC) and ν(CN); 766 w, 732 s, 705 w, δ(CF); further deformation vibrations at 600 w, 529 w, and 480 cm^{-1} m.

MS: m/e (rel. int. in %) = 231 (0.1) M^+; 212 (12.3) M^+-F; 169 (1.1) M^+-NSO, 162 (99) M^+-CF_3; 146 (1.0) CF_3CFNS^+; 143 (0.9) CF_3CNSO; 114 (5.1) CF_3CFN^+; 112 (38.0) CF_2NSO^+ ?; 96 (8.5) CF_2NS^+ ?; 76 (4.4) $F_2C_2N^+$; 69 (100) CF_3^+; 50 (7.4) CF_2^+; 48 (29.8) SO^+; 46 (18.2) SN^+; 32 (3.1) S^+; 31 (6.6) CF^+ [5].

Reaction of $O{=}S{=}NC_3F_7$-i with CsF (100% excess) at room temperature for 5 h yields the salt Cs^+i-$C_3F_7NSOF^-$ in equilibrium with the educts. Its oxidation with Cl_2 at low temperature yields the stable S^{VI} compound i-$C_3F_7NS(O)ClF$ [10]. When $((CH_3)_2N)_3S^+(CH_3)_3SiF_2^-$ ("TAS fluoride") is used as a fluoride ion donor instead of CsF, a stable salt $((CH_3)_2N)_3S^+$ i-$C_3F_7NSOF^-$ is obtained [6].

With $S({=}NSi(CH_3)_3)_2$, $O{=}S{=}NC_3F_7$-i reacts in diethylether at room temperature over a period of 3 d, mole ratio 2:1, in a twofold metathesis reaction. The end product III and also two intermediates, I and II, may be isolated [13, 14]:

$$2\,O{=}S{=}NC_3F_7\text{-i} + 2\,S({=}NSi(CH_3)_3)_2 \rightarrow$$

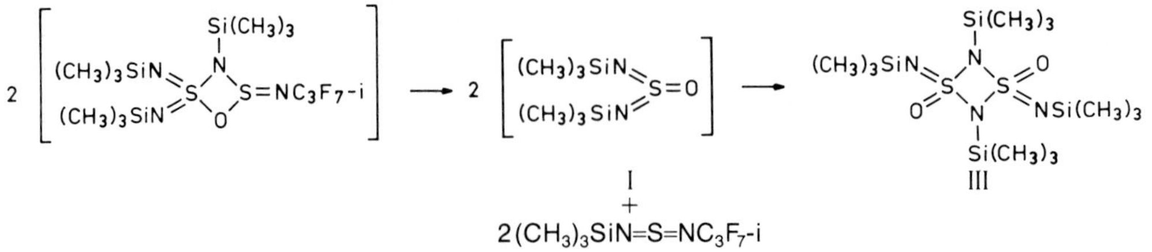

I + $2(CH_3)_3SiN{=}S{=}NC_3F_7$-i
II

III

2.6.22.8.5 N-Sulfinyl-chlorodifluoromethanamine, $O{=}S{=}NCF_2Cl$

When a mixture of ClC≡N and an excess of SOF_2 was heated in a metal tube for 11 h at 300°C $O{=}S{=}NCF_2Cl$ in 47% yield was obtained in addition to the main product, $C_3N_3Cl_3$ (trichloro-sym-triazine). The slightly yellowish liquid, b.p. 64°C, shows IR absorptions at 1420, 1345, 1310, 1235, 1088, and 815 cm^{-1}. It was identified by elemental analysis [11].

2.6.22.8.6 N-Sulfinyl-2-chloro-hexafluoro-2-propanamine, $O{=}S{=}NCCl(CF_3)_2$. N′-Sulfinyl-perfluoro(N-(1-methylethylidene)-2,2-propanediamine), $O{=}S{=}NC(CF_3)_2N{=}C(CF_3)_2$

To prepare the compounds excess $SOCl_2$ was condensed onto $(CF_3)_2C{=}NS(O)OLi$ at −196°C and warmed slowly to 25°C. After 12 h the products were separated by gas chromatography to give $O{=}S{=}NCCl(CF_3)_2$ (73% yield) and $O{=}S{=}NC(CF_3)_2N{=}C(CF_3)_2$ (16% yield) [7]. The latter compound was obtained in 47% yield from $SOCl_2$ and excess $LiN{=}C(CF_3)_2$ in the same manner as above, but some pentane had to be added as a heat sink to moderate the very exothermic, explosive reaction [12]. $O{=}S{=}NCCl(CF_3)_2$ is an intermediate product in the formation of $O{=}S{=}NC(CF_3)_2N{=}C(CF_3)_2$ [7, 12]. It also forms, when $(CF_3)_2C{=}NC(O)OLi$ reacts with $SOCl_2$ [7].

Boiling points of the colorless liquids, enthalpy and entropy of vaporization, vapor pressure equation, and ^{19}F NMR data are compiled in Table 48, p. 182. IR spectra are listed in the papers [7, 12]. The ^{19}F NMR spectrum of $O{=}S{=}NC(CF_3)_2N{=}C(CF_3)_2$ can be explained if the fluorine atoms in the terminal CF_3 groups are magnetically nonequivalent, even at temperatures as high as 90°C. At −60°C the isomerization is slowed sufficiently that on an NMR time scale the molecule exists in essentially one configuration [12].

MS of $O{=}S{=}NCCl(CF_3)_2$; ions (rel. int. in %): M^+-F (4); M^+-Cl (25); $C_3F_6ClN^+$(2); $C_3F_5ClN^+$ (37); M^+-CF_3 (100); $C_2F_4NSO^+$ (13); $SOCl^+$ (26); CF_3^+ (93); SCl^+ (25); SO_2^+ (33); CF_2^+ (13); SO^+ (73) [7].

MS of $O{=}S{=}NC(CF_3)_2N{=}C(CF_3)_2$; ions (rel. int. in %): M^+-F (<0.1); M^+-NSO (19); M^+-CF_3 (13); $M^+-C_3F_6N$ (98); $C_3F_3N_2SO^+$ (11); $C_2F_4NSO^+$ (22); $C_3F_4N^+$ (11); $C_3F_4^+$ (22); $C_2F_3NH^+$ (16); CF_3 (100); FSO^+ (55) [12].

2.6.22.8.7 N-Sulfinyl-trifluoroacetamide, $O{=}S{=}NC(O)CF_3$

The compound was prepared by dropping $SOCl_2$ onto $CF_3C({=}NSi(CH_3)_3)(OSi(CH_3)_3)$, mole ratio 1:1 at −30°C. Fractional distillation at normal pressure gives $O{=}S{=}NC(O)CF_3$ in 21% yield as a yellow liquid. Melting point, boiling point, enthalpy and entropy of vaporization, vapor pressure equation, and ^{19}F NMR data are compiled in Table 48, p. 182.

IR (gas): ν (in cm^{-1}) = 1785 vs, ν(CO); 1331 s, $\nu_{as}(N{=}S{=}O)$; 1266 vs, 1234 s, 1189 sh, $\nu(CF_3)$; 1179 vs, $\nu_s(N{=}S{=}O)$; 1029 w, 978 s, ν(CN) and ν(CC); further five deformation vibrations. The IR spectrum shows the compound to be acyclic.

MS: m/e (rel. int. in %) = 143 (0.1) CF_3CNSO^+; 97 (0.7) CF_3CO^+; 90 (18) $CONSO^+$; 76 (3.6) $C_2F_2N^+$; 69 (22.2) CF_3^+; 62 (0.6) NSO^+; 50 (5.8) CF_2^+; 48 (100) SO^+; 46 (6.0) SN^+; 32 (20.4) S^+; 31 (1.2) CF^+; 28 (3.6) CO^+, and the recombination fragments 47 FCO^+, 64 SO_2^+, 77 $FCNS^+$.

The yellow water-sensitive liquid smells like SO_2. When it stands at room temperature for a long time it turns light red [5].

References:

[1] Lustig, M. (Inorg. Chem. **5** [1966] 1317/9).
[2] Leidinger, W.; Sundermeyer, W. (Chem. Ber. **115** [1982] 2892/7).
[3] De Marco, R. A.; Shreeve, J. M. (J. Fluorine Chem. **1** [1971/72] 269/76).
[4] Abe, T.; Shreeve, J. M. (Inorg. Chem. **19** [1980] 3063/7).
[5] Von Halasz, S. P.; Glemser, O. (Chem. Ber. **103** [1970] 553/60).
[6] Heilemann, W.; Mews, R. (Chem. Ber. **121** [1988] 461/3).
[7] Swindell, R. F.; Shreeve, J. M. (J. Fluorine Chem. **2** [1972/73] 191/3).
[8] Tesky, F.-M.; Mews, R. (Chemiker-Ztg. **111** [1987] 345/6).
[9] Jäger, U.; Sundermeyer, W.; Pritzkow, J. (Chem. Ber. **120** [1987] 1191/5).
[10] Mews, R.; Kricke, P.; Stahl, J. (Z. Naturforsch. **36b** [1981] 1093/8).

[11] Vorob'ev, M. D.; Filatov, A. S.; Englin, M. A. (Zh. Obshch. Khim. **43** [1973] 1653; J. Gen. Chem. [USSR] **43** [1973] 1638).
[12] Swindell, R. F.; Babb, D. P.; Ouellette, Th. J.; Shreeve, J. M. (Inorg. Chem. **11** [1972] 242/5).
[13] Tesky, F. M.; Mews, R.; Krebs, B. (Angew. Chem. **91** [1979] 231/2; Angew. Chem. Intern. Ed. Engl. **18** [1979] 235).
[14] Tesky, F. M.; Mews, R.; Krebs, B. (Z. Naturforsch. **36b** [1981] 1465/78, 1470).

2.6.23 Reactions of N-Sulfinylamines (Sulfinyl Imides) with Organic Compounds and Derivatives

2.6.23.1 Reactions with Alkenes and Cycloalkenes

N-Sulfinylamines O=S=N–R show "enophilic" as well as "dienophilic" reactions (see p. 196). In the "enophilic" reaction the S=N part of the molecule (formally seen to be the S atom) adds to a double-bond C atom and the C=C double bond is shifted. Another type of reaction affords N-sulfinylamines which contain double bonds (π bonds) conjugated with the S=N bond (O=S=N–aryl, O=S=N–COR', O=S=N–COOR'). They may behave as "heterodienes" and form [2+4] cycloadducts with alkenes. A general rule for both types of reaction is that the reactivity of the (electrophilic) N-sulfinylamines O=S=N–R is enhanced by electron acceptor group R.

Reactions with olefinic enolizable aldehydes are not described here, but on pp. 224/6, reactions with olefinic organometallic compounds on pp. 244 and 249/50, reactions with olefinic alcohols on pp. 206/7.

Ene Reaction. O=S=N-alkyl compounds show no enophilic reactions [1].

Electron withdrawal by ester groups is sufficient to enable ene reactions of O=S=N–COOR' with reactive alkenes at room temperature [1] (see Table 49).

Table 49

Reaction Velocities of N-Sulfinylamines with Alkenes in $CDCl_3$ at 20°C.
Half-life time of educts $\tau_{1/2}$ in min if not stated otherwise.

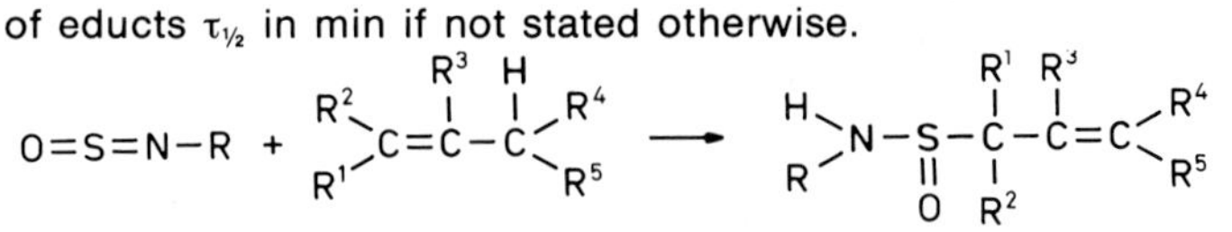

R	R^1	R^5	R^2	R^3	R^4	$\tau_{1/2}$	Ref.
$C(O)OCH_3$	CH_3	H	CH_3	H	H	60	[1]
	CH_3	H	CH_3	CH_3	H	60	[1]

Table 49 (continued)

R	R^1	R^5	R^2	R^3	R^4	$\tau_{1/2}$	Ref.
$C(O)OC_2H_5$	CH_3	H	CH_3	H	H	60	[1]
	CH_3	H	CH_3	CH_3	H	30	[1]
	$-[CH_2]_2-$		H	H	H	900	[1]
$C(O)OCH_2CCl_3$	CH_3	H	CH_3	CH_3	H	10	[1]
	H	CH_3	H	H	C_6H_5	>50 h	[1]
$SO_2C_6H_4CH_3$-4	H	CH_3	H	H	CH_3	180	[1, 6]
	H	CH_3	H	H	C_6H_5	600	[1, 6]
	H	H	CH_3	H	C_6H_5	30	[6]
	$-[CH_2]_2-$		H	H	H	180	[6]
	$-[CH_2]_3-$		H	H	H	600	[6]
	H	H	H	H	$CH_2C(O)OCH_3$	250 h	[1, 6]
	H	H	H	Cl	H	200 h	[1, 6]
	H	CH_3	Cl	CH_3	H	– (after 100 h)	[1, 6]
SO_2CCl_3	H	CH_3	H	H	CH_3	<1	[1]
	H	CH_3	H	H	C_6H_5	<1	[1]
	H	H	H	H	$CH_2C(O)OCH_3$	15	[1]
	H	H	H	Cl	H	25	[1]
	H	CH_3	Cl	CH_3	H	1	[1]
$SO_2C_4F_9$-n*)	H	CH_3	H	H	CH_3	<1	[1]
	H	CH_3	H	H	C_6H_5	<1	[1]
	$-[CH_2]_3-$		H	H	H	<1	[1, 25]
	H	H	H	H	$CH_2C(O)OCH_3$	5	[1, 25]
	H	H	H	Cl	H	5	[1, 25]
	H	CH_3	Cl	CH_3	H	<1	[1, 25]

*) Reactions of $O{=}S{=}N{-}SO_2C_4F_9$-n with 9 alkenes in CCl_4 at 0, −20, and −30°C are listed in [28].

The reactivity of N-sulfinyl-sulfonamides $O{=}S{=}N{-}SO_2R'$ is still higher [1 to 7]. The enophile most frequently used is $O{=}S{=}N{-}SO_2C_6H_4CH_3$-4 ("N-sulfinyl-p-toluenesulfonamide"), which is an effective reagent and easy to prepare. For instance, the reaction of $O{=}S{=}N{-}SO_2C_6H_4CH_3$-4 with alkenes $R^1CH_2{-}CR^2{=}CH_2$ ($R^1 = C_6H_5$, $R^2 = H$ or $R^1 = H$, $R^2 = C_6H_5$) in absolute ether at 0°C yields the stable amides $R^1CH{=}CR^2{-}CH_2{-}S(O){-}NH{-}SO_2{-}C_6H_4CH_3$-4 [2]. It should be remarked, however, that O=S=N–tos (tos = $SO_2C_6H_4CH_3$-4) reacts about ten times slower than the analogous diimide tos–N=S=N–tos [6]. The reactivity of the S=N group may be enhanced still further by variation of the group neighbored to the SO_2 group. $O{=}S{=}N{-}SO_2CCl_3$ reacts more quickly and efficiently than the tosyl derivative and $O{=}S{=}N{-}SO_2C_4F_9$-n even more so. Even cycloalkenes $HC{=}CH{-}[CH_2]_n{-}CH_2$ (n = 2, 3, 4, 5), which are not reactive when compared with acyclic alkenes, react with $O{=}S{=}N{-}SO_2CCl_3$ and $O{=}S{=}N{-}SO_2C_4F_9$-n within minutes, whereas $O{=}S{=}N{-}SO_2C_6H_4$-CH_3-4 needs hours [1, 25]. Similarly, the reactivity towards halogen-substituted alkenes rises. For the reaction of $CH_2{=}CClCH_3$ and $CH_2{=}CBrCH_3$ with O=S=N–R the estimated half-life time is $\tau_{1/2} \approx 12000$ min for $R = SO_2C_6H_4CH_3$-4, 25 min for $R = SO_2CCl_3$, and 5 min for $R = SO_2C_4F_9$-n. Alkenes with more halogen atoms in the molecule, however, such as 1,1- and 2,3-dichloropropene show no reaction with O=S=N–R [1]. With 3-halogenopropenes and O=S=N–R (R = SO_2-$C_6H_4CH_3$-4, SO_2CCl_3, $SO_2C_4F_9$-n) no ene products are obtained, but polymerization occurs. The same occurs with (E)-1-chloro-2-butene and 3-chloro-1-butene. It is suggested that the N-sulfinyl compounds behave in this case as Lewis acids [1]. O=S=N–C_6H_5, O=S=N– $C_6H_4NO_2$-4 and O=S=N–$SO_2C_6H_5$ react with perfluorinated isobutene only in the presence of CsF as a nucleophilic catalyst [8]:

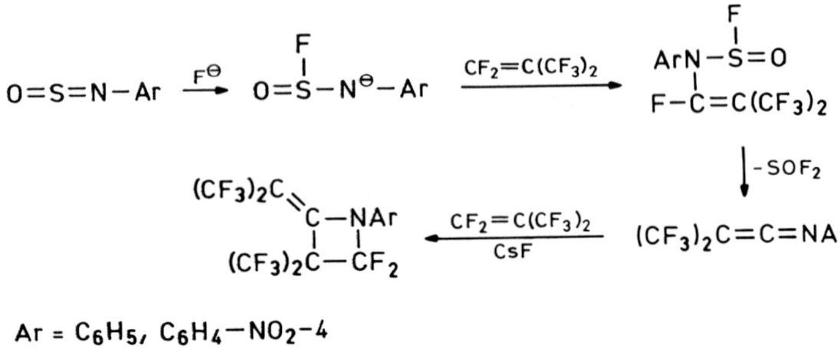

O=S=N–$SO_2C_6H_5$ yields the adduct $(CF_3)_2C$=C=$NSO_2C_6H_5$·CsF [8].

Allenylsilanes and O=S=N–$SO_2C_6H_5$ yield products with the acetylene structure (in C_6H_6; R = H or CH_3) [7]:

O=S=N–$SO_2C_6H_5$ + $(CH_3)_3$Si–CH=C=CR_2 → $(CH_3)_3$Si–C≡C–CR_2–SO–NH–$SO_2C_6H_5$.

The ene reaction occurs also with **cyclic alkenes** (as already mentioned above for C_nH_{2n-2}, n = 5 to 8, and shown in Table 49, pp. 186/7 and p. 187). The first reported ene reaction is that of 6,6-dimethyl-2-methylene-bicyclo[3.3.1]heptene (= β-pinene) with O=S=N–C_6H_5. After holding the substances together for many hours at 100°C, a 1:1 adduct was obtained in 38% yield [9]. The reaction runs better with N-sulfinyl-sulfonamides. E.g., with O=S=N–$SO_2C_6H_5$ and cyclohexene in CH_2Cl_2 an adduct in 33% yield was obtained after 48 h:

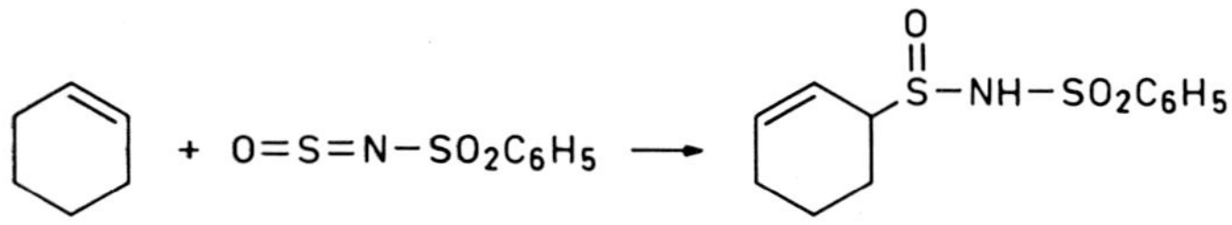

With β-pinene the reaction (in C_6H_6) occurred immediately with 85% yield:

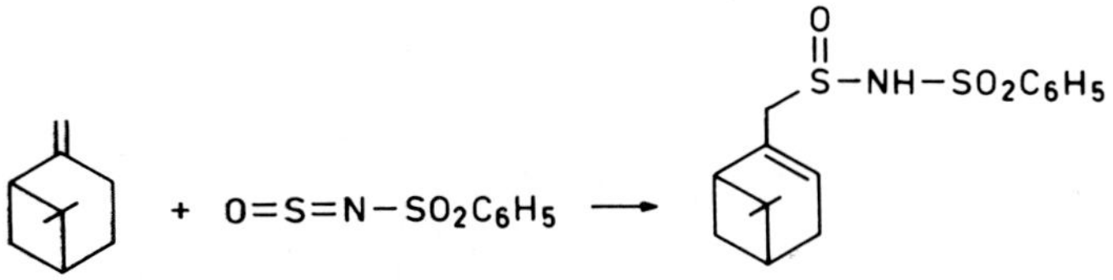

The yield with α-pinene is much lower (in C_6H_6 43% in 40 h; α-pinene can be regained quasi-quantitatively from an α-β-mixture and can be separated from β-pinene in this way) [3]:

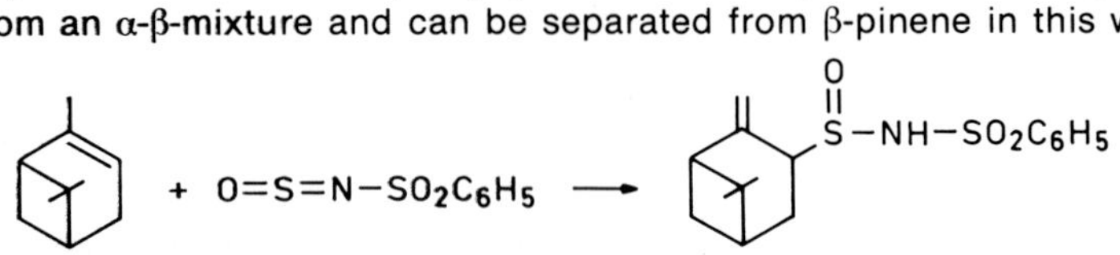

The reactivity of O=S=N–$SO_2C_6H_5$ is generally greater towards terminal or exocyclic double bonds than to internal double bonds [3].

1-Sila- and 1-germa-cyclopent-3-enes also undergo the ene reaction with $O{=}S{=}N{-}SO_2C_6H_5$ [7]:

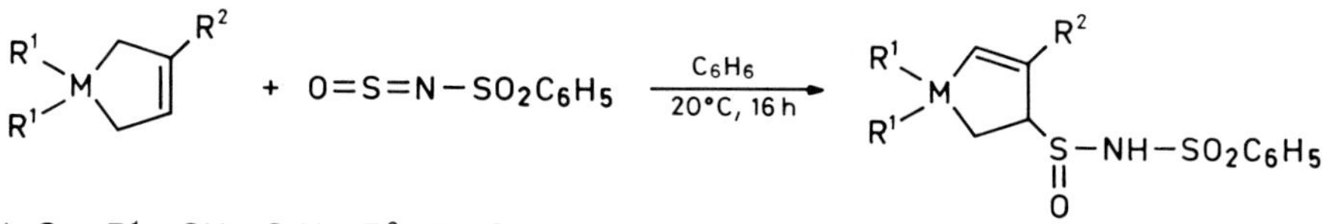

M = Si, Ge; $R^1 = CH_3$, C_6H_5; $R^2 = H$, CH_3

The reaction of N-sulfinyl-sulfonamides with alkenes is reversible and can be exploited to specifically introduce deuterium or tritium into the allylic position of an alkene. As an example, the deuteration of β-pinene is shown in the following scheme [5]:

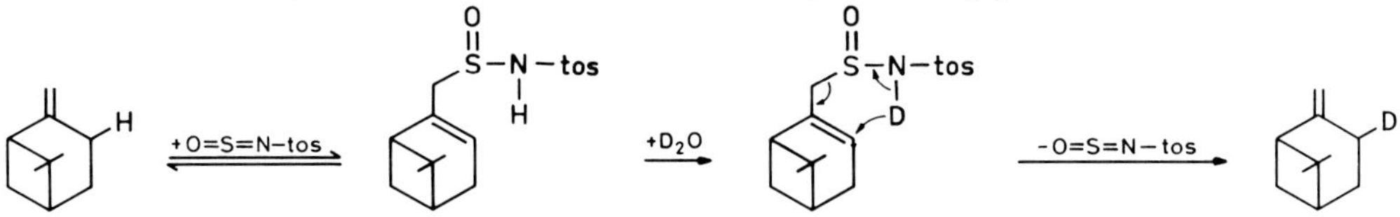

tos = $SO_2C_6H_4CH_3$-4

Similarly, the allylic chlorination of olefins in a special position succeeded with the aid of O=S=N–tos, e.g. [22]:

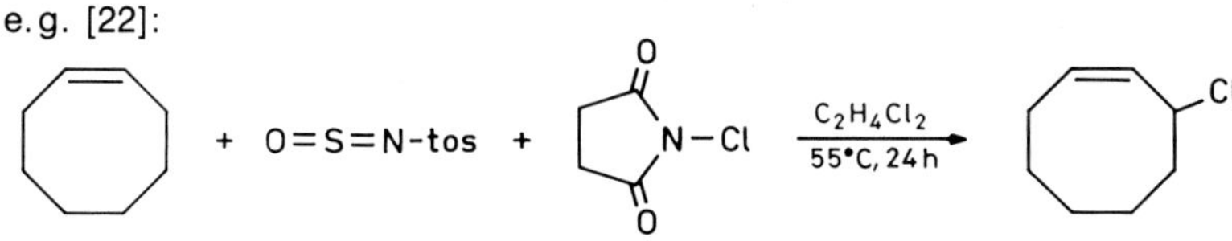

65 %

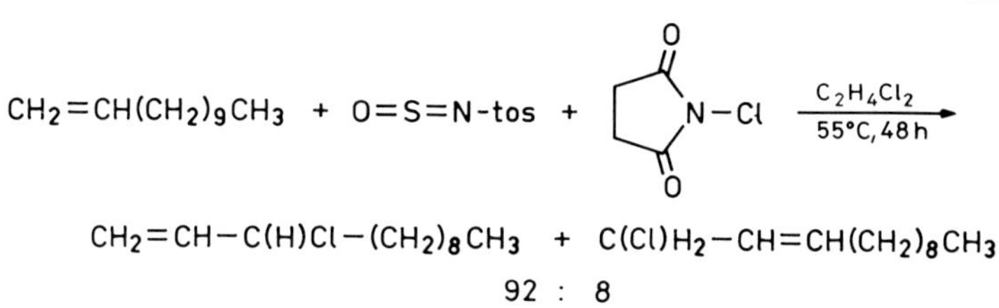

Reversible protection of amino groups with the sulfinyl group allows nucleophilic substitution, e.g. [24]:

$$O{=}S{=}N{-}C_6H_4CH_2Br\text{-}2 \xrightarrow{+ROH} O{=}S{=}N{-}C_6H_4CH_2OR\text{-}2 \xrightarrow{+H_2O} H_2N{-}C_6H_4{-}CH_2OR\text{-}2$$

The regiospecifity of the ene reactions with $O{=}S{=}N{-}SO_2C_6H_4CH_3$-4 holds strongly for the site of the molecule where the enophile attacks. The S atom of the NSO group always correlates with the less substituted C atom of the double bond. In general, the abstraction of the H atom also runs selectively at the neighbored CH_2 group and not at CH_3 [6]. The reactions are stereospecific, which means only one of the possible stereoisomers forms in the reactions with stereoisomers [6]:

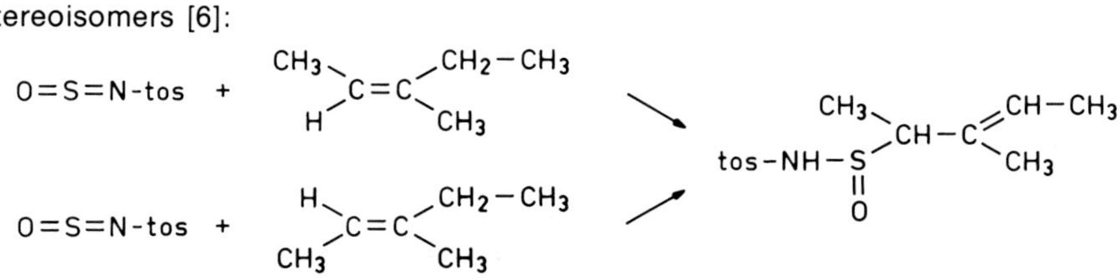

tos = $SO_2C_6H_4CH_3$-4

The ene reaction of the N-sulfinylcarbamate of *trans*-2-phenylcyclohexanol, O=S=N–COO-C_6H_{10}–C_6H_5-2, with simple *cis*-alkenes may be used to prepare allylic sulfinamides with high levels of both absolute stereochemical, as well as regiochemical control [30].

The regiospecific reaction of O=S=N–$SO_2C_6H_5$ with terpene hydrocarbons was used to alkylate in certain positions [4], to shift double bonds in ring systems (calarene→aristolene) [4] or to prepare certain thiols [23]:

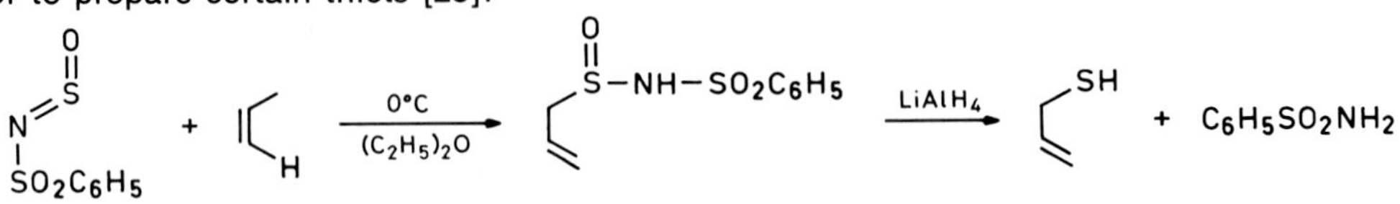

The ene reaction may be used to transfer chirality. E.g., the reaction of O=S=N–tos with (S)-(+)-3-phenyl-1-butene (in toluene, room temperature) yields an optically active (E)-2-alkene-sulfinamide:

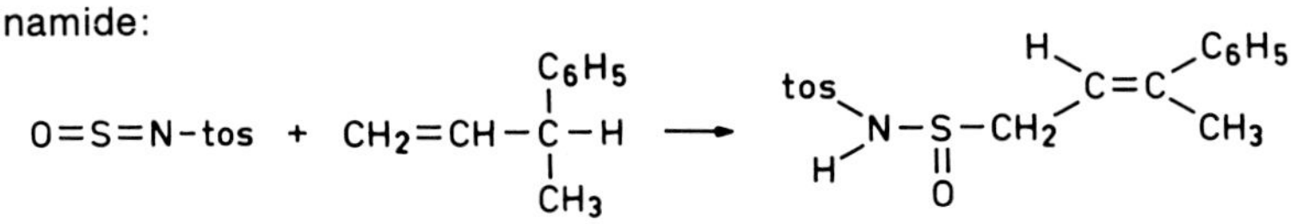

The results are in agreement with the formation of an orienting [2+2] complex between the reactants, preceeding the rate-determining H abstraction by the lone electron pair of the N atom (in principle other structures for the intermediate are possible) [10]:

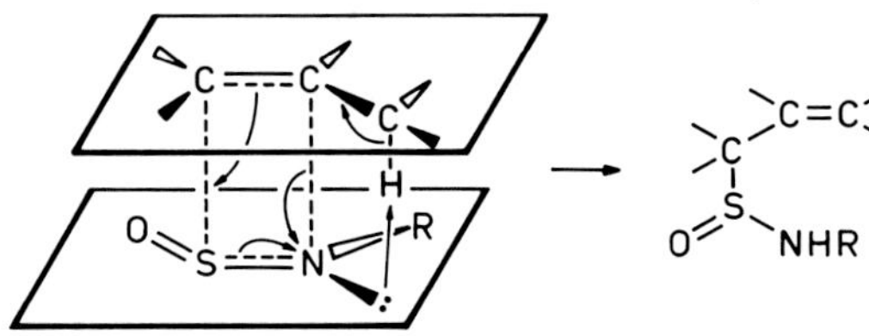

A kinetic, temperature-independent isotope effect for the intramolecular, as well as for the intermolecular hydrogen transfer, was observed for the reaction of O=S=N–tos with CH_2=CH–CHD–C_6H_5 (in $CHCl_3$, −10 to +30°C; k_H/k_D (intramolecular) = 2.8651 ± 0.0002 [11, 29], k_H/k_D (intermolecular) = 2.77 ± 0.15 [29]) and for the reaction of O=S=N–tos with methylenindan (in $CHCl_3$, 20°C; k_H/k_D (intermolecular) = 2.45 ± 0.5) [29].

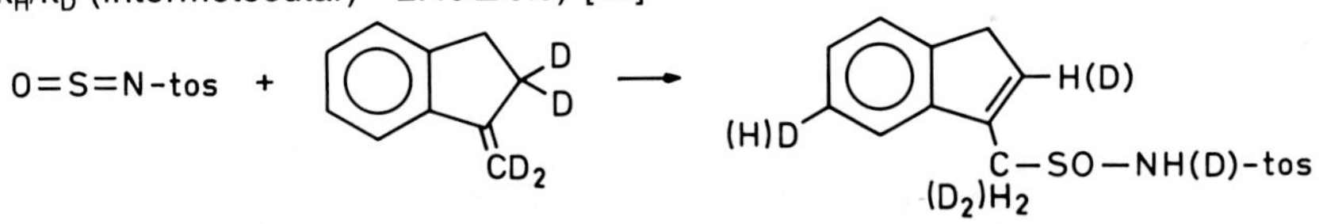

These results and the comparison with kinetic data of other heteroenophiles also point to a nonconcerted two-step ene reaction, namely complexation and hydrogen transfer with bond switching, with hydrogen transfer as the rate-determining step [29].

[2+4] Cycloaddition. When N-sulfinylamines O=S=N–R with R = aryl, R′CO or R′OOC react with alkenes a totally different reaction may occur. These N-sulfinylamines, which possess double bonds (π bonds) conjugated with the S=N bond, can act as heterodienes in [2+4] cycloadditions of Diels-Alder type. A systematic investigation of the reactions of alkenes with various N-sulfinylamines O=S=N–R showed that no "ene reaction" occurred with R = C(O)C_6H_5 and C(O)CF_3 [1], but the [2+4] cycloaddition is much faster than the "ene reaction".

The reaction of O=S=N–C(O)CH_3 and O=S=N–C(O)$C_6H_4CH_3$-4 was systematically studied with symmetrically substituted alkenes [12], and that of O=S=N–C(O)$C_6H_4CH_3$-4 with unsymmetrically substituted alkenes [13] (in CH_2Cl_2, ~0°C, 48 h). The low temperature was chosen to quench the retro-cycloaddition reaction shown by all adducts.

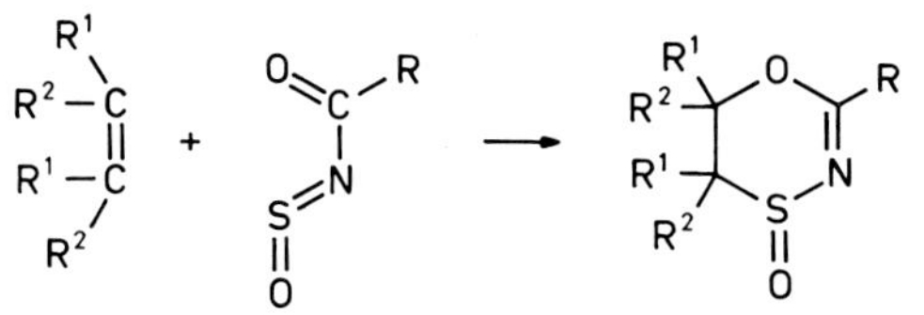

Formation of the cycloadduct is described for $O{=}S{=}N{-}C(O)C_6H_4CH_3$-4 with *cis*- and *trans*-2-butene ($R^1 = CH_3$, $R^2 = H$), *trans*-4-octene ($R^1 = n\text{-}C_3H_7$, $R^2 = H$), 2,3-dimethyl-2-butene ($R^1 = R^2 = CH_3$), cyclohexene ($(R^1)_2 = (CH_2)_4$, $R^2 = H$), similarly for $O{=}S{=}N{-}C(O)CH_3$ with *trans*-2-butene ($R^1 = CH_3$, $R^2 = H$) and cyclohexene ($(R^1)_2 = (CH_2)_4$, $R^2 = H$). The presence of electron-withdrawing groups in the alkene inhibits the reaction of the (electrophilic) N-sulfinylamines. No such reaction occurred between $O{=}S{=}N{-}C(O)C_6H_4CH_3$-4 and *trans*-1,2-dichloroethene, even after 20 d at room temperature with a large excess of the N-sulfinylamine [12]. The reaction with activated alkenes such as bicyclo[2.2.1]hept-2-ene (= norbornene) is much better than with the relatively inactive alkenes 1-octene and cyclohexene [1, 12].

Cycloaddition of $O{=}S{=}N{-}C_6H_5$ to bicyclo[2.2.1]hept-2-ene in refluxing toluene was the first reaction of this kind reported [17]. With CH_3CN as solvent, the same reaction occurs at room temperature [18]. $O{=}S{=}N{-}C(O)CH_3$ and $O{=}S{=}N{-}C(O)C_6H_5$ react with norbornene in ether at −15°C [14].

Acyl derivatives $O{=}S{=}N{-}C(O)OR'$ ($R' = CH_3$, C_2H_5, CH_2CH_2Cl) undergo [2 + 4] cycloaddition with activated alkenes such as bicyclo[2.2.1]hept-2-ene [15, 16] in exothermic reaction (in ether, room temperature). With *cis*- and *trans*-stilbene $C_6H_5(H)C{=}C(H)C_6H_5$ and maleic anhydride $C(O){-}CH{=}CH{-}C(O){-}O$ no reaction occurred even when the solution was warmed [16].

Heterocyclic N-sulfinylamines with the O=S=N group in 2-position, e.g., 2-$O{=}S{=}N{-}C_5H_4N$ (C_5H_5N = pyridine) or 5-$O{=}S{=}N{-}SN_2C_2{-}C_6H_5$ ($SN_2C_2H{-}C_6H_5$ = 3-phenyl-1,2,4-thiadiazole) also undergo [2 + 4] cycloaddition with norbornene [18].

α- and β-$O{=}S{=}N{-}C_{10}H_7$ ($C_{10}H_7$ = naphthyl) react with cyclopentene [26] and also with dienophiles that possess norbornene structure [27] at 100°C in a sealed tube (or 100 h at 50°C) with [2 + 4] cycloaddition. The stereoisomerism of the products was studied [26, 27]. When equimolar amounts of $O{=}S{=}N{-}C_6H_5$ and bicyclo[2.2.1]hepta-2,5-diene (= norbornadiene) are refluxed for 20 min, two products (I and II) form [19]:

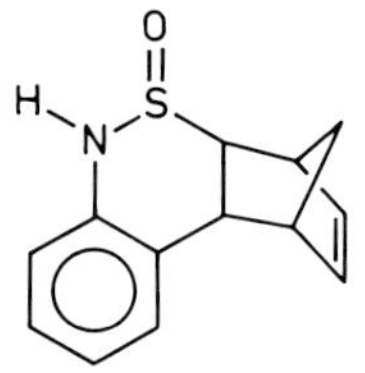

I

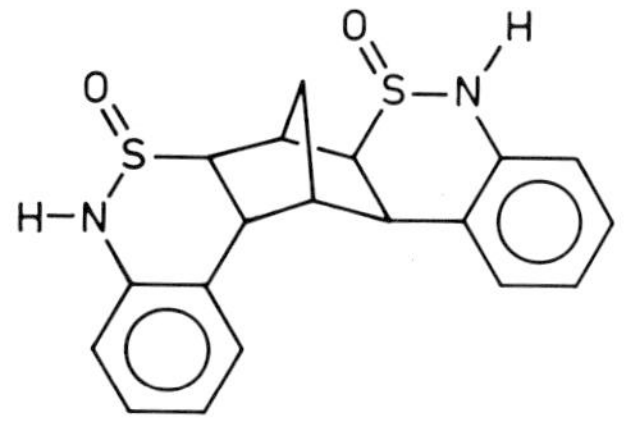

II

The tendency of the products to undergo retro-cycloaddition reactions was demonstrated by reacting the addition product of $CH_2{=}C(CH_3)C_6H_5$ and $O{=}S{=}N{-}C(O)C_6H_4CH_3$-4 with cyclohexene, which yielded the more stable adduct quantitatively with cyclohexene [13]:

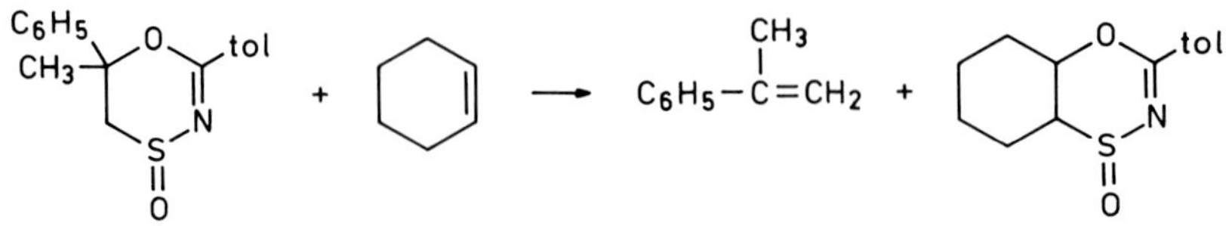

tol = $C_6H_4CH_3$-4

The regioselectivity of the [2+4] cycloaddition was demonstrated in the reaction of O=S=N–C(O)$C_6H_4CH_3$-4 with unsymmetrically substituted alkenes. Only products of one type were obtained:

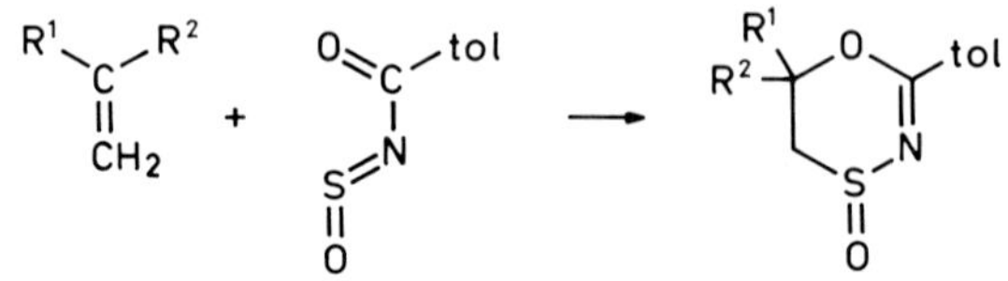

$R^1 = CH_3$	n-C_4H_9	C_6H_5	CH_2Cl	CH_3COO
R^2 = H or CH_3	H	H or CH_3	H	H or CH_3

The adducts were actually a mixture of two enantiomeric pairs [13].

For elucidation of the mechanism, the reaction of N-sulfinyl-anilines O=S=N–C_6H_4R (R = H, Cl, NO_2, OCH_3, CH_3) and N-sulfinylaminoazines O=S=N–R (R = pyridinyl or pyrimidinyl) with 1,4-epoxy-1,4-dihydronaphthalene (see the following scheme) was studied with varying temperature, solvent polarity, and substituent R. The second-order rate constant is only slightly affected by a change of the solvent. A two- or threefold increase was observed for a large range in solvent polarity, see Table 50. Therefore it appears unlikely that there is any high degree of charge separation in the transition state. The temperature dependence of the rate constants was measured in some cases and the activation parameters calculated (see Table 51, p. 194). The relatively low enthalpy of activation, and the strongly negative entropy of activation are consistent with a highly ordered transition state in which bond reorganization is synchronous. The influence of substituents in the N-sulfinyl-arenamines is important, see Table 52, p. 194. The correlation of the reaction rates of O=S=N–C_6H_4R in a dual-parameter analysis show that these cycloadditions can be expressed quantitatively in terms of the electronic character of the substituents [20]. The products show diastereomerism, owing to the chirality created at sulfur by the cycloaddition [21]. Steric hindrance by the substituent is of minor importance [20, 31]. E.g., the reaction of the 6-methyl-substituted pyridinyl derivative with the methyl-substituted 1,4-epoxy-1,4-dihydronaphthalene occurs in the same way as that of the unsubstituted pyridinyl derivative, in spite of the severe crowding of the methyl groups [31]. The experimental results indicate clearly that the [2+4]cycloadditions are pericyclic reactions of Diels-Alder type with inverse electron demand (where the N-sulfinylamine can be seen as a heterodiene and the alkene as a dienophile) proceeding via a normal suprafacial interaction [12, 13, 20, 21, 31]. The strong periselectivity of the cycloaddition of the unsymmetrical 2- and 3-(N-sulfinylamino)-pyridines, giving only the periisomers coupled via sulfur and ring positions 1 and 2, can be accounted for in terms of Hückel orbitals if all π interactions are considered. (The frontier orbital approximation fails on account of the close spacing of orbital energies that occurs in extensively conjugated compounds) [31].

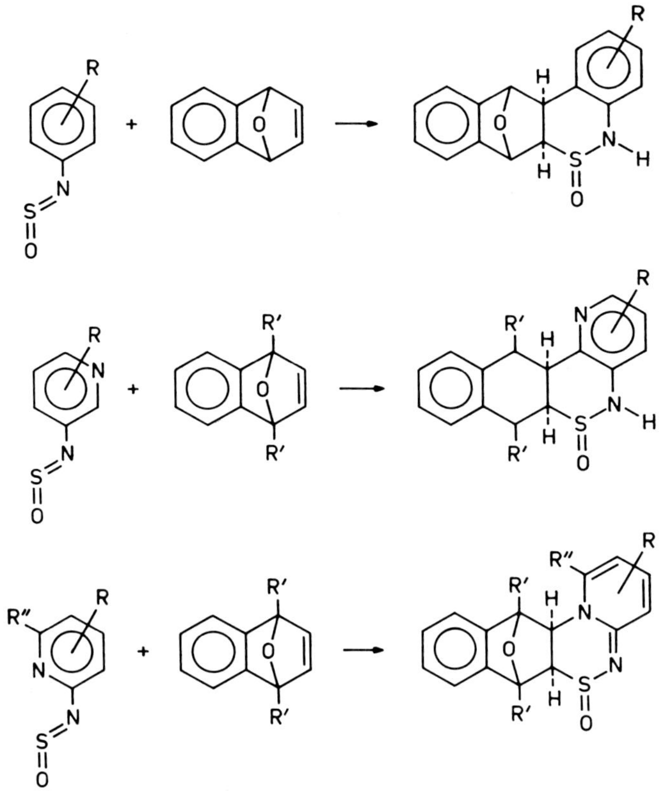

R′ = H, CH_3; R″ = H, CH_3; R is given in Table 50.

Table 50

Solvent Effects on Rates of Cycloaddition of N-Sulfinyl-arenamines, O=S=N–R to 1,4-Epoxy-1,4-dihydronaphthalene.

hetero-diene	t in °C	solvent	dielectric constant at 20°C	10^5 k in $L \cdot mol^{-1} \cdot s^{-1}$	Ref.
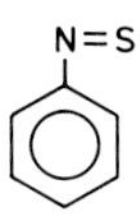	20	cyclohexane	2.023[a)]	4.7	[20]
	20	chlorobenzene	5.708[a)]	8.3	[20]
	20	acetonitrile	37.5[b)]	10.9	[20]
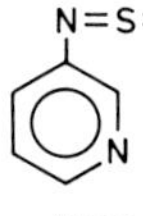	50.75	benzene	2.284[a)]	1.30 ± 0.04	[31]
	50.00	chlorobenzene	5.708[a)]	1.62 ± 0.02	[31]
	50.50	acetonitrile	37.5[b)]	1.80 ± 0.06	[31]
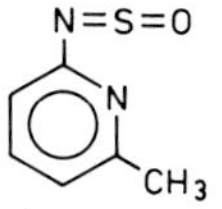	51.40	benzene	2.284[a)]	6.89 ± 0.48	[31]
	51.40	acetonitrile	37.5[b)]	14.10 ± 0.54	[31]

[a)] Weast, R. C. (Handbook of Chemistry and Physics, 63rd Ed., Chemical Rubber Co., Ohio 1982). – [b)] Bray, D. E. (American Institute of Physics Handbook, 2nd Ed., McGraw-Hill, New York 1963).

Table 51

Activation Parameters for the Cycloadditions of N-Sulfinyl-arenamines to 1,4-Epoxy-1,4-dihydronaphthalene.

hetero-diene	t in °C	10^5 k in $L \cdot mol^{-1} \cdot s^{-1}$	ΔH in kJ/mol	ΔS in $J \cdot mol^{-1} \cdot K^{-1}$	ΔG_{298} in kJ/mol	Ref.
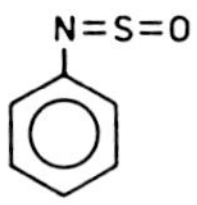 a)	80	3.61				
	90	8.31	80.2	−104.3	111.3	[20]
	100	14.72				
	110	37.74				
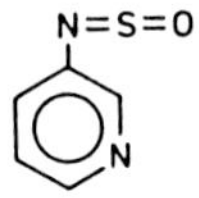 b)	50.7	1.30 ± 0.04				
	60.5	2.77 ± 0.08				
	60.7	2.82 ± 0.09	64 ± 2	−140 ± 6	106 ± 2	[31]
	70.0	5.66 ± 0.12				
	74.5	6.96 ± 0.16				
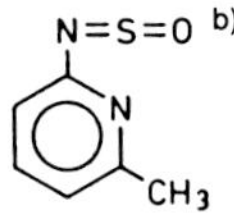 b)	40.2	3.43 ± 0.10				
	51.2	7.19 ± 0.25				
	51.4	6.89 ± 0.48	49 ± 2	−173 ± 6	101 ± 2	[31]
	60.3	12.56 ± 0.54				
	70.0	19.30 ± 1.36				

a) In chlorobenzene. – b) In benzene.

Table 52

Influence of Substituents on Rates of Cycloaddition of N-Sulfinyl-arenamines to 1,4-Epoxy-1,4-dihydronaphthalene.

heterodiene	10^5 k at 20°C in $L \cdot mol^{-1} \cdot s^{-1}$	Ref.
$O{=}S{=}N{-}C_6H_4OCH_3$-4	4.97	[20]
$O{=}S{=}N{-}C_6H_4Cl$-4	5.44	[20]
$O{=}S{=}N{-}C_6H_4CH_3$-4	7.19	[20]
$O{=}S{=}N{-}C_6H_5$	8.31	[20]
$O{=}S{=}N{-}C_6H_4NO_2$-3	9.53	[20]
$O{=}S{=}N{-}C_6H_4CH_3$-3	11.42	[20]
$O{=}S{=}N{-}C_6H_4Cl$-3	23.28	[20]
$O{=}S{=}N{-}C_6H_4OCH_3$-3	76.05	[20]

heterodiene	10^5 k at 50°C in $L \cdot mol^{-1} \cdot s^{-1}$	Ref.
$O{=}S{=}N{-}C_6H_5$ a)	0.29	[31]
$O{=}S{=}N{-}C_6H_3(CH_3)_2$-3,5	0.417 ± 0.002	[31]
4-$O{=}S{=}N{-}C_5H_4N$ b)	0.96 ± 0.06	[31]
3-$O{=}S{=}N{-}C_5H_4N$	1.25 ± 0.04	[31]
2-$O{=}S{=}N{-}C_5H_3N{-}CH_3$-6	6.47 ± 0.38	[31]
2-$O{=}S{=}N{-}C_5H_3N{-}CH_3$-5	21.9 ± 0.6	[31]
2-$O{=}S{=}N{-}C_5H_3N{-}CH_3$-4	29.3 ± 1.3	[31]
2-$O{=}S{=}N{-}C_5H_4N$	53.7 ± 1.6	[31]
2-$O{=}S{=}N{-}C_4H_3N_2$ c)	67.7 ± 3.4	[31]

a) Extrapolated from [20]. – b) C_5H_4N = pyridinyl. – c) $C_4H_3N_2$ = pyrimidinyl.

References:

[1] Bussas, R.; Kresze G. (Liebigs Ann. Chem. **1982** 545/63, 546/9, 552).
[2] Schönberger, N.; Kresze, G. (Liebigs Ann. Chem. **1975** 1725/31).

[3] Déléris, G.; Kowalski, J.; Dunoguès, J.; Calas, R. (Tetrahedron Letters **1977** 4211/4).
[4] Déléris, G.; Dunoguès, J.; Calas, R. (Tetrahedron Letters **1979** 4835/6).
[5] Hori, T.; Singer, S. P.; Sharpless, K. B. (J. Org. Chem. **43** [1978] 1456/9).
[6] Kresze, G.; Bussas, R. (Liebigs Ann. Chem. **1980** 843/57, 847, 852/3).
[7] Laporterie, A.; Dubac, J.; Manuel, G.; Déléris, G.; Kowalski, J.; Dunoguès, J.; Calas, R. (Tetrahedron **34** [1978] 2669/73).
[8] Zeifman, Yu. V.; Ter-Gabrielyan, E. G.; Del'tsova, D. P.; Gambaryan, N. P. (Izv. Akad. Nauk SSSR Ser. Khim. **1979** 396/8; Bull. Acad. Sci. USSR Div. Chem. Sci **1979** 366/8).
[9] Kataev, E. G.; Plemenkov, V. V. (Zh. Org. Khim. **2** [1966] 1124/5; J. Org. Chem. [USSR] **2** [1966] 1119).
[10] Bussas, R.; Muensterer, H.; Kresze, G. (J. Org. Chem. **48** [1983] 2828/32).

[11] Muensterer, H.; Kresze, G.; Brechbiel, M.; Kwart, H. (J. Org. Chem. **47** [1982] 2677/9).
[12] Carpanelli, C.; Gaiani, G.; Sancassan, F. (Gazz. Chim. Ital. **112** [1982] 469/73).
[13] Carpanelli, C.; Gaiani, G.; Sancassan, F. (Gazz. Chim. Ital. **114** [1984] 399/403).
[14] Scherer, O. J.; Schmitt, R. (Chem. Ber. **101** [1968] 3302/12, 3304, 3310).
[15] Hörhold, H.-H. (Angew. Chem. **79** [1967] 312/3; Angew. Chem. Intern. Ed. Engl. **6** [1967] 357/8).
[16] Hörhold, H.-H.; Eibisch, H. (Chem. Ber. **101** [1968] 3567/78, 3570).
[17] Collins, G. R. (J. Org. Chem. **29** [1964] 1688/91).
[18] Beecken, H. (Chem. Ber. **100** [1967] 2151/8, 2159/63).
[19] Macaluso, A.; Hamer, J. (J. Org. Chem. **32** [1967] 506/7).
[20] Hanson, P.; Lewis, R. J.; Stone, Th. W. (J. Chem. Soc. Perkin Trans. II **1983** 1719/23).

[21] Hanson, P.; Stone, Th. W. (J. Chem. Soc. Perkin Trans. I **1984** 2429/38).
[22] Hori, T.; Sharpless, K. B. (J. Org. Chem. **44** [1979] 4204/8).
[23] Gadras, A.; Dunoguès, J.; Calas, R.; Déléris, G. (J. Org. Chem. **49** [1984] 442/4).
[24] Beckert, R.; Mayer, R. (J. Prakt. Chem. **323** [1981] 511/5).
[25] Bussas, R.; Kresze, G. (Angew. Chem. **92** [1980] 748/9; Angew. Chem. Intern. Ed. Engl. **19** [1980] 732).
[26] Kataev, E. G.; Plemenkov, V. V. (Zh. Org. Khim. **4** [1968] 1094/5; J. Org. Chem. [USSR] **4** [1968] 1056/7).
[27] Kataev, E. G.; Plemenkov, V. V.; Pluzhnov, V. K. (Zh. Org. Khim. **3** [1967] 336/41; J. Org. Chem. [USSR] **3** [1967] 319/23).
[28] Ruminy, A. J.; Kresze, G. (Phosphorus Sulfur **29** [1987] 49/56).
[29] Starflinger, W.; Kresze, G.; Huss, K. (J. Org. Chem. **51** [1986] 37/40).
[30] Whitesell, J. K.; Carpenter, J. F. (J. Am. Chem. Soc. **109** [1987] 2839/40).

[31] Hanson, P.; Wren, S. A. C. (J. Chem. Soc. Perkin Trans. II **1987** 197/203).

2.6.23.2 Reactions with Alkoxyalkenes

With alkenes substituted by donor groups such as alkoxyalkenes, N-sulfinylsulfonamides react to form [2+2] cycloadducts. This was first reported in 1965 [1].

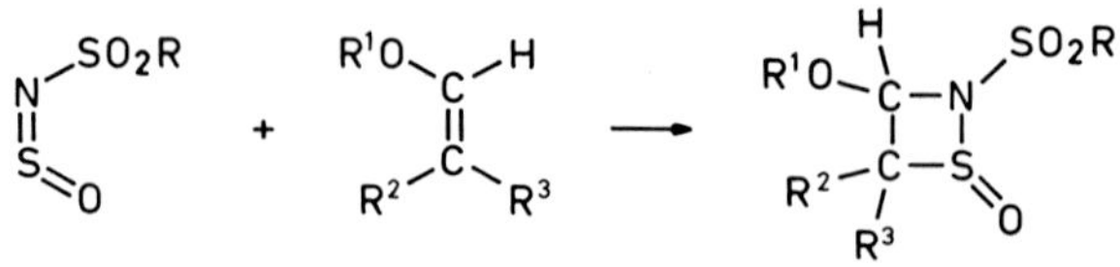

R = C_6H_5 [1], $C_6H_4CH_3$-4 [1, 2], C_6H_4Cl-4 [2]; $R^1 = C_2H_5$, $R^2 = R^3 = H$ [1, 2]; $R^1 = R^2 = C_2H_5$, $R^3 = H$ [2]; R^1, $R^2 = -[CH_2]_3-$, $R^3 = H$ [1, 2].

The S atom is attached to the β-C atom of the enol ether; this was proved by ring opening with Grignard reagents [2]. The reactions are exothermic [1] and occur at low temperatures (0°C in absolute $CHCl_3$ or ether) [2]. The cycloaddition is stereospecific as described for the reaction of O=S=N–$SO_2C_6H_4Cl$-4 with *cis*- and *trans*-C_2H_5O–CH=C(H)C_2H_5 in $CDCl_3$; the products do not isomerize [3]. The diastereomers obtained from O=S=N–$SO_2C_6H_4CH_3$-4 and alkylvinylethers form in nearly equal amounts, but their stability is different. The stereoisomer with the equatorial S=O group is more stable than that containing the axial S=O function [4]:

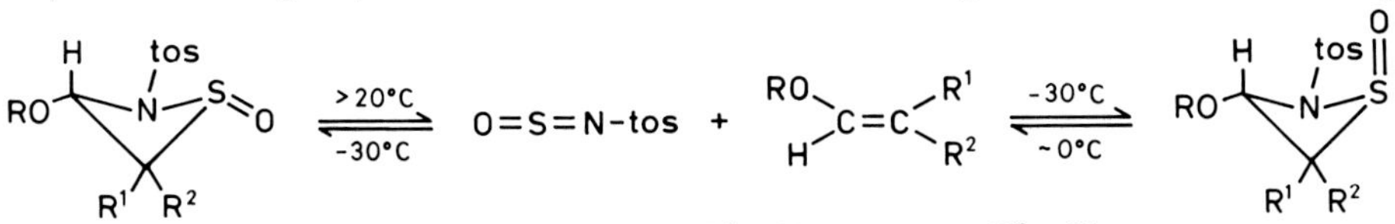

R	R^1	R^2
C_2H_5	H	H
C_6H_5	H	H
CH_3	n-C_4H_9	H
CH_3	H	n-C_4H_9
$-[CH_2]_3-$		H

References:

[1] Maccagnani, C. (Boll. Sci. Fac. Chim. Ind. Bologna **23** [1965] 399/400; C.A. **64** [1966] 15784).
[2] Effenberger, F.; Gleiter, R. (Chem. Ber. **99** [1966] 3903/9).
[3] Effenberger, F.; Kiefer, G. (Angew. Chem. **79** [1967] 936/7).
[4] Wucherpfennig, W. (Tetrahedron Letters **1971** 1891/4).

2.6.23.3 Reactions with Dienes

N-Sulfinylamines O=S=N–R are dienophiles and form [4+2] cycloadducts with conjugated dienes:

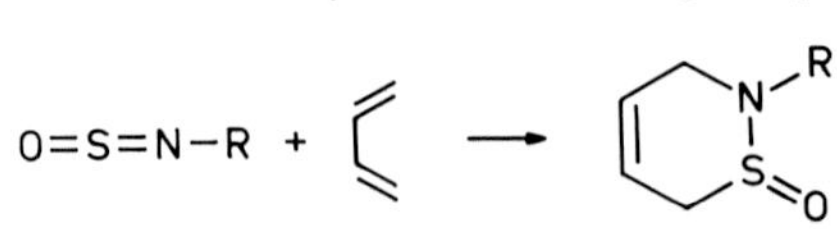

This was first published in 1953 [1] and has since become the most important reaction of N-sulfinylamines with respect to preparative use. The great number of publications is summarized in several reviews [2 to 5]. The short overview given here covers the chemical reactivity of the O=S=N group in various derivatives and recent considerations for the reaction mechanism.

Aliphatic N-sulfinylamines do not behave simply as dienophiles [1]. There was general agreement that only sulfinylamines with electron withdrawing groups at the nitrogen atom can be used in "Diels-Alder" diene syntheses [2, 4]. In a recent publication [25] however, it was reported that O=S=N–C_4H_9-n, in the presence of the Lewis acids BF_3, $SnCl_4$, $TiCl_4$ at −78°C, or under high pressures (6 and 12 kbar) at room temperature reacts with 2,3-dimethyl-1,3-butadiene CH_2=C(CH_3)–C(CH_3)=CH_2 and E,E-2,4-hexadiene CH_3–CH=CH–CH=CH–CH_3 to give [4+2] cycloadducts. Even with 1,3-cyclohexadiene the cycloadduct forms in 10% (BF_3-etherate) or 82% yield (12 kbar) [25]. Reactions of N-sulfinylamines with cyclobutadiene-AlX_3 complexes (X = Hal) were described earlier (see p. 202).

When compounds O=S=N–Ar (Ar = C_6H_5, $C_6H_4CH_3$-2, $C_6H_4CH_3$-4, $C_6H_4OCH_3$-4, $C_6H_4NO_2$-3, $C_6H_4NO_2$-4, C_6H_4Cl-3, $C_6H_4C(O)OCH_3$-4, α-$C_{10}H_7$, β-$C_{10}H_7$ ($C_{10}H_7$ ≙ naphthyl)) and 1,3-butadiene or 2,3-dimethyl-1,3-butadiene are heated on the steam bath for several hours, crystalline 1:1 adducts are obtained. In solution the reaction is usually very sluggish [1]. Substitution at the phenyl ring has no appreciable influence on the rate of formation and the yield of the adducts. The nitro-derivatives O=S=N–$C_6H_4NO_2$ with a somewhat higher binding order than the parent compounds react somewhat faster. However, even derivatives with electron-donating substituents (e.g. O=S=N–$C_6H_4OCH_3$) yield cycloadducts [1, 11]. The bis-sulfinyl derivative 1-O=S=N–C_6H_4–N=S=O-4 forms only one adduct; the second NSO group hydrolyzes readily to form the NH_2 group [1, 11]:

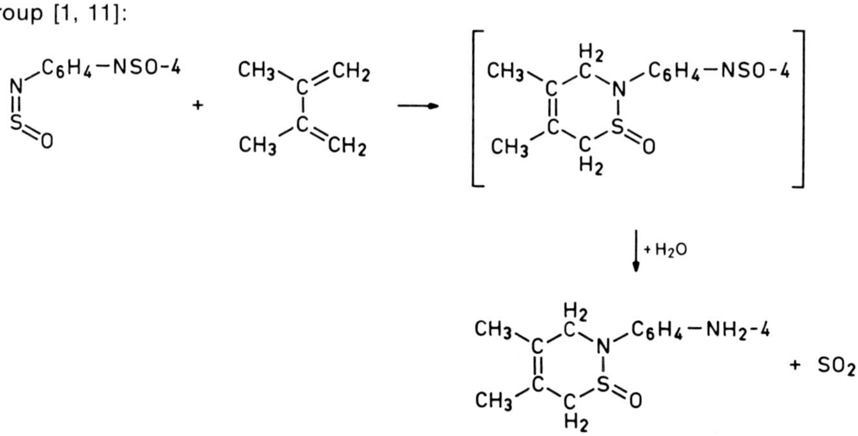

The reaction of N-sulfinyl-benzenamine with tetramethyldiallene in cyclohexane at room temperature for 17 h gives the 1:1 adduct in 64% yield [44]:

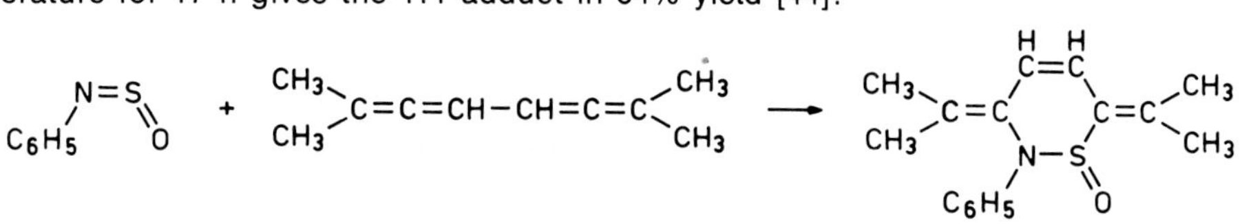

Substitution of the 1,3-dienes in 1- or 4-position may impede the addition reaction by steric hindrance and also by polar effects. For instance, 1-methyl- and 1-(4-nitrophenyl-)1,3-butadiene form adducts with N-sulfinylanilines [6]; 1-phenyl-1,3-butadiene gives small yields [4], but with 1-acetoxy-1,3-butadiene or 2,4-hexadienoic acid (=sorbic acid CH_3–CH=CH–CH=CH–COOH) or its ethyl ester, as well as with 1,3-cyclohexadiene, no adducts could be isolated [4]. O=S=N–C_6H_5 and cyclopentadiene under drastic conditions (when heated at 105°C for 21 h)

form a 1:2 adduct [6, 7]. The same adduct was obtained from O=S=N–C_6H_5 and dicyclopentadiene in refluxing toluene [7]:

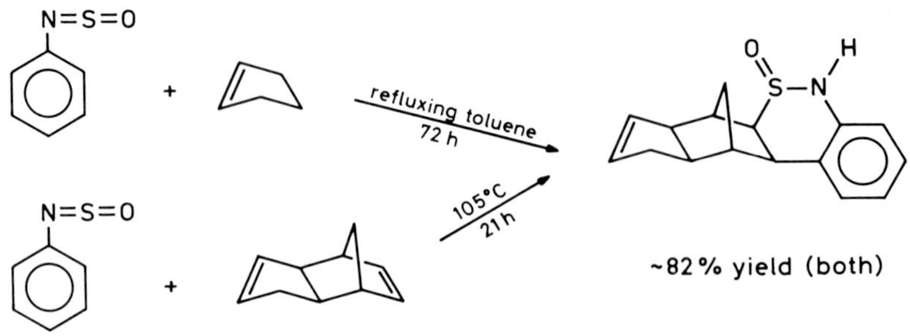

With N-sulfinylamines of the type O=S=N–C(O)Ar, the [4+2] cycloadditions are much better. O=S=N–C(O)C_6H_5 reacts at room temperature with 1,3-butadiene, 2-methyl-, 2-chloro-, or 2,3-dimethyl-1,3-butadiene [8]. O=S=N–CO–CO–N=S=O takes on two molecules of 2,3-dimethyl-1,3-butadiene [43]:

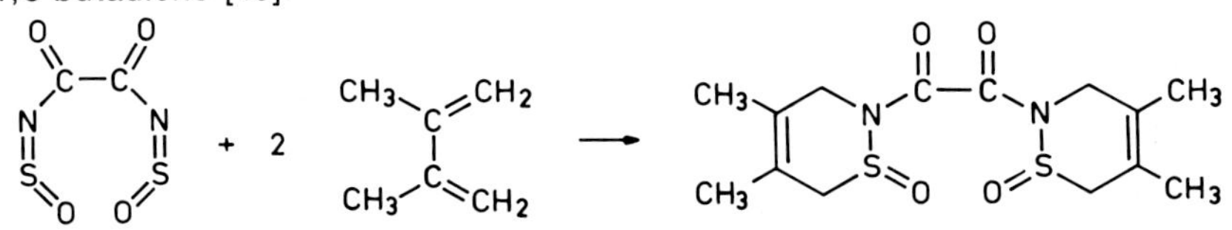

O=S=N–C(O)$C_6H_4CH_3$-4 reacts readily with 1- or 2-substituted 1,3-butadienes CH_2=CH–CH=CHR, CH_2=CH–CR=CH_2 (R = CH_3, C_6H_5, $C_6H_4OCH_3$-4, $C_6H_4NO_2$-4) in anhydrous benzene at room temperature with [4+2] cycloaddition, forming regioisomers. With products of 1-substituted 1,3-butadienes at temperatures higher than room temperature, retro cycloaddition occurs [9]. With disubstituted 1,3-butadienes, the tendency towards retro Diels-Alder reaction increases and several 1,4-disubstituted 1,3-butadienes do not yield isolable adducts [10]. Electron-withdrawing groups in the diene hinder the reaction. E.g., there is no reaction between O=S=N–C(O)C_6H_5 and 2,3-dichloro-1,3-butadiene [8].

N-Sulfinylamines of the type O=S=N–C(O)OR′ with R′ = alkyl or aryl form 1:1 adducts with dienes under mild conditions. The alkoxy-carbonyl group remains unchanged. E.g., O=S=N–C(O)OCH_3 [11, 31] as well as O=S=N–C(O)OC_6H_5 [12, 13] add to 1,3-butadiene [11, 13, 31] or 2,3-dimethyl-1,3-butadiene [12, 31], and even 2,3-dichloro-1,3-butadiene forms the adduct with O=S=N–C(O)OC_2H_5 in 70% yield when heated for 20 min at 100°C [14]:

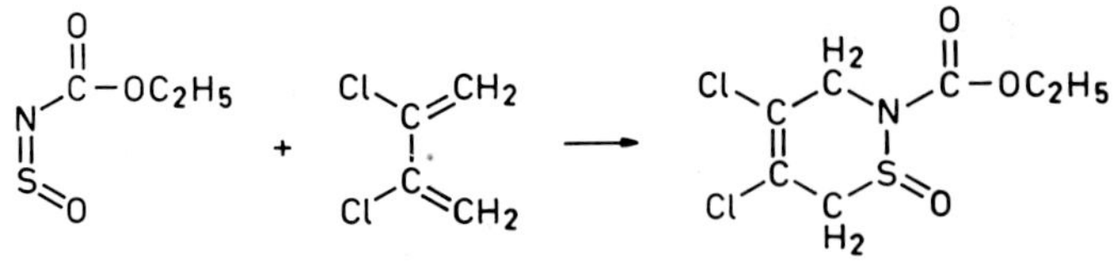

Similarly, O=S=N–C(O)OR′ with R′ = t-C_4H_9 or CH_2CCl_3 form 1:1 adducts with 1,3-butadiene and its derivatives CH_2=CR^1–CR^2=CH_2 (R^1 = R^2 = H, CH_3, C_6H_5, or Cl; R^1 = H, R^2 = CH_3 or Cl) [13].

Diene condensation with 1,2-dihydropyridines succeeds also [15]:

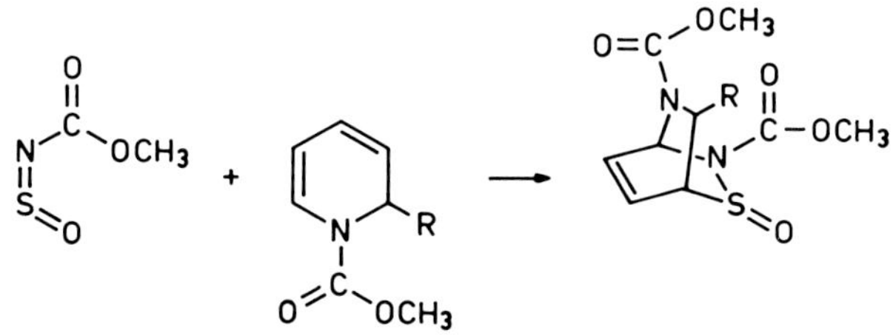

$R = H, CH_3, C_6H_5$

O=S=N–C≡N reacts exothermically with an ether solution of 2,3-dimethyl-1,3-butadiene, forming the colorless crystals of the [4+2] adduct in 90% yield [16].

The reactivity of the NSO group is still greater with the electron-withdrawing SO_2 group bound to the N atom. Condensation of $O{=}S{=}N{-}SO_2R'$ ($R' = C_6H_5$, C_6H_4Cl-4, C_6H_4Br-4, $C_6H_4NO_2$-2, $C_6H_4NO_2$-3, $C_6H_4NO_2$-4, $C_6H_4CH_3$-4) with 1,3-butadiene and 2,3-dimethyl-1,3-butadiene is exothermic, even in benzene solution [17, 18]. Without a solvent, explosion may occur [4, 18]. The substitution of the diene has an important influence on the reaction rate. Chloroprene $CH_2{=}CH{-}CCl{=}CH_2$ and 1,3-cyclohexadiene react fast, even at room temperature. 2,4-Hexadienoic acid $CH_3{-}CH{=}CH{-}CH{=}CH{-}COOH$ adds only sluggishly. 2-Chloro-1,3-cyclohexadiene needs a long time, and with hexachloro-cyclopentadiene (poor in electrons) no adduct is obtained, even after prolonged heating [2, 4]. [4+2] cycloaddition also occurs between $O{=}S{=}N{-}SO_2C_6H_4R$-4 (R = Cl or CH_3) and the siloxy-butadiene $CH_2{=}C(OSi(CH_3)_2C_4H_9\text{-t}){-}C(OSi(CH_3)_2C_4H_9\text{-t}){=}CH_2$ with yields of 56 (R = Cl) or 16% (R = CH_3) when the mixture is refluxed in benzene [40]. N-Sulfinyl-perfluoroorganylamines $O{=}S{=}N{-}R_f$ and 2,3-dimethyl-1,3-butadiene also form [4+2] cycloadducts, as shown from IR, ^{1}H, ^{13}C, and ^{19}F NMR spectra. $O{=}S{=}N{-}CF_3$ (neat) forms an isolable product at 0°C (yield 90%). $O{=}S{=}N{-}C_6F_5$ does not react spontaneously, and on heating the mixture it decomposes; refluxing for 5 h in CH_2Cl_2 affords the adduct (yield 75%). With $O{=}S{=}N{-}C(O)CF_3$ (neat) the addition product is obtained at room temperature in moderate yield (20%) [41]. Previously the formation of linear 1:1 adducts from $O{=}S{=}N{-}CF_3$ and 2,3-dimethyl-1,3-butadiene was claimed from IR, ^{1}H, and ^{19}F NMR spectra [42]. The cycloaddition is reversible. E.g., the adduct of $O{=}S{=}N{-}SO_2C_6H_5$ and cyclopentadiene may be isolated at low temperatures, but splits at room temperature because of the high ring tension in the bicyclic system [2, 4, and references therein].

The [4+2] cycloaddition of asymmetrically substituted 1,3-dienes is highly regioselective [30]. E.g., when $O{=}S{=}N{-}SO_2C_6H_4CH_3$-4 reacts with 2-substituted 1,3-butadienes, addition of the S atom occurs at C(1) [19].

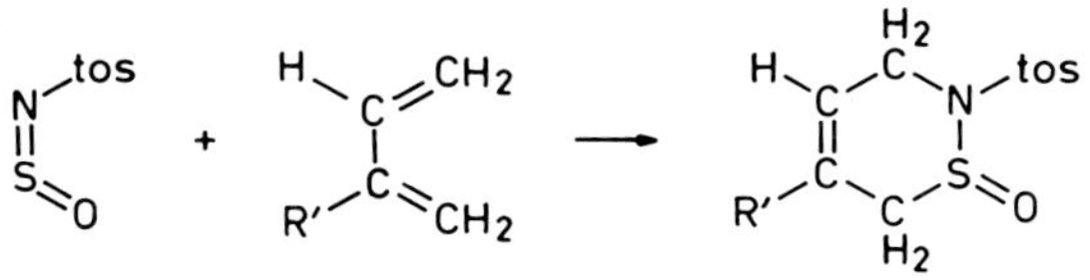

tos = $SO_2C_6H_4CH_3$-4

With 1-substituted 1,3-dienes in kinetically controlled reactions, the addition occurs at C(4) (with formation of, e.g., I), but when the reaction is thermodynamically controlled, the addition occurs at C(1) (with formation of, e.g., II).

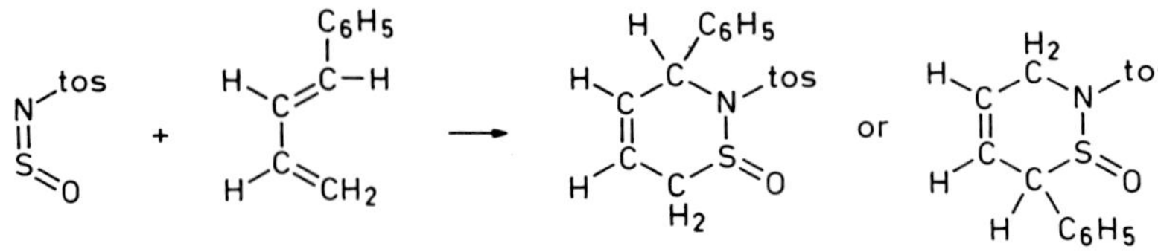

I II

When $O{=}S{=}N{-}SO_2C_6H_4CH_3$-4 reacts with $CH_2{=}CH{-}CH{=}CHC_6H_5$ in benzene at 5°C, only adduct I forms. In boiling benzene, however, II forms almost exclusively (>90%). I in boiling benzene transforms to II in a retro-diene reaction [19].

There is general agreement that the [4+2] cycloadditions of O=S=N–R with conjugated dienes are stereoselective and stereospecific. The stereospecific addition of $O{=}S{=}N{-}SO_2$-$C_6H_4CH_3$-4 to three isomeric 2,4-hexadienes (E,E; E,Z; Z,Z) stereoselectively produces all possible configurationally isomeric adducts [20]. Stereoselectivity and stereospecifity of the [4+2] cycloadditions allow synthetic application, e.g., in the production of complex organic substances and natural products (see for instance [21 to 24, 32 to 39]), or of pharmacologically active substances, for instance adamantane derivatives with antivirus activity [26].

The mechanism of [4+2] cycloadditions was supposed for a long time to be pericyclic, using the terminology "dienophile" and "Diels-Alder" (see e.g. [4, 27]), and was explicitely supposed by some authors [9, 10]. Other authors [20], however, opted for a nonconcerted (two-step) dipolar mechanism for all such reactions because the cycloaddition of $O{=}S{=}N{-}SO_2$-$C_6H_4CH_3$-4 with the highly hindered (Z,Z) isomer of hexa-2,4-diene excludes a concerted reaction. A recent kinetic investigation of the reaction of $O{=}S{=}N{-}C(O)OC_2H_5$ with 1,1′-bicyclohexenyl (I) in benzene at 281 to 318 K clearly indicates a pericyclic mechanism [28].

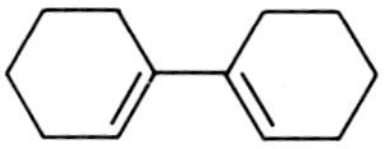

I

The entropy of activation is typically of large negative magnitude ($\Delta S = -176.9\ J \cdot mol^{-1} \cdot K^{-1}$), and the enthalpy of activation is small ($\Delta H = 30.3$ kJ/mol), amounting to 36% of the free energy of activation ($\Delta G = 83.0$ kJ/mol at 298 K). These imply a highly ordered early transition state in which a concerted making and breaking of bonds occurs. The solvent effect (C_6H_6 and $C_2H_4Cl_2$) is small. It is argued that N-sulfinylamino compounds, in general, cycloadd to dienes by a similar mechanism [28]. The orbital symmetry [29] and the regioselective behavior of N-sulfinylamino compounds is comprehensible in the Hückel frontier orbital approximation [28]. In some cases, however, as with (Z,Z)-hexa-2,4-diene [20] the pericyclic mechanism should change into another mechanism [28].

References:

[1] Wichterle, O.; Roček, J. (Chem. Listy **47** [1953] 1768/80, Collection Czech. Commun. **19** [1954] 282/97; C.A. **1955** 1053).

[2] Kresze, G. (in: Hamer, J., 1,4-Cycloaddition Reactions, Academic, New York – London 1967, pp. 453/71).

[3] Kresze, G.; Wucherpfennig, W. (Angew. Chem. **79** [1967] 109/27, 112/3; Angew. Chem. Intern. Ed. Engl. **6** [1967] 149/67).

[4] Kresze, G.; Maschke, A.; Albrecht, R.; Bederke, K.; Patzschke, H. P.; Smalla, H.; Trede, A. (Angew. Chem. **74** [1962] 135/44, 141/3; Angew. Chem. Intern. Ed. Engl. **1** [1962] 89/98).

[5] Ulrich, H. (Cycloaddition Reactions of Heterocumulenes, Organic Chemistry, A Series of Monographs, Vol. 9, Academic, New York 1967, pp. 306/27, 316/20).
[6] Kataev, E. G.; Plemenkov, V. V. (Zh. Obshch. Khim. **32** [1962] 3817/23; J. Gen. Chem. [USSR] **32** [1962] 3745/50).
[7] Collins, G. R. (J. Org. Chem. **29** [1964] 1688/91).
[8] Wucherpfennig, W. (Tetrahedron Letters **1967** 3235/7).
[9] Carpanelli, C.; Gaiani, G. (Gazz. Chim. Ital. **112** [1982] 187/90).
[10] Carpanelli, C.; Gaiani, G. (Gazz. Chim. Ital. **112** [1982] 191/4).

[11] Hörhold, H.-H.; Eibisch, H. (Chem. Ber. **101** [1968] 3567/78, 3570).
[12] Ichimura, K.; Ichikawa, S.; Imamura, K. (Bull. Chem. Soc. Japan **49** [1976] 1157/8).
[13] Wald, L.; Wucherpfennig, W. (Liebigs Ann. Chem. **746** [1971] 28/31).
[14] Hörhold, H.-H. (Angew. Chem. **79** [1967] 312/3; Angew. Chem. Intern. Ed. Engl. **6** [1967] 357/8).
[15] Pavlov, A. V.; Mochalin, V. B. (Zh. Org. Khim. **19** [1983] 234/5; J. Org. Chem. [USSR] **19** [1983] 216).
[16] Scherer, O. J.; Schmitt, R. (Angew. Chem. **79** [1967] 691; Angew. Chem. Intern. Ed. Engl. **6** [1967] 701/2).
[17] Levchenko, E. S.; Kirsanov, A. V. (Zh. Obshch. Khim. **32** [1962] 161/5; J. Gen. Chem. [USSR] **32** [1962] 157/61).
[18] Levchenko, E. S.; Bal'on, Ya. G.; Kirsanov, A. V. (Zh. Obshch. Khim. **33** [1963] 1579/84; J. Gen. Chem. [USSR] **33** [1963] 1540/4).
[19] Kresze, G.; Wagner, U. (Liebigs Ann. Chem. **762** [1972] 93/105, 94/5).
[20] Mock, W. L.; Nugent, R. M. (J. Am. Chem. Soc. **97** [1975] 6521/6).

[21] Weinreb, S. M.; Garigipati, R. S.; Gainor, J. A. (Heterocycles **21** [1984] 309/24).
[22] Garigipati, R. S.; Morton, J. A.; Weinreb, S. M. (Tetrahedron Letters **24** [1983] 987/90).
[23] Garigipati, R. S.; Weinreb, S. M. (J. Am. Chem. Soc. **105** [1983] 4499/501).
[24] Garigipati, R. S.; Freyer, A. J.; Whittle, R. R.; Weinreb, S. M. (J. Am. Chem. Soc. **106** [1984] 7861/7).
[25] Bell, S. I.; Weinreb, S. M. (Tetrahedron Letters **29** [1988] 4233/6).
[26] Sasaki, T.; Shimizu, K.; Ohno, M. (Chem. Pharm. Bull. [Tokyo] **32** [1984] 1433/40).
[27] Weinreb, S. M.; Staib, R. R. (Tetrahedron **38** [1982] 3087/3128, 3116/20).
[28] Hanson, P.; Stockburn, W. A. (J. Chem. Soc. Perkin Trans. II **1985** 589/95).
[29] Grunwell, J. R.; Danison, W. C., Jr. (Intern. J. Sulfur Chem. **8** [1976] 379/82).
[30] Levchenko, E. S.; Bal'on, Ya. G. (Zh. Org. Khim. **1** [1965] 150/5, 155/9; J. Org. Chem. [USSR] **1** [1965] 146/50, 151/5).

[31] Levchenko, E. S.; Bal'on, Ya. G.; Kirsanov, A. V. (Zh. Org. Khim. **3** [1967] 1883/9; J. Org. Chem. [USSR] **3** [1967] 1838/42).
[32] Whitesell, J. K.; James, D.; Carpenter, J. F. (J. Chem. Soc. Chem. Commun. **1985** 1449/50).
[33] Remiszewski, S. W.; Whittle, R. R.; Weinreb, S. M. (J. Org. Chem. **49** [1984] 3243/4).
[34] Remiszewski, S. W.; Stouch, T. R.; Weinreb, S. M. (Tetrahedron **41** [1985] 1173/8).
[35] Garigipati, R. S.; Cordova, R.; Parvez, M.; Weinreb, S. M. (Tetrahedron **42** [1986] 2979/83).
[36] Weinreb, S. M.; Gainor, J. A.; Joyce, R. P. (Bull. Soc. Chim. Belges **95** [1986] 1021/31).
[37] Remiszewski, S. W.; Yang, J.; Weinreb, S. M. (Tetrahedron Letters **27** [1986] 1853/6).
[38] Joyce, R. P.; Gainor, J. A.; Weinreb, S. M. (J. Org. Chem. **52** [1987] 1177/85).
[39] Butler, R. N.; O'Halloran, G. A. (Chem. Ind. [London] **1986** 750/1).
[40] Ried, W.; Reiher, U. (Chem. Ber. **121** [1988] 185/6).

[41] Jäger, U.; Sundermeyer, W. (Chem. Ber. **119** [1986] 3405/10).
[42] Roesky, H. W.; Holtschneider, G. (Z. Anorg. Allgem. Chem. **378** [1970] 168/76).
[43] Scherer, O. J.; Schmitt, R. (Chem. Ber. **101** [1968] 3302/12, 3305/6).
[44] Boan, C.; Skattebøl, L. (J. Chem. Soc. Perkin Trans. I **1978** 1568/72).

2.6.23.4 Reactions with Cyclobutadiene Aluminium Halide Complexes

Whereas aliphatically substituted N-sulfinylamines do not react with dienes in the absence of Lewis acids or high pressure, the reaction of O=S=N–C_4H_9-t with cyclobutadiene aluminium halide complexes occurs above −70°C to form a bicyclohexene-type sulfinamide I and a product II containing a bicyclobutane fragment [1].

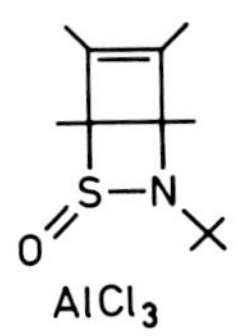

I

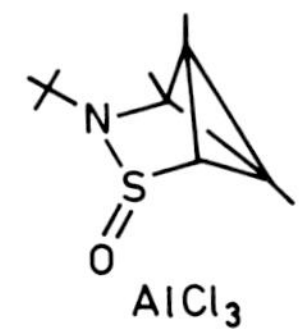

II

The reaction of N-sulfinyl-arenamines with cyclobutadiene aluminium halide σ complexes leads, depending on the temperature and the substitution pattern of the phenyl ring, to either monocyclic, bicyclic, or tricyclic products [2]:

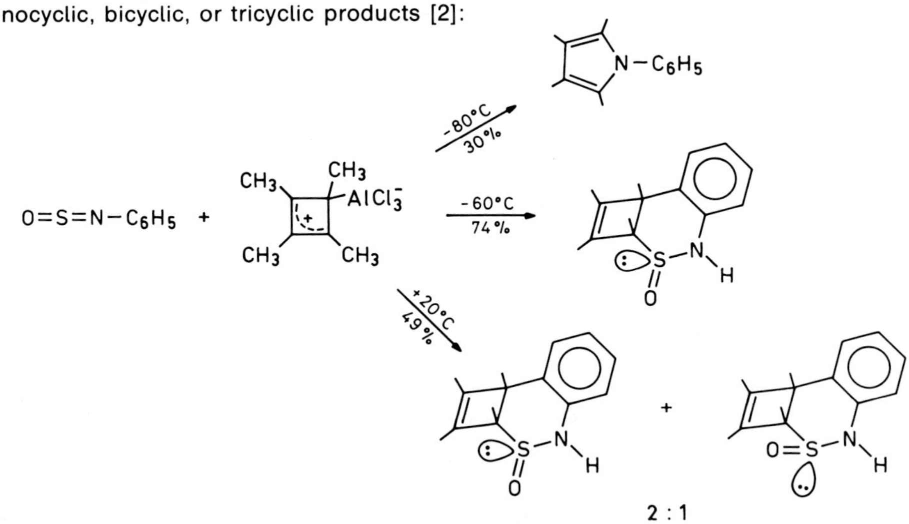

The reaction of O=S=N–C_6H_4R'-4 with R′ = Cl, OCH_3, or CH_3 with the cyclobutadiene complex $C_4H_8 \cdot AlCl_3$ yields the analogous monocyclic and tricyclic compounds. With trimethyl- or dimethyl-cyclobutadiene $AlBr_3$ complexes, the corresponding cyclic compounds form [1]. However, O=S=N–$C_6H_4NO_2$-4 with $C_4H_8 \cdot AlCl_3$ yields a bicyclus [1]:

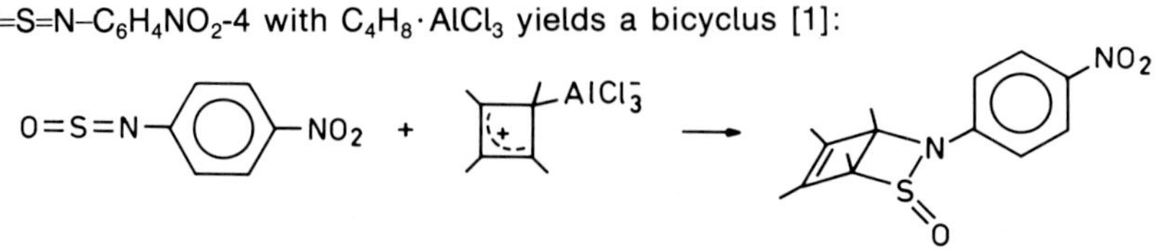

An analogous bicyclus forms with $O{=}S{=}N{-}C_6H_3(CH_3)_2$-2,6 [1].

References:

[1] Fongers, K. S.; Hogeveen, H.; Kingma, R. F. (J. Org. Chem. **48** [1983] 4275/80).
[2] Hogeveen, H.; Kingma, R. F.; Kok, D. M. (J. Org. Chem. **47** [1982] 1909/15).

2.6.23.5 Reactions with Alkines

N-Sulfinylaniline (I) adds to benzyne (dehydrobenzene) (II) prepared in situ by the oxidation of 1-aminobenzotriazole with $Pb(O_2CCH_3)_4$ in a dry solvent at 20°C. Aromatization to III is followed by oxidation to the sulfonamide IV under the oxidative conditions employed. IV is the only product isolated chromatographically in a low yield (4%) from the complex mixture of highly colored products [1].

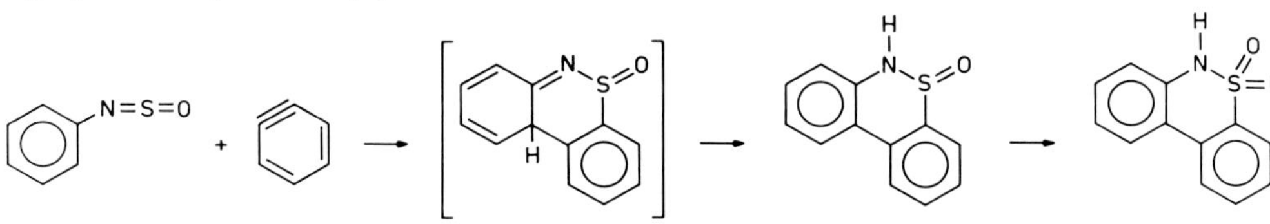

I II III IV

Addition with ring closure occurs similarly with alkoxy alkines (C_6H_6 or CH_3CN solution, room temperature) [2]:

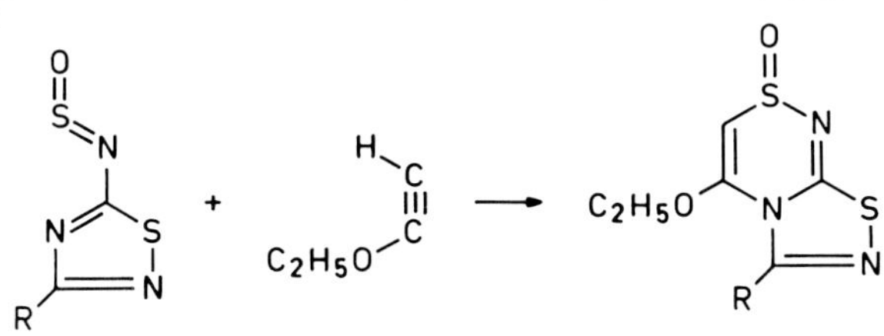

R = CH_3S, C_6H_5

The superreactive $O{=}S{=}NSO_2R$ compounds (R = CCl_3, n-C_4F_9) afford [2 + 2] cycloadducts even with 1-aryl-1-alkines, e.g., 1-propinylbenzene $CH_3{-}C{\equiv}C{-}C_6H_5$ [3]. More recently, the reaction of alkines with $O{=}S{=}N{-}C(O)C_6H_4CH_3$-4 was studied systematically (saturated solutions in dry CH_2Cl_2 at room temperature). In agreement with the previous observations, $O{=}S{=}N{-}C(O)$-$C_6H_4CH_3$-4 undergoes regiospecific [4 + 2] cycloadditions with 1-alkines, forming 6-substituted cycloadducts which are readily oxidized to form the corresponding 4,4-dioxides:

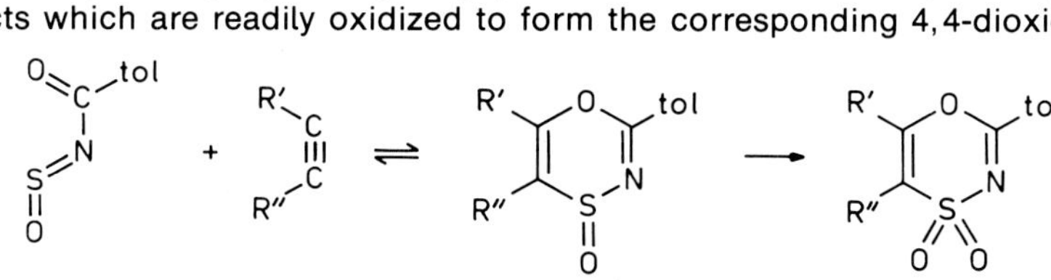

tol = $C_6H_4CH_3$-4

R′	R″	yield in %	R′	R″	yield in %
H	H	no reaction	CH_3OCH_2	H	20
CH_3	CH_3	40	CH_3OCH_2	CH_3OCH_2	no reaction
C_2H_5	C_2H_5	40	C_6H_5	H	80
n-C_3H_7	H	70	C_6H_5	C_6H_5	no reaction
n-C_4H_9	H	70	4-ClC_6H_4	H	50
n-C_5H_{11}	H	70	4-$CH_3C_6H_4$	H	50
$ClCH_2$	H	20			

The alkines bearing electron-withdrawing substituents (R′ = $ClCH_2$, CH_3OCH_2) react more slowly than the other members of this series. The tendency to retro-cycloaddition reactions is high, especially for the aryl-substituted adducts. Because of the similarity to the reactions of N-sulfinylamines with alkenes (see p. 190) a similar concerted mechanism is assumed. Thus, the cycloaddition appears to be a Diels-Alder reaction with inverse electron demand where the N-sulfinylamine acts as a heterodiene and the alkine as dienophile [4].

Inamines (V) combine with N-sulfinyl-arenamines or N-sulfinyl-sulfonamides to produce sulfine derivatives of type VII (in CH_2Cl_2 [6] or ether [5, 7], room temperature [6], 0 or −70°C [5, 7]). The formation of [2+2] cycloadducts (VI) is assumed, in analogy to related reactions of inamines. They are not detectable or separable, but the ring opens spontaneously to give VII [5, 6, 7]:

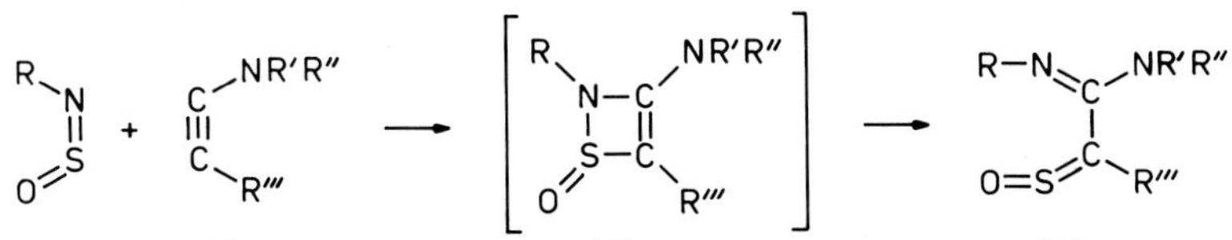

V VI VII

R = $SO_2C_6H_5$ [7], $SO_2C_6H_4OCH_3$-4 [7], $SO_2C_6H_4CH_3$-4 [5, 6, 7], $C_6H_4NO_2$-4 [6].
R′ = CH_3 [5, 7], C_2H_5 [5, 6, 7], C_6H_5 [5, 7]; R″ = C_2H_5 [5, 6, 7], C_6H_5 [5, 7].
NR′R″ = $N{-}CH_2{-}CH_2{-}O{-}CH_2{-}CH_2$ (morpholino) [6, 7], –N-2-C_6H_4–S–C_6H_4-2 [7].
R‴ = H, CH_3 [5, 7], C_6H_5 [5, 6, 7], $N(C_2H_5)_2$ [6], $C(O)C_6H_4Cl$-4, $C(O)C_6H_4CH_3$-4, $C(O)NHC_6H_5$, $C(O)NHC_6H_4OCH_3$-3, $C(O)NHC_6H_4CH_3$-4, $C(S)NHC_6H_5$ [7], $Si(C_6H_5)_3$ [7].

Reactions of N-sulfinyl-sulfonamides with metal compounds containing alkine groups are described on p. 250, reactions with acetylenic alcohols on p. 207.

References:

[1] Campbell, C. D.; Rees, C. W. (J. Chem. Soc. C **1969** 748/51).
[2] Beecken, H. (Chem. Ber. **100** [1967] 2159/63).
[3] Bussas, R. (unpublished results, in: Bussas, R.; Kresze, G.; Münsterer, H.; Schwöbel, A., Sulfur Rept. **2** [1983] 215/377, 272).
[4] Carpanelli, C.; Gaiani, G.; Sancassan, F. (Gazz. Chim. Ital. **115** [1985] 265/8).
[5] Kosack, S.; Himbert, G.; Maas, G. (Angew. Chem. **98** [1986] 443/4; Angew. Chem. Intern. Ed. Engl. **25** [1986] 459/60).
[6] Gotthard, H.; Löhr, T. (Liebigs Ann. Chem. **1987** 189/92).
[7] Kosack, S.; Himbert, G. (Chem. Ber. **121** [1988] 833/41).

2.6.23.6 Reactions with Alcohols

The aliphatic N-sulfinylamines O=S=N–R with R = t-C_4H_9, c-C_6H_{11}, n-C_7H_{15}, CH_2CH_2Cl, CH_2-CH_2CH_2Cl, showed no reaction when dissolved in anhydrous CH_3OH; the UV spectrum in solution remained unchanged [1]. Aromatic N-sulfinylamines react with alcohols, the ease of reaction depending on ring substitution and on the nature of the alcohol. N-Sulfinylaniline O=S=N–C_6H_5 decomposes rapidly and completely in anhydrous CH_3OH. The absorption line of $C_6H_5NH_2$ appears in the UV spectrum of the solution [2]. Similarly, complete decomposition is shown in 2-chloroethanol by means of UV spectrophotometry [2]. In contrast, O=S=N–C_6H_5 dissolves in absolutely anhydrous C_2H_5OH without decomposition, even at boiling temperature [3, 4]. Some authors report slow decomposition in C_2H_5OH in a first order reaction [5, 6, 7] or complete decomposition after 6 d [2]. But other authors state that there is no spontaneous reaction without a catalyst, that the discrepancies in the experimental results must be ascribed to the different methods of ethanol purification, and the stability of O=S=N–C_6H_5 is purely dependent on the dryness of ethanol [8]. Decomposition of O=S=N–C_6H_5 with formation of H_2S occurs only when it is heated with C_2H_5OH in a sealed tube up to 150°C [3]. With CH_2=CH–CH_2OH, $HO(CH_2)_2OH$, i-C_3H_7OH t-C_4H_9OH, $C_6H_5CH_2OH$, and $(C_2H_5)_3COH$ no reaction is observed [2]. However, in the presence of an acid catalyst such as trichloroacetic or acetic acid, i-C_3H_7OH reacts with O=S=N–C_6H_5 to give isopropylsulfite $OS(OC_3H_7\text{-i})_2$ [21]. With n-C_4H_9OH decomposition occurs on boiling, and $C_6H_5NH_2$, SO_2, and a polymeric product form. With phenol, O=S=N–C_6H_5 does not react even at 185°C [4].

Substituted N-sulfinylanilines react more easy with alcohols [9]. The UV spectra of solutions of O=S=N–$C_6H_4NO_2$-4 in CH_3OH, C_2H_5OH, n-C_3H_7OH, i-C_3H_7OH, n-C_4H_9OH, t-C_4H_9OH, and $HO(CH_2)_2OH$ show reactions with each of the alcohols [2]. Similarly, the UV spectra of 2- and 3-nitro-, 4-bromo-, 4-iodo-, 4-methyl-, 2,4-dimethyl-, 2,6-dimethyl-, and 4-sec-amyl-N-sulfinylanilines in CH_3OH indicate rapid and complete reaction. Even highly hindered N-sulfinylanilines containing electron donating groups, e.g., N-sulfinyl-2,6-dimethyl-benzenamine, can undergo rapid reaction [1]. The formation constants of charge-transfer complexes of C_2H_5OH with N-sulfinylaniline derivatives O=S=N–R (R = C_6H_5; 3- and 4-$CH_3C_6H_4$; 3- and 4-$CH_3OC_6H_4$; 3- and 4-ClC_6H_4; 4-$NO_2C_6H_4$; 4-$(CH_3)_2NC_6H_4$; 2,4,6-$(CH_3)_3C_6H_2$) were determined spectrophotometrically. The influence of structure and polarity of the substituents on the stability of the complexes indicates that the sulfur atom is nucleophilicly attacked by the alcoholic oxygen [7].

N-Sulfinylsulfonamides O=S=N–SO_2R, in general, react very vigorously with nondiluted alcohols [10]. But O=S=N–$SO_2C_6H_4CH_3$-4 does not react with benzoine $C_6H_5CH(OH)C(O)C_6H_5$ [11].

The reaction of N-sulfinylamines, e.g. O=S=N–C_6H_5, with alcohols, e.g. C_2H_5OH, is catalyzed by bases, e.g., triethylamine [6] or pyridine [12, 13, 14], and by metal (II) chlorides of Zn, Cu, Ni, Co, and Mn, resulting in the precipitation of the corresponding bisaniline complex $(C_6H_5NH_2)_2MCl_2$ and $(C_2H_5)_2S{=}O$ [15]. The formation of the amine results in autocatalysis of the reaction [9].

Kinetic measurements of the reaction of O=S=N–C_6H_5 with C_2H_5OH in the presence of $N(C_2H_5)_3$ [5, 6, 7] or $CuCl_2$ [8, 15, 16] by spectrophotometry indicate a first-order reaction with respect to N-sulfinylaniline and the catalyst. Similarly the ethanolysis of O=S=N–C_6H_4Cl-4 and O=S=N–$C_6H_4NO_2$-4 in the presence of $CuCl_2$ obeys a first-order rate law [9, 16]. The activation energy at 30 to 40°C is $\Delta H^{\neq} = 14.25 \pm 1.5$ kcal/mol [16], Arrhenius factor 8.0×10^9 [16] (2.83×10^{10} for the $N(C_2H_5)_3$-catalyzed reaction [6]), and $\Delta S^{\neq} = -15.3 \pm 2.3$ e.u. (on the basis of transition state theory) [16]. The ethanolysis reaction is accelerated by electron-withdrawing substituents as shown by measurements of O=S=N–C_6H_4R' with R' = H, Cl, or NO_2. The relation between reaction rate and Hammett σ constant of the N-sulfinylaniline indicates that no complex forms between the N-sulfinylaniline and catalyst (as was suggested earlier [6, 7]), but that the catalyst activates ethanol, converting it to a more effective nucleophile which then attacks the electron-deficient sulfur atom in the N-sulfinylaniline in the rate-determining step [9]. Calculations of the electron densities in O=S=N–C_6H_4R' (R' = H, Cl, or NO_2) by means of the CNDO/2 method corroborate this suggestion. The electron densities at the sulfur atom are proportional to the Hammett σ values of the substituents and also to log k_c for the alcoholysis [16]. A mechanism was proposed [9].

Kinetic studies on the reaction of O=S=N–C_6H_5 and O=S=N–$C_6H_4CH_3$-4 with CH_3OH were also made [12]. In most of the experiments, pyridine or a similar base was added as a catalyst [12, 13, 14]. The rate of the O=S=N–C_6H_5 reaction fits a first-order kinetic equation with $k_1 \times 10^3$ (in min^{-1}) = 1.26 (without base) [12, 14]; 1.2 (2-bromopyridine); 1.6 (2-methylpyridine); 1.7 (N,N-dimethylaniline); 1.9 (2,6-dimethylpyridine); 2.8 (3-bromopyridine); 3.5 (pyridine); 3.8 (quinoline); 4.5 (3-methylpyridine); 5.0 (4-methylpyridine) [14]. With pyridine, $\Delta H^{\neq} = 5.9$ kcal/mol, $\Delta S^{\neq} = -58$ e.u. [12]. The catalytic activity of the tertiary amine is governed by both its basicity and steric configuration [14]. The catalytic activity of pyridine is enhanced by the presence of benzoic acid [12]. The mechanism of the catalytic reaction is proposed to be similar to that in the presence of metal chlorides. The amine should enhance the nucleophilic properties of the alcohol by formation of an H-bonded complex [12, 13, 14]. In the reaction of O=S=N–C_6H_5 with CH_3OH in the presence of pyridine or 4-methylpyridine, the solvent may change the kinetic order with respect to CH_3OH. In THF, $(n\text{-}C_4H_9)_2O$, and CH_3CN the order of the reaction is close to 1; in the less polar solvents CCl_4, C_6H_6, and $C_2H_4Cl_2$ the order is between 1 and 2, dependent on the nature of the added base. The reason is assumed to be self-association of CH_3OH [13].

The type of products formed in the reaction of N-sulfinylamines with alcohols depends on the type of alcohol used. An investigation of the products formed from O=S=N–$SO_2C_6H_4CH_3$-4 (= O=S=N–tos) and 12 different alcohols in benzene or toluene enabled the authors to write the reaction mechanism shown in the following scheme.

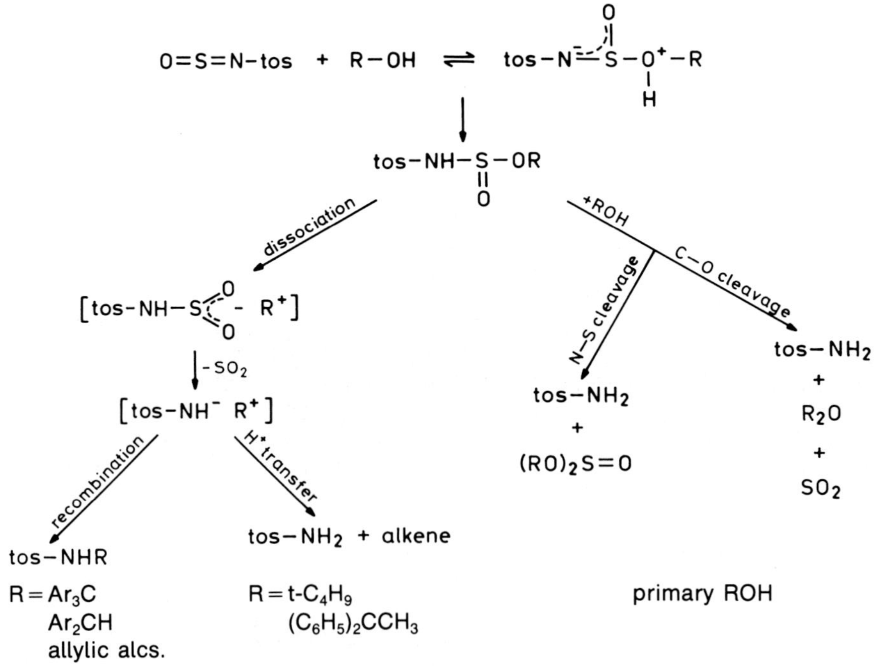

tos = 4-$CH_3C_6H_4SO_2$–
Ar = aryl

The primary product in all cases should be a 1:1 adduct, a thionamide acid O-ester. If the incipient carbonium ion R^+ is sufficiently stable, the adduct can dissociate into an ion pair, and the anion should easily lose SO_2. The new ion pair may recombine to give N-substituted sulfonamides. Alternatively, the carbonium ion in the ion pair may release a proton to produce an alkene. In those cases where relatively stable carbonium ions R^+ are virtually impossible, a second molecule of alcohol attacks the primary intermediate, resulting either in cleavage of the N–S bond and formation of dialkylsulfite, or cleavage of the C–O bond with formation of R_2O and SO_2 [17]. With acetylenic alcohols 1,2,3-oxathiazolidines form, presumably by cyclization of the primary product [18, 19]:

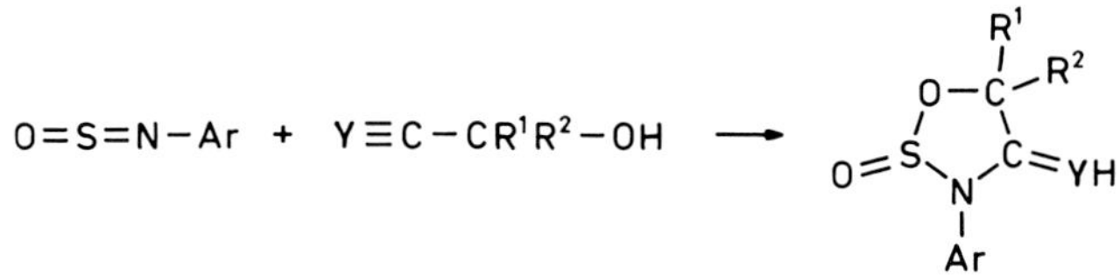

$Ar = C_6H_5$, $4\text{-}ClC_6H_4$, $4\text{-}NO_2C_6H_4$
R^1, $R^2 = H$, CH_3
$Y = CH$, N

N-Sulfinylamines are condensing agents for hydroquinones and isoprenoid alcohols:

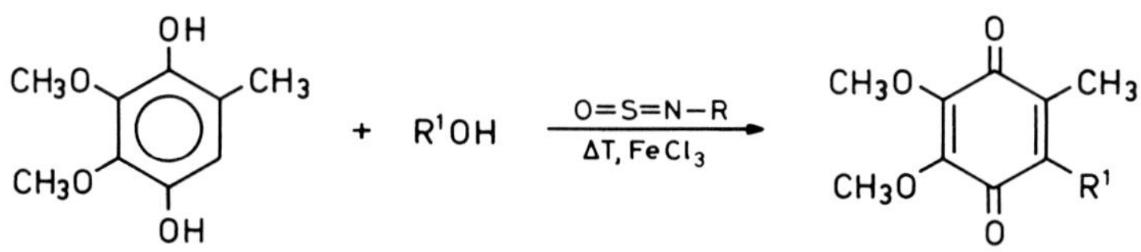

R = aryl, tosyl; R^1 = geranyl, farnesyl, neryl, phytyl.

The primary formation of a thionamide acid O-ester is assumed [20].

Reactions with α-hydroxy carboxylic acids are described on p. 228.

References:

[1] Smith, W. T., Jr.; Grinninger, L. D. (J. Org. Chem. **26** [1961] 2133/4).
[2] Smith, W. T., Jr.; Trimnell, D.; Grinninger, L. D. (J. Org. Chem. **24** [1959] 664/6).
[3] Michaelis, A. (Ber. Deut. Chem. Ges. **24** [1891] 745/55, 747).
[4] Carré, P.; Libermann, D. (Compt. Rend. **194** [1932] 2218/20).
[5] Jannelli, L.; Senatore, L.; Carpanelli, C. (Ann. Chim. [Rome] **53** [1963] 1150/60).
[6] Senatore, L.; Jannelli, L. (Ann. Chim. [Rome] **55** [1965] 980/90).
[7] Senatore, L.; Dell'Atti, A. (Ann. Chim. [Rome] **57** [1967] 177/90).
[8] Collins, N. C.; Glass, W. K. (J. Chem. Soc. B **1971** 2156/9).
[9] Collins, N. C.; Glass, W. K. (J. Chem. Soc. Perkin Trans. II **1974** 713/6).
[10] Kresze, G.; Maschke, A.; Albrecht, R.; Bederke, K.; Patzschke, H. P.; Smalla, H.; Trede, A. (Angew. Chem. **74** [1962] 135/44, 143; Angew. Chem. Intern. Ed. Engl. **1** [1962] 89/98).

[11] Kresze, G.; Sommerfeld, D.; Albrecht, R. (Chem. Ber. **98** [1965] 601/3).
[12] Mironova, D. F.; Konoplya, O. Ya. (Ukr. Khim. Zh. **42** [1976] 504/8; Soviet Progr. Chem. **42** No. 5 [1976] 52/5).
[13] Mironova, D. F.; Konoplya, O. Ya. (Ukr. Khim. Zh. **42** [1976] 629/32; Soviet Progr. Chem. **42** No. 6 [1976] 69/72).
[14] Mironova, D. F.; Konoplya, O. Ya. (Ukr. Khim. Zh. **42** [1976] 960/1; Soviet Progr. Chem. **42** No. 9 [1976] 63/4).

[15] Collins, N. C.; Glass, W. K. (Chem. Commun. **1967** 211).
[16] Glass, W. K.; King, I. J.; Shiels, A. (Inorg. Chim. Acta **25** [1977] 157/8).
[17] McFarland, J. W.; Schut, D.; Zwanenburg, B. (Tetrahedron **37** [1981] 389/93).
[18] Dovlatyan, V. V.; Mirzoyan, R. S. (Arm. Khim. Zh. **29** [1976] 959/62; C.A. **86** [1977] No. 171346).
[19] Dovlatyan, V. V.; Mirzoyan, R. S. (U.S.S.R. 516689 [1975/76] from C.A. **85** [1976] No. 177 431).
[20] Sugihara, H.; Sasaki, M.; Kawamatsu, Y.; Miromoto, H. (Liebigs Ann. Chem. **763** [1972] 121/7).

[21] Shin, J. M.; Kim, Y. H. (Tetrahedron Letters **27** [1986] 1921/4).

2.6.23.7 Reactions with Thiols

N-Sulfinylanilines react with thiols in ether–acetone to give di- and trisulfides [1, 2]. Intermediate products are unstable, but in some cases isolable [2]:

$$O{=}S{=}NR + R'SH \rightleftharpoons R\text{–}NH\text{–}S(O)\text{–}SR'$$

$$R = n\text{-}C_4H_9,\ C_6H_5,\ 4\text{-}CH_3C_6H_4;\ R' = C_6H_5,\ 4\text{-}NO_2C_6H_4$$

The intermediates decompose to give RNH_2, R′–SS–R′, R′–SSS–R′, and H_2O [1] or the hydrolysis products $RNH_3^+HSO_3^-$ [2].

The reaction with 2-HS–$C_6H_4NH_2$ is described on p. 214.

References:

[1] Holmberg, B. (Ber. Deut. Chem. Ges. **43** [1910] 226/7).
[2] Kresze, G.; Patzschke, H.-P. (Chem. Ber. **93** [1960] 380/6).

2.6.23.8 Reactions with Amines, Amides, Imines, and Related Compounds

Primary amines may undergo "transsulfinylation" when reacted with N-sulfinylamines:

$$O{=}S{=}N\text{–}R + R'NH_2 \rightleftharpoons O{=}S{=}N\text{–}R' + RNH_2$$

The reaction may be used as a method for preparation of aza analogs of SO_2 which otherwise are accessible only with difficulty [1]. The equlibrium constant for

$$O{=}S{=}N\text{–}C_6H_5 + 4\text{-}ClC_6H_4NH_2 \rightleftharpoons O{=}S{=}N\text{–}C_6H_4Cl\text{-}4 + C_6H_5NH_2$$

is K = 0.82 in absolute benzene at 20°C [2]. O=S=N–$C_6H_4CH_3$-4 does not react with aniline to form N-sulfinylaniline O=S=N–C_6H_5, however, the latter reacts with 4-$CH_3C_6H_4NH_2$ to form O=S=N–$C_6H_4CH_3$-4 because of the stronger nucleophilicity of 4-$CH_3C_6H_4NH_2$ [41].

The reaction of N-sulfinylamines with 2-substituted anilines depends on the nucleophilicity of the substituent. With 2-HO–$C_6H_4NH_2$ in dry ether and cooling transsulfinylation occurs [3]:

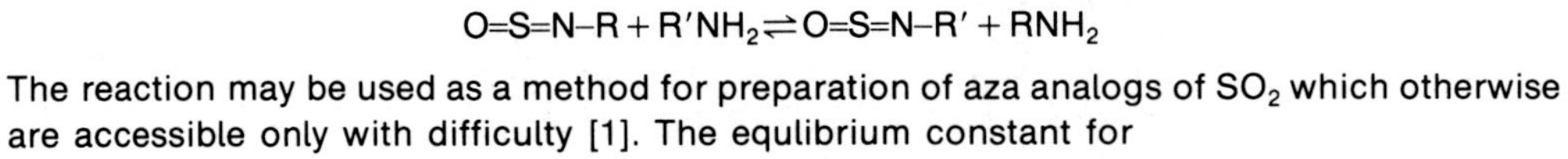

The hydroxyl group is not attacked. This quality enables $O{=}S{=}N{-}SO_2C_6H_4CH_3$-4 and $O{=}S{=}N{-}C(O)OC_2H_5$ to be used as reagents for steroid aminoalcohols with unprotected OH groups [32].

When $O{=}S{=}N{-}SO_2C_6H_5$ reacts with phenylenediamine, mole ratio 1:1, with cooling ($t<25°C$) in dry benzene, only one NH_2 group is substituted by the NSO group to form compound I. With the mole ratio 2:1 in dry $CHCl_3$ at room temperature it forms II. Compound I decomposes spontaneously on standing with warming to release SO_2 [3]:

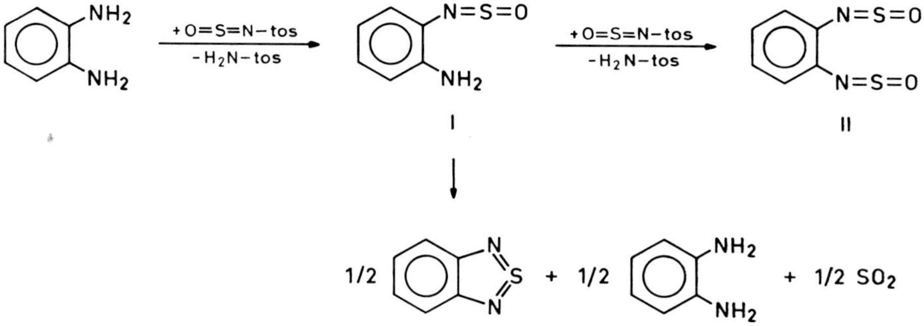

With aromatic o-diamines at 60 to 100°C cyclization occurs, e.g.,

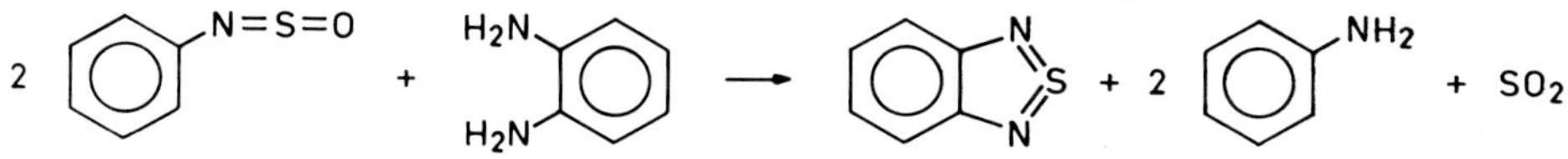

[4, 5, 6] (see "Sulfur-Nitrogen Compounds" Part 3, 1987, p. 192). A great number of substituted 2,1,3-benzothiadiazoles were prepared in this way [5, 6]. In a number of cases the reaction is vigorous and rapid, even at room temperature [6].

$O{=}S{=}N{-}C_6H_5$ reacts similarly with 2,3-diaminonaphthalene (mole ratio 3:1, 95% yield) in pyridine at 95°C [7]:

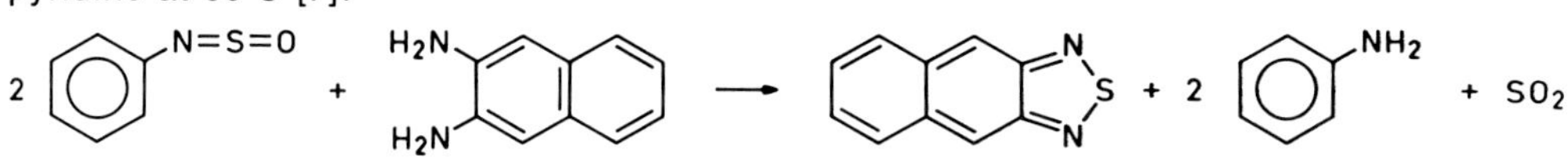

With 1,8-diaminonaphthalene and $O{=}S{=}N{-}C_6H_5$, bridging of the double ring yields the thiadiazole (IV). The intermediate, sulfinyldiamide (III), could be isolated:

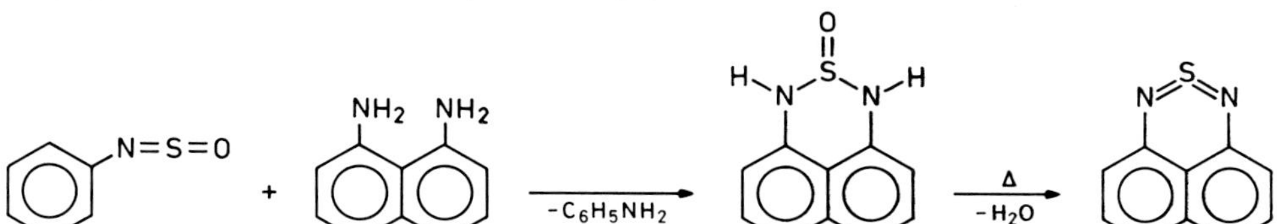

III IV

[8, 9] (see "Sulfur-Nitrogen Compounds", Part 4, 1987, pp. 66/7). The reaction of $O{=}S{=}N{-}C_6H_5$ with 1,8-diaminonaphthalene in $N(C_2H_5)_3$ yields the thiadiazole IV, but in the less basic solvent, pyridine, only the oxide III forms. The reaction of the more strongly electrophilic $O{=}S{=}N{-}SO_2$-C_6H_5 with 1,8-diaminonaphthalene results in both $N(C_2H_5)_3$ and pyridine in the formation of the thiadiazole IV [10].

When O=S=N–C_6H_5 is added to an equimolar amount of 9-amino-1-phenalenimine (V) in benzene at ~50°C phenaleno[1,9-cd] [1,2,6]thiadiazine 2-oxide (VI) is obtained [11]:

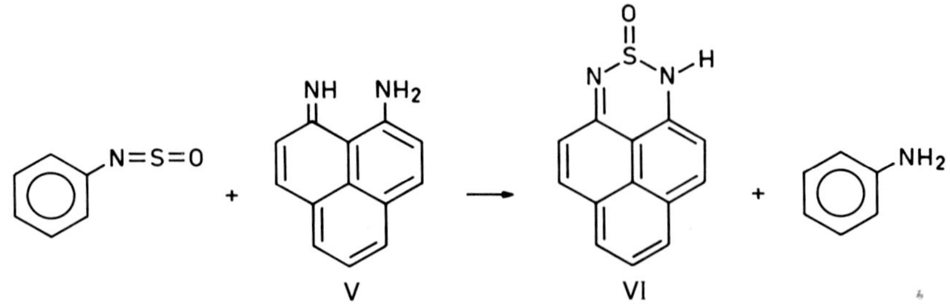

With 2,1,3-benzothiadiazole-4,5-diamine (VII) O=S=N–C_6H_5 reacts in refluxing xylene to form benzo[1,2-c:3,4-c′]bis[1,2,5]thiadiazole (VIII) (see "Sulfur-Nitrogen Compounds", Part 3, 1987, p. 222) [12].

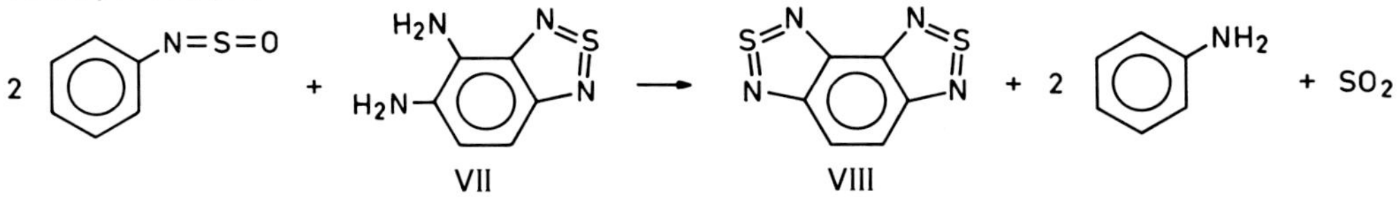

The reaction of O=S=N–C_6H_5 with compound IX in pyridine at 130 to 135°C yields compound X [13]:

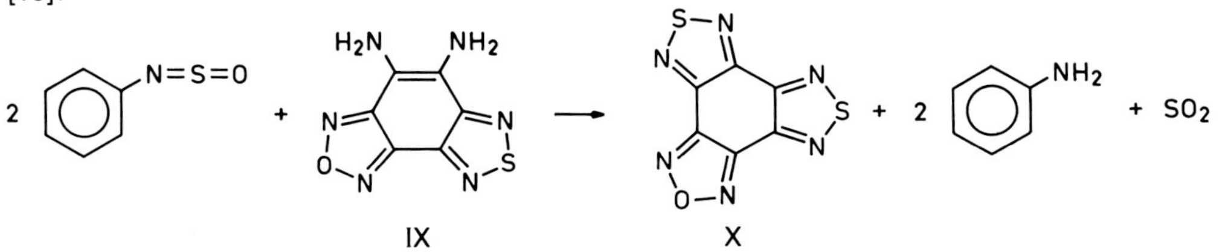

Polymeric amorphous products with electric conductivities of >10^{-6} $\Omega^{-1}\cdot cm^{-1}$ at 30°C and 3×10^5 Hz are obtained when bis-N-sulfinylbenzidine XI reacts with benzidine XII in N-methyl-pyrrolidone at 180°C for 7 h. The polymer contains –NH–SO–NH– groups XIII in the backbone [30], see also [44].

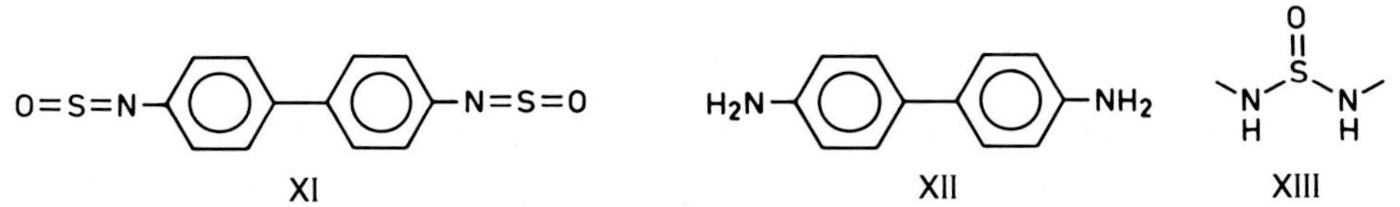

Aromatic diamines, which contain two amino groups with different reactivity, yield products depending on the basicity of the amino groups. The following scheme shows the possibilities of reactions (in $CHCl_3$, room temperature):

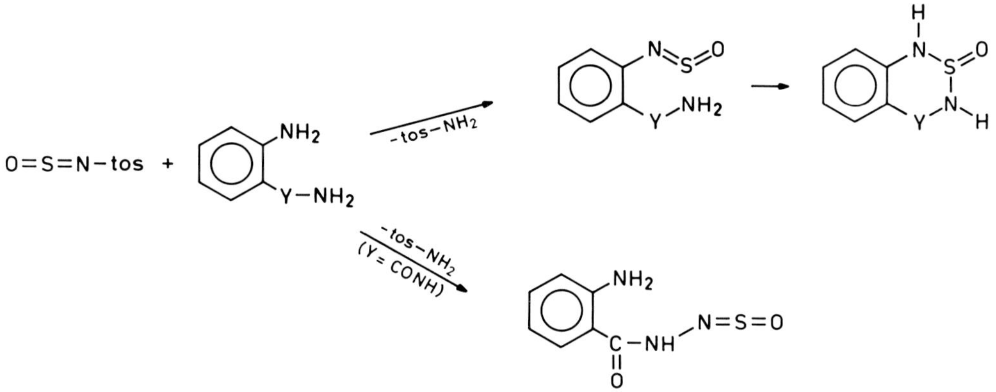

Y = CO, SO_2, CS, CONH

When Y = CO, SO_2, or CS transsulfinylation of the more basic NH_2 group occurs, and with Y = CO and CS cyclization follows immediately. When Y = SO_2 the primary product is stable and cyclization occurs only above the melting point. The urea derivative becomes substituted at the more basic CONH group. Cyclization is not observed. The primary product decomposes on heating with N_2 evolution [26]. $O{=}S{=}N{-}SO_2CH_3$ is used for cyclization of 5-amino-4-methyl-2,1,3-benzothiadiazole and 4-amino-5-methyl-2,1,3-benzothiadiazole [15]:

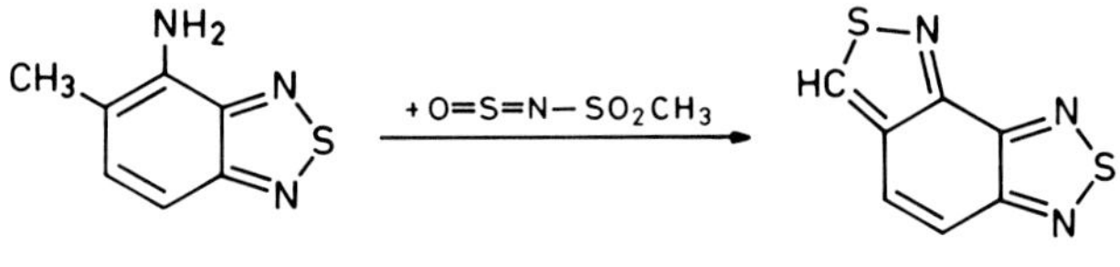

(40% yield)

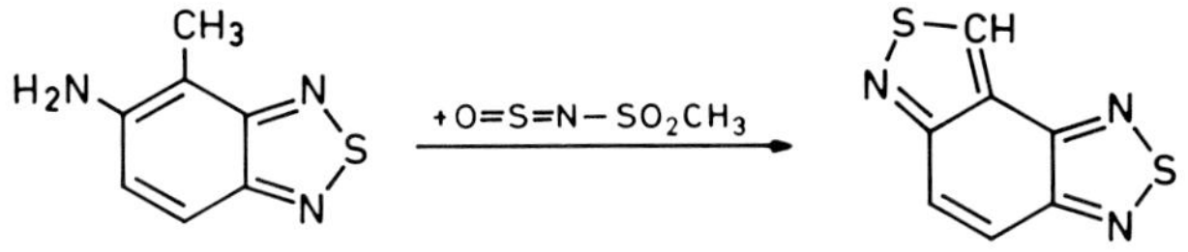

(39% yield)

In a similar way, $O{=}S{=}N{-}SO_2CH_3$ reacts with o-toluidines or N-sulfinyl-o-toluidines in a boiling mixture of pyridine and benzene forming 2,1-benzisothiazoles:

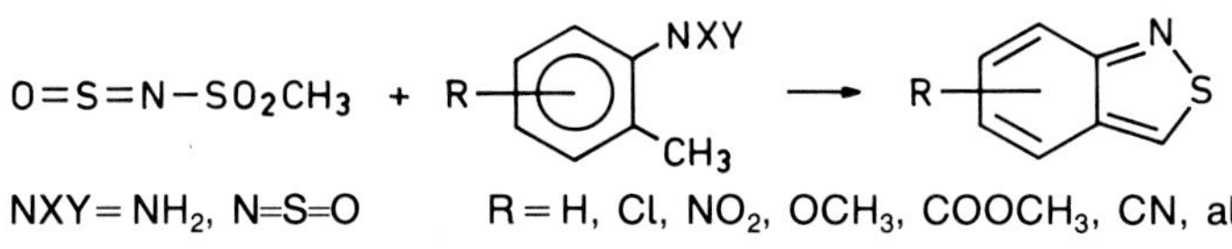

NXY = NH_2, N=S=O R = H, Cl, NO_2, OCH_3, $COOCH_3$, CN, alkyl

$O{=}S{=}N{-}SO_2C_6H_5$ and $O{=}S{=}N{-}SO_2C_6H_4CH_3$-4 react in the same manner. The unclear mechanism of this preparatively interesting reaction is discussed. It is clear that $O{=}S{=}N{-}C_6H_4CH_3$-2 does not cyclize [14]. The syntheses of all three possible angular benzo[c]bis-isothiazoles (I to III) and of the symmetrical benzo[c]tris-isothiazole (IV) by means of the same method (with $O{=}S{=}N{-}SO_2CH_3$) was successful [15].

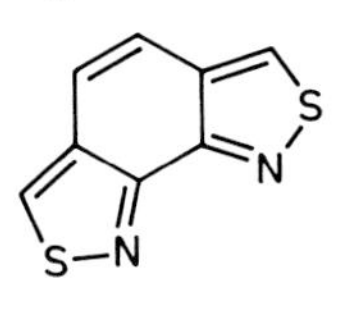

I

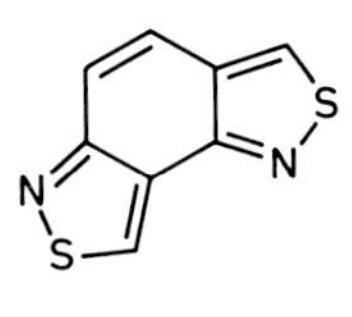

II

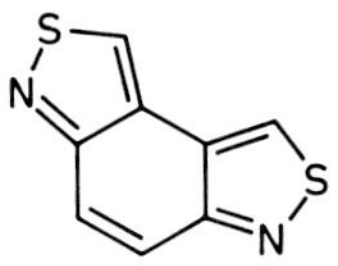

III

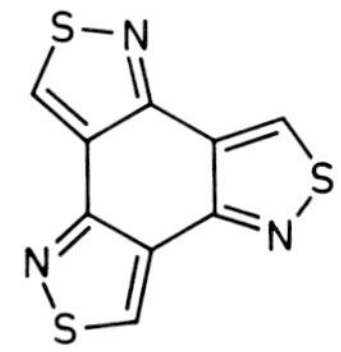

IV

$O{=}S{=}N{-}SO_2CH_3$, reacting by the same method with V, yields the tautomeric compounds VI and VII [16]:

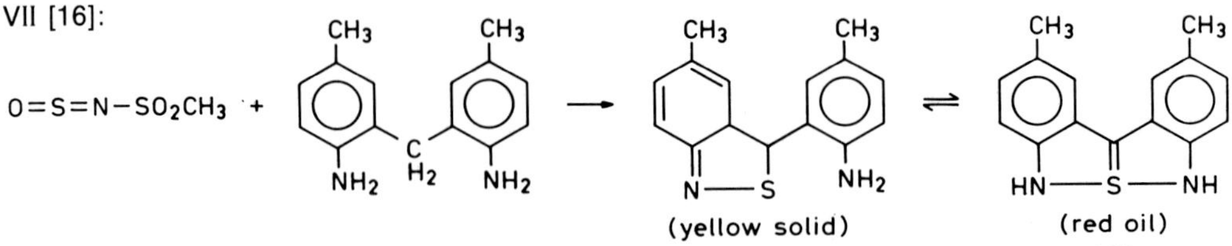

V VI VII

o-Diamines of heterocyclic compounds, e.g., derivatives of pyrimidine, are cyclized by $O{=}S{=}N{-}C_6H_5$ to form 1,2,5-thiadiazoles, see e.g. [17, 18].

$O{=}S{=}N{-}SO_2C_6H_5$ reacts with N-phenyl-phenylmethanimine in boiling benzene with exchange of the NSO group. The author assumes the intermediate formation of a four-membered ring [31]:

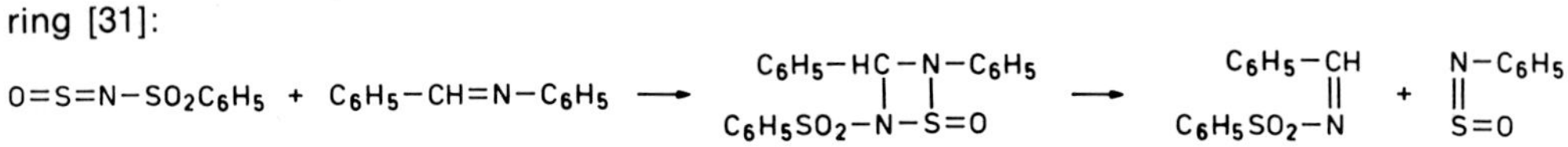

Benzylidene anilines undergo cycloaddition with N-sulfinylanilines, which means the latter behave as "dienes" (mole ratio 1:1, neat, room temperature, 7 to 10 d, quantitative yield) [20]:

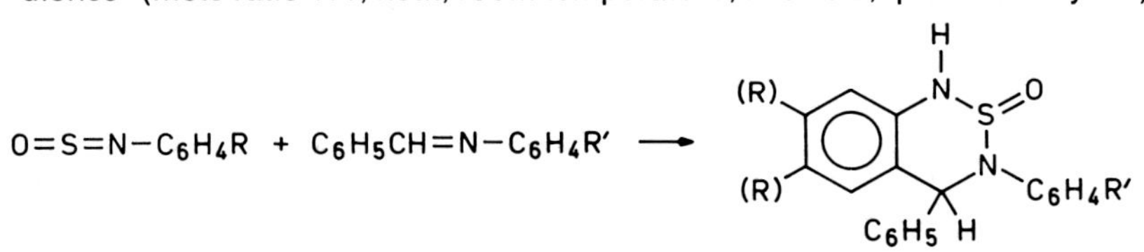

$R = R' = H$, 3-CH_3, 4-CH_3

With N-alkyl-substituted Schiff bases $C_6H_5CH{=}N{-}R$ (R = alkyl), no cyclic compounds form, but exchange occurs [20]:

$O{=}S{=}N{-}C_6H_5 + C_6H_5CH{=}N{-}C_2H_5 \rightarrow C_6H_5CH{=}N{-}C_6H_5 + [O{=}S{=}N{-}C_2H_5]$ (not isolated).

Tetraphenyloxalamidine I and O=S=N–tos, when refluxed 5 h in dry $CHCl_3$, yield the yellow adduct II (12% yield) and the light yellow heterocycle III (5% yield) [33]:

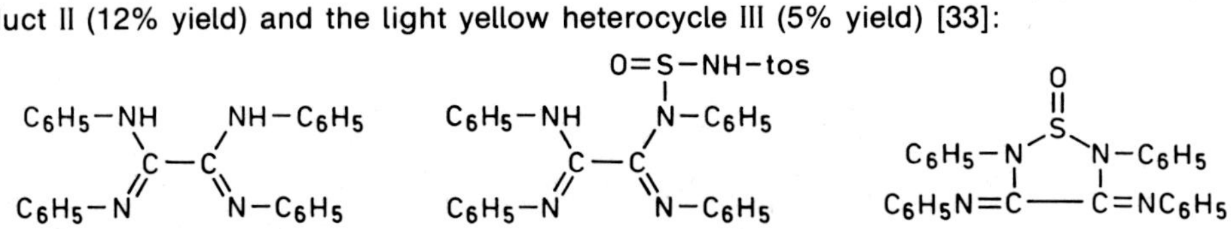

I II III

N-Aryl-substituted amidines (IV) react with N-sulfinylsulfonamides $O{=}S{=}N{-}SO_2R$ (R = CH_3 or 4-$CH_3C_6H_4$) readily at room temperature, forming compounds V in high yields:

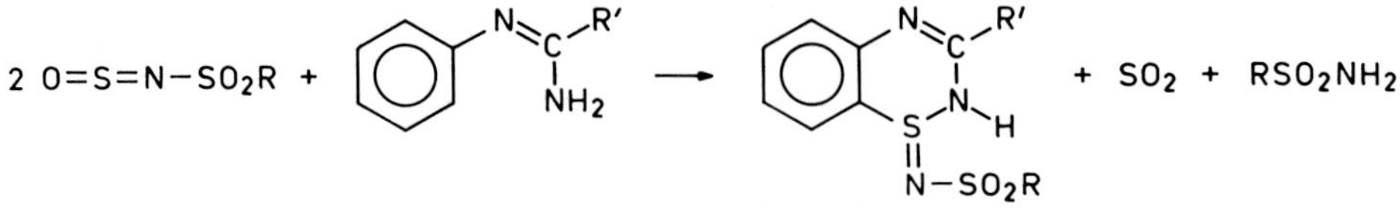

IV V

Compounds of similar structure, e.g., N-aryl-guanidines and -iso-thioureas react with N-sulfinylsulfonamides in the same way [19].

When amidinoyl isothiocyanates VI and N-sulfinylamines in DMF are heated at 140 to 150°C for 1 to 2 h, derivatives of 4-anilinoquinazoline VII are obtained in 20 to 40% yield. A mechanism is proposed [34].

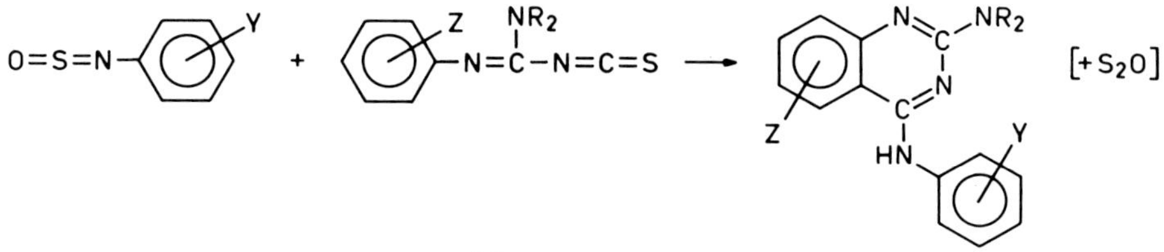

VI VII

Y = H, 4-NO_2, 2-NO_2, 4-CH_3
Z = H, 4-Cl, 4-Br
NR_2 = $N(C_2H_5)_2$, morpholino, N-phenylpiperazino

N-Sulfinylsulfonamides O=S=N–SO_2R (R = C_6H_5, 4-ClC_6H_4, 4-$CH_3C_6H_4$) are used in the synthesis of 1λ[4],2,4,6-thiatriazines from compounds containing an N–C–N–C–N sequence, e.g. [21, 22]:

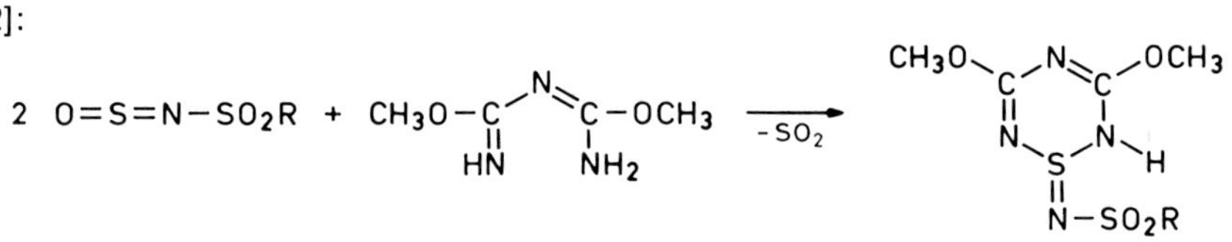

Synthesis of 1λ[4],2,4,6-thiatriazino[4,3-a]benzimidazoles (VIII), 1λ[4],2,4,6-thiatriazino[2,3-a]-benzimidazoles (IX) [23], and pyrazolo[3,2-c][1λ[4],2,4,6]thiatriazines (X) [24] succeeded, using O=S=N–SO_2R (R = C_6H_5, 4-ClC_6H_4, 4-$CH_3C_6H_4$):

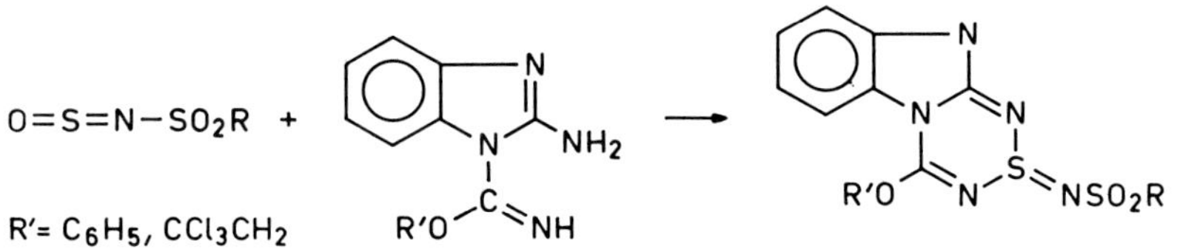

R′ = C_6H_5, CCl_3CH_2

VIII

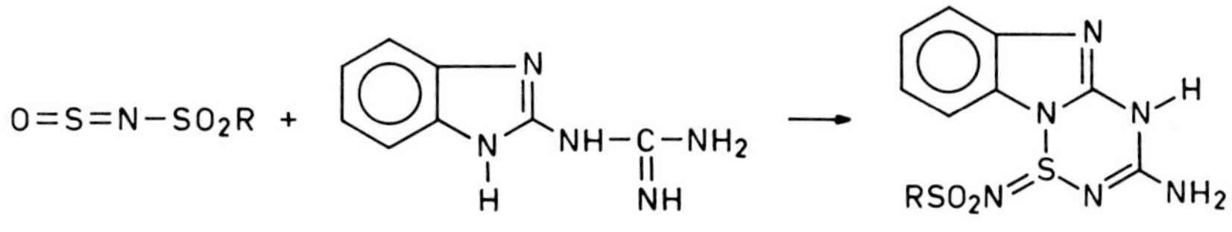

IX

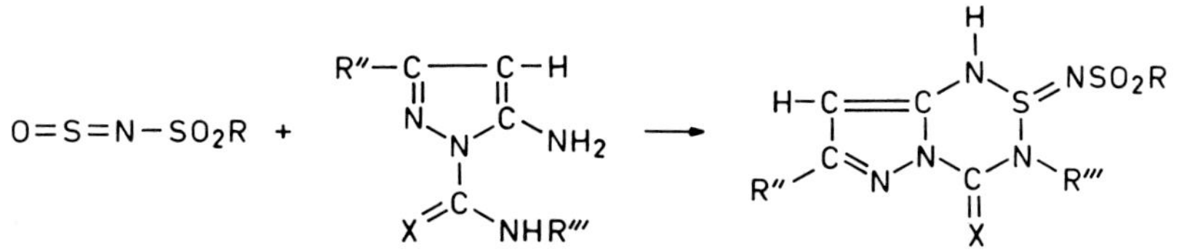

X = S, O; R″ = C_6H_5; R‴ = H, C_3H_7, C_6H_5

X

The mechanism of the reactions is discussed [21, 22, 24].

Guanidinobenzimidazoles containing another functional group capable of cyclization are also cyclized by N-sulfinylsulfonamides at the biguanide moiety because of the high nucleophilicity of exocyclic nitrogen [42]:

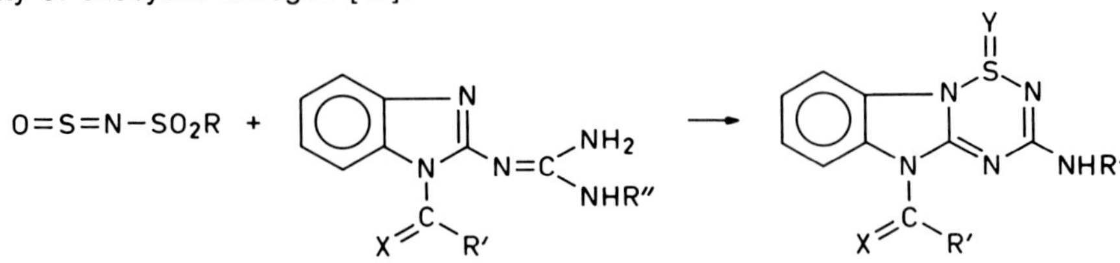

$R = 4\text{-}CH_3C_6H_4$, $4\text{-}ClC_6H_4$; X = O, NH; Y = O, N–tos
$R' = C_6H_5O$, $4\text{-}CH_3C_6H_4O$, $4\text{-}CH_3OC_6H_4O$, C_6H_5NH
$R'' = H$, $4\text{-}ClC_6H_4$

When N-sulfinylanilines O=S=NR ($R = C_6H_5$, $2\text{-}ClC_6H_4$, $3\text{-}ClC_6H_4$, $4\text{-}NO_2C_6H_4$) react with $H_2NC(CH_3)_2CN$ in an organic solvent, e.g., absolute ether, acetone, containing pyridine at 0 to 60°C cyclization occurs [25]:

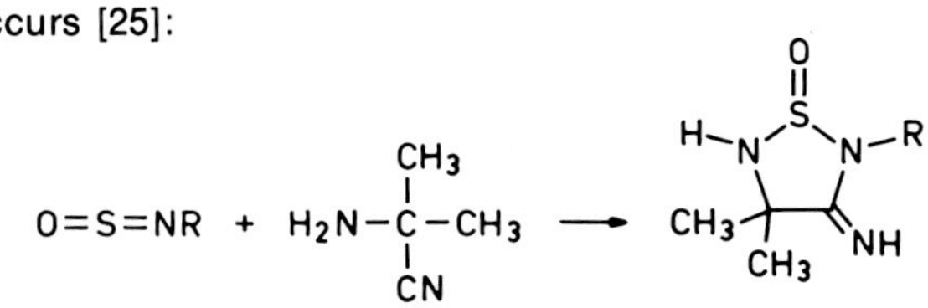

Reactions with inamines $R_2N–C≡C–R$ are described on p. 204, reactions with conjugated –N=C–C=P bonds on p. 240.

2-Mercaptoaniline $2\text{-}HS–C_6H_4NH_2$ and O=S=N–tos yield many compounds even when cooled, which could not be separated by column chromatography. It is suggested that the strongly nucleophilic SH group reacts immediately, presumably before the NH_2 group [3].

$O=S=N–C_6H_5$ reacts with α-amino acids or their sodium salts at room temperature or at 80°C in DMSO or in benzene, to give either N-phenylimines or the corresponding carbonyl compounds:

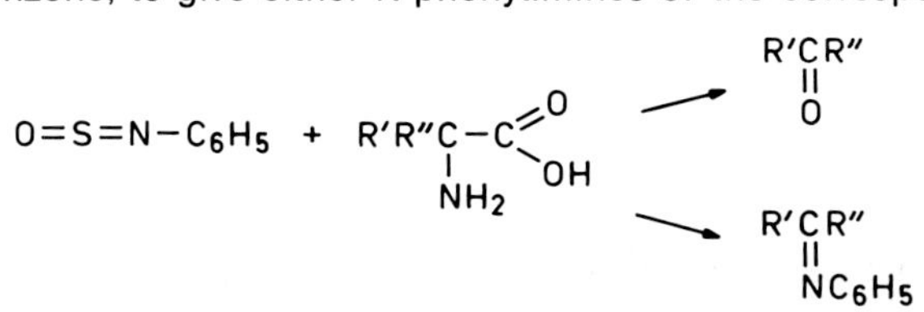

The imine forms when $R' = R'' = H$ (9% yield; Na salt 35%), $R' = C_6H_5$, $R'' = H$ (57%, Na salt 76%), and $R' = C_6H_5$, $R'' = CH_3$ (Na salt 72%). The carbonyl compound forms when $R' = CH_3$, $R'' = H$ (Na salt 72%), $R' = CH(CH_3)_2$, $R'' = H$ (24%, Na salt 54%), and also with the ester $C_6H_5CH(NH_2)$-$COOC_2H_5$ (31%). Glycyl ethyl ester in DMSO, benzene, or ether yield a 2,1,3-thiadiazole (24 to 30%):

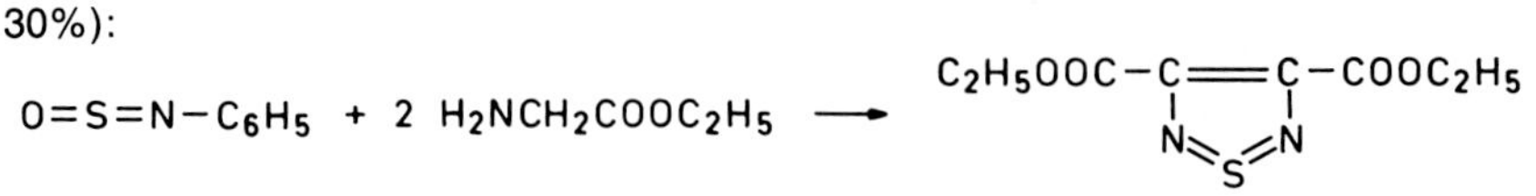

Reaction mechanisms are discussed. An initial sulfonyl group transfer to give $C_6H_5NH_2$ and the type O=S=N–C(R',R")COOH as intermediate products is proposed [38]. Subsequent decarboxylation of such intermediates and formation of a 4π-electron species, believed to be sulfinylaminomethanide anions I, is assumed by other authors. These anions could be trapped by N-substituted maleimides II in 4π + 2π cycloaddition to form III. Indeed, when $O=S=N–C_6H_5$, II, and α-amino acids in DMSO at 80°C are heated, the adducts III are obtained in 60 to 90% yield. In toluene at 110°C α-amino acid esters react analogously [40, 43].

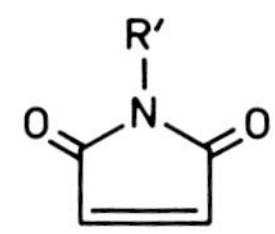
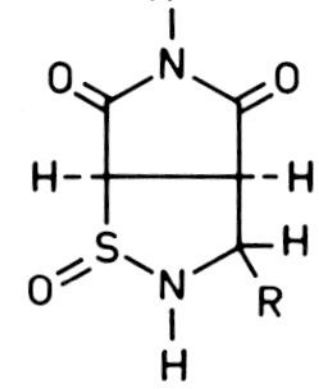

I II III

$R = i\text{-}C_3H_7$, $s\text{-}C_4H_9$, $i\text{-}C_4H_9$, $(CH_2)_2SCH_3$, C_6H_5, $CH_2C_6H_5$, CH_2-Ind (Ind = indol-3-yl)
$R' = CH_3$, $CH_2CH{=}CH_2$

Formamide $HCONH_2$ and O=S=N–tos react with elimination of H_2O and formation of HC≡N [35]. Higher carboxylic acid amides $RCONH_2$ and N-sulfinyl-arensulfonamides $O{=}S{=}N{-}SO_2Ar$ (formed initially from the corresponding sulfur diimide) yield nitriles:

$R{-}CONH_2 + O{=}S{=}N{-}SO_2Ar \rightarrow R{-}C{\equiv}N + ArSO_2NH_2 + SO_2.$

Presumably the N-sulfinyl derivative, formed initially by transsulfinylation, loses SO_2 to give the corresponding nitrile [1, p. 257], [27]. N,N-disubstituted formamides react with O=S=N–SO_2Ar in the same way as aromatic aldehydes (see p. 222), with elimination of SO_2 and formation of formamidines [35, 36, 37] e.g.,

$O{=}S{=}N{-}SO_2Ar + O{=}CH{-}N(CH_3)_2 \rightarrow ArSO_2{-}N{=}CHN(CH_3)_2 + SO_2.$

The highly reactive azirine IV reacts with $O{=}S{=}N{-}SO_2C_6H_4Cl\text{-}4$ to give the corresponding 1,2,5-thiadiazoline V (in ether, room temperature) [39]:

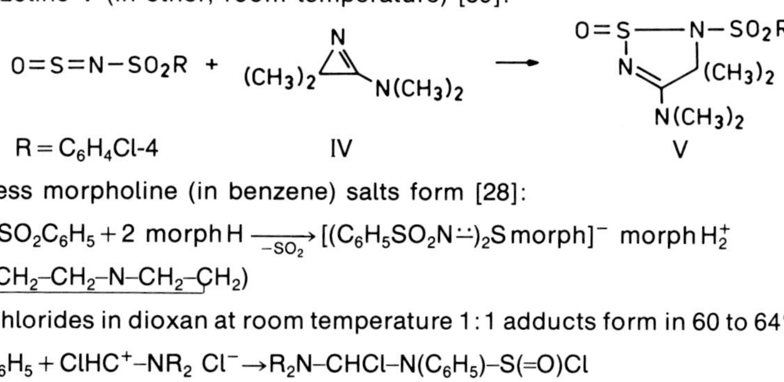

$R = C_6H_4Cl\text{-}4$ IV V

With excess morpholine (in benzene) salts form [28]:

$2\,O{=}S{=}N{-}SO_2C_6H_5 + 2\ \text{morphH} \xrightarrow[-SO_2]{} [(C_6H_5SO_2N{\because})_2S\,\text{morph}]^-\ \text{morphH}_2^+$

(morph = O–CH₂–CH₂–N–CH₂–CH₂)

With amide chlorides in dioxan at room temperature 1:1 adducts form in 60 to 64% yield [29]:

$O{=}S{=}N{-}C_6H_5 + ClHC^+{-}NR_2\ Cl^- \rightarrow R_2N{-}CHCl{-}N(C_6H_5){-}S({=}O)Cl$

$R_2 = (CH_3)_2$, $-(CH_2)_5-$ (piperidino)

References:

[1] Bussas, R.; Kresze, G. ; Münsterer, H.; Schwöbel, A. (Sulfur Rept. **2** [1983] 215/377, 255).
[2] Czerwiec, Z. (Roczniki Chem. **50** [1976] 2197/200).
[3] Wucherpfennig, W. (Chem. Ber. **101** [1968] 371/2).
[4] Michaelis, A. (Liebigs Ann. Chem. **274** [1893] 173/266, 182, 262/3).
[5] Khaletskii, A. M.; Pesin, V. G.; Chi-Chun Chao [Chzhao Chzi-Chzhun] (Dokl. Akad. Nauk SSSR **106** [1956] 88/91; Proc. Acad. Sci. USSR Chem. Sect. **106/111** [1956] 31/4).
[6] Pesin, V. G.; Khaletskii, A. M.; Chzhao Chzi-Chzhun [Chi-Chun Chao] (Zh. Obshch. Khim. **27** [1957] 1570/5; J. Gen. Chem. [USSR] **27** [1957] 1643/7).
[7] Cava, M. P.; Schlessinger, R. H. (Tetrahedron Letters **1964** 3815/8).

[8] Dietz, R. (Chem. Commun. **1965** 57).
[9] Behringer, H.; Leiritz, K. (Chem. Ber. **98** [1965] 3196/9).
[10] Beecken, H. (Chem. Ber. **100** [1967] 2164/9; 2170/7).

[11] Kaplan, M. L.; Haddon, R. C.; Hirani, A. M.; Schilling, F. C.; Marshall, J. H. (J. Org. Chem. **46** [1981] 675/8).
[12] Pesin, V. G.; Khaletskii, A. M.; Sergeev, V. A. (Zh. Obshch. Khim. **32** [1962] 181/6; J. Gen. Chem. [USSR] **32** [1962] 177/81).
[13] Uno, T.; Takagi, K.; Tomoeda, M. (Chem. Pharm. Bull. [Tokyo] **28** [1980] 1909/12).
[14] Singerman, G. M. (J. Heterocycl. Chem. **12** [1975] 877/82).
[15] Danylec, B.; Davis, M. (J. Heterocycl. Chem. **17** [1980] 533/6, 537/9).
[16] McKinnon, D. M; Duncan, K. A.; McKinnon, A. M.; Spevack, P. A. (Can. J. Chem. **63** [1985] 882/6).
[17] Shealy, Y. T.; Clayton, J. D.; Montgomery, J. A. (J. Org. Chem. **27** [1962] 2154/60).
[18] Pesin, V. G.; Khaletskii, A. M.; Zolotova-Zolotukhina, L. V. (Zh. Obshch. Khim. **29** [1959] 3214/9; J. Gen. Chem. [USSR] **29** [1959] 3178/81).
[19] Kresze, G.; Seyfried, C.; Trede, A. (Tretrahedron Letters **1965** 3933/6).
[20] Zoller, U.; Rona, P. (Tretrahedron Letters **26** [1985] 2813/4).

[21] Fischer, E.; Rembarz, G.; Teller, M. (J. Prakt. Chem. **324** [1982] 920/4).
[22] Fischer, E.; Teller, M.; Martin, D. (Wiss. Z. Wilhelm-Pieck-Univ. Rostock Naturwiss. Reihe **31** [1982] 1/4).
[23] Fischer, E.; Teller, M.; Kálmán, A.; Argay, G. (Tetrahedron **40** [1984] 385/90).
[24] Fischer, E.; Teller, M.; Schroeder, H. (Sulfur Letters **1** [1983] 119/25).
[25] Dovlatyan, V. V.; Mirzoyan, R. S. (U.S.S.R. 530029 [1975/76] from C.A. **86** [1977] No. 89832).
[26] Grill, H.; Kresze, G. (Liebigs Ann. Chem. **749** [1971] 171/7, 172).
[27] Carpanelli, C.; Gaiani, G. (Gazz. Chim. Ital. **100** [1970] 689/92).
[28] Levchenko, E. S.; Ugarov, B. N.; Sheinkman, I. E. (Zh. Org. Khim. **2** [1966] 1388/92; J. Org. Chem. [USSR] **2** [1966] 1378/81).
[29] Ito, Y.; Katsuragawa, S.; Okano, M.; Oda, R. (Tetrahedron **23** [1967] 2159/67, 2162).
[30] Ghosh, M.; Rahman, Md. S.; Maiti, S. (J. Polymer Mater. **2** [1985] 29/34).

[31] Senning, A. (Acta Chem. Scand. **18** [1964] 1958/60).
[32] Sólyom, S.; Szilágyi, K.; Toldy, L. (Liebigs Ann. Chem. **1983** 1001/19, 1002).
[33] Beckert, R.; Mayer, R. (J. Prakt. Chem. **322** [1980] 273/80).
[34] Stankovský, Š.; Burkovska, V. (Chem. Zvesti **37** [1983] 831/6).
[35] Albrecht, R.; Kresze, G.; Mlakar, B. (Chem. Ber. **97** [1964] 483/9).
[36] Davis, M.; Paproth, T. G.; Stephens, L. J. (J. Chem. Soc. Perkin Trans. I **1973** 2057/9).
[37] Levchenko, E. S.; Ugarov, B. N. (Zh. Org. Khim. **4** [1968] 1413/20; J. Org. Chem [USSR] **4** [1968] 1361/7, 1363/4).
[38] Taguchi, T.; Morita, S.; Kawazoe, Y. (Chem. Pharm. Bull. [Tokyo] **23** [1975] 2654/9).
[39] Schaumann, E.; Kausch, E.; Walter, W. (Chem. Ber. **107** [1974] 3574/88, 3576, 3582).
[40] Grigg, R.; Dowling, M.; Sridharan, V. (J. Chem. Soc. Chem. Commun. **1986** 1777/9).

[41] Shin, J. M.; Kim, Y. H. (Tetrahedron Letters **27** [1986] 1921/4).
[42] Teller, M.; Fischer, E.; Martin, D.; Graubaum, H. (J. Prakt. Chem. **328** [1986] 199/204).
[43] Dorrity, M. R. J.; Dowling, M.; Grigg, R.; Malone, J. F.; Sridharan, V. (Tetrahedron **44** [1988] 4941/52).
[44] El-Sonbati, A. Z.; Ahmed, M. A.; Reicha, F. M.; Diab, M. A. (Eur. Polym. J. **25** [1989] 125/7).

2.6.23.9 Reactions with Nitrogen Compounds with N–N and N–N–N Groups

$O{=}S{=}N{-}SO_2C_6H_4CH_3$-4 and the hydrazine derivative I in dry $CHCl_3$ undergo a transsulfinylation reaction at room temperature:

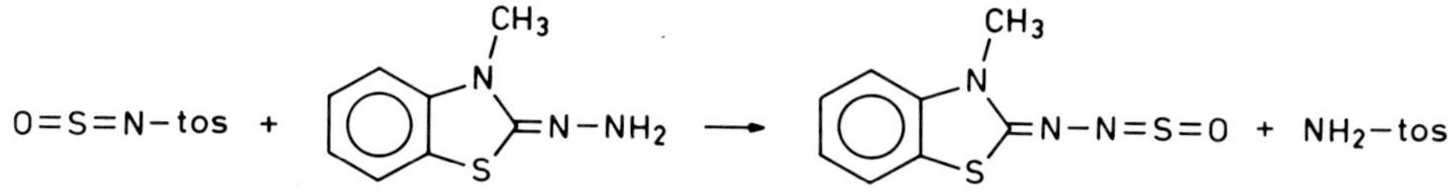

I

tos = $SO_2C_6H_4CH_3$-4

Preparation of N-sulfinylhydrazones was done in this way [1].

Dibenzalhydrazine (benzalazine) and $O{=}S{=}N{-}SO_2C_6H_5$ react in the same manner [2]:

$$O{=}S{=}N{-}SO_2C_6H_5 + C_6H_5{-}CH{=}N{-}N{=}CH{-}C_6H_5 \rightarrow C_6H_5{-}CH{=}N{-}SO_2C_6H_5$$
$$(+\,C_6H_5{-}CH{=}N{-}N{=}S{=}O)\ \text{(unstable)}$$

Transsulfinylation of (the acetate of) phenylhydrazine $C_6H_5NH{-}NH_2$ and other aromatic hydrazines by $O{=}S{=}N{-}OCH_2C_6H_5$ in absolute alcohol was observed by Michaelis [8, 9].

Azoalkenes react with N-sulfinylamines (mole ratio 1:1) in benzene at room temperature, forming six-membered ring compounds by [4+2] cycloaddition:

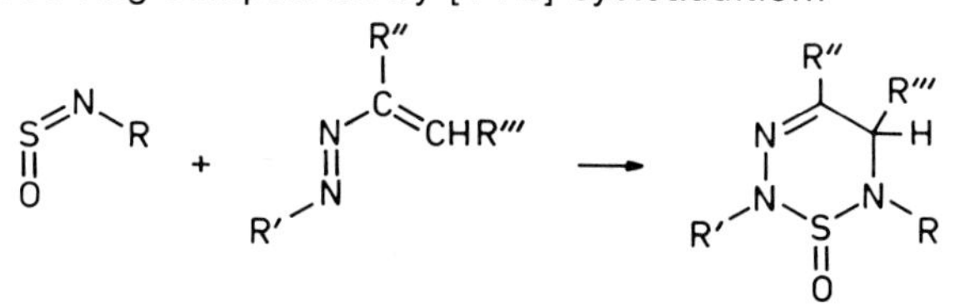

R = SO_2CH_3, $SO_2C_6H_4CH_3$-4, $C(O)OC_2H_5$, $C(O)C_6H_5$, C_6H_5; R′, R″, R‴ = alkyl, aryl

The rate of addition increases with increasing electron acceptor strength of R and increasing donor strength of the diene substituents. E.g., the reaction of $O{=}S{=}N{-}SO_2C_6H_4CH_3$-4 and the diene with $R' = CH_3$; $R''R''' = -(CH_2)_4-$ is rapid and very exothermic whereas the reaction of $O{=}S{=}N{-}C_6H_5$ with the same diene needs 5 d for completion. Reaction of $O{=}S{=}N{-}SO_2C_6H_4CH_3$-4 with the diene where $R' = R'' = CH_3$, $R''' = CO{-}NHC_6H_5$, requires 1 d [7].

With 2-(1,3-alkadienyl)-diazene carboxylic acid esters at 20°C O=S=N–tos undergoes [4+2] cycloaddition to the C=C–N=N system, but not to the C=C–C=C unit [3]:

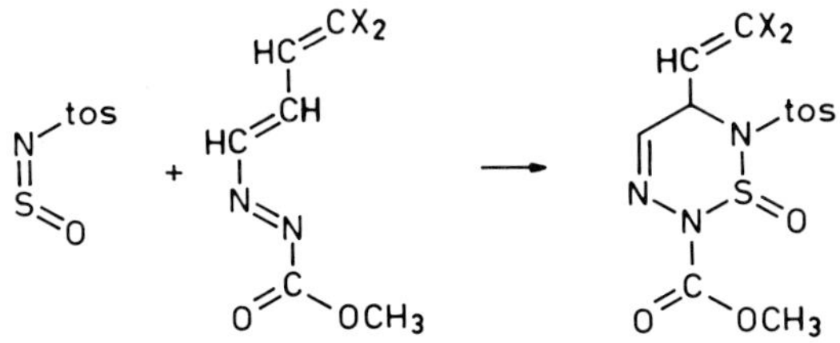

X = H, Cl

Reactions with $C_6H_5{-}C{\equiv}N^{\oplus}{-}N^{\ominus}{-}C_6H_5$ are described on pp. 242/3. Aryl azides $Ar'N_3$ ($Ar' = C_6H_5$, $4\text{-}ClC_6H_4$, $4\text{-}CH_3OC_6H_4$, $4\text{-}CH_3C_6H_4$) and N-sulfinyl-arenamines O=S=N–Ar (Ar = C_6H_5, $4\text{-}ClC_6H_4$, $4\text{-}CH_3OC_6H_4$, $4\text{-}CH_3C_6H_4$) react in boiling chlorobenzene, and also photochemically when irradiated at room temperature, to yield several products:

$$O{=}S{=}N{-}Ar + Ar'N_3 \rightarrow Ar{-}N{=}N{-}Ar + Ar'{-}N{=}N{-}Ar' + Ar{-}N{=}N{-}Ar' + ArNH_2$$
$$+\,Ar'NH_2 + O{=}S{=}N{-}Ar' + S + SO_2$$

The main product is Ar–N=N–Ar′. A three-membered SN_2 ring I is suggested to form as an intermediate and to yield the different products by decomposition [4]:

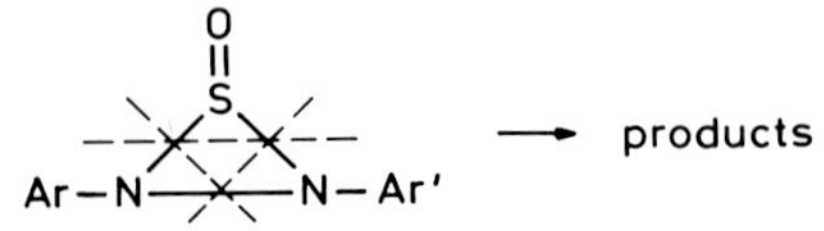

I

With the 1,2,4,5-tetrazine II, N-sulfinylanilines O=S=N–C_6H_4R' (R′ = H, CH_3, C_2H_5) presumably undergo primarily a [4 + 2] cycloaddition. Then the 1,2,4-triazole III forms by ring contraction [5]:

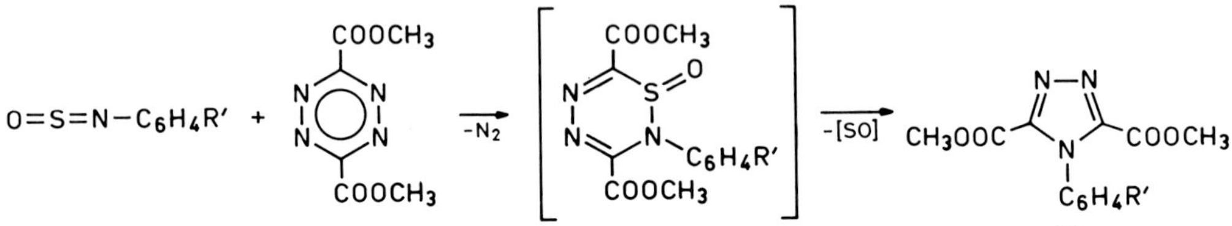

II III

Substituted triazolium imides IV and O=S=N–Ar, when refluxed in benzene, give 8π-2,4,5,6-tetra-substituted 1,2,3,5-tetrazines V [6]:

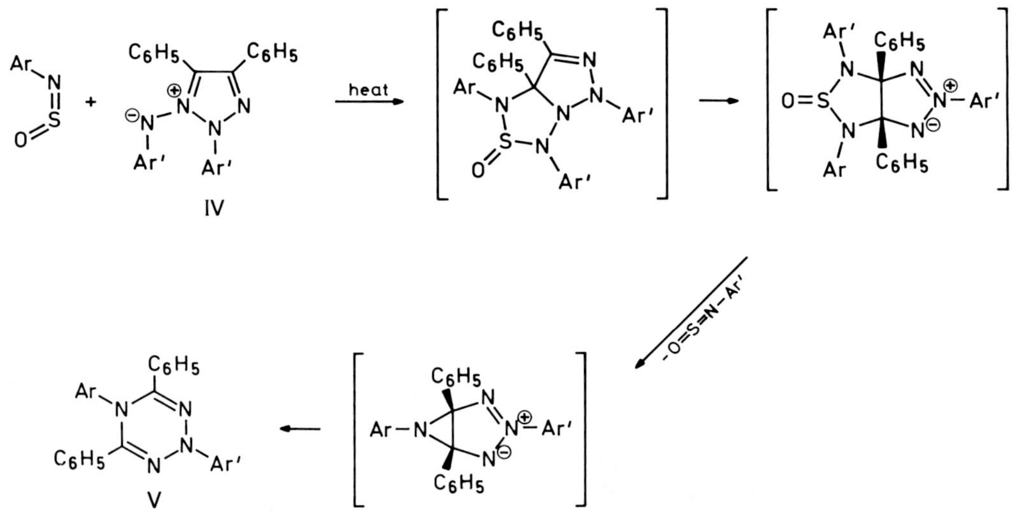

Ar = C_6H_5, 4-BrC_6H_4, 4-$NO_2C_6H_4$, 5-methyloxazol-3-yl; Ar′ = C_6H_5, 4-$NO_2C_6H_4$

References:

[1] Mayer, R.; Bleisch, S.; Heller, V. (Z. Chem. [Leipzig] **20** [1980] 185).
[2] Senning, A. (Recl. Trav. Chim. **82** [1963] 790).
[3] Sommer, W. (unpublished results in: Bussas, R.; Kresze, G.; Münsterer, H.; Schwöbel, A., Sulfur Rept. **2** [1983] 215/377, 295/6).
[4] Benati, L.; DeLuca, G.; Maccagnani, G.; Tundo, A. (J. Chem. Soc. Chem. Commun. **1972** 702/3).
[5] Seitz, G.; Kämpchen, T. (Arch. Pharm. **310** [1977] 269/71).

[6] Butler, R. N.; Cunningham, D.; McArdle, P.; O'Halloran, G.A. (J. Chem. Soc. Chem. Commun. **1988** 232/4).
[7] Sommer, S. (Synthesis **1977** 305/7).
[8] Michaelis, A. (Ber. Deut. Chem. Ges. **24** [1891] 745/55, 751/2).
[9] Michaelis, A.; Schröter, G. (Ber. Deut. Chem. Ges. **26** [1893] 2155/7).

2.6.23.10 Reactions with N-Oxides and NO Compounds

$O{=}S{=}N{-}C_6F_5$ and $O{=}S{=}N{-}CF(CF_3)_2$ are oxidized by the N-oxides $ON(CH_3)_3$ and $C_7H_{13}NO$ ($C_7H_{13}N$ = quinuclidine = 1-azabicyclo[2.2.2]octane) in CH_2Cl_2 at low temperature (−78°C). The sulfimides which formed are stabilized as adducts of the tertiary amines [1]:

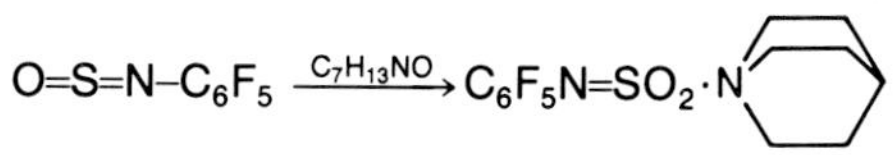

(85%)

$$O{=}S{=}N{-}CF(CF_3)_2 \xrightarrow{ON(CH_3)_3} (CF_3)_2CF{-}N{=}SO_2 \cdot N(CH_3)_3 \quad (44\%)$$

Pyridine N-oxide reacts with $O{=}S{=}N{-}SO_2R'$ ($R' = C_6H_5$, $C_6H_4CH_3$-4) in benzene exothermically to give stable zwitterionic adducts [2, 3]:

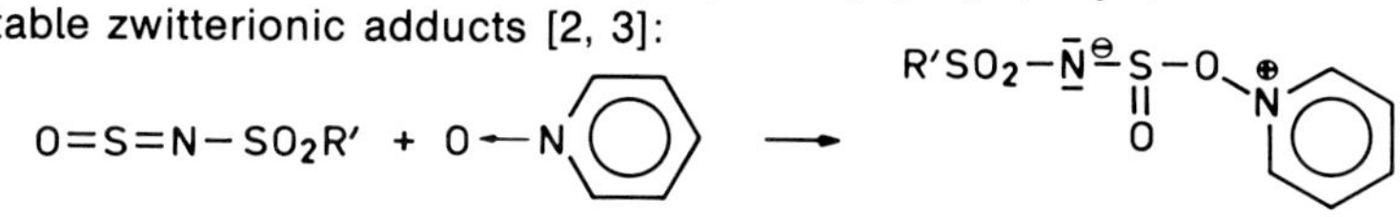

With C,N-diphenylnitrone (in benzene) quite another reaction occurs with warming and SO_2 evolution:

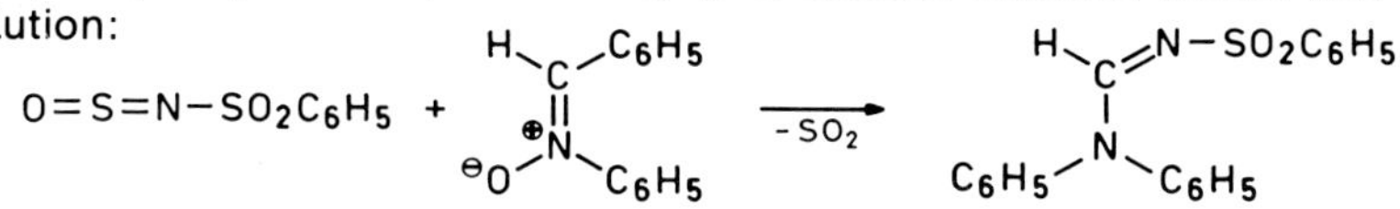

The intermediary formation of an unstable, not isolable [2+3] cycloadduct, which rearranges to the corresponding amidine, is proposed [2, 4]. Analogous reactions were observed with $O{=}S{=}N{-}SO_2C_6H_5$ and otherwise substituted nitrones [4].

With cyclic nitrones [5, 6, 7] under mild conditions, the corresponding substituted products are obtained, e.g.,

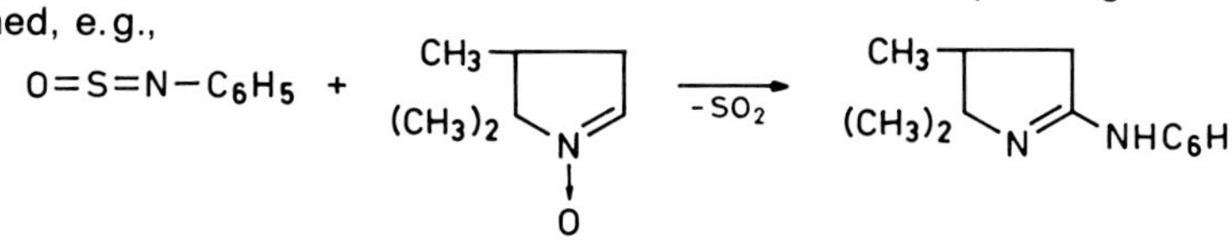

[5]

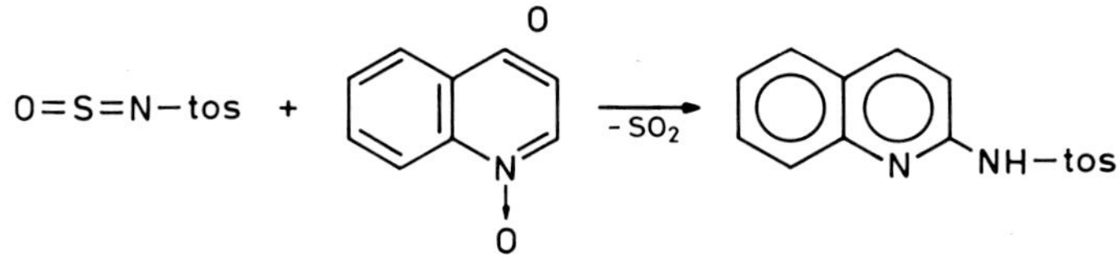

[6]

Nitrosobenzene and p-dimethylaminonitrosobenzene do not react with $O{=}S{=}N{-}C_6H_5$ at room temperature. However, when boiled in toluene it was possible to isolate a small yield of the azo compounds. Their formation was explained by assuming a [2+2] cycloaddition mechanism [8]:

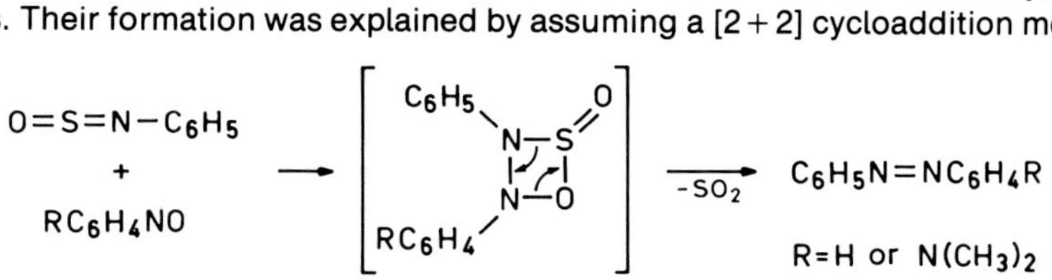

N-Sulfinylamines O=S=N–R (R = aryl, SO_2Ar, or alkyl) react with nitrile oxides $R'C{\equiv}N^+{-}O^-$ (R′ = aryl, $Si(CH_3)_3$) to form [3 + 2] cycloadducts at room temperature [9 to 15]. Higher temperatures lead to decomposition with SO_2 evolution and rearrangement to carbodiimides [9, 10, 13, 15]:

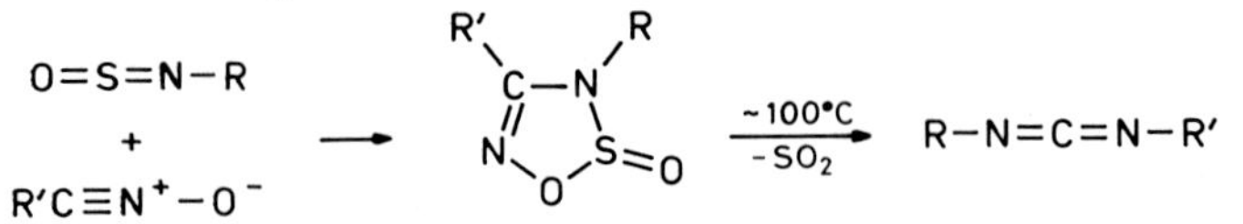

Kinetic investigations of the reaction in various solvents give evidence for a concerted mechanism with a dipolar transition state [12, 13]. The reactivity of the N-sulfinylamines O=S=N–R depends on R. For instance, $4\text{-}NO_2C_6H_4{-}C{\equiv}N^{\oplus}{-}O^{\ominus}$ and N-sulfinylaniline $O{=}S{=}N{-}C_6H_5$ yield the [3 + 2] cycloadduct quantitatively, whereas the aliphatic substituted $O{=}S{=}N{-}C_4H_9$-n yields only 9% (benzene, room temperature) [11]. At higher temperatures N-sulfinylanilines O=S=N–R and $(CH_3)_3SiCNO$ give carbodiimides $(CH_3)_3Si{-}N{=}C{=}N{-}R$ (and SO_2) in 79% yield for R = C_6H_5, ca. 60% for R = $C_6H_4NO_2$-4, and 30% for R = $C_6H_4CH_3$-4 [15].

References:

[1] Jäger, U.; Sundermeyer, W.; Pritzkow, H. (Chem. Ber. **120** [1987] 1191/5).
[2] Albrecht, R.; Kresze, G. (Chem. Ber. **98** [1965] 1205/9).
[3] Bussas, R.; Kresze, G.; Münsterer, H.; Schwöbel, A. (Sulfur Rept. **2** [1983] 215/377, 271).
[4] Stark, B. P.; Ratcliffe, M. H. G. (J. Chem. Soc. **1964** 2640/1).
[5] Tsuge, O.; Tashiro, M.; Mataka, Sh. (Tetrahedron Letters **1968** 3877/80).
[6] Onaka, T. (Itsuu Kenkyusho Nempo No. 15 [1968] 29/39 from C.A. **72** [1970] No. 90231).
[7] Kreher, R.; Morgenstern, H. (Tetrahedron Letters **21** [1980] 2141/4).
[8] Pozdnyakova, T. M.; Zefirov, N. S. (Zh. Org. Khim. **8** [1972] 1107; J. Org. Chem. [USSR] **8** [1972] 1120).
[9] Rajagopalan, P.; Daeniker, H. U. (Angew. Chem. **75** [1963] 91; Angew. Chem. Intern. Ed. Engl. **2** [1963] 46).
[10] Rajagopalan, P.; Advani, B. G. (J. Org. Chem. **30** [1965] 3369/71).

[11] Eloy, F.; Lenaers, R. (Bull. Soc. Chim. Belges **74** [1965] 129/35).
[12] Beltrame, P.; Comotti, A.; Veglio, C. (Chem. Commun. **1967** 996/7).
[13] Beltrame, P.; Vintani, C. (J. Chem. Soc. B **1970** 873/6).
[14] Schmitt, E. E.; American Cyanamide (U.S. 3118902, 3118903 [1962] from C.A. **60** [1964] 10691, 10692).
[15] DeSarlo, F.; Brandi, A.; Goti, A.; Guarna, A.; Rovero, P. (Heterocycles **20** [1983] 511/8, 512, 517).

2.6.23.11 Reactions with Sulfinyl Derivatives

Aromatic N-sulfinylamines react with sulfinic acids at low temperature (−15°C) to form the corresponding sulfinylsulfones [1]:

$$O{=}S{=}N{-}R + 2\,R'SO_2H \rightarrow RNH_2 + R'S(O)SO_2R' + SO_2$$
$$R = C_6H_4CH_3\text{-}4,\ C_6H_4NO_2\text{-}4;\ R' = C_6H_4CH_3\text{-}4$$

N-Sulfinylsulfonamides react with sulfonic acids at 40 to 150°C to give disulfonimides in high yields (mole ratio 1:1; neat components or dissolved in inert solvents, e.g., $C_6H_5CH_3$, C_6H_5Cl, $C_6H_4Cl_2$) [2]:

$$O{=}S{=}N{-}SO_2R + R'SO_3H \rightarrow RSO_2{-}NH{-}SO_2R' + SO_2$$
$$R = CH_3,\ C_6H_5,\ C_6H_4NO_2\text{-}3;\ R' = C_6H_4CH_3\text{-}4$$

N-Sulfinylaniline O=S=N–C_6H_5 or its nitro derivative O=S=N–$C_6H_4NO_2$-4 do not react with aliphatic or aromatic sulfinyl derivatives [3]. N-Sulfinylamines containing electron-withdrawing groups, e.g., O=S=N–$SO_2C_6H_4CH_3$-4, O=S=N–C(O)C_6H_5, react readily and exothermically with aliphatic sulfinyl derivatives R′S(O)R″ in inert solvents, e.g., benzene, forming iminosulfuranes RN=SR′R″. Warming at 80 to 100°C for several hours is necessary for a reaction with aromatic sulfinyl derivatives in benzene or without a solvent [3]. Electron donating groups R′, R″ attached to the SO group increase the reactivity [4, p. 276]. Reactions of O=S=N–SO_2CH_3 [3], O=S=N–$SO_2C_6H_4CH_3$-4 [3, 5, 6, 7], O=S=N–SO_2F [8], O=S=N–SO_2CF_3 [8, 16], O=S=N–$SO_2C_4F_9$-n [8], and O=S=N–SO_2CCl_3 [9] with aliphatic sulfinyl derivatives, yielding compounds RN=SR′R″, are described.

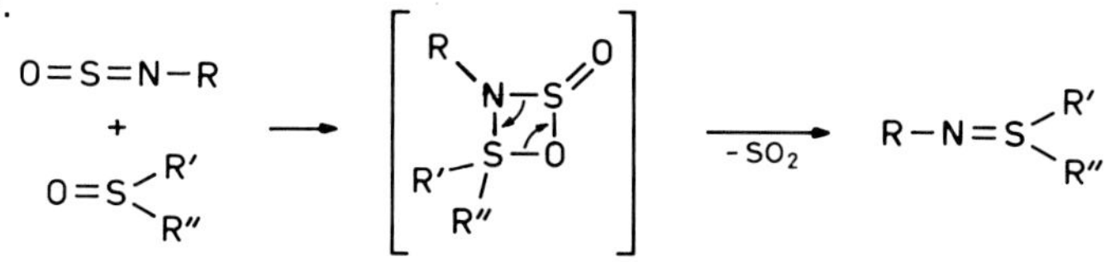

R = SO_2CH_3, $SO_2C_6H_4CH_3$-4, SO_2F, SO_2CF_3, $SO_2C_4F_9$-n, SO_2CCl_3; R′, R″ = alkyl, aryl

Sulfinic acid amides R′SO_2NH_2 [10, 11] and N-sulfinylbenzamidine O=S=N–C(=NH)C_6H_5 [10] react in the same manner.

Reactions of sulfinyl compounds containing electron-withdrawing substituents are more complex. N-Sulfinylamines O=S=N–R (I, R = SO_2CH_3, $SO_2C_6H_4CH_3$-4, or C(O)C_6H_5), when refluxed in dry benzene with $CH_3S(O)CH_2R'$ (II, R′ = C_6H_5CO, 4-$CH_3C_6H_4CO$, CH_3COO), give compounds of the types $CH_3SCH(NHR)R'$ (III), $(RHN)_2CHR'$ (IV), $(CH_3S)_2CHR'$ (V), and CH_3SCH_2R' (VI). For instance, O=S=N–$SO_2C_6H_4CH_3$-4 and $CH_3S(O)CH_2COC_5H_{11}$-n yield only the type IV compound in 58% yield. O=S=N–$SO_2C_6H_4CH_3$-4 and $CH_3S(O)CH_2CN$ yield compounds of type IV (61%), V (53%), and VI (4%) whereas O=S=N–C(O)C_6H_5 and $CH_3S(O)CH_2OC_6H_4OCH_3$-4 yield only type III (49%). The first step in the complex reaction is assumed to be the formation of a [2 + 2] cycloadduct [4, p. 277], which rearranges and decays in several steps [12]:

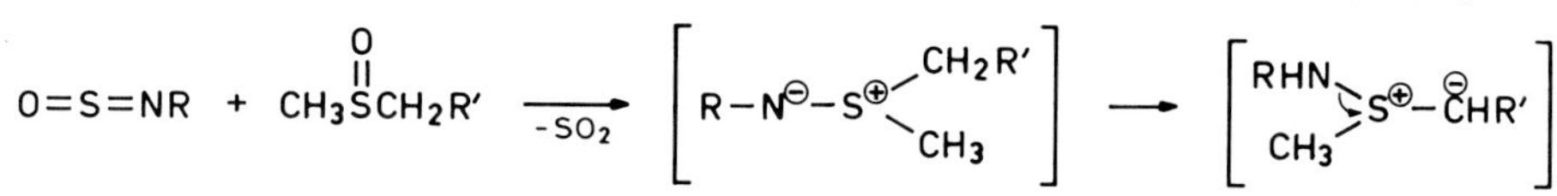

I II

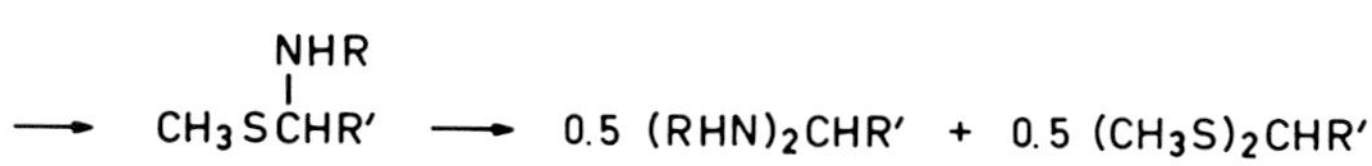

III IV V

But it should be noted that the mechanism of the reactions of O=S=N–R with sulfinyl derivatives is unclear up to date [4, p. 278]. The reaction of O=S=N–$SO_2C_6H_4CH_3$-4 with 4-$CH_3C_6H_4$-(CH_3)S=O in pyridine at 25°C appears to be of third-order (second-order in O=S=N–SO_2-$C_6H_4CH_3$-4 and first-order in sulfoxide); $k^3 = 0.317 \pm 0.002\ M^{-2} \cdot min^{-1}$ [13]. The reports are discrepant about the stereochemistry of the processes [5, 6, 7, 13, 14], and the solvent seems to play an important role. E.g., O=S=N–$SO_2C_6H_4CH_3$-4 reacts with $CH_3S(O)C_4H_9$-n in benzene, predominantly with retention in configuration, and in pyridine with inversion of configuration [14].

O=S=N–CF_3 and O=S=N–C_2F_5 are methylated at the N atom when they react with net CH_3O-$SO^+SbF_6^-$ at room temperature [15]:

$$O{=}S{=}N{-}R + CH_3OSO^+SbF_6^- \xrightarrow[-SO_2]{} [O{=}S{\overset{..}{-}}N(CH_3)R]^+SbF_6^-$$

$$R = CF_3,\ C_2F_5$$

References:

[1] Bederke, K. (Dipl.-Arbeit, TU Berlin 1961 from [4. p. 259]).
[2] Bentz, F.; Nischk, G.-E. (Angew. Chem. **82** [1970] 80; Angew. Chem. Intern. Ed. Engl. **9** [1970] 66).
[3] Schulz, G.; Kresze, G. (Angew. Chem. **75** [1963] 1022/3; Angew. Chem. Intern. Ed. Engl. **2** [1963] 736).
[4] Bussas, R.; Kresze, G.; Münsterer, H.; Schwöbel, A. (Sulfur Rept. **2** [1983] 215/377).
[5] Day, J.; Cram, D. J. (J. Am. Chem. Soc. **87** [1965] 4398/9).
[6] Johnson, C. R.; Rigau, J. J. (J. Org. Chem. **33** [1968] 4340/3).
[7] Christensen, B. W.; Kjaer, A. (Chem. Commun. **1969** 934/5).
[8] Roesky, H. W.; Holtschneider, G.; Wiezer, H.; Krebs, B. (Chem. Ber. **109** [1976] 1358/61).
[9] Buchholt, H. C.; Senning, A. (Acta Chem. Scand. **24** [1970] 2255/6).
[10] Seyfried, C. (Diss. TU München 1966 from [4, p. 276]).

[11] Trede, A. (Diss. TU München 1964 from [4, p. 276]).
[12] Minami, T.; Tsumori, Y.; Yoshida, K.; Agawa, T. (J. Org. Chem. **39** [1974] 3412/5).
[13] Cram, D. J.; Day, J.; Rayner, D. R.; von Schriltz, D. M.; Duchamp, D. J.; Garwood, D. C. (J. Am. Chem. Soc. **92** [1970] 7369/84, 7370/3, 7382).
[14] Christensen, B. W. (J. Chem. Soc. D **1971** 597/8).
[15] Henle, H.; Mews, R. (Chem. Ber. **115** [1982] 2935/42, 2939).
[16] Roesky, H. W.; Holtschneider, G. (Z. Anorg. Allgem. Chem. **378** [1970] 168/76, 170, 174).

2.6.23.12 Reactions with Carbonyl Compounds

N-Sulfinylamines undergo heterometathesis with nonenolizable aldehydes in a "Quasi-Wittig Reaction" to form SO_2 and the corresponding imines (Schiff's bases) in high yields:

$$O{=}S{=}N{-}Ar + O{=}CHR' \rightarrow ArN{=}CHR' + SO_2$$

Ar = C_6H_5, 4-$CH_3C_6H_4SO_2$;
R' = H (the product is presumably trimeric), CCl_3, C_6H_5, 4-$NO_2C_6H_4$, α-furyl, α-pyridyl

N-Sulfinylanilines O=S=N–Ar need several hours at boiling temperature before the reaction is complete. N-Sulfinylsulfonamides O=S=N–SO_2R undergo the same reaction within some minutes. The components are heated in the mixture or in an inert solvent. In the latter case the reaction may be catalyzed by HCl, $AlCl_3$, BF_3, or similar compounds [1, 2], [3, pp. 115, 117]. O=S=N–$SO_2C_4F_9$-n, when refluxed at 120°C with O=$CHCCl_3$ in the presence of $AlCl_3$, yields n-$C_4F_9SO_2$–N=$CHCCl_3$ (69%) and SO_2 [31].

Aliphatic aldehydes [3, p. 115], and also the polymeric trifluoroacetaldehyde [3, p. 115], [4] produce the 1,1-bis(arylsulfonylamido)alkanes $R'CH_2CH(NHSO_2Ar)_2$ in poor yield. The first step in the heterometathesis is presumably the formation of a [2 + 2] cycloadduct [5, p. 279]:

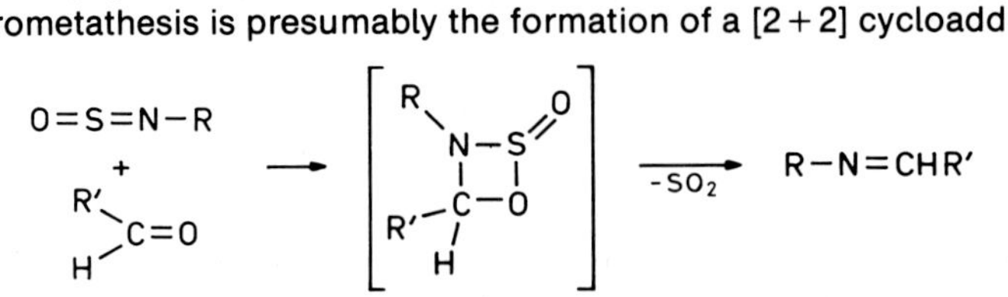

The isolation of unstable solid [2 + 2] cycloadducts was reported [6], but other authors found no such adducts in aprotic solvents. Only in the presence of water did they obtain ammonium α-aminosulfonate salts $R_2'NH_2^+$ Ar–NH–CH(Ar)–OSO_2^-, formed by hydrolysis of O=S=N–Ar [7].

Catalyzation of the process occurs in the case of Lewis acids such as $AlCl_3$, presumably via complexation with an oxygen atom of the reactants [5, p. 280].

O=S=N–tos and glyoxylic acid ethyl ester, when refluxed in benzene in the presence of some $AlCl_3$, react at the aldehyde group [8]:

$$O{=}S{=}N{-}tos + OCH{-}COOC_2H_5 \rightarrow tos{-}N{=}CH{-}COOC_2H_5 + SO_2$$

O=S=N–tos and glyoxylic acid butyl ester O=CH–$COOC_4H_9$-n, under the same conditions, react similarly, with formation of tos–N=CH–$COOC_4H_9$ [9]. When O=CH–$COOC_4H_9$-n is added dropwise to O=S=N–$SO_2C_4F_9$-n, instant formation of n-$C_4F_9SO_2$–N=CH–$COOC_4H_9$-n takes place if warmed (hot water) [31].

α-Halogenated aldehydes also react with N-sulfinylamines by metathesis of the aldehyde O atom. E.g., when O=S=N–tos and $ClCH_2$–CCl_2–CHO are refluxed in benzene, $ClCH_2$–CCl_2–CH=N–tos is obtained [20]. O=S=N–tos reacts similarly with aromatic carboxylic acid chlorides to give imidochlorides, which decompose on heating:

$$O{=}S{=}N{-}tos + O{=}C(Cl)R' \xrightarrow[-SO_2]{} tos{-}N{=}C(Cl)R' \xrightarrow{\Delta} tos{-}Cl + N{\equiv}CR'$$

The reaction of O=S=N–tos with aliphatic carboxylic acid chlorides is rather complex [21].

N-Sulfinylaniline and the α-pyrone derivative I undergo heterometathesis to give compound II:

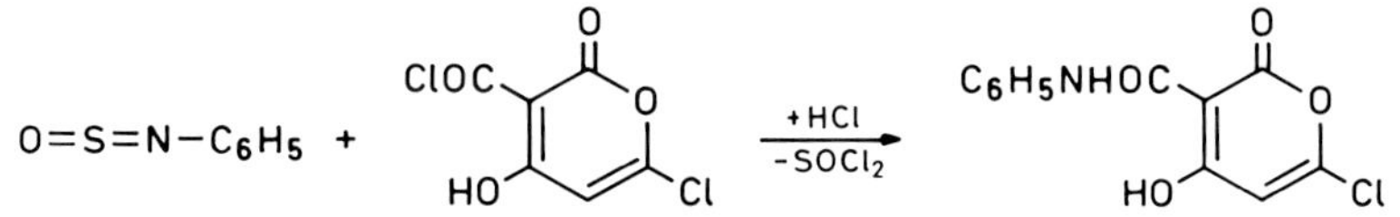

I II

Compounds of the same type II are obtained when N-sulfinylanilines O=S=N–C_6H_4R' (R' = H, 4-Cl, 4-CH_3, 2- or 4-NO_2, 2- or 4-OCH_3) react with malonyl dichloride $CH_2(COCl)_2$. The latter reactant presumably dimerizes with cyclization to I [22].

O=S=N–tos does not react with benzophenone $(C_6H_5)_2CO$, even in the presence of $AlCl_3$ [10]. With cyclohexanone, only tos–NH_2 was isolated [3, p. 122]. O=S=N–C_6H_5 does not react with $(CF_3)_2CO$ even when heated (100°C) for long periods, however in the presence of CsF in CH_3CN this reaction goes readily at 0 to 20°C. The catalytic action of the fluoride ion is assumed to consist of converting the electrophilic sulfinylamine into a nucleophilic anion O=S(F)–$\underline{\overline{N}}^{\ominus}$–Ar, which is able to add to the carbonyl group. The end product ArN=$C(CF_3)_2$ should form by cyclization and decomposition of the [2+2] cycloadduct [11].

With benzil in boiling benzene in the presence of $AlCl_3$ heterometathesis occurs easily:

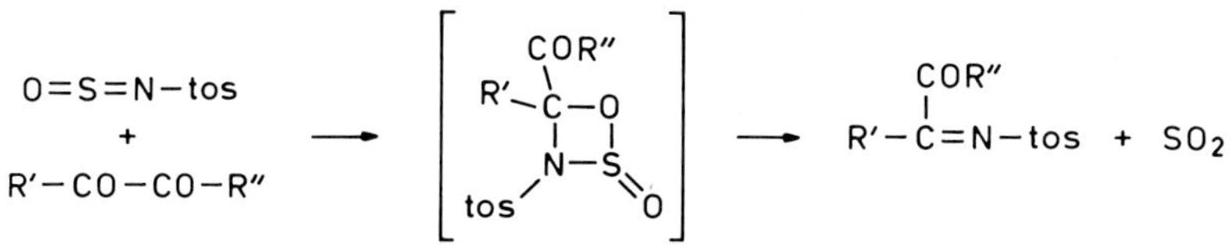

R' = R'' = C_6H_5

Any further reactions with formation of diimide derivatives of benzil were not observed, even with excess O=S=N–tos. The 4-chloro- and the 4,4'-dichloro-derivative (R' = 4-ClC_6H_4, R'' = C_6H_5; R' = R'' = 4-ClC_6H_4) react like benzil. With the dimethyl derivative CH_3–CO–CO–CH_3, the tautomeric form H_2C=C(NH–tos)–CO–CH_3 is produced, however there is no reaction with 4-methoxy- and 4,4'-dimethoxybenzil (R' = OCH_3, R'' = C_6H_5; R' = R'' = OCH_3) [3, pp. 117/8], [10].

With **enolizable carbonyl compounds** the reaction is more complex. The reaction of O=S=N–tos with 1-aryl-1-ethanones gives low yields of 1-aryl-2,2-bis(tosylamino)-1-ethanones [12]:

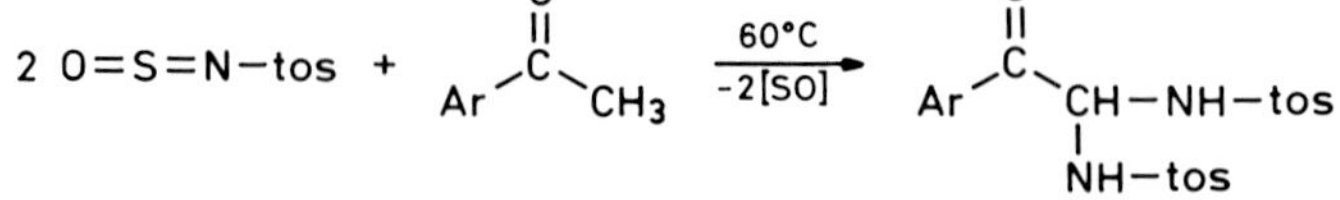

With malonic ester $CH_2(COOC_2H_5)_2$, or desoxybenzoin $C_6H_5C(O)CH_2C_6H_5$ and its derivatives in the presence of $AlCl_3$, tetrasubstituted ethylenes [13] or their precursors [14] are obtained [5, p. 308]:

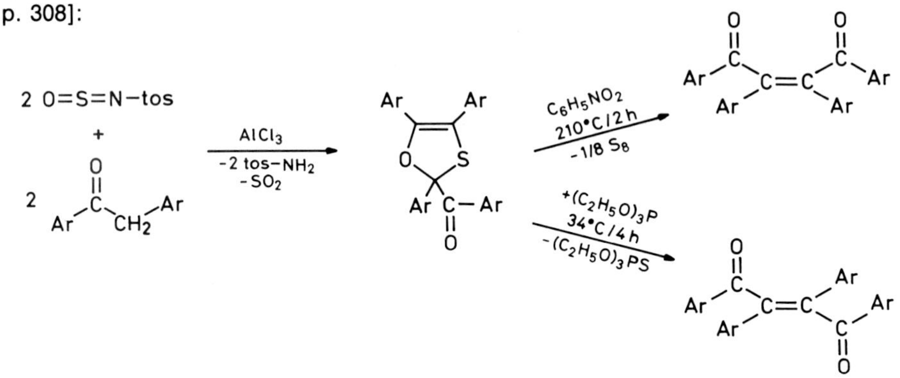

O=S=N–$SO_2C_6H_5$ reacts with 5,5-dimethyl-1,3-dioxocyclohexane (dimedone) in benzene at room temperature to form a product of complex tricyclic structure [15]:

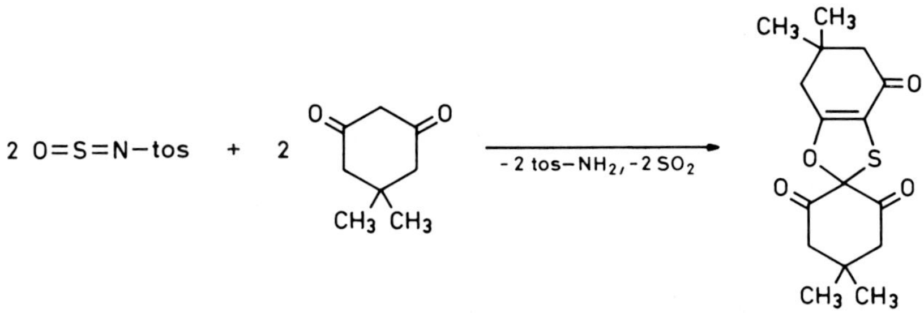

O=S=N–$SO_2C_4F_9$-n reacts at 20°C with alkyl-substituted alkanones and cycloalkanones within minutes, with 1-aryl-substituted alkanones within some hours to form stable adducts. The alkanone probably reacts in the enolic form, for the acidity of the C–H group has no decisive influence [16]:

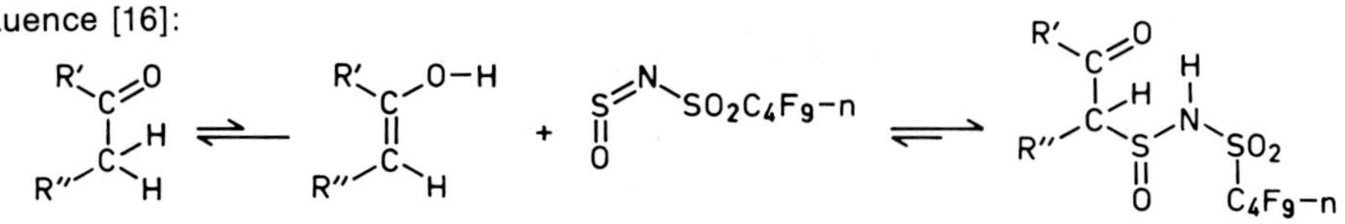

R′ = alkyl, aryl; R″ = H, Br, alkyl

Variation of the organic part of an N-sulfinylamine in reactions with the same alkenone may result in various products. The cyclohexyl derivative O=S=N–C_6H_{11}-c and diphenylcyclopropenone give a [2+3] cycloaddition product I when heated in DMF at 60 to 65°C in the presence of $Ni(CO)_4$. On the other hand, the phenyl and the tolyl derivative O=S=N–Ar (Ar = C_6H_5 or $C_6H_4CH_3$-4), under the same conditions, undergo exchange of C=O for S=O and yield pyrroline-2,5-diones II:

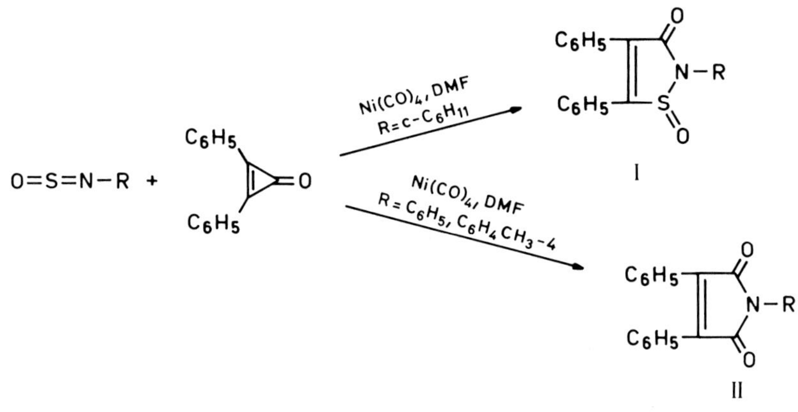

Isolated I does not react with $Ni(CO)_4$. Therefore the exchange of C=O for S=O takes place before the formation of I. A 1:1:1 complex of the reactants and the $Ni(CO)_2$ moiety is assumed to be the intermediate [17, 18].

With 1,3-dioxalan-2-one (ethylene carbonate, III) and LiBr or $(C_2H_5)_4N^+Br^-$ as catalysts, the reaction takes another course and N,N'-diarylhexahydro-1,4-diazines VI form in 11 to 47% yield (reaction temperature 140 to 170°C). The order of reactivity of the sulfinylanilines O=S=N–C_6H_4R' is R' = Cl > H > CH_3. The formation of VI was interpreted as below:

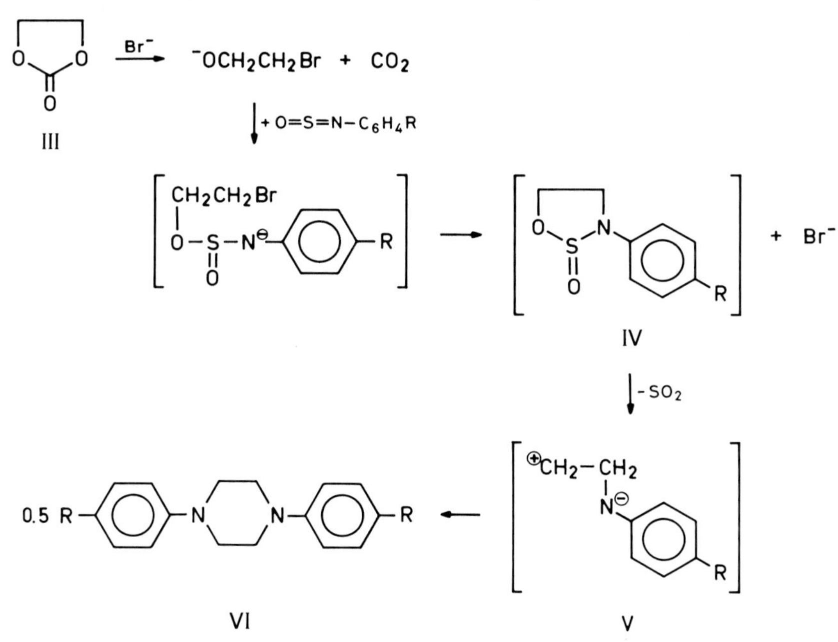

The 1,2,3-oxathiazolidine 2-oxide IV forms, subsequent elimination of SO_2 from IV yields the intermediate V, which dimerizes to form the product VI [19].

Olefinic enolizable aldehydes and O=S=N–tos, in the presence of promoting boron trifluoride etherate, presumably form N-tosyliminum complexes at first. Subsequent intramolecular electrophilic cyclization affords homoallylic amines. Such treatment of VII with O=S=N–tos in benzene at 0°C in the presence of BF_3–$(C_2H_5)_2O$ yields a 1:1 mixture of XII and XIII. A Lewis acid complexed N-tosyliminum intermediate (VIII, IX), which cyclizes to form carbonium ions (X, XI), followed by deprotonation, yields the observed ene-like products.

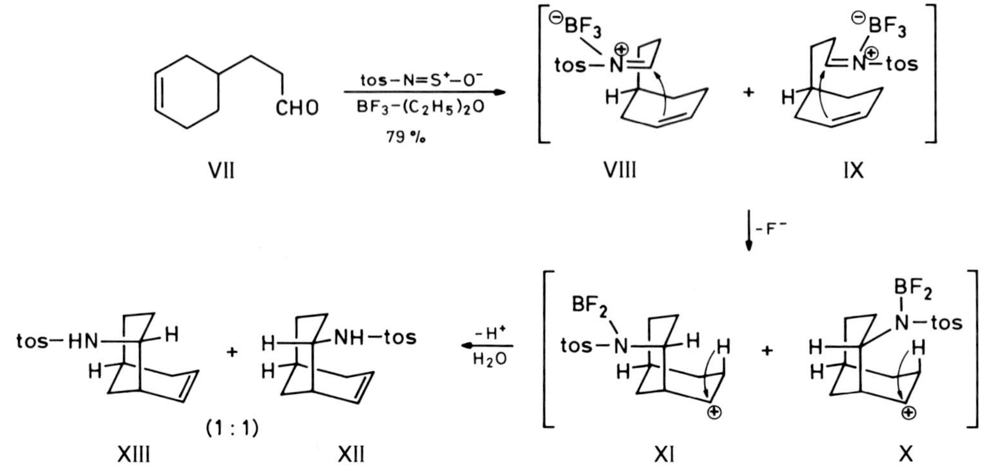

Two further examples of stereoselective cyclization in CH_2Cl_2 at 0°C are given:

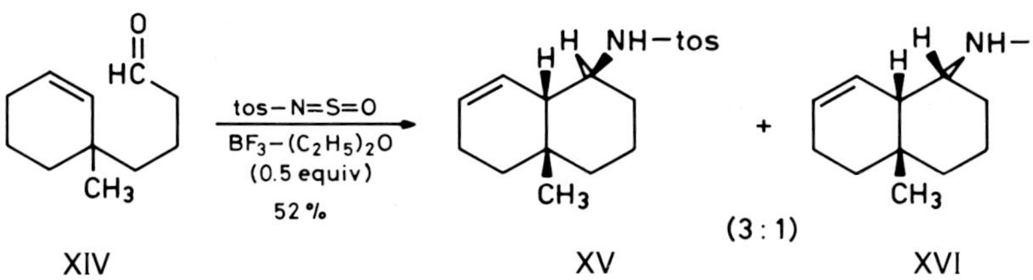

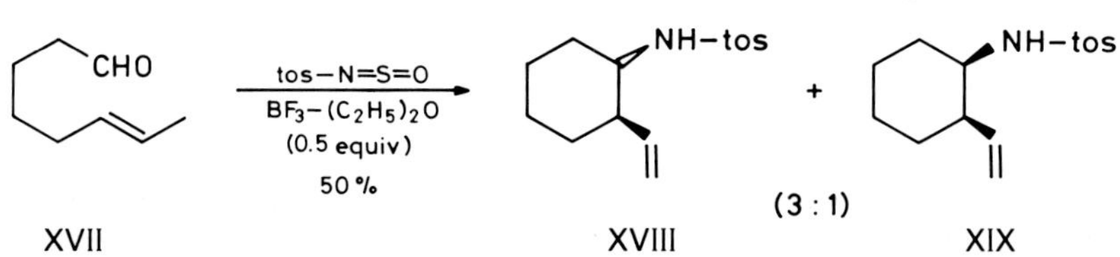

The cyclization of aldehyde XX by O=S=N–tos in the presence of BF_3–$(C_2H_5)_2O$ or $FeCl_3$ (1 equivalent, in CH_2Cl_2) takes a slightly different course. When only 0.2 equivalent BF_3–$(C_2H_5)_2O$ is taken, the products are XXI and XXII in equivalent yields. With $FeCl_3$ as catalyst XXIII is formed as a single stereoisomer [29].

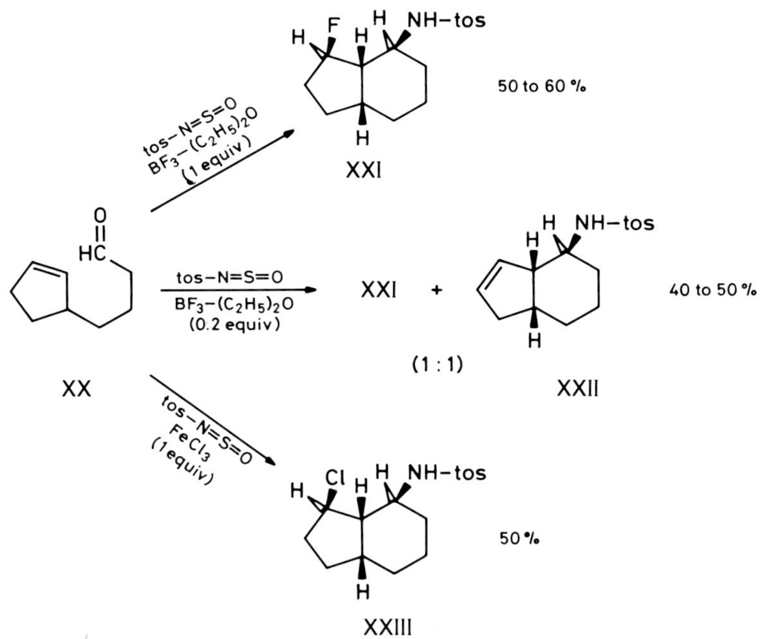

The reaction of O=S=N–C_6H_5 with **carboxylic acids** is similar to that of aldehydes and affords the corresponding anilides and SO_2 [23]:

$$O{=}S{=}N{-}C_6H_5 + R'COOH \rightarrow C_6H_5{-}NH{-}COR' + SO_2$$

Formic acid reacts at ambient temperature with warming to form C_6H_5–NH–COH. The other carboxylic acids, e.g., acetic, butanoic, or stearic acid, need warming to temperatures near the decomposition point (110 to 180°C) to afford the corresponding anilides [23]. Other authors state that traces of HCl or $SOCl_2$ are necessary for the reaction of sulfinylamines O=S=N–R (R = C_6H_5, 4-$NO_2C_6H_4$, c-C_6H_{11}) with carboxylic acids, e.g., formic, acetic, or chloroacetic acid. The reaction of O=S=N–$C_6H_4NO_2$-4 with acetic acid is more vigorous than that of O=S=N–C_6H_5 [24]. When dissolved in benzene, N-sulfinylalkan- and N-sulfinylarenamines O=S=N–R do not react with formic or acetic acid [30].

Aliphatic and aromatic N-sulfinyl-sulfonamides O=S=N–SO_2R react with aliphatic and aromatic carboxylic acids to give the corresponding N-sulfonyl-carboxamides and SO_2:

$$O{=}S{=}N{-}SO_2R + R'COOH \rightarrow RSO_2{-}NH{-}COR' + SO_2$$

(e.g., R = 3-$NO_2C_6H_4$, R' = C_2H_5 (yield 84%); R = C_6H_5, R' = $ClCH_2$ (67%); R = 3-$NO_2C_6H_4$, R' = Cl_2CH (63%); R = C_6H_5, R' = CCl_3 (61%); R = 3-$NO_2C_6H_4$, R' = C_6H_5 (73%)) [25, 30]. The sometimes vigorous reactions are carried out in the melt or in an inert solvent such as benzene. Disubstituted compounds form with dicarboxylic acids, e.g., from hexanedioic acid (adipic acid HOOC–$(CH_2)_4$–COOH) and O=S=N–$SO_2C_6H_4NO_2$-3 in 82% yield [25].

Kinetics of the reaction of O=S=N–R (R = C_6H_5, 4-$CH_3C_6H_4$, 4-ClC_6H_4) with HCOOH were studied in the temperature range 40 to 90°C. The reaction is facilitated by the introduction of an electron-withdrawing substituent into the phenyl group (e.g. Cl), but is retarded by that of an electron-releasing substituent (e.g. CH_3). Activation energies (in kcal/mol) are E_a = 14.6 (R = C_6H_5), 15.1 (4-$CH_3C_6H_4$), 10.1 (4-ClC_6H_4). An intermediate was detected by ^1H NMR spectroscopy, and two possible structures were proposed:

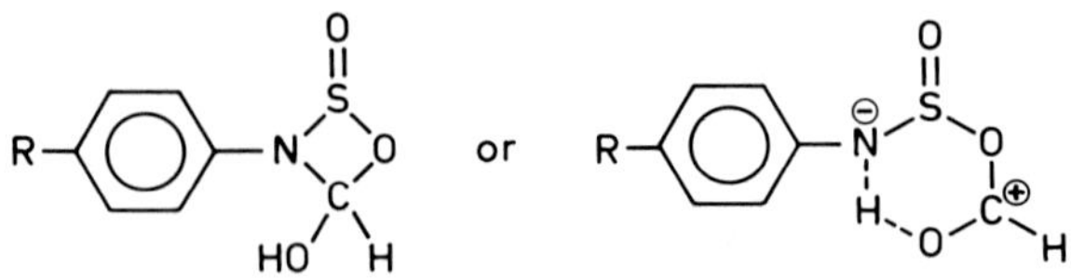

Presumably the reaction is initiated by the interaction between a sulfur atom of the O=S=N group and a carbonyl oxygen of formic acid with formation of a 1:1 complex as the rate-determining process. The decay of the complex yields anilides and SO_2 [26].

With α-hydroxycarboxylic acids R′CHOH-COOH (R′ = H, CH_3, C_2H_5, C_6H_5) and $HOOC(CHOH)_2$-COOH (d-tartaric acid) N-sulfinylanilines O=S=NR (R = C_6H_5, 4-ClC_6H_4, 4-$CH_3C_6H_4$) react to give the corresponding α-hydroxyanilides R′CH(OH)–C(O)NHR in quantitative yield under mild conditions (in CH_3CN, 25°C). Presumably the reaction begins with the protonation of the N-sulfinylamine by the COOH group, followed by the attack of the α-OH group onto the NSO group; the formation of the cyclic intermediate and the aniline was indirectly proved [32].

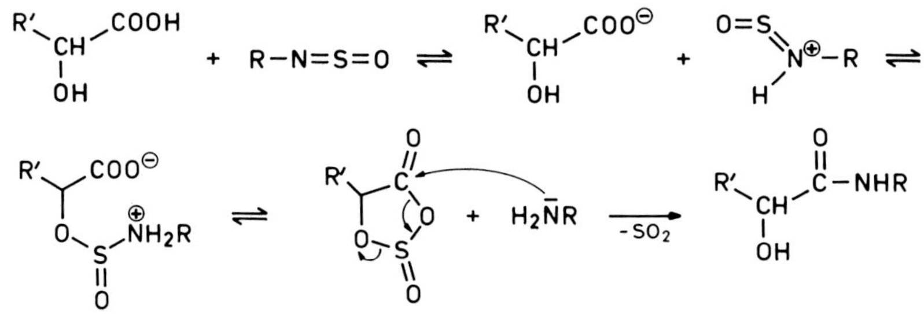

The important role of the α-carboxylic acid moiety in this reaction becomes evident in the fact that the esters, e.g., C_6H_5–CHOH–$COOC_2H_5$ do not react with N-sulfinylanilines [32].

The reaction of N-sulfinyl-sulfonamides O=S=N–SO_2Ar with benzoic or o-phthalic acid anhydride results in a displacement of oxygen by NSO_2Ar [27]:

$$O{=}S{=}N{-}SO_2Ar + (C_6H_5CO)_2O \longrightarrow (C_6H_5CO)_2NSO_2Ar + SO_2$$

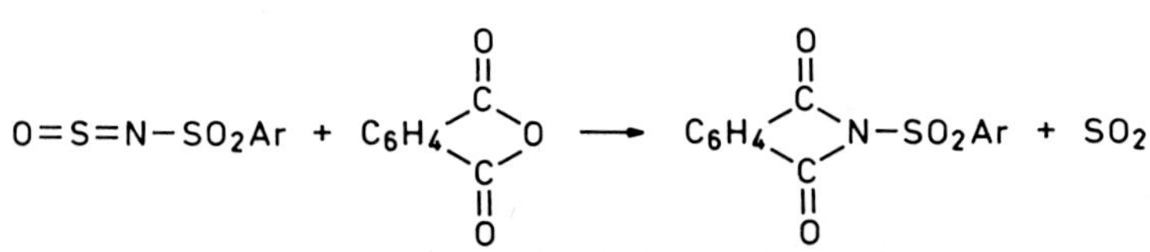

The condensation reactions of N-sulfinylamines O=S=N–Ar with aldehydes, catalyzed by organic bases such as piperidine, may be used for preparation of polymers [28].

Reactions of N-sulfinylamines with α-amino acids and carboxylic acid amides are described on pp. 214/5, reactions with 1-alkenones (ketenes) on pp. 234/6, reactions with compounds containing conjugated O=C–C=P bonds on pp. 238/9.

References:

[1] Kresze, G.; Albrecht, R. (Angew. Chem. **74** [1962] 781/2; Angew. Chem. Intern. Ed. Engl. **1** [1962] 595).
[2] Albrecht, R.; Kresze, G.; Mlakar, B. (Chem. Ber. **97** [1964] 483/9).
[3] Kresze, G.; Wucherpfennig, W. (Angew. Chem. **79** [1967] 109/27; Angew. Chem. Intern. Ed. Engl. **6** [1967] 149/67).
[4] Kresze, G.; Albrecht, R. (Chem. Ber. **97** [1964] 490/3).

[5] Bussas, R.; Kresze, G.; Münsterer, H.; Schwöbel, A. (Sulfur Rept. **2** [1983] 215/377).
[6] Zefirov, N. S.; Pozdnyakova, T. M. (Zh. Org. Khim. **7** [1971] 951/3; J. Org. Chem. [USSR] **7** [1971] 966/8).
[7] Mirek, J.; Rachwat, St. (Phosphorus Sulfur **3** [1977] 333/8).
[8] Tschaen, D. M.; Turos, E.; Weinreb, S. M. (J. Org. Chem. **49** [1984] 5058/64, 5061).
[9] Albrecht, R.; Kresze, G. (Chem. Ber. **98** [1965] 1431/4).
[10] Kresze, G.; Sommerfeld, D.; Albrecht, R. (Chem. Ber. **98** [1965] 601/3).

[11] Zeifman, Yu. V.; Ter-Gabrielyan, E. G.; Del'tsova, D. P.; Gambaryan, N. P. (Izv. Akad. Nauk SSSR Ser. Khim. **1979** 396/8; Bull. Acad. Sci. USSR Div. Chem. Sci. **1979** 366/8).
[12] Berger, M. (unpublished results in [5, p. 308]).
[13] Sommerfeld, D. (Diss. TH München 1966 from [3, pp. 122/3]).
[14] Jacobsson, U.; Kempe, T.; Novin, T. (J. Org. Chem. **39** [1974] 2722/8, 2726).
[15] Levchenko, E. S.; Gaidamaka, S. N.; Kalinin, V. N.; Budnik, L. V. (Zh. Org. Khim. **17** [1981] 990/6; J. Org. Chem. [USSR] **17** [1981] 871/5).
[16] Kresze, G.; Bussas, R. (Angew. Chem. **92** [1980] 750/1; Angew. Chem. Intern. Ed. Engl. **19** [1980] 737).
[17] Baba, A.; Ohshiro, Y.; Agawa, T. (Chem. Letters **1976** 11/2).
[18] Ohshiro, Y.; Nanimoto, H.; Baba, A.; Tone, T.; Tanaka, H.; Komatsu, M.; Agawa, T. (Heterocycles **9** [1978] 123).
[19] Tsuge, O.; Mataka, S.; Tashiro, M.; Mashiba, F. (Bull. Chem. Soc. Japan **40** [1967] 2709/11).
[20] Zinner, H.; Siems, W. E.; Erfurt, G. (J. Prakt. Chem. **316** [1974] 698/702).

[21] Kresze, G.; Wucherpfennig, W. (Chem. Ber. **101** [1968] 365/70).
[22] Al-Ajely, M. S.; Al-Rawi, J. M. A.; Elvidge, J. A. (J. Chem. Soc. Perkin Trans. I **1982** 1575/8).
[23] Carré, P.; Libermann, D. (Compt. Rend. **194** [1932] 2218/20).
[24] Smith, W. T.; King, G. G. (J. Org. Chem. **24** [1959] 976/8).
[25] Bentz, F.; Nischk, G.-E. (Angew. Chem. **82** [1970] 80; Angew. Chem. Intern. Ed. Engl. **9** [1970] 66).
[26] Yamada, F.; Nishiyama, T.; Emi, T. (Bull. Chem. Soc. Japan **45** [1972] 271/4).
[27] Levchenko, E. S.; Ugarov, B. N. (Zh. Org. Khim. **4** [1968] 1413/20; J. Org. Chem. [USSR] **4** [1968] 1361/7).
[28] Hörhold, H. H. (Z. Chem. [Leipzig] **12** [1972] 41/52).
[29] Melnick, M. J.; Freyer, A. J.; Weinreb, S. M. (Tetrahedron Letters **29** [1988] 3891/4).
[30] Markovskii, L. N.; Fedyuk, G. S.; Levchenko, E. S. (Zh. Org. Khim. **6** [1970] 2426/31; J. Org. Chem. [USSR] **6** [1970] 2438/42).

[31] Braxmeier, H.; Kresze, G. (Synthesis **1985** 683/4).
[32] Shin, J. M.; Kim, Y. H. (Tetrahedron Letters **27** [1986] 1921/4).

2.6.23.13 Reactions with Thiocarbonyl Compounds

Reactions with compounds containing the –N=C=S group are described on p. 233.

$O{=}S{=}N{-}SO_2C_6H_4CH_3$-4 reacts with thioamides (including thiocarbamides), dithioesters and their sulfonyl analogs under mild conditions [1]:

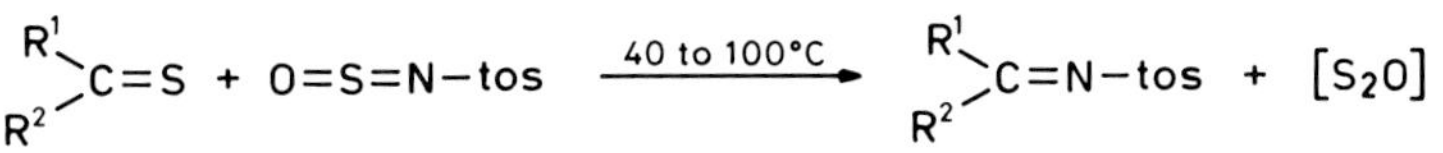

R^1, R^2 = Ar, R^3NH, R^3R^4N, R^5S, R^6SO_2

Other N-sulfinyl-sulfonamides, e.g., $O{=}S{=}N{-}SO_2CH_3$ may be used for syntheses with similar efficiency, but N-sulfinylanilines need much stronger conditions. For instance, $O{=}S{=}N{-}SO_2{-}C_6H_4CH_3$-4 reacts with $C_6H_5{-}C(S)NH_2$ in boiling $CHCl_3$ within 2 h, yielding 87% of compound I ($R = C_6H_5$). With $O{=}S{=}N{-}C_6H_5$ 48 h at 100°C were necessary to obtain a detectable amount of I ($R = C_6H_5$) [2]. The nature of the end products depends on the structure of the thio compound and the reaction conditions. With O=S=N–tos (tos = $SO_2C_6H_4CH_3$-4), besides tos–NH_2, SO_2, and elemental sulfur, the three types of main products are [2 to 7]:

I II III

R = H, alkyl, aryl, NR′R″, O-alkyl, aryl-SO_2

The first intermediate product was assumed to be a zwitterion IV [2, 3]:

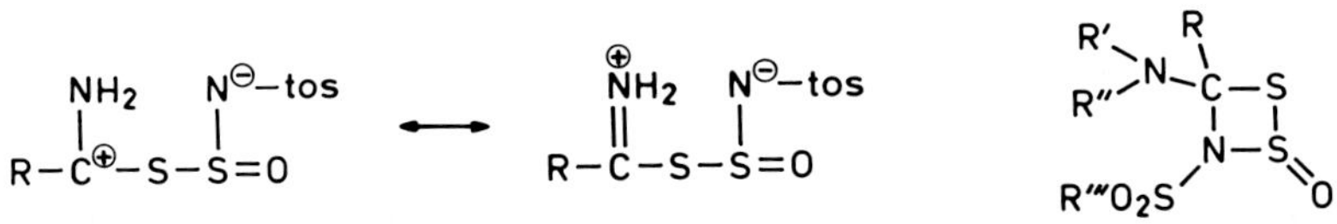

IV V

Other authors suggest a four-membered ring compound, e.g., V [5]. The reaction with $H_2NC_6H_4{-}C(S)NH_2$-2 is described on p. 211.

References:

[1] Bussas, R.; Kresze, G.; Münsterer, H.; Schwöbel, A. (Sulfur Rept. **2** [1983] 215/377, 282).
[2] Kresze, G.; Horn, A.; Philippson, R.; Trede, A. (Chem. Ber. **98** [1965] 3401/9).
[3] Kresze, G.; Horn, A. (Chem. Ber. **100** [1967] 1655/60).
[4] Kresze, G.; Maschke, A.; Albrecht, R.; Bederke, K.; Patzschke, H. P.; Smalla, H.; Trede, A. (Angew. Chem. **74** [1962] 135/44, 144).
[5] Senning, A. (Acta Chem. Scand. **18** [1964] 1958/60).
[6] Senning, A. (Acta Chem. Scand. **21** [1967] 1293/6).
[7] Nilsson, N. H.; Jacobsen, C.; Sørensen, O. N.; Haunsøe, N. K.; Senning, A. (Chem. Ber. **105** [1972] 2854/71, 2861).

2.6.23.14 Reactions with Oxiranes and Aziridines

N-Sulfinylanilines $O{=}S{=}N{-}C_6H_4R'$ react with oxirane and its 2-methyl or 2-chloromethyl derivative to give [2+3] cycloadducts [1, 2, 3]. The presence of catalysts, $(C_2H_5)_4NBr$ at 95 to 100°C [1], LiCl or LiBr or $(C_2H_5)_4NBr$ in DMF at 40 to 65°C [3] gives the adducts as a pair of cis and trans isomers at the S=O bond in good yields:

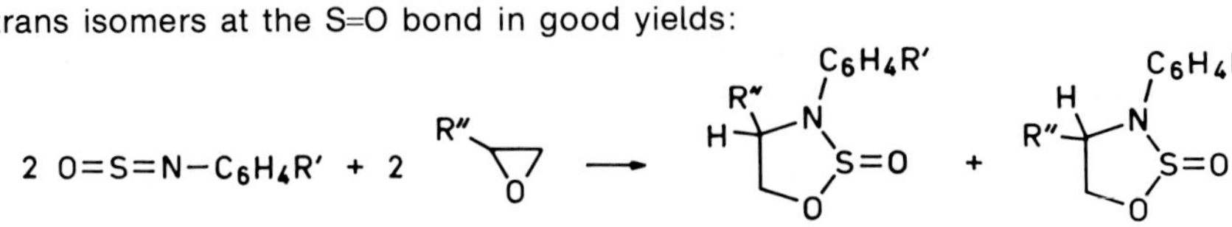

R′ = H, CH_3, Cl, NO_2; R″ = H, CH_3, CH_2Cl

The ratio of the stereoisomer mixture depends on R′ and R″, but is independent of the catalyst. The net reactants O=S=N–C_6H_4R' and oxirane at 15 to 65°C yield mainly a resinous product, which decomposes with evolution of SO_2 when heated at 160°C to N,N′-diaryl-hexahydro-1,4-diazines. The following mechanism is proposed [2]:

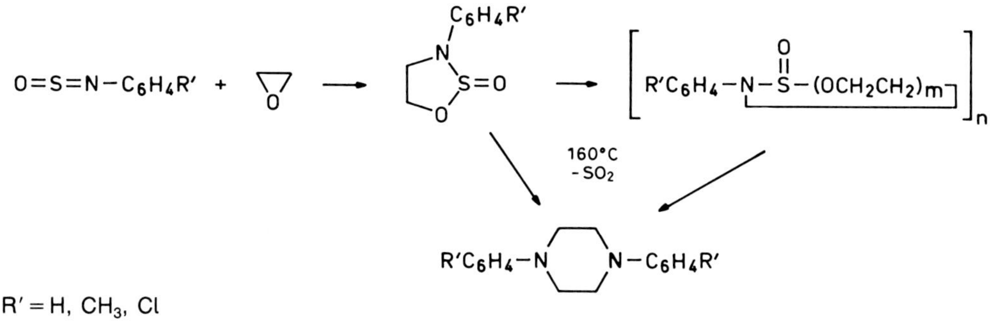

R′ = H, CH_3, Cl

Variation of the type of N-sulfinylamine in the reaction with 2-phenyloxirane in the presence of $(C_2H_5)_4NBr$ yields various end products as shown in the following schemes 1) [4], 2) [5], 3) [6]:

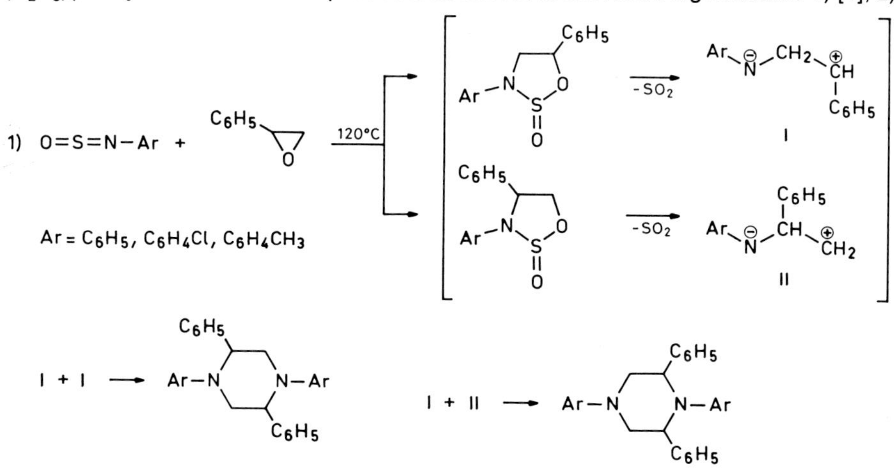

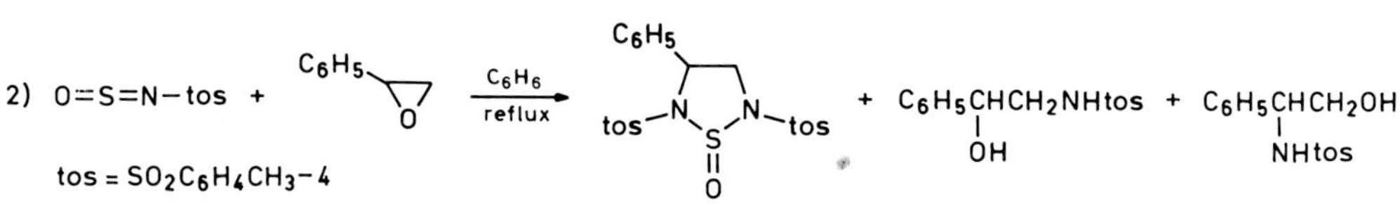

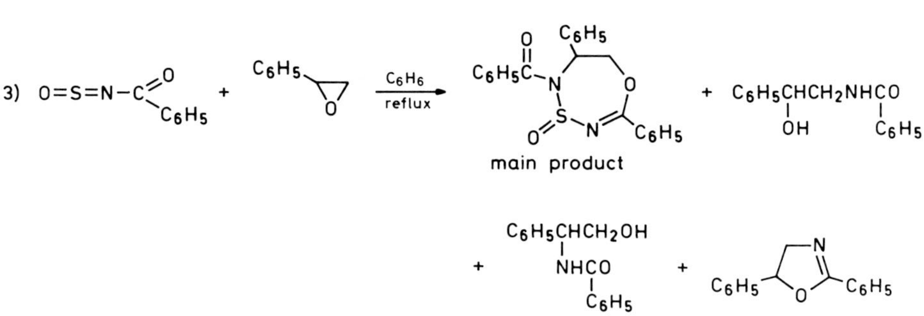

Quite another type of reaction seems to occur between N-sulfinylaniline (I) and 1,3-diphenylisobenzofuran (II) in the presence of BF_3-etherate. After 13 h refluxing, III was isolated in 78% yield. When I and II without a solvent are heated at 84°C for 36 h, III is obtained in 46% yield.

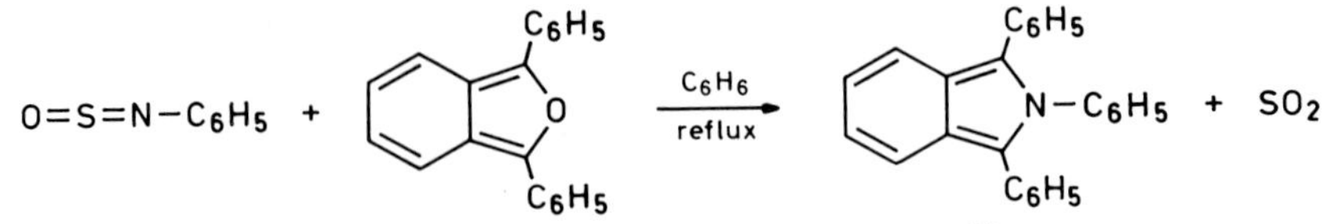

I II III

When I and II are mixed in approximately equimolar amounts, a red color appears immediately, indicating a charge-transfer complex (V). Rearrangement of the supposed primary [2+3] adduct IV could lead to the sultam VI, which then collapses to III [7]:

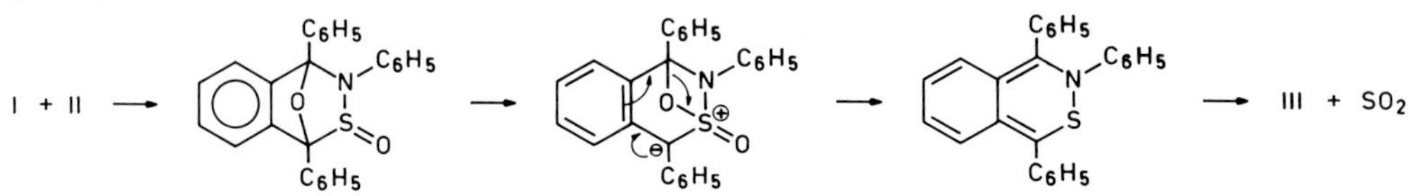

IV V VI

When O=S=N–tos reacts with N-tosylaziridine in the presence of $(C_2H_5)_4NBr$, products analogous to those from 2-phenyloxirane are obtained [5]:

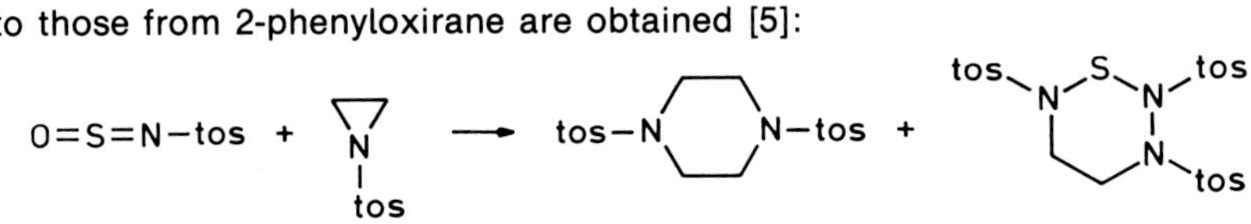

References:

[1] Etlis, V. S.; Sineokov, A. P.; Sergeeva, M. E. (Khim. Geterosikl. Soedin **1966** No. 2, pp. 682/6; Chem. Heterocycl. Compounds [USSR] **1966** No. 2, pp. 521/8).

[2] Yamada, F.; Nishiyama, T.; Kinugasa, M.; Nakatani, M. (Bull. Chem. Soc. Japan **43** [1970] 3611/3).

[3] Nishiyama, T.; Yamada, F. (Bull. Chem. Soc. Japan **44** [1971] 3073/7).

[4] Tsuge, O.; Mataka, S. (Bull. Chem. Soc. Japan **44** [1971] 1896/900).

[5] Tsuge, O.; Mataka, S. (Nippon Kagaku Zasshi **92** [1971] 543/7; C.A. **76** [1972] No. 99574).

[6] Tsuge, O.; Mataka, S. (Bull. Chem. Soc. Japan **44** [1971] 2836/9).

[7] Cava, M. P.; Schlessinger, R. H. (J. Org. Chem. **28** [1963] 2464/5).

2.6.23.15 Reactions with Heterocumulenes

With Isocyanates. $O{=}S{=}N{-}C_4H_9$-t undergoes exchange reactions with halogenosulfonylisocyanates $X{-}SO_2{-}N{=}C{=}O$ (X = F, Cl) at −78°C. At higher temperatures, when refluxed in CH_2Cl_2, the diimides form.

$$O{=}S{=}N{-}C_4H_9\text{-t} + X{-}SO_2{-}N{=}C{=}O \xrightleftharpoons{-78^\circ C} XSO_2{-}N{=}S{=}O + \text{t-}C_4H_9{-}N{=}C{=}O$$

$$\downarrow \text{reflux in } CH_2Cl_2$$

$$XSO_2{-}N{=}S{=}N{-}C_4H_9\text{-t} + CO_2$$

$O{=}S{=}N{-}C_4H_9$-t does not react with organylsulfonylisocyanates $R'{-}SO_2{-}N{=}C{=}O$ when refluxed in CH_2Cl_2 for several days. N-Sulfinylaniline $O{=}S{=}N{-}C_6H_5$ is less reactive and does not react with

XSO_2–N=C=O (X = F, Cl) at relatively low temperatures [1]. At higher temperatures in the reactions with isocyanates, exchange reactions and formation of azo compounds take place at the same time. The initial formation of a [2 + 2] cycloadduct is assumed:

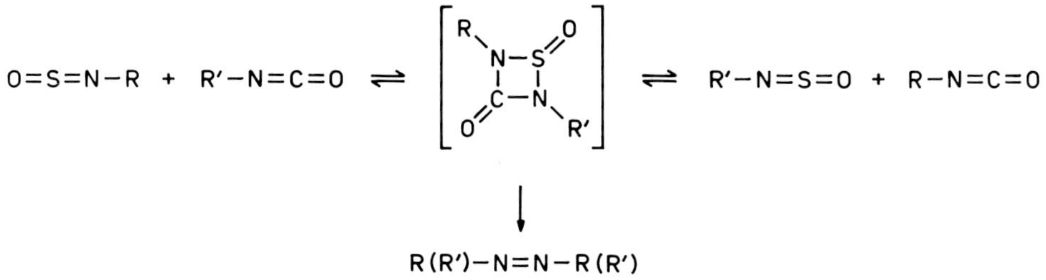

E. g., when O=S=N–tos reacts with R′–N=C=O at 110 to 130°C, the tosyl isocyanate tos–N=C=O is obtained in varying amounts depending on the isocyanate used (36.3% from C_6H_5N=C=O; 62.5% from c-C_6H_{11}–N=C=O; 16.2% from n-C_4H_9–N=C=O). N-Sulfinylanilines undergo exchange reactions at 180 to 200°C. Aromatic derivatives yield azo by-products. E. g., O=S=N–C_6H_4Cl-4, when stirred with C_6H_5N=C=O at 195°C for 5 h, gives 4-ClC_6H_4–N=C=O (37.3%), 4-ClC_6H_4–N=N–C_6H_4Cl-4 (10.2%), C_6H_5–N=N–C_6H_5 (3%), and 4-ClC_6H_4–N=N–C_6H_5 (3%) [2]. In the presence of metal carbonyls $Co_2(CO)_8$ or $Fe(CO)_5$ at 150°C, a cyclocondensation product forms in low yield [3]:

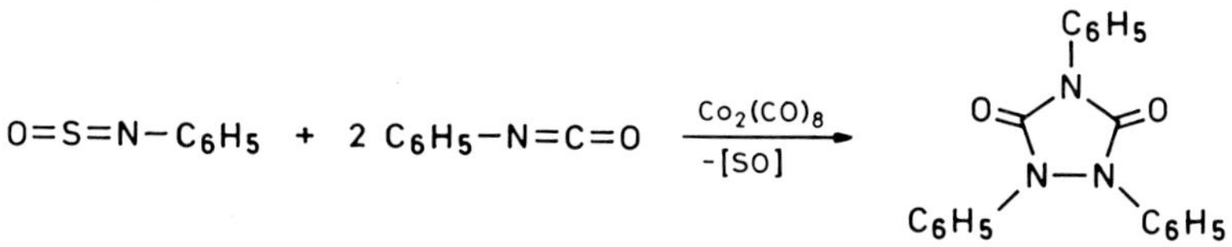

With Isothiocyanates. N-Sulfinylamines and aromatic isothiocyanates in equimolar amounts at 170 to 200°C yield exchange products O=S=N–R′ and R–N=C=S, and also carbodiimides. Addition of the C–S bond as well as of the N=C bond to the S=N bond of the sulfinylamine explains the formation of various products:

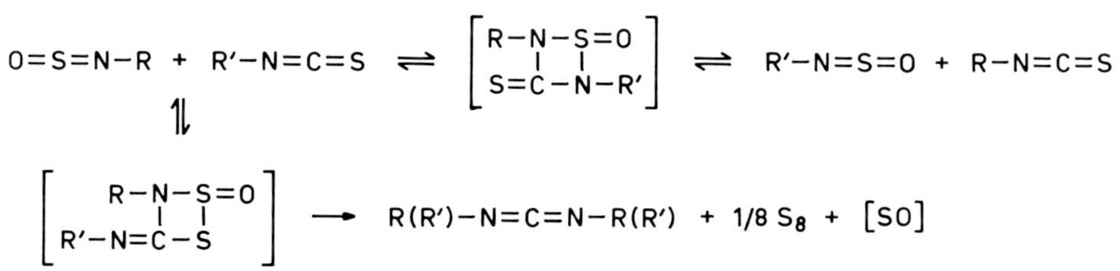

R = C_6H_5, 2-, or 4-$CH_3C_6H_4$, 4-$CH_3OC_6H_4$, 2-ClC_6H_4, 4-$CH_3C_6H_4SO_2$; R′ = C_6H_5
R = R′ = 2-$CH_3C_6H_4$

With R′–C(O)N=C=S in equimolar amounts the products vary:

$$O{=}S{=}N{-}C_6H_5 + R'{-}C(O)N{=}C{=}S \rightarrow C_6H_5{-}N{=}C{=}S + R'CN + S + SO_2$$

With R′ = C_6H_5 at 170°C, 8.5 h, C_6H_5–N=C=S (8.2% yield), C_6H_5CN (18.5%), and sulfur (75.8%) form. With R′ = CH_3 at 145°C, 5.0 h, sulfur (50.0%) and traces of other products form [4].

Reactions with amidinoyl-isothiocyanates are described on p. 213.

With Carbodiimides. N-Sulfinylalkanamines and N-sulfinylarenamines O=S=N–R (R = alkyl, aryl) give with R′–N=C=N–R′ no cycloadducts. N-Sulfinyl-sulfonamides O=S=N–SO_2R, however, form remarkably stable [2 + 2] cycloadducts at room temperature or even lower temperatures in ether. These undergo thermal decomposition only at 80 to 180°C to give exchange products.

Alkaline or acidic hydrolysis in ethanol, or reduction with Raney nickel in the same solvent yield guanidine derivatives [6]:

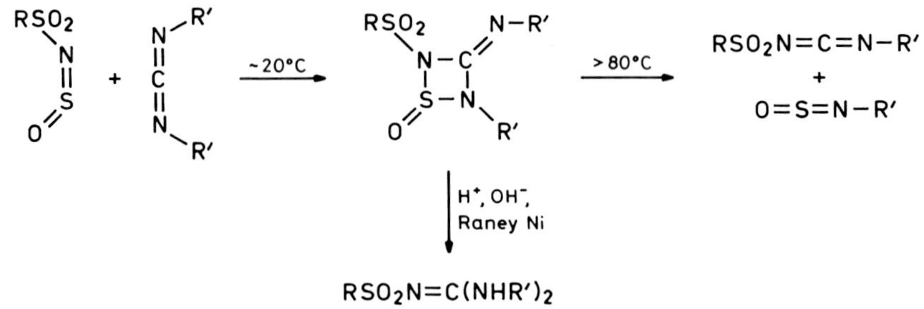

$R = CH_3$, $4\text{-}CH_3C_6H_4$
$R' = (CH_3)_2CH$, $n\text{-}C_4H_9$, $c\text{-}C_6H_{11}$, C_6H_5

In the absence of a solvent O=S=N–tos reacts with CH_3–N=C=N–C_4H_9-t exothermically to form the expected exchange products O=S=N–CH_3 and tos–N=C=N–C_4H_9-t [7].

N-Sulfinyl-carboxamides O=S=N–C(O)R and carbodiimides yield a complex mixture of products. The authors assume the initial formation of [2+2] and [4+2] cycloadducts. Their thermal decomposition gives isocyanates, nitriles, and amides as major products [6]:

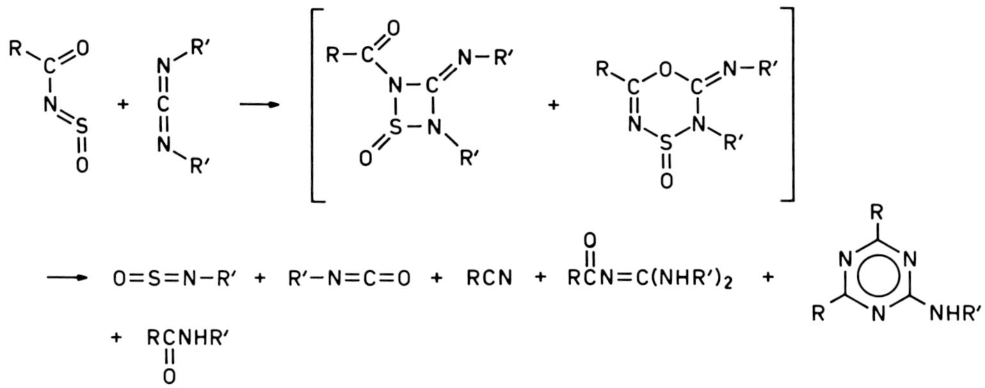

$R = CH_3$, C_6H_5; $R' = c\text{-}C_6H_{11}$, C_6H_5

[2+3] Cycloaddition of O=S=N–R to the 1,3-dipolar species produced by thermolysis of compound I enables trapping of the intermediary product II:

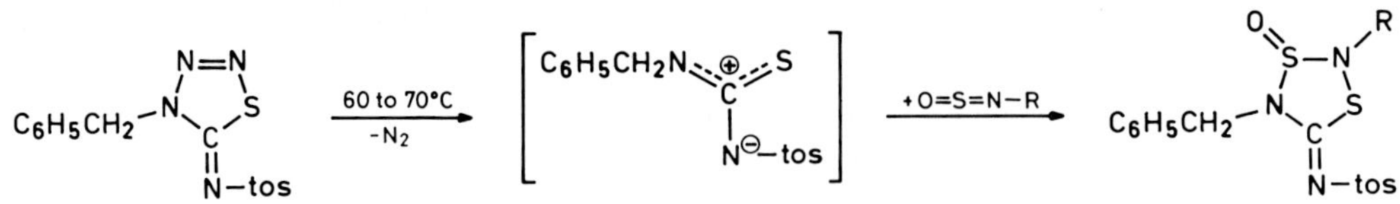

I II

$R = 4\text{-}ClC_6H_4$, $4\text{-}NO_2C_6H_4$, $4\text{-}CH_3C_6H_4$, $4\text{-}CH_3C_6H_4SO_2$, $CH_3CH_2CH_2$, CH_3CO [5]

With Ketenes. The reaction of unsubstituted gaseous ketene H_2C=C=O with O=S=N–C_6H_5 in dry acetone at −78°C is reported to yield an unstable [2+2] cycloadduct, which adds aromatic amines [12, 13]:

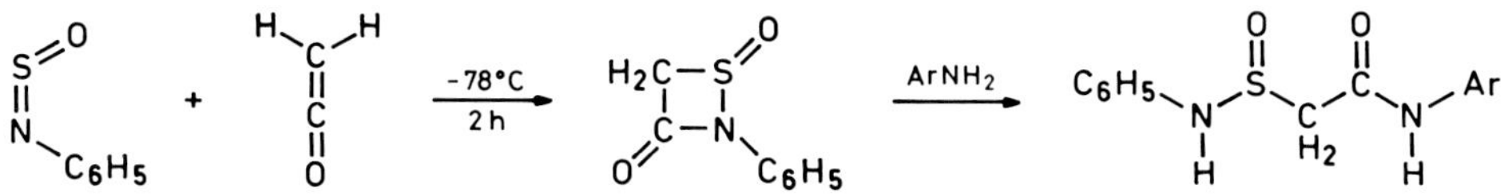

Diphenylketene reacts with aliphatic and aromatic O=S=N–R compounds forming [2+2] cycloadducts nearly quantitatively at room temperature or even higher temperatures [8, 14]:

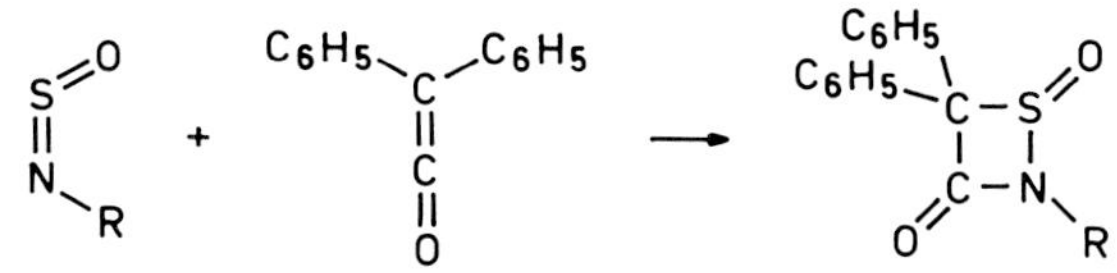

R = c-C_6H_{11}, C_6H_5, C_6H_4Cl-4, C_6H_4Cl-3, C_6H_4Cl-2, $C_6H_3Cl_2$-2,4, $C_6H_4NO_2$-4 [14], $SO_2C_6H_4CH_3$-4 [8]

With the cyclohexyl derivative, the cycloadduct forms immediately at room temperature [14]. O=S=N–$SO_2C_6H_4CH_3$-4 reacts smoothly under conditions where O=S=N–C_6H_5 yields no product (room temperature, ether–petroleum ether solution) [8]. For the reaction of O=S=N–C(O)OR (R = CH_3, C_2H_5) with diphenylketene in ether, the first step is assumed to be a [4+2] cycloaddition, followed by CO_2 elimination and further reaction and transformation [18]:

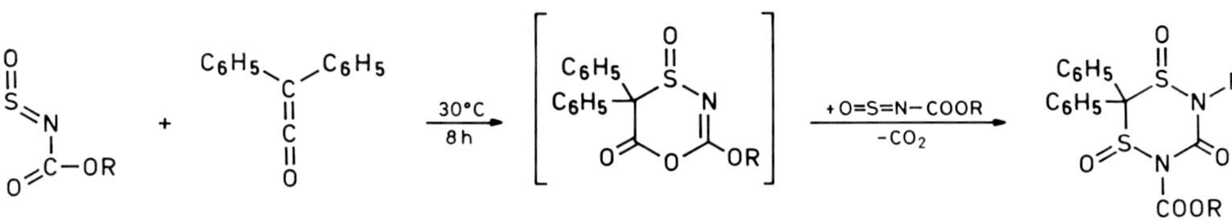

Bis(trifluoromethyl)ketene and N-sulfinylamines form [2+2] cycloadducts:

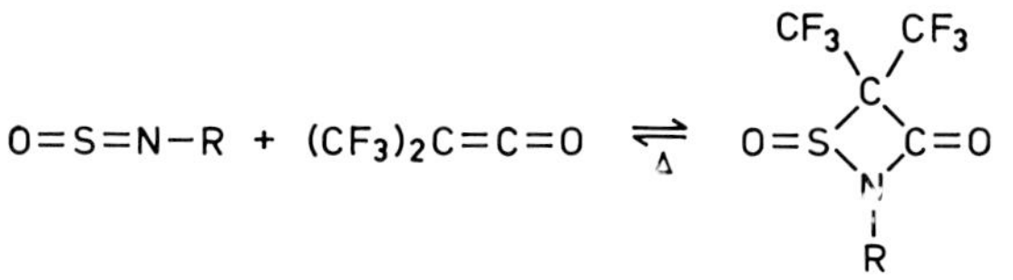

R = CH_3, c-C_6H_{11}, C_6H_5, CF_3, C_6F_5

The reaction with O=S=N–CH_3 proceeds quantitatively without a solvent at −20°C. With O=S=N–C_6H_{11}-c and O=S=N–C_6H_5 the reaction occurs readily in $CHCl_3$ at room temperature. With O=S=N–CF_3 the adduct forms only in $CHCl_3$ solution with $(C_2H_5)_3N$ as a catalyst at −35°C. With O=S=N–C_6F_5 the presence of $(C_2H_5)_3N$ is also necessary for [2+2] cycloaddition at room temperature without a solvent [19].

[2+2] Cycloaddition occurs readily with biphenyleneketene and O=S=N–R (R = C_6H_5, C_6H_4Cl-3, $C_6H_3Cl_2$-2,4) in petroleum ether [14]:

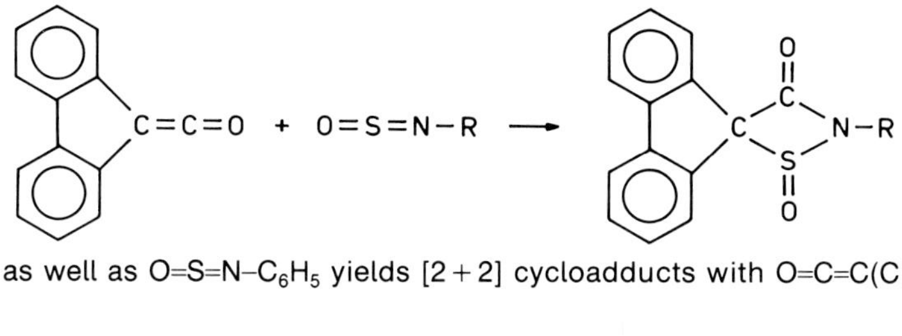

O=S=N–C_4H_9-t as well as O=S=N–C_6H_5 yields [2+2] cycloadducts with O=C=C(CN)C_4H_9-t [15].

[4 + 2] Cycloaddition occurs with acyl ketenes, even with alkyl derivatives O=S=N–R (in refluxing benzene) [16]:

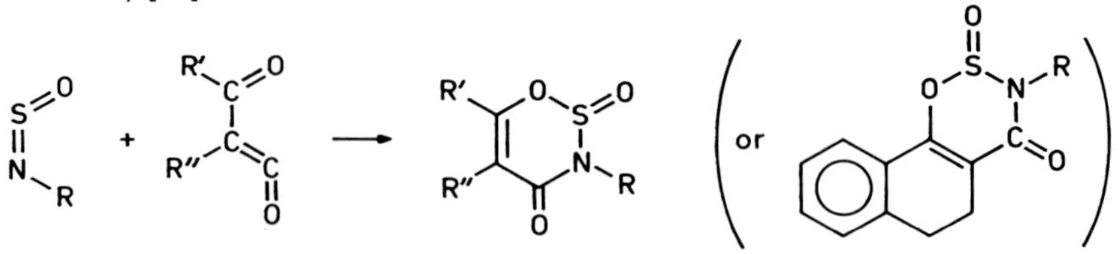

R = c-C_6H_{11}, C_6H_5, C_6H_4Cl-4, $C_6H_4CH_3$-4, $SO_2C_6H_4CH_3$-4; R′ = C_6H_5, R″ = H; R′, R″ = 2-$(CH_2)_2C_6H_4$–

Acyl ketenes generated in situ from α-diazoalkanes form [4 + 2] and sometimes [2 + 2] cycloadducts (in boiling toluene) [17]:

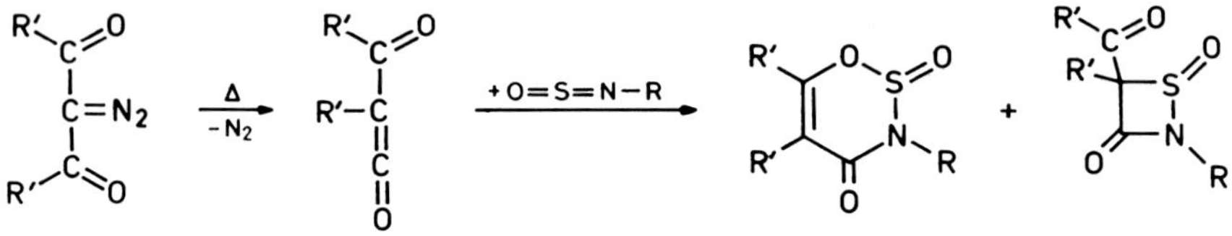

I II

R	R′	product
n-C_3H_7	C_6H_5	I (20%)
c-C_6H_{11}	C_6H_5	I (40%) + II (9%)
C_6H_5	C_6H_5	I (38%)
4-$CH_3C_6H_4$	C_6H_5	I (30%)
4-$CH_3OC_6H_4$	C_6H_5	I (29%)
4-$CH_3OC_6H_4$	4-$CH_3OC_6H_4$	I (16%)
4-$NO_2C_6H_4$	C_6H_5	I (31%)
2,6-$(CH_3)_2C_6H_3$	C_6H_5	I (18%) + II (not separated)

When the tosyl derivative O=C=C($C_6H_4CH_3$-4)$SO_2C_6H_4CH_3$-4 reacts with O=S=N–$C_6H_4NO_2$-4 the only product is the [2 + 2] cycloadduct II in 24% yield [17].

With Ketenimines. N-Sulfinyl-alkanamines or N-sulfinyl-arenamines O=S=N–R (R = alkyl, aryl) and ketenimines (1-alken-1-imines) R′R″C=C=N–R‴ afford no cycloadduct. The more electrophilic O=S=N–tos or O=S=N–SO_2CH_3 and ketenimines in dry ether at room temperature form an unstable, partly isolable [2 + 2] cycloadduct [9].

The structure assumed initially is wrong. IR and NMR spectra and chemical reactions show that the [2 + 2] adduct contains an SNC_2 ring [10]. The adducts may form by heterometathesis and subsequent [2 + 2] cylcoaddition [11]:

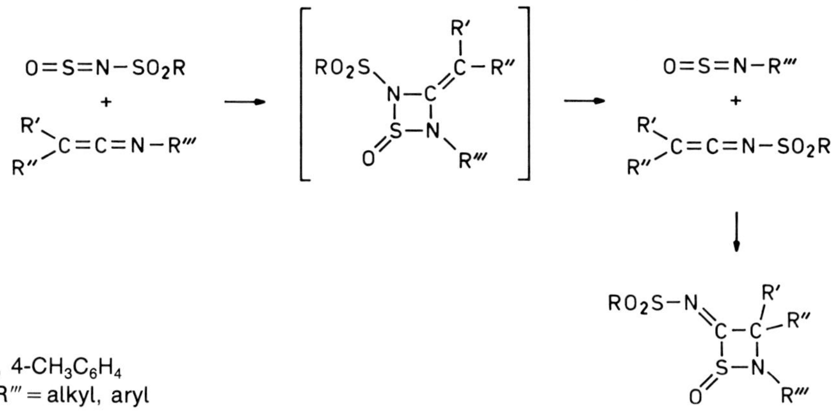

$R = CH_3$, $4\text{-}CH_3C_6H_4$
R′, R″, R‴ = alkyl, aryl

With Sulfur Diimides, Imidothionyl Chlorides (Iminosulfuranes). Exchange reactions of $O{=}S{=}NSC_6H_4R$ ($R = 2\text{-}NO_2$, 4-Cl) with $S({=}N{-}SO_2R')_2$ ($R' = C_6H_4CH_3\text{-}4$) in dry benzene, in the presence of pyridine, yield the wanted N-sulfonyl-N′-sulfenyl sulfur diimides:

$$O{=}S{=}NSC_6H_4R + R'SO_2{-}N{=}S{=}N{-}SO_2R' \rightarrow O{=}S{=}N{-}SO_2R' + RC_6H_4{-}S{-}N{=}S{=}N{-}SO_2R'$$

With $R = 2\text{-}NO_2$ refluxing for 6 h gives 10% yield. With R = 4-Cl stirring at room temperature for 24 h gives 17% yield [20].

N-Sulfinyl-sulfonamides $O{=}S{=}N{-}SO_2R$ react with N-substituted imidothionyl chlorides of the type $RSO_2N{=}SCl_2$ at about 140 to 145°C for 50 min, mole ratio 1:1, to give symmetrically or asymmetrically substituted sulfur diimides. Intermediate formation of an adduct is assumed, which splits off $SOCl_2$ [21]:

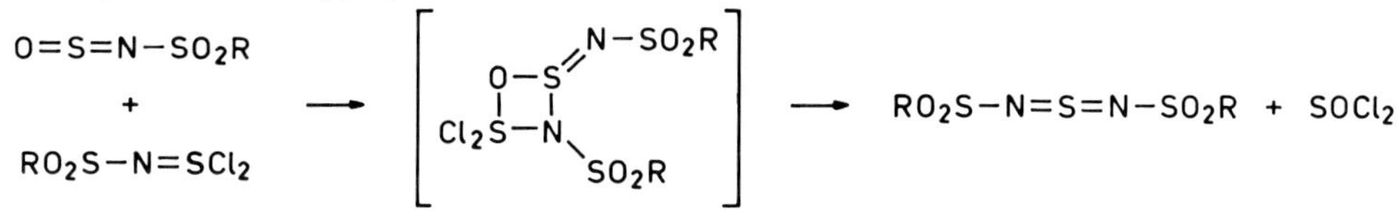

$R = C_6H_5$, $4\text{-}CH_3C_6H_4$

References:

[1] Appel, R.; Montenarh, M. (Chem. Ber. **109** [1976] 2437/41).
[2] Minami, T.; Miki, H.; Agawa, T. (Kogyo Kagaku Zasshi **70** [1967] 1831/3; C.A. **69** [1968] No. 18790).
[3] Drapier, J.; Hubert, A. J.; Teyssié, P. (Synthesis **1975** 649).
[4] Minami, T.; Agawa, T. (Tetrahedron Letters **1968** 2651/3).
[5] L'Abbé, G.; Van Asch, A.; Declercq, J. P.; Germain, G.; Van Meerssche, M. (Bull. Soc. Chim. Belges **87** [1978] 285/92).
[6] Minami, T.; Fukuda, M.; Abe, M.; Agawa, T. (Bull. Chem. Soc. Japan **46** [1973] 2156/9).
[7] Ulrich, H.; Tucker, B.; Sayigh, A. A. R. (J. Am. Chem. Soc. **94** [1972] 3484/7).
[8] Kresze, G.; Maschke, A.; Albrecht, R.; Bederke, K.; Patzschke, H. P.; Smalla, H.; Trede, A. (Angew. Chem. **74** [1962] 135/44, 143; Angew. Chem. Intern. Ed. Engl. **1** [1962] 89/98).
[9] Minami, T.; Takimoto, F.; Agawa, T. (Bull. Chem. Soc. Japan **48** [1975] 3259/61).
[10] L'Abbé, G.; Van Asch, A.; Dekerk, J.-P.; Minami, T. (Tetrahedron Letters **22** [1981] 583/4).

[11] Bussas, R.; Kresze, G.; Münsterer, H.; Schwöbel, A. (Sulfur Rept. **2** [1983] 215/377, 286/7).
[12] Bohen, J. M.; Joullié, M. M. (J. Org. Chem. **38** [1973] 2652/7).

[13] Semple, J. E.; Joullié, M. M. (J. Org. Chem. **43** [1978] 3066/8).
[14] Beeken, H.; Korte, F. (Tetrahedron **18** [1962] 1527/31).
[15] Goldish, D. M.; Axon, B. W.; Moore, H. W. (Heterocycles **20** [1983] 187/92).
[16] Minami, T.; Yamauchi, Y.; Ohshiro, Y.; Agawa, T.; Murai, S.; Sonoda, N. (J. Chem. Soc. Perkin Trans. I **1977** 904/8).
[17] Capuano, L.; Urhahn, G.; Willmes, A. (Chem. Ber. **112** [1979] 1012/22).
[18] Hörhold, H. H.; Eibisch, H. (Chem. Ber. **101** [1968] 3567/78, 3571, 3575/6).
[19] Jäger, U.; Schwab, M.; Sundermeyer, W. (Chem. Ber. **119** [1986] 1127/32).
[20] Buchholt, H. C.; Senning, A.; Kelly, P. (Acta Chem. Scand. **23** [1969] 1279/87).

[21] Markovskii, L. N.; Fedyuk, G. S.; Levchenko, E. S. (Zh. Org. Khim. **8** [1972] 286/91; J. Org. Chem. [USSR] **8** [1972] 287/91).

2.6.23.16 Reactions with Organo-Phosphorus Compounds

The products of the reactions of alkylidenephosphoranes with N-sulfinylamines depend on the type of the N-sulfinylamine used [1]:

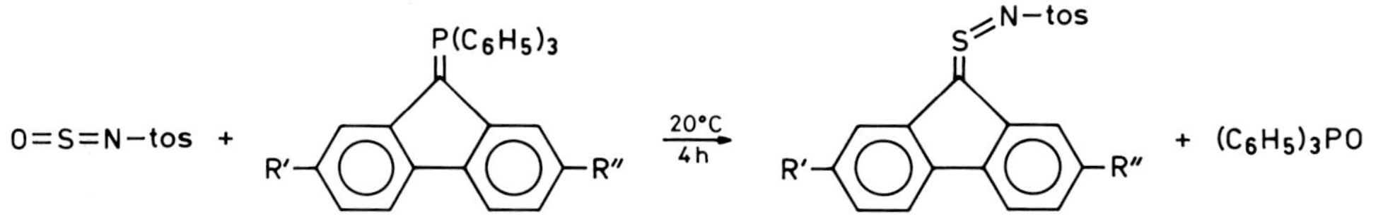

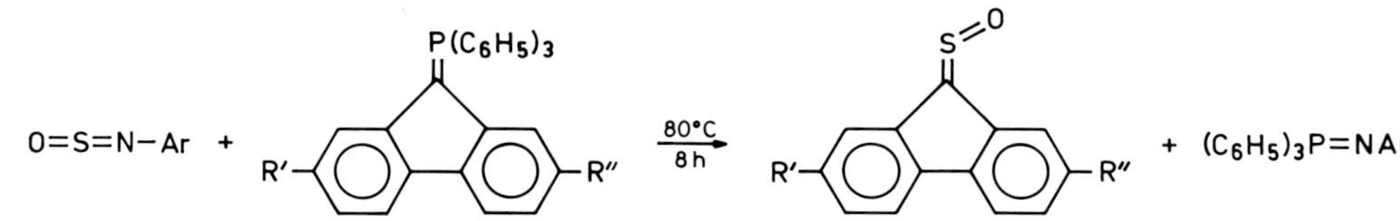

tos = $SO_2C_6H_4CH_3$-4; Ar = $C_6H_4NO_2$-4; R′, R″ = H, Br, NO_2

The mechanism of these metathesis reactions is not clear. Either different bonds, S=O or S=N, undergo the reaction, or the sulfine is the primary product in each case and reacts further with O=S=N–tos to give the thionimide [2, p. 283].

The Wittig-type reactions of O=S=N–tos with O=C-, C=C-, and N=C-conjugated phosphorus ylides afford quite different products, as shown in the following three schemes. When (acyl arylmethylene)triphenyl phosphoranes I are refluxed with O=S=N–tos in benzene, both functional sites, namely the ylide and the acyl moiety, react to form tos–N=C-conjugated thione S-imides II, which undergo intramolecular 1,3-dipolar cycloaddition of the CSN group with the aromatic C=C bond of the tosyl group to give a structurally unique spiro-fused tricyclic adduct III. Besides III, compounds IV and a small amount of blue V form [3]. The previously [1] assumed structure VI for the reaction product was revised [3]. The structure of one compound III (R′ = C_6H_5, R″ = 4-$CH_3OC_6H_4$) was proved by means of X-ray structure analysis. The structure of the other compounds III was confirmed by comparison of their spectroscopic data (IR, MS, ^{1}H and ^{13}C NMR) with those of the X-ray-studied compound. The orange-colored compounds II, which formed at the first stage of the reaction, are labile, however, they could be isolated (in the case of R′ = C_6H_5, R″ = 4-$CH_3C_6H_4$ or R′ = C_6H_5, R″ = 4-$CH_3OC_6H_4$) when the reaction was performed at room temperature or below. The IR, ^{1}H, and ^{13}C NMR spectra were measured [3].

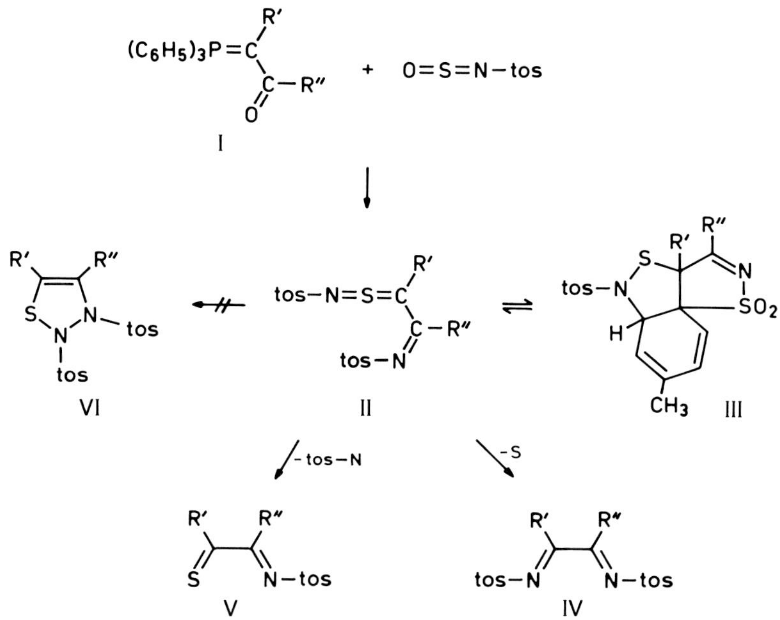

tos = 4-$CH_3C_6H_4SO_2$; R′ = C_6H_5, 4-$CH_3OC_6H_4$, 4-$CH_3C_6H_4$; R″ = C_6H_5, 4-ClC_6H_4, 4-$CH_3C_6H_4$

The reaction of O=S=N–tos with the C=C-conjugated phosphorus ylide VIIa (R = H) (when warmed in benzene until the solution becomes clear) yield the 1-azadiene X, which might form from the conjugated thione S-imide IX by sulfur extrusion via a thiaziridine ring. Neither IX nor its cyclization product XI were obtained. On the other hand, the reaction of O=S=N–tos with VIIb (R = C_6H_5) under the same conditions stops at the stage of betaine VIII, but this could not be isolated.

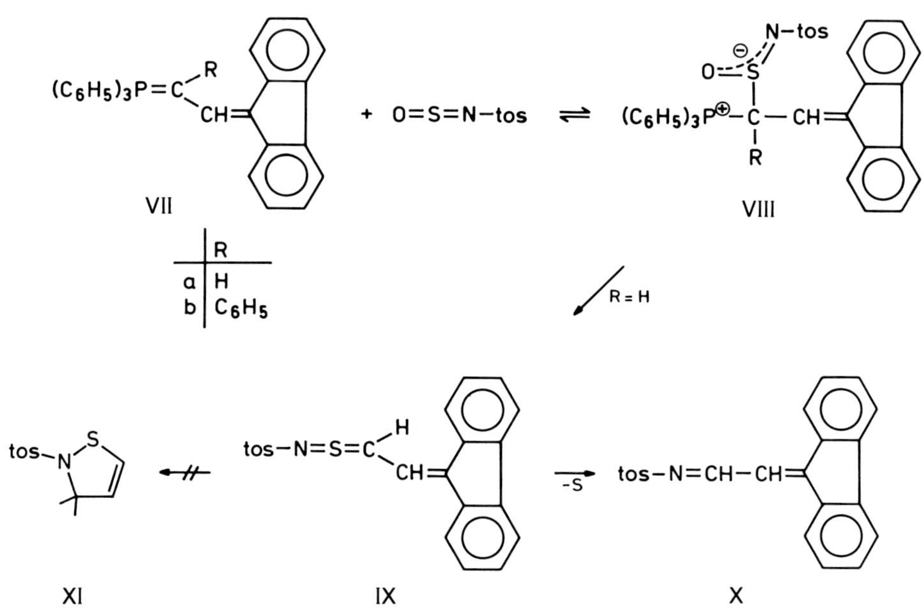

The reaction of O=S=N–tos with XII, having an N=C bond in conjugation with the ylide function (in benzene, ~10°C), affords the hygroscopic betaine XIII, which on standing in air is hydrolyzed to give XIV. The corresponding conjugated thione S-imides XV and S-oxide XVI, respectively, could not be obtained [3].

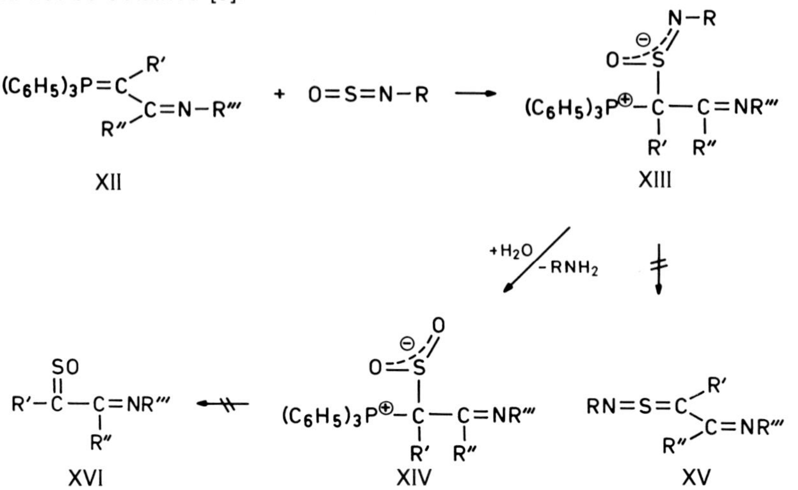

$R = C_6H_5$, $4\text{-}CH_3C_6H_4SO_2$; $R' = H$, C_6H_5; $R'' = C_6H_5$, $C_6H_5CH_2$; $R''' = C_6H_5$, $(CH_3)_3C_6H_2$

N-Sulfinyl-sulfonamides $O{=}S{=}N{-}SO_2R$ react with 1,3,2λ⁵-dioxaphosphole I in CH_2Cl_2 when mixed with ice-cooling and then refluxed for 1 h with elimination of the corresponding phosphate III and formation of the 3-sulfonyl-3H-1,2,3-oxathiazole 2-oxide II [4, 5]:

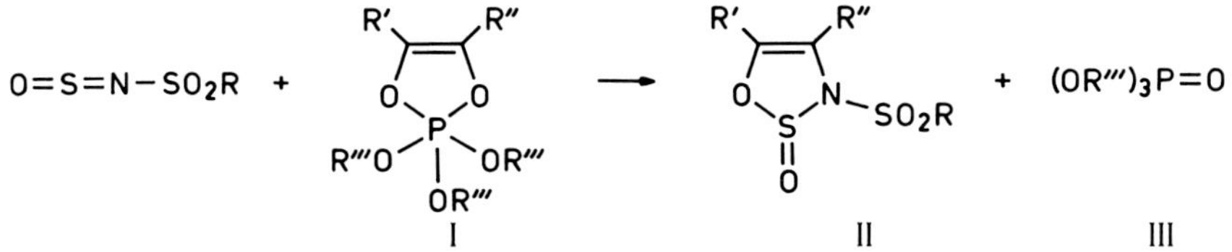

$R = C_2H_5$, C_6H_5, $4\text{-}ClC_6H_4$, $4\text{-}BrC_6H_4$ [4], $4\text{-}CH_3C_6H_4$ [4, 5]; R′, $R'' = H$, CH_3, C_2H_5, C_6H_5 [4], $(-CH_2-)_n$, n = 4, 5, 6 [5]; $R''' = CH_3$ [4, 5], C_2H_5 [5]

When R′ ≠ R″ isomeric mixtures form. N-Sulfinyl-arenamines O=S=N–Ar do not react under these conditions [4].

References:

[1] Saito, T.; Motoki, Sh. (J. Org. Chem. **42** [1977] 3922/5).
[2] Bussas, R.; Kresze, G.; Münsterer, H.; Schwöbel, A. (Sulfur Rept. **2** [1983] 215/377, 283).
[3] Saito, T.; Nakane, M.; Watanabe, T.; Motoki, Sh.; Kobayashi, K.; Sakurai, T. (Bull. Chem. Soc. Japan **58** [1985] 2882/8).
[4] Neidlein, R.; Mosebach, R. (Arch. Pharm. **307** [1974] 291/301).
[5] Neidlein, R.; Friederich, W. (Arch. Pharm. **310** [1977] 622/35, 625, 631/2).

2.6.23.17 Reactions with Oxonium Salts

N-Sulfinyl-alkanamines react with oxonium salts (neat, 15°C) to give the corresponding N-sulfinyl-aminium salts in nearly quantitative yield [1, 2]:

$$O{=}S{=}NR + R'_3O^+BF_4^- \rightarrow [O{=}S{\doteq}NR'R]^+BF_4^- + R'_2O$$

$$R = CH_3,\ C_2H_5;\ R' = CH_3,\ C_2H_5$$

References:

[1] Kresze, G.; Rössert, M. (Angew. Chem. **90** [1978] 61; Angew. Chem. Intern. Ed. Engl. **17** [1978] 63).
[2] Kresze, G.; Rössert, M. (Liebigs Ann. Chem. **1981** 58/64).

2.6.24 Reactions of N-Sulfinylamines (Sulfinyl Imides) with Radicals, Carbenes, Ylides, and Carbanions

The reaction of N-sulfinylaniline $O{=}S{=}NC_6H_5$ with dibenzoyl peroxide $(C_6H_5CO_2)_2$ in boiling benzene (under nitrogen atmosphere), i.e. with the $C_6H_5^{\bullet}$ radical forming by thermolysis, yields $C_6H_5N{=}NC_6H_5$ (30%), $C_6H_5SSO_2C_6H_5$ (20%), $C_6H_5{-}C_6H_5$ (20%), $C_6H_5SSC_6H_5$ (10%), and C_6H_5-$SO_2OSO_2C_6H_5$ (5%). Obviously the addition occurs at the sulfur atom to form I:

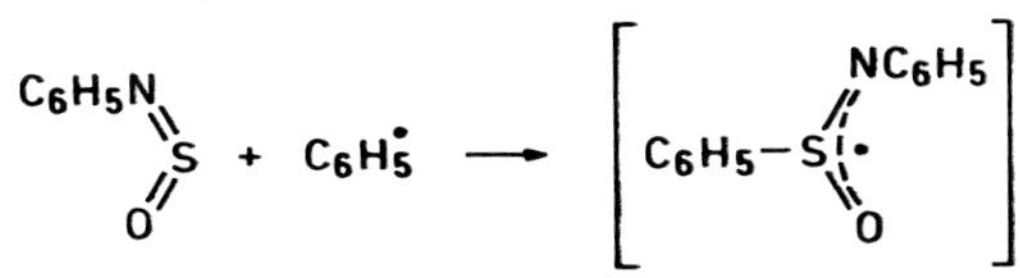

I

Decomposition, combination, and rearrangement of this intermediate product (I) gives the end products cited above. The reaction of N-sulfinyl-arenamines $O{=}S{=}NAr$ with arylazides $Ar'N_3$ in boiling chlorobenzene, i.e. with triplet arylnitrenes $Ar'\dot{N}^{\bullet}$, yields $ArN{=}NAr'$ (~30%), $ArN{=}NAr$ (~15%), $Ar'N{=}NAr'$ (~15%), SO_2, sulfur, and other products. Here also the addition should occur at the sulfur atom to give radicals such as I. Cleavage of the SN bond and combination of the fragments follow [1].

1,8-Dehydronaphthalene (III), generated via $Pb(O_2CCH_3)_4$ oxidation of 1-aminonaphtho[1,8-de]triazine (II), in the presence of N-sulfinylaniline or its 4-chloro derivative in CH_2Cl_2, yields a complex mixture of colored products from which IV (4%) and V (20%) were isolated [2]:

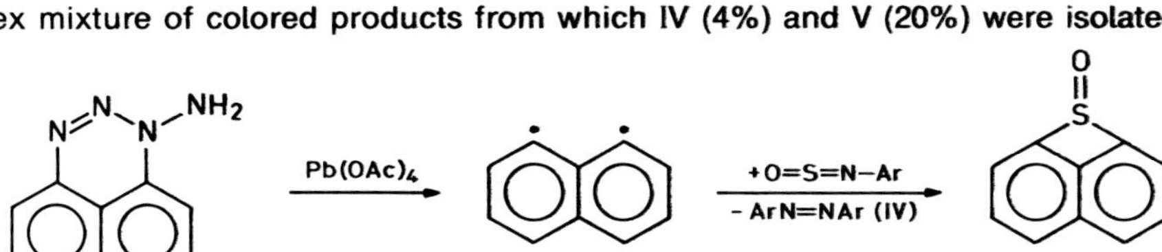

II III V

$Ar = C_6H_5$, 4-ClC_6H_4

ESR studies were made on the addition of photochemically produced alkoxy or alkyl radicals to N-sulfinylamines in cyclopropane solution. Radicals form which are characterized by a g factor of ca. 2.0042 and nitrogene splittings of a(N) = 9 to 10 G:

$$O{=}S{=}NR + X^{\bullet} \rightarrow RN^{\bullet}{=}S(O)X$$

$R = C_2H_5$, c-C_6H_{11}, t-C_4H_9

$X = CH_3$ (from $CH_3N{=}NCH_3$), t-C_4H_9 (from t-$C_4H_9N{=}NC_4H_9$-t), t-C_4H_9O (from t-$C_4H_9OOC_4H_9$-t), $(CH_3)_3SiO$ (from $(CH_3)_3SiOOSi(CH_3)_3$)

Irradiation of a mixture of $O{=}S{=}NC_4H_9\text{-t}$ and $\text{t-}C_4H_9N{=}NC_4H_9\text{-t}$ yields another signal tentatively ascribed to a radical $\text{t-}C_4H_9N{=}S(O)(C_4H_9\text{-t}){-}\dot{N}C_4H_9\text{-t}$. The first step in its formation might be the photolysis of $O{=}S{=}NC_4H_9\text{-t}$:

$$O{=}S{=}NC_4H_9\text{-t} \xrightarrow{h\nu} [\text{t-}C_4H_9N|] + [SO]$$

$$[\text{t-}C_4H_9N|] + O{=}S{=}NC_4H_9\text{-t} \rightarrow [\text{t-}C_4H_9N{=}S(O){=}NC_4H_9\text{-t}]$$

$$\text{t-}C_4H_9^{\bullet} + [\text{t-}C_4H_9N{=}S(O){=}NC_4H_9\text{-t}] \rightarrow [\text{t-}C_4H_9N{=}S(O)(C_4H_9\text{-t}){-}\dot{N}C_4H_9\text{-t}]$$

Photolysis of a mixture containing $O{=}S{=}NC_4H_9\text{-t}$ and $\text{t-}C_4H_9OOC_4H_9\text{-t}$ in isobutane gave rise to the spectrum of $\text{t-}C_4H_9\dot{N}{-}S(O)C_4H_9\text{-t}$ and $\text{t-}C_4H_9N{=}S(O)(C_4H_9\text{-t}){-}\dot{N}C_4H_9\text{-t}$. In propene solution, $\text{t-}C_4H_9\dot{N}{-}S(O)C_3H_5$ forms. Photolysis of the same mixture in $\text{i-}C_3H_7OH$ gives rise to the spectrum of another radical ascribed to $\text{t-}C_4H_9N{=}SOH$. This radical presumably arises by a direct hydrogen transfer from $(CH_3)_2\dot{C}OH$ (equation 2), or by an addition-elimination sequence (equations 3 and 4) [3]:

$$(1)\ \text{t-}C_4H_9O^{\bullet} + (CH_3)_2CHOH \rightarrow (CH_3)_2\dot{C}OH + \text{t-}C_4H_9OH$$

$$(2)\ O{=}S{=}NC_4H_9\text{-t} + (CH_3)_2\dot{C}OH \rightarrow (CH_3)_2CO + \text{t-}C_4H_9N{=}\dot{S}OH$$

$$(3)\ O{=}S{=}NC_4H_9\text{-t} + (CH_3)_2\dot{C}OH \rightarrow \text{t-}C_4H_9\dot{N}S(O)C(OH)(CH_3)_2$$

$$(4)\ \text{t-}C_4H_9\dot{N}S(O)C(OH)(CH_3)_2 \rightarrow \text{t-}C_4H_9N{=}\dot{S}OH + (CH_3)_2CO$$

The photolysis of an equimolar mixture of $O{=}S{=}NC_6H_5$ and diphenyldiazomethane in hexane under nitrogen gives $C_6H_5N{=}C(C_6H_5)_2$ (67% yield), $(C_6H_5)_2C{=}O$, $(C_6H_5)_2C{=}C(C_6H_5)_2$, $(C_6H_5)_2C{=}N{-}N{=}C(C_6H_5)_2$ and minor products. The reaction is assumed to proceed by a carbene addition to the SN double bond via a three-membered cyclic intermediate [4]:

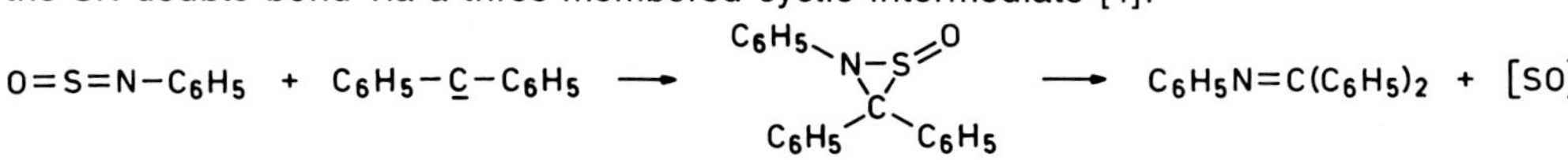

A carbene addition as an intermediate process should be involved in the preparation of isocyanides by the reaction of N-sulfinylanilines or N-sulfinylcyclohexanamine with $CHCl_3{-}KOH$ in the presence of crown ethers or $(C_4H_9)_4NCl$ or other phase-transfer catalysts [5, 6]. Yield 70 to 96% [5].

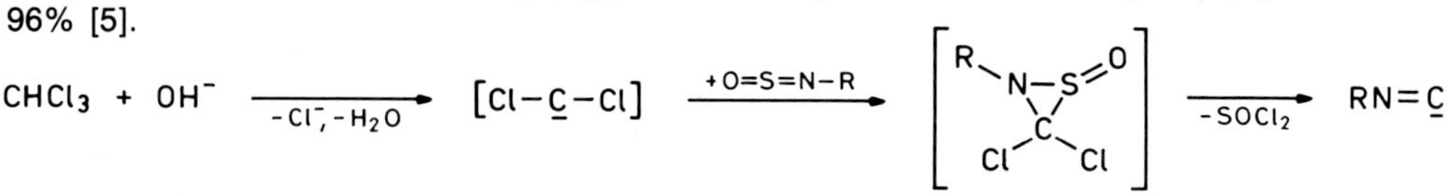

$R = C_6H_5$, $4\text{-}CH_3C_6H_4$, $4\text{-}ClC_6H_4$, $2,4,6\text{-}(CH_3)_3C_6H_2$, $c\text{-}C_6H_{11}$

Nitrile ylides $(CF_3)_2C^{\ominus}{-}N^{\oplus}{\equiv}CR$ (produced from I) react with $O{=}S{=}NC_6H_5$ in boiling benzene to give [2+3] cycloadducts [7]:

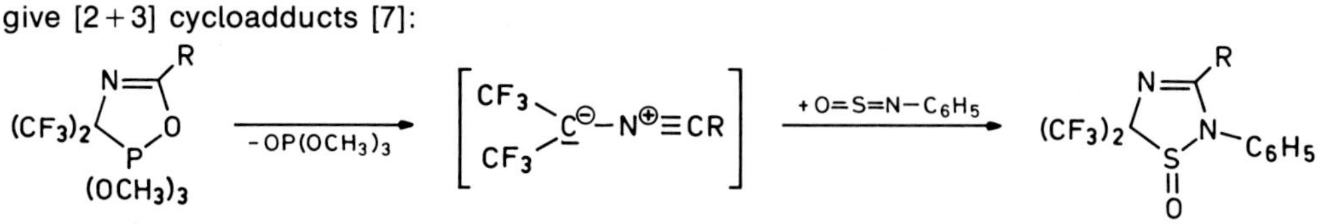

I

$R = C(CH_3)_3$ (yield 54%), C_6H_5 (45%), $4\text{-}CH_3C_6H_4$ (49%), $4\text{-}FC_6H_4$ (57%), $4\text{-}ClC_6H_4$ (29%)

$O{=}S{=}N{-}C_6H_5$ undergoes [2+3] cycloaddition with diphenylnitrile imine prepared in situ in the presence of $N(C_2H_5)_3$ in boiling benzene. The new heterocycle formed in 87% yield [9]:

N-Sulfinylanilines react with α-silylcarbanions (produced from the corresponding neutral compound and LiC_4H_9-n in situ) to give thione S-imides:

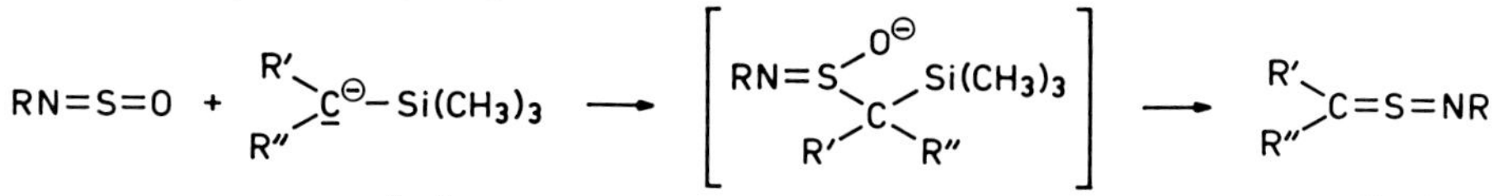

Ia,b IIa,b

$R = 2,4,6\text{-}(t\text{-}C_4H_9)_3C_6H_2$

R′, R″ =

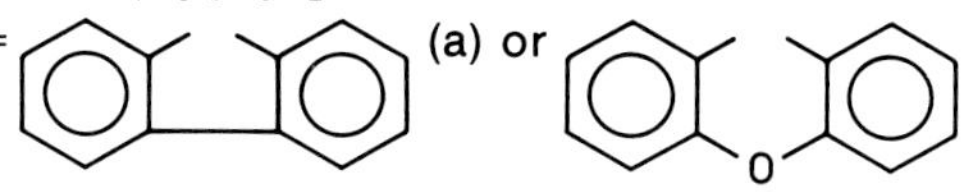

(a) or (b)

77% yield 68% yield

With $R = 2\text{-}CH_3\text{-}4,6\text{-}(t\text{-}C_4H_9)_2C_6H_2$ steric hindrance prevents the formation of the imide II.

The N-sulfinyl-sulfonamide $O{=}S{=}N{-}SO_2C_6H_4CH_3\text{-}4$ and Ia do not yield the expected thione S-imide IIa, but $4\text{-}CH_3C_6H_4{-}SO_2{-}NH_2$ [8].

References:

[1] De Luca, G.; Renzi, G.; Cipollini, R.; Pizzabiocca, A. (Gazz. Chim. Ital. **107** [1977] 357/8).
[2] De Luca, G.; Pizzabiocca, A.; Renzi, G. (Chem. Ind. [London] **1983** 679).
[3] Baban, J. A.; Roberts, B. P. (J. Chem. Soc. Perkin Trans. II **1978** 678/83).
[4] Stoffer, J. O.; Musser, H. R. (J. Chem. Soc. D **1970** 481).
[5] Domschke, G.; Beckert, R.; Mayer, R. (Synthesis **1977** 275/6).
[6] Bussas, R.; Kresze, G.; Münsterer, H.; Schwöbel, A. (Sulfur Rept. **2** [1983] 215/377, 313/4).
[7] Burger, K.; Goth, H.; Daltrozzo, E. (Z. Naturforsch. **37b** [1982] 473/85, 474).
[8] Porskamp, P. A. T. W.; Zwanenburg, B. (J. Org. Chem. **48** [1983] 4582/5).
[9] Huisgen, R.; Grashey, R.; Seidel, M.; Knupfer, H.; Schmidt, R. (Liebigs Ann. Chem. **658** [1962] 169/80, 174, 180).

2.6.25 Reactions of N-Sulfinylamines (Sulfinyl Imides) with Organometallic Compounds and Metal Complexes

2.6.25.1 Reactions with Organometallic Compounds of Group I, II, and III Metals

Lithium aryl compounds react vigorously with N-sulfinylaniline and related compounds in ethereal solutions. Hydrolysis of the primary adducts with aqueous NH_4Cl solution yields anilides of the aryl sulfinic acids [1]:

$$O{=}S{=}NC_6H_5 + LiAr \xrightarrow{\text{ether}} [\text{adduct}] \xrightarrow[H_2O]{NH_4Cl} C_6H_5NHS(O)Ar + LiOH$$

$$Ar = C_6H_5,\ C_{10}H_7\ (\text{naphthyl})$$

Condensation of lithium-metalated esters $LiC(R')R''{-}CO_2R'''$ (R′, R″ = H, CH_3; R‴ = C_2H_5, t-C_4H_9) and O=S=NR (R = c-C_6H_{11}, C_6H_5) in benzene at room temperature, followed by hydrolysis with aqueous NH_4Cl solution, yields $R{-}NH{-}SO{-}C(R')R''{-}CO_2R'''$ [2].

In the same way, N-sulfinylamines react exothermically with Grignard reagents in inert solvents, yielding the corresponding sulfinamides after hydrolysis in aqueous solution [3 to 11]:

$$O{=}S{=}NR + R'MgX \rightarrow [\text{adduct}] \xrightarrow{H_2O} RNHS(O)R' + Mg(OH)_2$$

R = alkyl, aryl [3 to 9, 11], $SO_2C_6H_5$ [9], $(CH_2)_nCl$ (n = 2, 3) [10]; R′ = alkyl, aryl [3 to 7], vinyl and substituted vinyl [11], allyl, benzyl [7], R″C≡C [8], C_nF_{2n+1} (n = 2, 4, 6, 8) [9]; X = Cl, Br

Chemical behavior of the primarily formed adducts indicates that the MgX group has added to the oxygen and not to the nitrogen atom [4]. In agreement therewith, steric hindrance at the N atom has no deciding influence on the Grignard addition. Thus $O{=}S{=}NC_4H_9$-t and C_6H_5MgBr yield only 7% t-C_4H_9–NH–S(O)C_6H_5; while 2,4,6-$(CH_3)_3C_6H_2N{=}S{=}O$ and C_6H_5MgBr yield 71% 2,4,6-$(CH_3)_3C_6H_2$–NH–S(O)C_6H_5 [6].

Bromo-zinc esters BrZnC(R′)R″–CO_2R‴ (R′, R″ = H or CH_3; R‴ = C_2H_5, t-C_4H_9) and O=S=NR (R = c-C_6H_{11}, C_6H_5) yield the corresponding β-aminosulfinyl compound R–NH–SO–C(R′)R″–CO_2R‴ [2].

$O{=}S{=}N{-}SO_2CF_3$ and $Hg(CH_3)_2$ in CH_2Cl_2 at room temperature give an N–S=N derivative in 58% yield [12] (i.e., $Hg(CH_3)_2$ catalyzes the conversion of N-sulfinylamine to the corresponding diimide [13]):

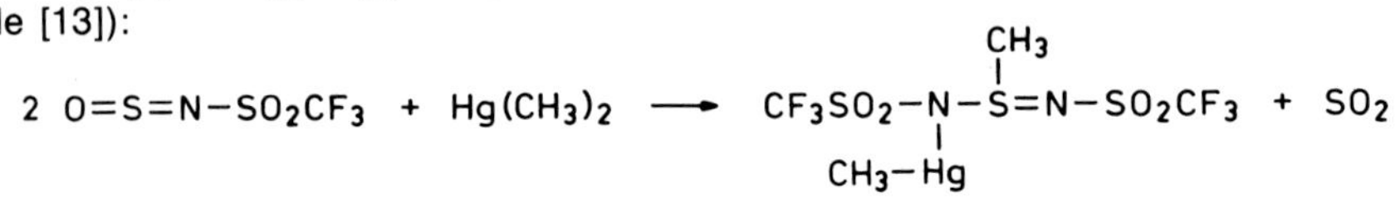

With $[Al(CH_3)_3]_2$ in pentane at 0°C 1:1 adducts, $[(CH_3)_2Al\{RNS(CH_3)O\}]_n$, form. The abrupt color change of the orange-red solution to white allows the use of the reaction for titration of $[Al(CH_3)_3]_2$. Temperature and concentration dependence of the ^{1}H NMR spectrum of $[(CH_3)_2Al\{RNS(CH_3)O\}]_n$ in $CDCl_3$ shows the existence of an equilibrium of two dimeric and one monomeric species. Possible structures of the complexes are proposed [14]:

$$2\ O{=}S{=}N{-}R\ +\ [Al(CH_3)_3]_2\ \longrightarrow$$

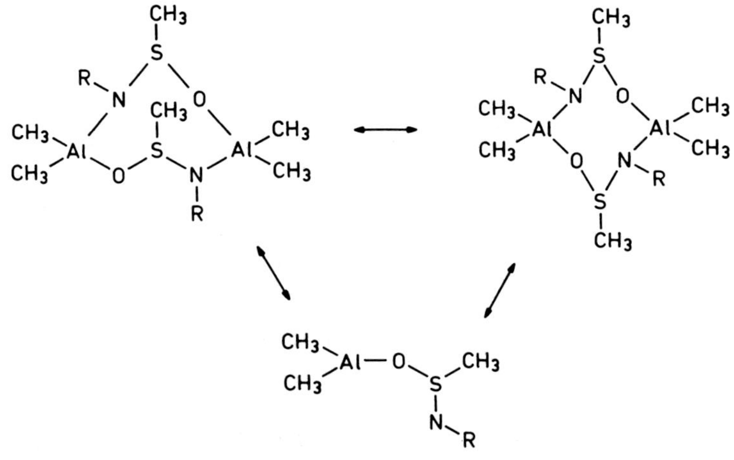

References:

[1] Schönberg, A.; Stephenson, A.; Kaltschmitt, H.; Petersen, E.; Schulten, H. (Ber. Deut. Chem. Ges. **66** [1933] 237/44).
[2] De Blic, A.; Cazaux, L.; Gorrichon-Guigon, L.; Perry, M. (Synthesis **1982** 281/3).
[3] Sonn, A.; Schmidt, E. (Ber. Deut. Chem. Ges. **57** [1924] 1355/6).
[4] Gilman, H.; Morris, H. L. (J. Am. Chem. Soc. **48** [1926] 2399/404).
[5] Klamann, D.; Sass, C.; Zelenka, M. (Chem. Ber. **92** [1959] 1910/3).
[6] Smith, W. T., Jr.; Thio, P. A.; Grasley, M. (J. Org. Chem. **27** [1962] 692/3).
[7] Bowlus, S. B.; Katzenellenbogen, J. A. (Syn. Commun. **4** [1974] 137/41).
[8] Selling, H. A.; Mak, H. J. (Syn. Commun. **6** [1976] 129/34).

[9] v. Werner, K. (J. Fluorine Chem. **8** [1976] 451/5).
[10] Smith, W. T., Jr.; Grasley, M. (Chim. Ther. **3** [1968] 266/9).

[11] Baudin, J.-B.; Julia, S. A. (Tetrahedron Letters **27** [1986] 837/40).
[12] Roesky, H. W.; Schönfelder, L. (Chem. Ber. **115** [1982] 1460/6).
[13] Bussas, R.; Kresze, G.; Münsterer, H.; Schwöbel, A. (Sulfur Rept. **2** [1983] 215/377, 322/3).
[14] Klerks, J. M.; van Vliet, R.; van Koten, G.; Vrieze, K. (J. Organometal. Chem. **214** [1981] 1/14, 3/4, 12).

2.6.25.2 Reactions with Organometallic Compounds of Group IV Metals

O=S=N–tos (tos = $SO_2C_6H_4CH_3$-4) reacts with $Sn(CH_3)_4$ with elimination of SO_2 (conditions are not reported) [1]:

$$2\,O{=}S{=}N{-}tos + Sn(CH_3)_4 \rightarrow (CH_3)_3Sn{-}N(tos){-}S(CH_3){=}N{-}tos + SO_2$$

With bis-trialkyl tin oxides N-sulfinylamines react immediately and exothermically to give 1:1 adducts [2, 3]:

$$O{=}S{=}NR + (R'_3Sn)_2O \rightarrow R'_3Sn{-}N(R){-}S(O)OSnR'_3$$

$$R = C_2H_5,\ n\text{-}C_3H_7,\ C_6H_5,\ 4\text{-}CH_3C_6H_4,\ 4\text{-}NO_2C_6H_4,\ 4\text{-}CH_3C_6H_4SO_2$$
$$R' = n\text{-}C_4H_9,\ C_6H_5$$

When R is an ethyl or propyl group, IR spectra and chemical reactions (with CCl_3CHO) indicate that this adduct is stable in the mixture (R = C_2H_5) or the CCl_4 solution (R = C_3H_7-n). The adduct (n-$C_4H_9)_3$Sn–N(C_3H_7-n)–S(O)OSn(C_4H_9-n$)_3$ could be isolated, and decomposed at 75°C/0.05 Torr, yielding ((n-$C_4H_9)_3Sn)_2O$ and O=S=NC_3H_7-n.

When R is an aryl group, the 1:1 adduct is not stable, but appears to be in equilibrium with the corresponding N,N′-sulfinyl-bis-stannylamine and sulfite [2]:

$$2\,[R'_3Sn{-}N(Ar){-}S(O)OSnR'_3] \rightleftharpoons R'_3Sn{-}N(Ar){-}S(O){-}N(Ar){-}SnR'_3 + R'_3Sn{-}OS(O)O{-}SnR'_3$$

$$Ar = C_6H_5,\ 4\text{-}CH_3C_6H_4,\ 4\text{-}NO_2C_6H_4$$
$$R' = n\text{-}C_4H_9,\ C_6H_5$$

When R = tos the reaction with ((n-$C_4H_9)_3Sn)_2O$ is very exothermic to give the 1:1 adduct as a viscous substance, which is almost glass at room temperature. The heat of formation of the amidosulfite is sometimes sufficient to induce partial loss of SO_2 [3]:

$$O{=}S{=}N{-}tos + ((n\text{-}C_4H_9)_3Sn)_2O \rightarrow (n\text{-}C_4H_9)_3Sn{-}N(tos){-}S(O)OSn(C_4H_9\text{-}n)_3$$
$$\downarrow \Delta$$
$$((n\text{-}C_4H_9)_3Sn)_2N{-}tos + SO_2$$

With (n-$C_4H_9)_3SnOCH_3$ N-sulfinyl-arenamines O=S=N–Ar react rapidly and exothermically at room temperature to yield a 1:1 adduct in equilibrium with the starting compounds. If an attempt is made to distil the product, the equilibrium is shifted wholly to the left. Addition of a further mole of N-sulfinylamine at room temperature causes a slow reaction during a period of 4 to 8 months to give a mixture of diimide and sulfite:

$$O{=}S{=}N{-}Ar + (n\text{-}C_4H_9)_3SnOCH_3 \rightleftharpoons (n\text{-}C_4H_9)_3Sn{-}N(Ar){-}S(O)OCH_3$$
$$\downarrow + O{=}S{=}N{-}Ar$$
$$Ar{-}N{=}S{=}N{-}Ar + (n\text{-}C_4H_9)_3Sn{-}O{-}S(O)OCH_3$$

$Ar = C_6H_5,\ 4\text{-}CH_3C_6H_4,\ 4\text{-}NO_2C_6H_4$

The same end products form when an equilibrium 1:1 mixture of (n-$C_4H_9)_3SnOCH_3$ and O=S=N–Ar is set aside at room temperature. With Ar = C_6H_5 and 4-$CH_3C_6H_4$ this reaction takes place slowly during a period of some months, whereas with Ar = 4-$NO_2C_6H_4$ the diimide separates within 4 d [2].

O=S=N–tos reacts with $(n\text{-}C_4H_9)_3SnOCH_3$ very exothermically, yielding $(n\text{-}C_4H_9)_3Sn$–N(tos)–$S(O)OCH_3$. With the less reactive $(n\text{-}C_4H_9)_3SnOC_6H_5$, the analogous reaction occurs only on warming. Thermal decomposition of the O=S=N–tos·$(n\text{-}C_4H_9)_3SnOCH_3$ adduct does not yield the same products as the corresponding adducts of O=S=N–Ar, but instead products similar to that of O=S=N–Ar·$((n\text{-}C_4H_9)_3Sn)_2O$ adducts [3]:

$$2(n\text{-}C_4H_9)_3Sn\text{–}N(tos)\text{–}S(O)OCH_3 \xrightarrow{\Delta} ((n\text{-}C_4H_9)_3Sn(tos)N)_2SO + (CH_3O)_2SO$$

When O=S=NC_6H_5 in light petroleum was added dropwise to $(CH_3)_3SnN(CH_3)_2$ an immediate exothermal reaction resulted with formation of a 1:1 adduct, characterized by elemental analysis and IR spectra [4]:

$$O{=}S{=}NC_6H_5 + (CH_3)_3SnN(CH_3)_2 \rightarrow (CH_3)_3Sn\text{–}N(C_6H_5)\text{–}S(O)N(CH_3)_2$$

Another way of reaction was assumed for O=S=N–C_4H_9-t and $(CH_3)_3SnN(CH_3)_2$ without giving experimental details. A 1:1 adduct, formed unusually at the S=O double bond, disproportionates to give t-C_4H_9–N=S=N–C_4H_9-t and $(CH_3)_3SnOS(=O)N(CH_3)_2$ [1]. The reaction of O=S=N–tos with $(CH_3)_3SnN(CH_3)_2$ is unclear. With certainty $O{=}S(N(CH_3)_2)_2$ and tos–$N(Sn(CH_3)_3)_2$ were identified as products [1].

$((n\text{-}C_4H_9)_3Sn)_2NC_6H_5$ and O=S=N–tos, when mixed at room temperature, undergo exothermic exchange of the electronegative groups. The reaction proceeds via a 1:1 adduct [3]:

$$O{=}S{=}N\text{–}tos + ((n\text{-}C_4H_9)_3Sn)_2NC_6H_5 \rightarrow [(n\text{-}C_4H_9)_3Sn\text{–}N(tos)\text{–}S(O)\text{–}N(C_6H_5)\text{–}Sn(C_4H_9\text{-}n)_3] \rightarrow$$
$$O{=}S{=}N\text{–}C_6H_5 + ((n\text{-}C_4H_9)_3Sn)_2N\text{–}tos$$

Similarly, $((CH_3)_3Sn)_3N$ and O=S=N–C_6F_5 in CH_2Cl_2 give the exchange products in a moderately exothermic reaction:

$$O{=}S{=}N\text{–}C_6F_5 + ((CH_3)_3Sn)_3N \rightarrow O{=}S{=}N\text{–}Sn(CH_3)_3 + ((CH_3)_3Sn)_2N\text{–}C_6F_5$$

The yields are 88 and 55% of theory. This reaction is said to be the best method for preparation of $((CH_3)_3Sn)_2N$–C_6F_5 [5].

When $(CH_3)_3SnN(Si(CH_3)_3)_2$ is refluxed with O=S=NR (R = aryl) in CH_3CN for 2 h (mole ratio 1:1) different products (dependent on R) form: RN=S=NR, RN=S=$NSi(CH_3)_3$, RN=S=$NSn(CH_3)_3$, $((CH_3)_3SiN{=})_2S$, $(CH_3)_3SiOSn(CH_3)_3$, and $((CH_3)_3Si)_2O$ [8].

The cyclic Sn–N compound $[\text{–}(CH_3)_2Sn\text{–}N(SO_2CF_3)\text{–}]_3$ and excess O=S=N–SO_2CF_3, when refluxed in CH_2Cl_2 for 15 h, yield an Sn–N–S ring in a complex reaction [6]:

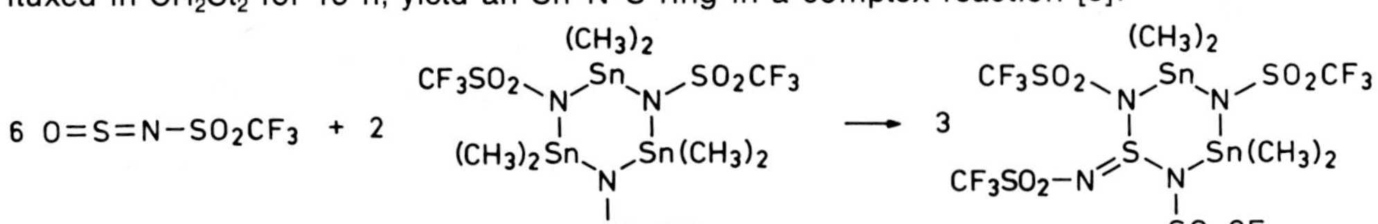

For the reaction of O=S=N–C≡N with $H_3CN(Sn(CH_3)_3)_2$, see p. 179.

When $(\eta^5\text{-}C_5H_5)_2TiCl_2$ is added to a stirred solution of equimolar amounts of $(O{=}S{=}N)_2S$ in liquid NH_3 at −78°C $(\eta^5\text{-}C_5H_5)_2Ti(N{=}S{=}O)_2$ is formed [7].

References:

[1] Hänssgen, D.; Roelle, W. (J. Organometal. Chem. **63** [1973] 269/78, 270/2).
[2] Davies, A. G.; Kennedy, J. D. (J. Chem. Soc. C **1968** 2630/40, 2633/5).

[3] Davies, A. G.; Kennedy, J. D. (J. Chem. Soc. C **1970** 1570/6).
[4] George, T. A.; Jones, K.; Lappert, M. F. (J. Chem. Soc. **1965** 2157/65, 2158, 2164).
[5] Roesky, H. W.; Ambrosius, K. (Israel J. Chem. **17** [1978] 132/6).
[6] Fuess, H.; Bats, J. W.; Diehl, M.; Schönfelder, L.; Roesky, H. W. (Chem. Ber. **114** [1981] 2369/74).
[7] Parkin, J. P.; Slawin, A. M. Z.; Williams, D. J.; Woollins, J. D. (Polyhedron **8** [1989] 835/9).
[8] Miller, A. O.; Zibarev, A. V.; Fedotov, M. A.; Furin, G. G. (Zh. Obshch. Khim. **59** [1989] 586/90; J. Gen. Chem. [USSR] **59** [1989] 518/21).

2.6.25.3 Reactions with Complexes and Organometallic Compounds of Group VI, VII, and VIII Metals

N-Sulfinylanilines O=S=N–Ar (Ar = C_6H_5, 4-$CH_3C_6H_4$, 4-ClC_6H_4, 3,5-$(CH_3)_2C_6H_3$, 2,4,6-$(CH_3)_3$-C_6H_2) can enter into transition metal complexes as unchanged ligands. Preparation and properties of such complexes are described on pp. 270/88. Metal complexes with N-sulfinyl-p-toluenesulfonamide O=S=N–$SO_2C_6H_4CH_3$-4 are described on pp. 268/9. In other cases the reaction with complexes and organometallic compounds causes metathesis with decay of the OSN group, oxidation, or ring closure including the SN group.

O=S=N–$C_6H_4CH_3$-4 converts molybdenum-dioxygen complexes into the imino derivatives [1]:

$$O{=}S{=}N{-}C_6H_4CH_3\text{-}4 + cis\text{-}Mo(chel)_2O_2 \rightarrow Mo(chel)_2(O)(NC_6H_4CH_3\text{-}4) + SO_2$$

chel = $S_2CN(C_2H_5)_2$ (stirring in benzene for 64 h at room temperature, 6.5% yield)
chel = $(CH_3C(O))_2CH$ (stirring of the suspended mixture for 7 d at room temperature, 83% yield)

The reaction of O=S=N–Ar with a rhenium-oxygen complex in boiling benzene or a benzene–n-hexane mixture is similar, producing good yields [2]:

$$O{=}S{=}NAr + ReCl_3(O)(P(C_6H_5)_3)_2 \rightarrow ReCl_3(NAr)(P(C_6H_5)_3)_2 + SO_2$$
$$Ar = C_6H_5,\ 4\text{-}CH_3C_6H_4,\ 4\text{-}CH_3OC_6H_4$$

With [Re(O)Cl_2(S–C(OC_2H_5)=NC_6H_5)(P$(C_6H_5)_3$)] excess O=S=N–$C_6H_4CH_3$-4 in non-anhydrous toluene at reflux temperature reacts (2 h) to give the mixed imido, amine rhenium (V) complex [$ReCl_3(NH_2C_6H_4CH_3$-4)(=$NC_6H_4CH_3$-4)(P$(C_6H_5)_3$)] in poor yield [11].

$(O{=}S{=}N)_2S$ in liquid NH_3 at −78°C reacts with equimolar amounts of $PtCl_2(PR_3)_2$ (PR_3 = $P(CH_3)_3$, $P(C_2H_5)_3$, $P(C_3H_7\text{-}n)_3$, $P(C_4H_9\text{-}n)_3$, $P(CH_3)_2C_6H_5$, $PCH_3(C_6H_5)_2$, $P(C_6H_5)_3$, and $(C_6H_5)_2PCH_2$-$CH_2P(C_6H_5)_2$) and $PdCl_2(C_6H_5)_2PCH_2CH_2P(C_6H_5)_2$ to give $Pt(N{=}S{=}O)_2(PR_3)_2$ in good yield (~60%) and minor amounts of $Pt(S_2N_2)(PR_3)_2$ (5%) and $Pd(N{=}S{=}O)_2(C_6H_5)_2PCH_2CH_2P(C_6H_5)_2$, respectively. Using a slight excess of $PtCl_2(PR_3)_2$ to $(O{=}S{=}N)_2S$ (mole ratio 1.2:1) ~25% of $Pt(S_2N_2)(PR_3)_2$ is observed by ^{31}P NMR spectroscopy. Further increasing of the mole ratio to 2:1 gives $Pt(N{=}S{=}O)_2(PR_3)_2$ (25%), $Pt(N{=}S{=}O)Cl(PR_3)_2$ (50%), and $Pt(S_2N_2)(PR_3)_2$ (25%) [15].

$Pt(P(C_6H_5)_3)_2O_2$ oxidizes O=S=N–C_6H_5 in benzene to afford the sulfato group in $Pt(P(C_6H_5)_3)_2SO_4$ [8]. Previously, in the same reaction, the formation of the 1:1 adduct $Pt(P(C_6H_5)_3)_2O_2$·O=S=N–C_6H_5 was erroneously claimed [12]. The reactions of $Pt(P(C_6H_5)_3)_2O_2$ and $Pd(P(C_6H_5)_3)_2O_2$ with N-sulfinylanilines O=S=N–C_6H_4R-4 (R = H, CH_3) proceed even at low temperature (~5°C, in benzene) leading to the sulfato derivatives $Pt(P(C_6H_5)_3)_2SO_4$ and $Pd(P(C_6H_5)_3)_2SO_4$ [13].

IrCl(CO)$(P(C_6H_5)_3)_2$ (Vaska's complex) reacts with O=S=N–C_6H_4R-4 (R = H, CH_3, NO_2) to give IrCl(CO)$(P(C_6H_5)_3)_2$·SO_2 [14]. In the absence of moisture and O_2 only O=S=$NC_6H_4NO_2$-4 reacted with *trans*-$IrX(CO)(P(C_6H_5)_3)_2$ (X = Cl, Br) to give the complexes $IrX(CO)(P(C_6H_5)_3)_2 \cdot O{=}S{=}N{-}C_6H_4NO_2$-4 (see p. 282) [8]. $IrCl(CO)(P(C_6H_5)_3)_2O_2$ (dioxygen adduct of Vaska's complex) does not react with $O{=}S{=}NC_6H_5$ when moisture and O_2 are excluded, but it reacts with $O{=}S{=}N{-}C_6H_4NO_2$-4 to give a mixture of unreacted dioxygen complex, $IrCl(CO)(P(C_6H_5)_3)_2SO_4$ and the $O{=}S{=}N{-}C_6H_4NO_2$-4 adduct of Vaska's complex (see p. 282) [8]. In a previous short publication, the reaction of $O{=}S{=}N{-}C_6H_5$, $O{=}S{=}N{-}C_6H_4CH_3$-4 as well as $O{=}S{=}N{-}C_6H_4NO_2$-4 with $IrCl(CO)(P(C_6H_5)_3)_2O_2$ and formation of $IrCl(CO)(P(C_6H_5)_3)_2SO_4$ was reported without giving experimental details [14]. Other authors assume that this adduct is accounted for by reaction with adventitious SO_2 rather than with N-sulfinylanilines [8].

O=S=N–tos reacts with osmium-alkylidene complexes $L_nOs{=}CHR$ ($L_nOs = Cl(NO)P(C_6H_5)_3Os$; R = H, $COOC_2H_5$) in toluene or THF, respectively, at room temperature to give the cycloadducts $L_n\overline{Os{-}CHR{-}S(O)}{=}N{-}tos$ in 80% yield. $O{=}S{=}N{-}C_4H_9$-t and $O{=}S{=}N{-}C_6H_4CH_3$-4 react with $L_nOs{=}CH_2$ to give $L_n\overline{Os{-}CH_2{-}S(O)}{=}O$ [16].

The behavior towards cyclopentadienyl-carbonyl-metal compounds has been studied in more detail. The N-sulfinylamines $O{=}S{=}N{-}C_6H_{11}$-c and $O{=}S{=}N{-}C_6H_5$ do not react with $(\eta^5\text{-}C_5H_5)Fe(CO)_2CH_3$ and $(\eta^5\text{-}C_5H_5)W(CO)_3CH_3$ at ambient temperatures in C_6D_6. In the presence of BF_3, $O{=}S{=}N{-}C_6H_{11}$-c reacts with both compounds, but no stable adducts could be isolated and fully characterized. By contrast, under comparable conditions N-sulfinyl-sulfonamides $O{=}S{=}N{-}SO_2R$ insert into the Fe–R′ σ bond in an electrophilic cleavage process to yield Fe–N-bonded products I ($CHCl_3$ or toluene–CH_2Cl_2 solution, room temperature). The products rearrange on storage at room temperature for 4 months or on heating the solid (50°C) or refluxing the CH_2Cl_2 solution, forming Fe–S bonds to give compounds II [3, 4]:

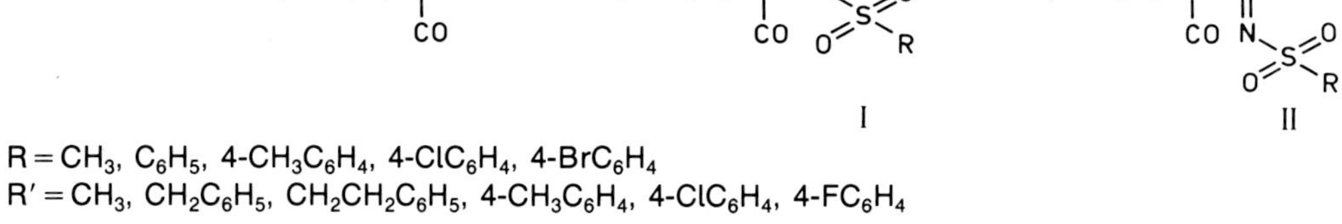

I II

R = CH_3, C_6H_5, 4-$CH_3C_6H_4$, 4-ClC_6H_4, 4-BrC_6H_4
R′ = CH_3, $CH_2C_6H_5$, $CH_2CH_2C_6H_5$, 4-$CH_3C_6H_4$, 4-ClC_6H_4, 4-FC_6H_4

Similarly, $O{=}S{=}N{-}SO_2CH_3$ reacts with the monocarbonyl-Fe complexes (in CH_2Cl_2, room temperature) to form unstable, not isolated N-bonded insertion products I, which transform rapidly into the S-bonded species II [9]:

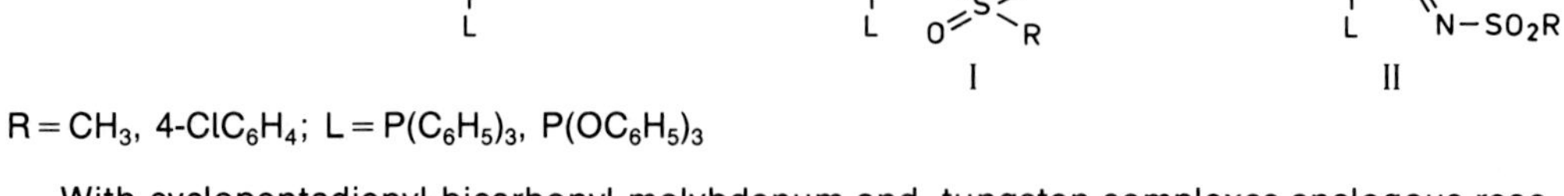

I II

R = CH_3, 4-ClC_6H_4; L = $P(C_6H_5)_3$, $P(OC_6H_5)_3$

With cyclopentadienyl-bicarbonyl-molybdenum and -tungsten complexes analogous reactions occur (in CH_2Cl_2, room temperature) [9]:

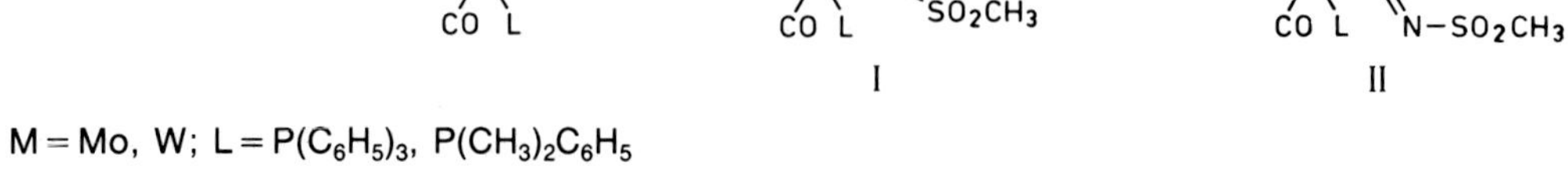

I II

M = Mo, W; L = $P(C_6H_5)_3$, $P(CH_3)_2C_6H_5$

With M = Mo the species I is unstable and transforms rapidly to species II. With M = W the species I was isolated in the crystalline state, and transforms to II only on reflux in CH_2Cl_2 (when $L = P(C_6H_5)_3$) or decomposes gradually at room temperature (when $L = P(CH_3)_2C_6H_5$). The complexes I and II have been characterized by chemical reactions, IR, ^1H, and ^{13}C NMR spectroscopy [3, 4, 9]; for one member of species II $(\eta^5\text{-}C_5H_5)Fe(CO)_2\{S[NS(O)_2CH_3](O)CH_2\text{-}C_6H_5\} \cdot \frac{1}{2}CH_2Cl_2$ the crystal and molecular structure was determined by X-ray diffraction [9].

Cyclopentadienyl-tricarbonyl-molybdenum or -tungsten complexes of the type $(\eta^5\text{-}C_5H_5)M(CO)_3R$ form N- and O-bonded species III and IV at −25°C in $CDCl_3$ and CH_2Cl_2 which decompose into intractable materials on warming [9]:

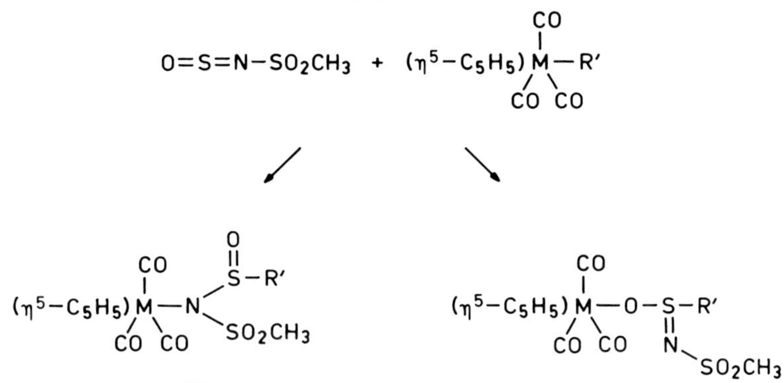

III IV

M = Mo, W; R' = CH_3, 4-$ClC_6H_4CH_2$

O=S=N–SO_2CH_3 and the cyclopentadienyl-nitrosyl-chromium complex $(\eta^5\text{-}C_5H_5)Cr(NO)_2CH_3$ form the N-bonded insertion product V (type I) (in CH_2Cl_2, room temperature), which did not convert to the S-bonded compound (type II).

O=S=N–SO_2CH_3 and the penta-carbonyl-manganum complex $Mn(CO)_5CH_3$ form the N-bonded species VI (in $CDCl_3$, −25°C; not isolated at room temperature). With $Mn(CO)_4(P(C_6H_5)_3)CH_3$ quite another product VII forms, presumably by disproportionation of O=S=N–SO_2CH_3 to SO_2 and $(CH_3SO_2N)_2S$, followed by insertion of the latter into the Mn–CH_3 bond [9]:

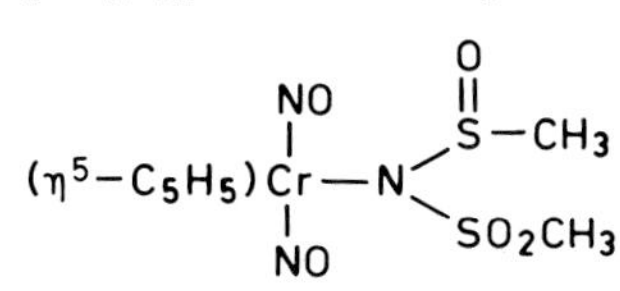

V

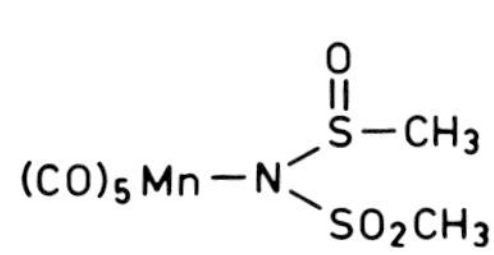

VI

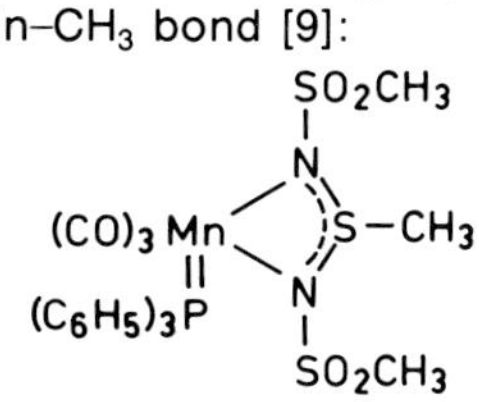

VII

The mechanism of the insertion reaction was studied by ^1H NMR spectroscopy using O=S=N–SO_2R (R = CH_3, C_6H_5) and $(\eta^5\text{-}C_5H_5)Fe(CO)_2CHDCHDC_6H_5$ as reactants. The insertion must proceed with a high degree of inversion of configuration at the α-carbon atom of R' (R' = $CHDCHDC_6H_5$). Therefore an electrophilic attack of the sulfur atom and subsequent recombination of the resulting ion pair is probable [4].

Quite other reaction products form from N-sulfinylsulfonamides and carbonyl- and cyclopentadienyl-carbonyl metal compounds containing unsaturated C=C or C≡C bonds. From allyl derivatives, stable solid [2 + 3] cycloadducts VIII are obtained in 63 to 67% yield (in CH_2Cl_2, room temperature) [5, 10]:

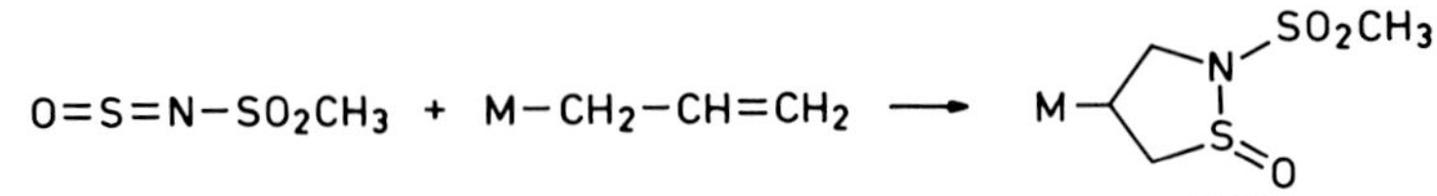

VIII

M = $(\eta^5\text{-}C_5H_5)Fe(CO)_2$, $(\eta^5\text{-}C_5H_5)Mo(CO)_3$, $Mn(CO)_5$

Substituted allyl complexes $(\eta^5\text{-}C_5H_5)Fe(CO)_2(CH_2C(CH_3){=}CH_2)$ and $(\eta^5\text{-}C_5H_5)Mo(CO)_3$-$(CH_2C(CH_3){=}CH_2)$ with N-sulfinyl-sulfonamides $O{=}S{=}N{-}SO_2R$ yield brown non-carbonyl decomposition materials [10].

$O{=}S{=}N{-}C_6H_5$ [7] and N-sulfinyl-sulfonamides and cyclopentadienyl-carbonyl-2-propinyl metal compounds form stable solid [2+3] cycloadducts IX in 50 to 77% yield [6, 7, 10]:

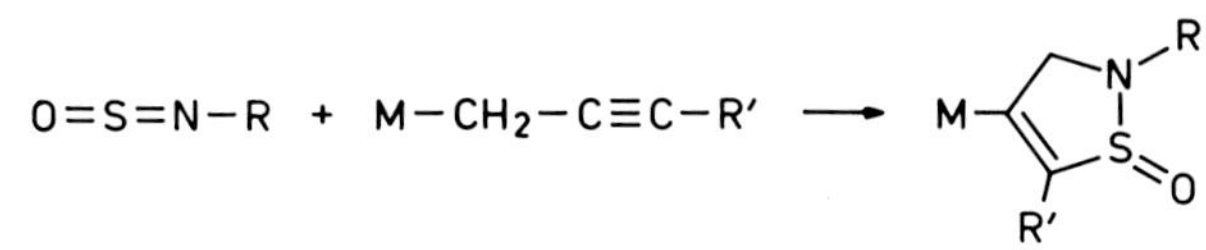

IX

M = $(\eta^5\text{-}C_5H_5)Fe(CO)_2$, $(\eta^5\text{-}C_5H_5)Mo(CO)_3$, $Mn(CO)_5$ (no elemental analysis)
R = C_6H_5, SO_2CH_3, $SO_2C_6H_5$, $SO_2C_6H_4CH_3$-4; R′ = CH_3, C_6H_5, $CH_2Fe(CO)_2(\eta^5\text{-}C_5H_5)$

The cycloadducts were identified by IR and NMR spectra [5, 6, 7, 10]. The reaction of $O{=}S{=}N{-}SO_2R$ with transition metal-η^1-allyl and -propinyl complexes probably proceeds via dipolar metal-η^2-alkene intermediates X or XI, respectively,

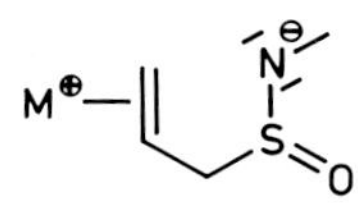

X

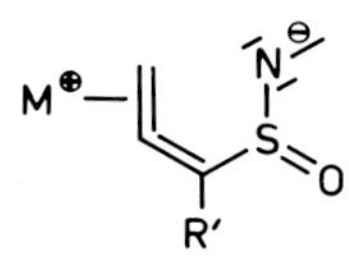

XI

which collapse to VIII or IX, respectively, by nucleophilic attack of the N atom at the η^2-alkene terminus [10].

The reaction of $O{=}S{=}N{-}SO_2CH_3$ with the cyclopentadienyl-carbonyl-cyclopropyl Fe-complex $(\eta^5\text{-}C_5H_5)Fe(CO)_2CH_2\overline{CHCH_2CH_2}$ (in CH_2Cl_2, room temperature) affords both the red insertion product XII and the yellow cycloaddition product XIII (obtained as a mixture of diastereomers) in 19 and 50% yields, respectively [10]:

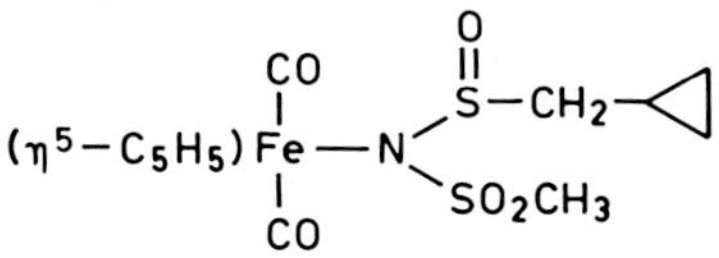

XII

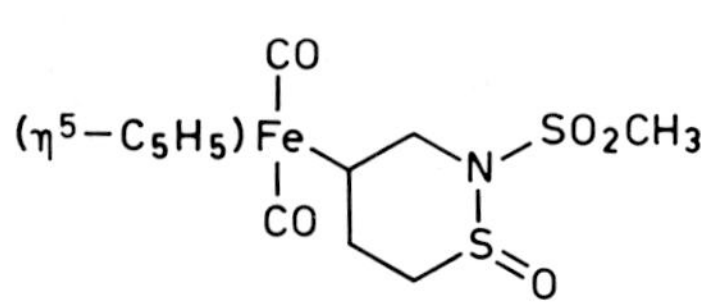

XIII

References:

[1] Cenini, S.; Pizzotti, M. (Inorg. Chim. Acta **42** [1980] 65/8).
[2] La Monica, G.; Cenini, S. (Inorg. Chim. Acta **29** [1978] 183/7).
[3] Severson, R. G.; Wojcicki, A. (J. Organometal. Chem. **149** [1978] C66/C70).
[4] Severson, R. G.; Leung, T. W.; Wojcicki, A. (Inorg. Chem. **19** [1980] 915/23).
[5] Leung, T. W.; Christoph, G. G.; Wojcicki, A. (Inorg. Chim. Acta **76** [1983] L281/L282).
[6] Robinson, P. W.; Wojcicki, A. (J. Chem. Soc. D **1970** 951/2).

[7] Chen, L. S.; Lichtenberg, D. W.; Robinson, P. W.; Yamamoto, Y.; Wojcicki, A. (Inorg. Chim. Acta **25** [1977] 165/72).
[8] Blake, D. M.; Reynolds, J. R. (J. Organometal. Chem. **113** [1976] 391/6).
[9] Leung, T. W.; Christoph, G. G.; Gallucci, J.; Wojcicki, A. (Organometallics **5** [1986] 366/74).
[10] Leung, T. W.; Christoph, G. G.; Gallucci, J.; Wojcicki, A. (Organometallics **5** [1986] 846/53).

[11] Rossi, R.; Marchi, A.; Duatti, A.; Magon, L.; Casellato, U.; Graziani, R. (Inorg. Chim. Acta **142** [1988] 23/5).
[12] Kolomnikov, I. S.; Koreshkov, Yu. D.; Lobeeva, T. S.; Vol'pin, M. E. (Izv. Akad. Nauk SSSR Ser. Khim. **1972** 1181/3; Bull. Acad. Sci. USSR Div. Chem. Sci. **1972** 1132/3).
[13] La Monica, G.; Pizzotti, M.; Cenini, S. (Gazz. Chim. Ital. **108** [1978] 611/3).
[14] De Luca, G.; Panattoni, C.; Toniolo, L. (Chem. Ind. [London] **1973** 742).
[15] Parkin, I. P.; Slawin, A. M. Z.; Williams, D. J.; Woollins, J. D. (Polyhedron **8** [1989] 835/9).
[16] Herberhold, M.; Hill, A. F. (J. Organometal. Chem. **354** [1988] 227/31).

2.7 Transition Metal Complexes of Sulfinyl Imides and the Sulfinyl Imide Ion (1−)

Transition metal complexes of $[O{=}S{=}N]^-$, O=S=NR (R = H, n-C_4H_9, c-C_6H_{11}, aryl), and $(O{=}S{=}N)_2S$ have been known only a few years and the investigation of the chemistry is still in progress. The ligands coordinate via N, S, or via π-N=S double bond and via O (in the case of $(O{=}S{=}N)_2S$ and probably in the case of O=S=NH). In the structurally investigated complexes of Ti(II), Pt(II), and Rh(I) the $[O{=}S{=}N]^-$ ligands are coordinated via N. The crystal structure analysis of the Zn and Ag complexes confirm that the $(O{=}S{=}N)_2S$ ligand is bonded via O, forming chains between the Zn and Ag atoms. In the few O=S=NH complexes which have been prepared, reliable structural information is not available. O=S=NH is probably coordinated via N or O. In the structurally studied O=S=NR (R = aryl) complexes η^1-S- and η^2-N=S coordination was observed. In Rh(I) and Ir(I) complexes both forms have been detected in solid and solution side by side. The O=S=NAr ligand has been found in the cis conformation. For reviews see [1 to 4].

References:

[1] Vrieze, K.; van Koten, G. (Recl. Trav. Chim. **99** [1980] 145/53).
[2] Chivers, T.; Edelmann, F. (Polyhedron **5** [1986] 1661/99, 1676/7).
[3] Kelly, P. F.; Woollins, J. D. (Polyhedron **5** [1986] 607/32, 620/1).
[4] Roesky, H. W. (Chem. Soc. Rev. **15** [1986] 309/34, 324).

2.7.1 Complexes of Sulfinyl Imide, O=S=NH

$[Mn(CO)_5(O{=}S{=}NH)]^+AsF_6^-$

The complex was prepared by reacting $[Mn(CO)_5(SO_2)]^+AsF_6^-$ and a twofold excess of O=S=NH in SO_2 at −80 to −50°C. After evaporation of solvent and excess O=S=NH, the analytically pure title compound remained in quantitative yield. It decomposes at 117°C. The proposed coordination of the O=S=NH ligand is via nitrogen.

IR (Nujol): 15 bands in the range 3222 to 398 cm^{-1}; ν (in cm^{-1}) = 3222 ν(NH); 2170, 2075, 2055 ν(CO); 1251, 1194 ν_{as}(N=S=O); Raman: ν (in cm^{-1}) = 1048 vs, 1012 sh ν_s(N=S=O) [1].

$[Re(CO)_5(O{=}S{=}NH)]^+AsF_6^-$

The title compound was synthesized analogously to $[Mn(CO)_5(O{=}S{=}NH)]^+AsF_6^-$ (see above) and also by condensing stoichiometric amounts of $(CH_3)_3SnOH$ on $[Re(CO)_5(NSF)]^+AsF_6^-$ in SO_2 at −196°C. After warming to −30°C, and stirring for 2 h, the complex was isolated in quantitative yield. The formation of the title compound by direct hydrolysis of $[Re(CO)_5$-$(NSF)]^+AsF_6^-$ with H_2O in SO_2 at −60°C is difficult to control. The compound decomposes at 102°C. The proposed coordination of the O=S=NH ligand is via nitrogen.

IR (Nujol): 16 bands ranging from 3180 to 397 cm^{-1}; ν (in cm^{-1}) = 3180 ν(NH); 2175, 2068, 2007 ν(CO); 1260, 1238 ν_{as}(N=S=O); Raman: 14 bands ranging from 2175 to 272 cm^{-1}; ν (in cm^{-1}) = 2172, 2166, 1996 ν(CO); 1024 ν_s(N=S=O) [1].

$[Ni(O{=}S{=}NH)_4]^{2+}(AsF_6^-)_2$

The complex was prepared analogously to $[Mn(CO)_5(O{=}S{=}NH)]^+AsF_6^-$ (see above) from $[Ni(SO_2)_2]^{2+}(AsF_6^-)_2$ and O=S=NH in SO_2 at ca. −50°C (quantitative yield). The product decomposes at 108°C. On the basis of the IR spectrum the presence of two different O=S=NH groups was deduced. Two of the O=S=NH groups act as bidentate ligands.

IR (Nujol): 9 bands in the range 3238 to 397 cm^{-1}; ν (in cm^{-1}) = 3238, 3203 ν(NH); 1252, 1237, 1169, 1145 ν_{as}(N=S=O) [1].

$Pt(O{=}S{=}NH)_2(P(C_6H_5)_3)_2$

The title compound was reportedly obtained by the reaction of about equimolar amounts of $S_4N_4H_4$ in acetone and $Pt(P(C_6H_5)_3)_4$ in benzene at ambient temperature (94% yield) [2, 5]. Other workers were not able to reproduce the synthesis. They obtained (in acetone in the presence of air or under nitrogen in dry solvents) $Pt(P(C_6H_5)_3)_2(S_2N_2)$ in high yield [3] or found that the reaction in toluene leads to $Pt(P(C_6H_5)_3)_2(S_2N_2)$ along with $(C_6H_5)_3PS$, $(C_6H_5)_3PO$, and $(C_6H_5)_3P$ [4]. In an earlier paper, the reaction product was identified to be $Pt(P(C_6H_5)_3)_2(S_2N_2H_2)$ based on analytical methods [5], but an X-ray structure analysis revealed the presence of the title compound [2,6]. Probably the compound was formed under the presence of catalytic quantities of impurities [4], or may have been the result of the degradation of the initial reaction product, $Pt(P(C_6H_5)_3)_2(S_2N_2)$ by atmospheric moisture and/or oxygen [3].

The complex, $Pt(O{=}S{=}NH)_2(P(C_6H_5)_3)_2 \cdot 0.5\,H_2O$ forms triclinic crystals, space group $P\bar{1}$-C_i^1 (No. 2) with two molecules in the unit cell of dimensions a = 10.082(3), b = 10.258(3), c = 17.812(6) Å, α = 86.05(3)°, β = 81.02(3)°, and γ = 72.45(3)°; V = 1734.4 Å^3; D_x = 1.637, D_m = 1.628 g/cm^3 [2, 6]. R = 0.048 for 4082 independent reflections. Atomic and thermal parameters are tabulated in the paper [6]. The title compound exhibits a square-planar structure with a cis arrangement of the ligands, see **Fig. 17** [2, 6]. The O=S=NH ligands are coordinated via O to Pt. However, the positions of the hydrogen atoms could not be determined. Two isomeric forms are possible: thionyl imide (O=S=NH) and thiazyl-S-hydroxide (HOS≡N) form. Their presence at the terminus of the O=S=N chain is assumed. But in the absence of a definite location for each H atom it is difficult to distinguish whether this is a compound of Pt (II) or Pt (0). The geometry of the two ligands is somewhat divergent. The bond lengths are remarkably short compared to those obtained by microwave spectroscopy for the free ligand. Selected bond lengths and bond angles are given in Table 53 [6], see also [2].

The yellow-colored compound melts at 220°C. UV absorption maxima have been observed at λ_{max} = 470, 365, and 325 nm and a shoulder at 405 nm, due to transitions $^1A_{1g} \rightarrow {}^1A_{2g}$, $^1A_{1g} \rightarrow {}^1B_{1g}$, and $^1A_{1g} \rightarrow {}^1E_g$. The IR spectrum shows 35 absorption bands in the range 3150 to

280 cm^{-1}; ν (in cm^{-1}) = 3150 ν(NH); 1660; 1030 ν(SO); 925 ν(NS); 545, 525 ν(PtO); 445, 422 ν(PtP). The mass spectrum shows no parent peak but fragment peaks at m/e = 63 $OSNH^+$, 48 SO^+, and 47 SNH^+ [2].

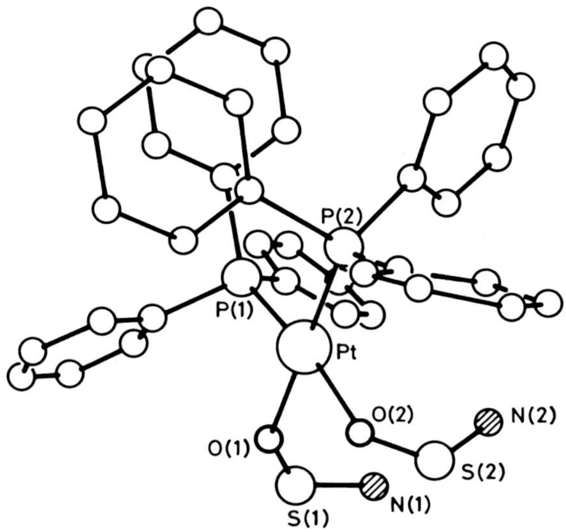

Fig. 17. Molecular Structure of $Pt(O{=}S{=}NH)_2(P(C_6H_5)_3)_2$ [2, 6].

Table 53
Bond Lengths and Bond Angles of $Pt(O{=}S{=}NH)_2(P(C_6H_5)_3)_2$ [6]; see also [2].

bond length	in Å	bond angle	in °
S(1)–N(1)	1.48(2)	O(1)–S(1)–N(1)	121.9(12)
S(2)–N(2)	1.45(2)	O(2)–S(2)–N(2)	122.0(9)
S(1)–O(1)	1.26(1)	Pt–O(2)–S(2)	129.5(9)
S(2)–O(2)	1.34(2)	Pt–O(1)–S(1)	138.1(10)
Pt–O(1)	2.173(11)	O(1)–Pt–O(2)	86.0(4)
Pt–O(2)	2.088(12)	P(2)–Pt–O(2)	90.8(4)
Pt–P(1)	2.288(4)	P(2)–Pt–O(1)	175.2(3)
Pt–P(2)	2.264(4)	P(1)–Pt–O(2)	171.1(4)
		P(1)–Pt–O(1)	85.1(3)
		P(1)–Pt–P(2)	98.0(1)

$Pt(O{=}S{=}NH)_2(S_2N_2H_2)$

The complex was prepared by adding an aqueous solution of K_2PtCl_4 to a suspension of $S_4N_4H_4$ in acetone (mole ratio 1:1.1). The product precipitated within 5 min and after cooling the mixture; yield 95% (based on K_2PtCl_4). The red-brown solid melts in the range 133 to 141°C with decomposition.

The ^1H NMR spectrum exhibits two singlets at δ = 3.45 and 0.75 ppm (1:1), which have been tentatively assigned to the disulfur diimide and thionyl imide protons, respectively. The UV spectrum in DMF shows three bands at λ_{max} (in nm) (ε in $L \cdot mol^{-1} \cdot cm^{-1}$) 450 (939.0), 310 (3883.5), and 300 (3982.9), possibly due to electronic transitions $^1A_{1g} \rightarrow {}^1A_{2g}$, $^1A_{1g} \rightarrow {}^1B_{2g}$, and $^1A_{1g} \rightarrow {}^1E_g$ (typical for square-planar d^8 configuration). The IR spectrum (KBr) shows 13 bands in the range 3300 to 360 cm^{-1}; ν (in cm^{-1}) = 3300 ν(NH); 1040 ν(SO); 470 ν(PtN); 420 ν(PtO); 360 ν(PtS). On the basis of spectroscopic data, the geometrical structure below was proposed [7].

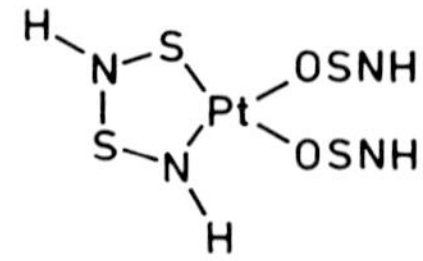

References:

[1] Hartmann, G.; Hoppenheit, R.; Mews, R. (Inorg. Chim. Acta **76** [1983] L201/L202).
[2] Bhattacharyya, A. A.; Turner, A. G.; Holt, E. M.; Alcock, N. W. (Inorg. Chim. Acta **44** [1980] L185/L186).
[3] Chivers, T.; Edelmann, F.; Behrens, U.; Drews, R. (Inorg. Chim. Acta **116** [1986] 145/51).
[4] Jones, R.; Kelly, P. F.; Williams, D. J.; Woollins, J. D. (Polyhedron **5** [1986] 607/32).
[5] Bhattacharyya, A. A.; McLean, J. A., Jr.; Turner, A. G. (Inorg. Chim. Acta **34** [1979] L199/L200).
[6] Holt, E. M.; Alcock, N. W.; Bhattacharyya, A. A.; Turner, A. G. (Inorg. Chim. Acta **47** [1981] 255/9).
[7] Bhattacharyya, A. A.; Turner, A. G. (Inorg. Chim. Acta **53** [1981] L89/L90).

2.7.2 Complexes of the Sulfinyl Imide Ion (1−), $[O{=}S{=}N]^-$

$(\eta^5\text{-}C_5H_4R)_2Ti(N{=}S{=}O)_2$, R = H, CH_3

$(\eta^5\text{-}C_5H_5)_2Ti(N{=}S{=}O)_2$ was synthesized and characterized at the same time by two different working groups. It was prepared in 80% yield by reaction of $(\eta^5\text{-}C_5H_5)_2TiCl_2$ with two equivalents of K^+ $[O{=}S{=}N]^-$ in CH_3CN at 0°C [1] or at −40°C and then at room temperature overnight [11]. Treatment of $(\eta^5\text{-}C_5H_5)_2TiCl_2$ or $(\eta^5\text{-}C_5H_5)_2Ti(CO)_2$ with $(O{=}S{=}N)_2Hg$ also produced the bis(thionyl imide) complex [1, 13]. $(\eta^5\text{-}C_5H_5)_2Ti(N{=}S{=}O)_2$ was obtained in 64% yield by passing a solution of $(\eta^5\text{-}C_5H_5)_2Ti(N{=}S{=}NC_4H_9\text{-}t)_2$ (obtained from $(\eta^5\text{-}C_5H_5)_2TiCl_2$ and $K^+t\text{-}C_4H_9N{=}S{=}N^-$) in n-pentane through a column containing silica [2]. The title compound was prepared in 75% yield by adding $(\eta^5\text{-}C_5H_5)_2TiCl_2$ to a stirred solution of equimolar amounts of $(O{=}S{=}N)_2S$ in liquid NH_3 at −78°C. Then the reaction mixture was stirred for 30 min. After removal of NH_3 the compound is extracted into CH_2Cl_2 and filtered through a celite/glass wool plug. Crystals were grown by slow diffusion of n-hexane into the CH_2Cl_2 solution at room temperature [10].

$(\eta^5\text{-}C_5H_5)_2Ti(N{=}S{=}O)_2$ crystallizes in the orthorhombic space group Fdd2–C_{2v}^{19} (No. 43) with 8 molecules in the unit cell [1, 2]; a = 29.170(6), b = 14.139(2), and c = 6.188(2) Å. R = 0.035 for 1540 unique reflections [1]; a = 29.219(5), b = 14.153(3), and c = 6.189(1) Å, V = 2559.4 $Å^3$, D_x = 1.568 g/cm^3, R = 0.0334, R_w = 0.0450 for 620 unique reflections. Positional and equivalent isotropic displacement parameters are given in the paper [2]. The crystal structure confirms that the O=S=N units are bonded over nitrogen (see **Fig. 18**). The atoms of the $Ti(N{=}S{=}O)_2$ moiety are nearly coplanar; deviations from the N–Ti–N′ plane: S = 0.03, O = 0.02 Å. The N=S=O angle and the N=S distance are similar to the isoelectronic SO_2 [1, 2]. The molecular structure with selected bond lengths and bond angles (from [1]) is shown in Fig. 18; nearly identical bond lengths and bond angles are given in [2]. No isomerization was found in the NMR spectra. Therefore the configuration of the N=S=O ligands in solution is likely to be Z, Z, as found for the solid state [2].

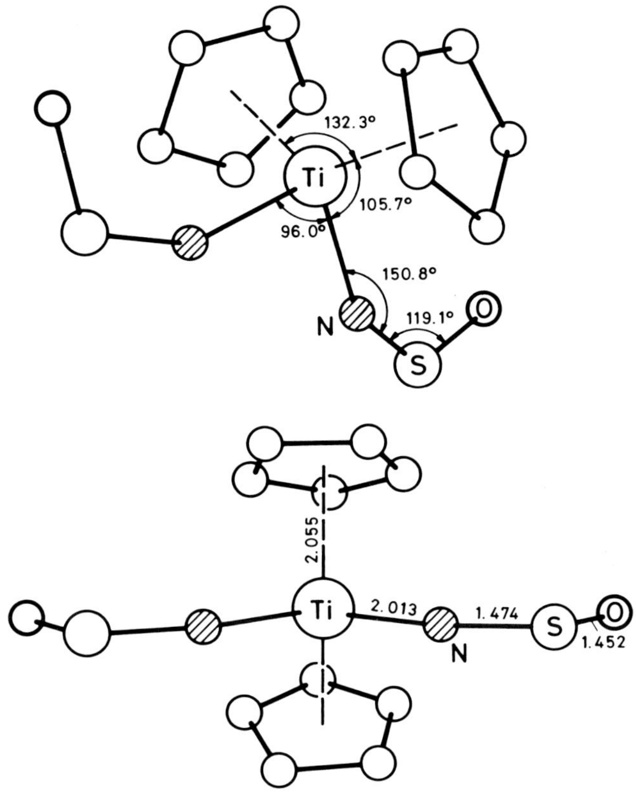

Fig. 18. Molecular structure of $(\eta^5\text{-}C_5H_5)_2Ti(N{=}S{=}O)_2$; bond lengths in Å, bond angles in ° [1].

The yellow crystals decompose upon heating over a wide temperature range without melting [2]; decomposition >230°C [11]. ^{1}H NMR ($CDCl_3$/TMS): δ = 6.38 [1] or 6.37 ppm (s, C_5H_5) [2, 10]; $^{13}C\{^1H\}$ NMR ($CDCl_3$/TMS): δ = 114.4 ppm (C_5H_5) [2]; IR (KBr): ν (in cm^{-1}) = 3110; 1235 vs [1, 10], 1232 [2] ν_{as}(N=S=O); 1075 s [1, 10], 1076 [2] ν_s(N=S=O); 830; 525 m δ(N=S=O) [1, 10]; 430; 400 [1]. MS: m/e = 302 M^+ [2, 10]; 246 (M^+–NSO) [10].

The complex was described to be stable towards air and atmospheric moisture by [11], whereas air-sensitivity was observed by [2]; solutions of the complex exposed to air decompose rapidly [11]. The reaction of $(\eta^5\text{-}C_5H_5)_2Ti(N{=}S{=}O)_2$ with $LiN(Si(CH_3)_3)_2$ (mole ratio 1:2) yields yellow crystals of $(\eta^5\text{-}C_5H_5)_2Ti(N{=}S{=}NSi(CH_3)_3)_2$ and $LiOSi(CH_3)_3$ [1]; 65% after recrystallization [11].

$(\eta^5\text{-}C_5H_4CH_3)_2Ti(N{=}S{=}O)_2$ was formed in a 12% yield by passing a THF solution containing $(\eta^5\text{-}C_5H_4CH_3)_2TiCl_2$ and $K^+t\text{-}C_4H_9N{=}S{=}N^-$ (mole ratio 1:2) that had been refluxed for 3 h through a column containing silica.

The yellow, needle-like, air-sensitive crystals melt at 127 to 130°C with decomposition. ^{1}H NMR ($CDCl_3$/TMS): δ (in ppm) = 6.18 (m, C_5H_4), 2.16 (s, CH_3); $^{13}C\{^1H\}$ NMR ($CDCl_3$/TMS): δ (in ppm) = 129.9, 117.6, 109.6 (C_5H_4), 15.9 (CH_3). No isomerization was found in the NMR spectra. Therefore the configuration of the N=S=O ligand in solution is likely to be Z,Z, as found for $(\eta^5\text{-}C_5H_5)_2Ti(N{=}S{=}O)_2$.

IR (KBr): ν (in cm^{-1}) = 1241 ν_{as}(N=S=O); 1069 ν_s(N=S=O).

MS: m/e = 330 M^+ [2].

References on p. 263

(η^5-C_5H_5)Ti(N=S=O)$_3$

The complex was synthesized by adding a solution of O=S=NSn(CH_3)$_3$ in toluene to a solution of (η^5-C_5H_5)$TiCl_3$ in toluene at 0°C (mole ratio ~3:1). The complex was isolated as a pale yellow solid in 80% yield.

The compound decomposes at >55°C. IR (Nujol): ν (in cm^{-1}) = 3075 m, 1260, 1220 vs, 1090 m, 1035 m, 845 s, 740 m, 620 m, and 540 w.

The solid is insoluble in noncoordinating solvents; slightly soluble in CH_3CN. Such a solution decomposed within a few minutes. It was not possible to obtain NMR data or to recrystallize the compound [11].

(η^5-C_5H_5)(η^5-$C_5(CH_3)_5$)Zr(N=S=O)$_2$

The complex was prepared by adding slowly with stirring CH_3CN to a mixture of (η^5-C_5H_5)-(η^5-$C_5(CH_3)_5$)$ZrCl_2$ and K^+[O=S=N]$^-$ (mole ratio 1:2.1) at −40°C. After warming up to room temperature overnight the title compound was isolated as bright yellow solid in 56% yield (after recrystallization).

The compound crystallizes in the monoclinic space group C2/m–C^3_{2h} (No. 12) with a = 15.185(8), b = 14.768(7), c = 9.312(5) Å, β = 98.58(2)°; V = 2065 Å^3, Z = 4, D_x = 1.483 g/cm^3, R = 0.036, R_w = 0.06 for 1348 reflections. Atomic coordinates were listed in the paper. The structure of (η^5-C_5H_5)(η^5-$C_5(CH_3)_5$)Zr(N=S=O)$_2$ is comparable to that of (η^5-C_5H_5)$_2$Ti(N=S=O)$_2$ with the central metal atom in a pseudo-tetrahedral environment. The N=S=O groups are coordinated via the N atoms, both reveal cis orientation. The Zr(N=S=O)$_2$ unit is nearly planar. Selected bond lengths and bond angles are shown in **Fig. 19**.

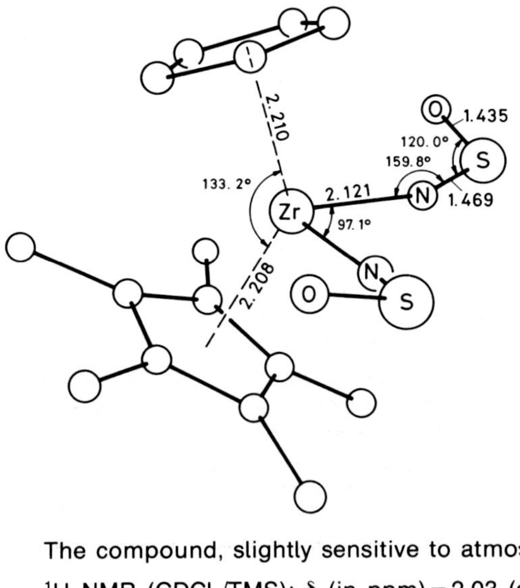

Fig. 19. Molecular structure of (η^5-C_5H_5)-(η^5-$C_5(CH_3)_5$)Zr(N=S=O)$_2$; bond lengths in Å (±0.003 to 0.004 Å), bond angles in ° (±0.2°) [11].

The compound, slightly sensitive to atmospheric moisture, decomposes at >215°C.

^{1}H NMR ($CDCl_3$/TMS): δ (in ppm) = 2.03 (s, η-$C_5(CH_3)_5$), 6.22 (s, η-C_5H_5). IR (Nujol): ν (in cm^{-1}) = 3100 w, 1280 s, 1210 vs, 1090 m, 1080 m, 1030 m, 1010 m, 840 m, 825 s, 730 m, 515 m, 380 s. EIMS (70 eV): m/e = 414 (M^+) [11].

(η^5-$C_5(CH_3)_5$)$_2$Zr(N=S=O)$_2$

The complex was prepared analogously to (η^5-C_5H_5)(η^5-$C_5(CH_3)_5$)Zr(N=S=O)$_2$ in 43% yield, see above.

The bright yellow solid, slightly sensitive to atmospheric moisture, decomposes at >220°C.

^{1}H NMR ($CDCl_3$/TMS): δ (in ppm) = 1.95 (η^5-$C_5(CH_3)_5$). IR (Nujol): ν (in cm^{-1}) = 1290 sh, 1260 vs, 1090 s, 1035 m, 735 m, 525 w. EIMS (70 eV): m/e = 484 (M^+) [11].

(η^5-C_5H_5)(η^5-$C_5(CH_3)_5$)Hf(N=S=O)$_2$

The complex was prepared analogously to (η^5-C_5H_5)(η^5-$C_5(CH_3)_5$)Zr(N=S=O)$_2$ in 40% yield, see above.

The pale yellow solid, slightly sensitive to atmospheric moisture, decomposes at >155°C.

^{1}H NMR ($CDCl_3$/TMS): δ (in ppm) = 2.06 (s, η^5-$C_5(CH_3)_5$), 6.14 (s, η^5-C_5H_5). IR (Nujol): ν (in cm^{-1}) = 3100 w, 1290 sh, 1260 s, 1090 s, 1085 sh, 1020 m, 835 s, 730 m, 520 m. EIMS (70 eV): m/e = 504 (M^+) [11].

OsX_2(N=S=O)$_2$(P(C_6H_5)$_3$)$_2$, X = Cl, Br

$OsCl_2$(N=S=O)$_2$(P(C_6H_5)$_3$)$_2$ was prepared by the reaction of Os(NO)Cl_3(P(C_6H_5)$_3$)$_2$ with sulfur (mole ratio 4:1) in benzene (4 h refluxing). The green solid was recrystallized from benzene–hexane (1:3), yield 30%.

It is stable under nitrogen atmosphere; m.p. 210°C. The magnetic moment μ, determined by Gouy's method (t = 25°C), is 1.70 B.M. IR (KBr): ν (in cm^{-1}) = 1237 ν_{as}(N=S=O); 1052 ν_s(N=S=O); 575 δ(N=S=O); 325 ν(Os–Cl). UV: λ_{max} (in nm) = 505, 465, 390, 275 [3].

$OsBr_2$(N=S=O)$_2$(P(C_6H_5)$_3$)$_2$ was synthesized analogously to the corresponding complex above. The green solid was obtained in 22% yield.

It is stable under nitrogen atmosphere and melts at 215°C. The magnetic moment μ is 1.72 B.M. (Gouy's method, t = 25°C). IR (KBr): ν (in cm^{-1}) = 1235 ν_{as}(N=S=O); 1055 ν_s(N=S=O); 575 δ(N=S=O) [3].

(RhCl(N=S=O)$_2$(P(C_6H_5)$_3$)(H_2O))$_2$

The dimeric complex was synthesized by the reaction of RhCl(P(C_6H_5)$_3$)$_3$ or Rh(CO)Cl(P(C_6H_5)$_3$)$_2$ with "NSOH" in $CHCl_3$ at room temperature under argon atmosphere for 2 h. The complex was recrystallized from CH_2Cl_2–hexane (1:3), yielding a brown crystalline solid which melts at 165 to 166°C.

It is stable in air, soluble in $CHCl_3$, CH_2Cl_2, CH_3NO_2, C_6H_6, and insoluble in hexane and ether. The conductivity in $C_6H_5NO_2$ is low, ~4 $\Omega^{-1}\cdot cm^2\cdot mol^{-1}$. The complex is diamagnetic. The proposed structure is based only on the IR spectrum: ν (in cm^{-1}) = 1118, 1060, 970, 630 (N=S=O); 3400, 1630, 930 (H_2O); 260 ν(Rh–Cl, bridging Cl); additionally the characteristic bands of P(C_6H_5)$_3$ were observed. The dimeric nature of the complex was confirmed by measurement of the molecular weight in benzene. It reacts with excess of (C_6H_5)$_3$P to give the monomeric complex, see below [4].

RhCl(N=S=O)$_2$(P(C_6H_5)$_3$)(H_2O)

The complex was obtained by refluxing the dimeric complex (RhCl(N=S=O)$_2$(P(C_6H_5)$_3$)(H_2O))$_2$ with excess of (C_6H_5)$_3$P in CH_2Cl_2.

The red-brown complex shows an IR absorption at 335 cm^{-1} and no bands in the range 230 to 280 cm^{-1}, which was interpreted as the absence of a Cl bridge. The monomeric structure was confirmed by the measurement of the molecular weight in benzene [4].

References on p. 263

***trans*-Rh(CO)(N=S=O)(P(C_6H_5)$_3$)$_2$**

The complex was obtained in 56% yield by adding a solution of $CsNSO_2$ in CH_3CN to an equimolar amount of RhH(CO)(P(C_6H_5)$_3$)$_3$ in CH_3CN at room temperature. Subsequently the reaction mixture was refluxed for 4 h, filtered and the filtrate concentrated in vacuum. Yellow crystals formed on cooling, m.p. 190°C with decomposition.

The crystals are triclinic, space group $P\bar{1}-C_i^1$ (No. 2) with a = 9.248(2), b = 9.502(2), c = 10.567(2) Å, α = 71.52(2)°, β = 89.71(2)°, and γ = 69.80(2)°; V = 820.8 Å^3; Z = 1; D_m = 1.43(1), D_x = 1.427 g/cm^3; R_w = 0.047 and R = 0.045 for 3377 observed reflections. The crystal structure contains an arrangement of mononuclear square-planar molecules with a trans configuration of the P(C_6H_5)$_3$ ligands. A view of the complex molecule is presented in **Fig. 20**; selected bond distances and bond angles are given in Table 54; atomic coordinates are listed in the paper. The N=S=O ligand shows bond lengths close to double bonds and the N–S–O angle is also consistent with a double-bond valence structure. The arrangement of the molecules in the crystal is disordered, with a superposition of two half-occupied CO and N=S=O ligand sites in the electron-density map, simulating the crystallographic inversion center at the Rh site.

IR (Nujol): 18 absorptions between 3060 and 510 cm^{-1}; ν (in cm^{-1}) = 1985 s ν(CO); 1240 s ν_{as}(N=S=O); 1050 s ν_s(N=S=O); 580 s δ(N=S=O) [5].

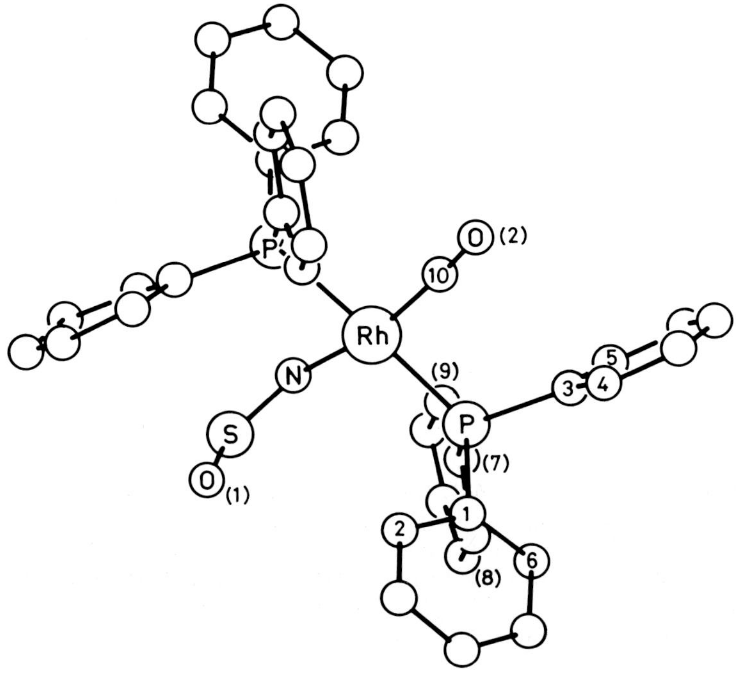

Fig 20. Molecular structure of *trans*-Rh(CO)(N=S=O)(P(C_6H_5)$_3$)$_2$ [5].

Table 54
Selected Bond Lengths and Bond Angles of *trans*-Rh(CO)(N=S=O)(P(C_6H_5)$_3$)$_2$ [5].

bond length	in Å	bond angle	in °	bond angle	in °
Rh–P	2.330(1)	P–Rh–N	93.6(2)	Rh–P–C(3)	116.5(1)
Rh–N	1.989(9)	P'–Rh–N	86.4(2)	C(1)–P–C(7)	105.8(2)
Rh–C(10)	1.843(6)	P–Rh–C(10)	86.3(2)	C(1)–P–C(3)	103.7(2)

Table 54 (continued)

bond length	in Å	bond angle	in °	bond angle	in °
N–S	1.480(12)	P′–Rh–C(10)	93.7(2)	C(7)–P–C(3)	104.0(2)
S–O(1)	1.509(10)	N–Rh–C(10)	164.4(3)	P–C(1)–C(2)	118.3(3)
C(10)–O(2)	1.137(10)	Rh–N–S	135.1(8)	P–C(1)–C(6)	122.8(3)
P–C(1)	1.818(4)	N–S–O(1)	122.5(5)	P–C(7)–C(8)	122.9(3)
P–C(7)	1.823(4)	Rh–C(10)–O(2)	170.8(7)	P–C(7)–C(9)	118.2(3)
P–C(3)	1.823(4)	Rh–P–C(1)	112.3(1)	P–C(3)–C(4)	118.8(3)
		Rh–P–C(7)	113.5(1)	P–C(3)–C(5)	122.1(3)

$IrCl_2(CO)(N{=}S{=}O)(P(C_6H_5)_3)_2$

The complex was prepared by the reaction of $S_3N_3Cl_3$ with $IrCl(CO)(P(C_6H_5)_3)_2$ in THF in the presence of oxygen (40% yield), and also by the reaction of $IrCl(CO)(O_2)(P(C_6H_5)_3)_2$ with $S_3N_3Cl_3$ in THF. The compound melts at 115°C.

IR absorptions at 1160, 1060, and 565 cm^{-1} were assigned to ν_{as}(N=S=O), ν_s(N=S=O), and δ(N=S=O), respectively. Two further IR bands at 2060 ν(CO) and at ~330 cm^{-1} ν(IrCl), indicating trans-Cl atoms, are reported. The complex is diamagnetic and a nonelectrolyte [6].

$Pd(N{=}S{=}O)_2((C_6H_5)_2PCH_2CH_2P(C_6H_5)_2)$

The title compound formed in 32% yield by adding $PdCl_2((C_6H_5)_2PCH_2CH_2P(C_6H_5)_2)$ to an equimolar amount of $(O{=}S{=}N)_2S$ in liquid NH_3 at −78°C and stirring the reaction mixture for 30 min. After evaporation of NH_3, the compound was extracted with CH_2Cl_2, filtered through a celite/glass wool plug and layered with petroleum ether.

^1H NMR (solvent?/TMS): δ (in ppm) = 2.4 (m, 4H), 7.3 (m, 12H), and 7.6 (m, 8H).

^{31}P NMR (solvent?/external 85% H_3PO_4): δ = 53.4 ppm [10].

$PtCl(N{=}S{=}O)(PR_3)_2$, $PR_3 = P(C_2H_5)_3$, $P(C_4H_9\text{-}n)_3$, **and $PtCl(N{=}S{=}O)((C_6H_5)_2PCH_2CH_2P(C_6H_5)_2)$**

The title compounds (50%) were observed by ^{31}P NMR spectroscopy along with $Pt(N{=}S{=}O)_2(PR_3)_2$ (25%) and $Pt(S_2N_2)(PR_3)_2$ (25%) in the reaction of $PtCl_2(PR_3)_2$ with $(O{=}S{=}N)_2S$ (mole ratio 2:1) in liquid NH_3.

The ^{31}P NMR spectra, listed in the table below, are of the AX type with $^2J(^{31}P,^{31}P)$ couplings of 20 Hz. The average $^{195}Pt,^{31}P$ coupling constants of 3650 and 2850 Hz are typical of phosphorus trans to chlorine and nitrogen in Pt(II) complexes.

^{31}P NMR (solvent?/external 85% H_3PO_4, δ in ppm, J in Hz) [10]:

complex	δ_A	δ_X	$^1J_A(^{195}Pt,^{31}P)$	$^1J_X(^{195}Pt,^{31}P)$	$^2J_{AX}(^{31}P,^{31}P)$
$PtCl(N{=}S{=}O)(P(C_2H_5)_3)_2$	7.4	0.2	3640	2734	20
$PtCl(N{=}S{=}O)(P(C_4H_9\text{-}n)_3)_2$	−0.6	−6.3	3695	2944	20
$PtCl(N{=}S{=}O)((C_6H_5)_2PCH_2)_2$	39.8	33.8	3700	3003	8

$Pt(N{=}S{=}O)_2(PR_3)_2$, $PR_3 = P(CH_3)_3$, $P(C_2H_5)_3$, $P(C_3H_7\text{-}n)_3$, $P(C_4H_9\text{-}n)_3$, $P(CH_3)_2C_6H_5$, $PCH_3(C_6H_5)_2$, $P(C_6H_5)_3$, **and $Pt(N{=}S{=}O)_2((C_6H_5)_2PCH_2CH_2P(C_6H_5)_2)$**

Preparation. Properties. *cis*-$Pt(N{=}S{=}O)_2(PR_3)_2$ complexes (see p. 260) with the ligands $P(CH_3)_3$, $P(C_2H_5)_3$, $P(C_3H_7\text{-}n)_3$, $P(C_4H_9\text{-}n)_3$, $P(CH_3)_2C_6H_5$, $PCH_3(C_6H_5)_2$, $P(C_6H_5)_3$, and $(C_6H_5)_2PCH_2CH_2P(C_6H_5)_2$ were prepared by adding *cis*-$PtCl_2(PR_3)_2$ to a solution of equimolar amounts of $(O{=}S{=}N)_2S$ in liquid NH_3 at −78°C. After stirring the reaction mixture for 30 min NH_3 was evaporated and the residue extracted with CH_2Cl_2, filtered through a celite/glass wool plug, and layered with petroleum ether [10]. *cis*-Pt(N=S=O)$_2$(PR$_3$)$_2$ (R = ?) was also synthesized by adding SOCl$_2$ and Pt(PR$_3$)$_2$Cl$_2$ to liquid NH$_3$ [12]. *cis*-Pt(N=S=O)$_2$(P(C$_6$H$_5$)$_3$)$_2$ was prepared in ~67% yield by reacting solutions of C$_6$H$_5$As(N=S=O)$_2$ and (η^2-C$_2$H$_4$)Pt(P(C$_6$H$_5$)$_3$)$_2$ (mole ratio ~1:1) in toluene at −17°C, followed by stirring at 23°C for 17 h [8]. The reaction of (η^2-C$_2$H$_4$)Pt-(P(C$_6$H$_5$)$_3$)$_2$ with 2 mole equivalents of O=S=NSi(CH$_3$)$_3$ also yielded *cis*-Pt(N=S=O)$_2$(P(C$_6$H$_5$)$_3$)$_2$ [9]. The formation of the cis complex was also assumed on the basis of ^{31}P NMR data by reaction of *cis*-PtCl$_2$(P(C$_6$H$_5$)$_3$)$_2$ with (O=S=N)$_2$Hg, but the product was not isolated [7, 13]. *trans*-Pt-(N=S=O)$_2$(P(C$_6$H$_5$)$_3$)$_2$ was prepared in 40% yield by adding (η^2-C$_2$H$_4$)Pt(P(C$_6$H$_5$)$_3$)$_2$ to an equimolar amount of (O=S=N)$_2$Hg in THF. After stirring for 2 h the solvent was removed and the residue extracted with C$_6$H$_6$. Concentration of this solution and addition of hexane gave the product as yellow crystals [7, 13]. Yields, melting points, IR and mass spectral data of the complexes are summarized in Table 55; ^{1}H and ^{31}P NMR data are compiled in Table 56, p. 262.

Structure. The ^{31}P NMR spectra of the complexes consist of singlets with platinum satellites indicating that the phosphines are equivalent. The magnitude of the ^{195}Pt,^{31}P coupling constant (~3100 Hz) is appropriate for Pt(II) and suggests nitrogen rather than oxygen bound NSO$^-$ ligands. The ^{1}H NMR spectra of the complexes with PR$_3$ = P(CH$_3$)$_3$, P(CH$_3$)$_2$C$_6$H$_5$, and PCH$_3$(C$_6$H$_5$)$_2$ reveal CH$_3$ resonances split by both ^{2}J(P,H) of 11 Hz and ^{3}J(Pt,H) of ~34 Hz. This again indicates identical phosphine groups in a symmetric environment [10].

X-ray crystal structure analyses have been performed for *cis*-Pt(N=S=O)$_2$(P(CH$_3$)$_3$)$_2$ [10, 13] and *trans*-Pt(N=S=O)$_2$(P(C$_6$H$_5$)$_3$)$_2$ [7]. Crystal data are given in Table 57, p. 263. The molecular structures of the complexes, along with selected bond lengths and bond angles are shown in **Fig. 21** and **Fig. 22**, p. 263. *cis*-Pt(N=S=O)$_2$(P(CH$_3$)$_3$)$_2$ has a slightly distorted cis square-planar geometry with local noncrystallographic C$_2$ symmetry. The NSO$^-$ ligands are bonded via nitrogen to platinum. The Pt–N(1)–N(2) and Pt–P(1)–P(2) planes are rotated ~3° with respect to each other about an axis bisecting the P(1)–Pt–P(2) angle. The two Pt–N–S–O planes are rotated 85° from the coordination plane [10]. *trans*-Pt(N=S=O)$_2$(P(C$_6$H$_5$)$_3$)$_2$ is square-planar and the anionic NSO$^-$ ligands are also σ-bonded via nitrogen to platinum. The S–N and S–O distances indicate double bonds [7].

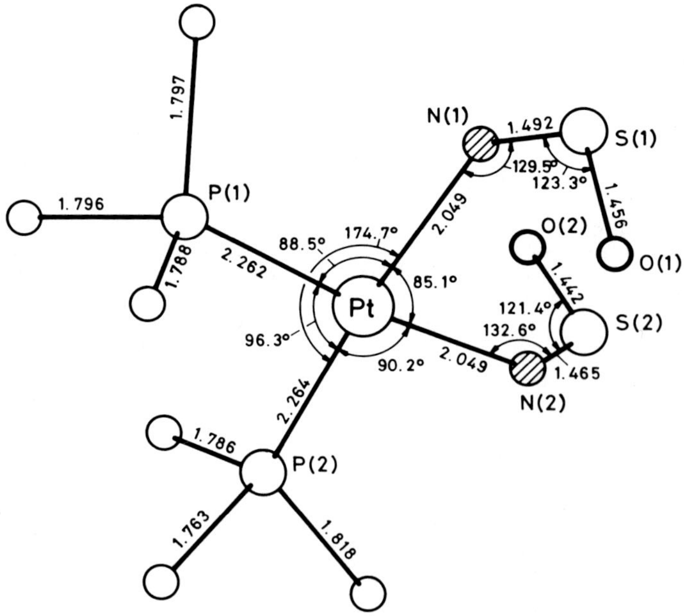

Fig. 21. Molecular structure of *cis*-Pt(N=S=O)$_2$(P(CH$_3$)$_3$)$_2$; bond lengths in Å, bond angles in ° [10].

Table 55

Yields, Melting Points, IR Frequencies of O=S=N Vibrations, and Mass Spectral Data of *cis*-$Pt(N{=}S{=}O)_2(PR_3)_2$ Complexes and *trans*-$Pt(N{=}S{=}O)_2(P(C_6H_5)_3)_2$.

PR_3	yield in %	m.p. in °C	IR (KBr), ν in cm^{-1} ν_{as}(OSN)	ν_s(OSN)	δ(OSN)	MS, m/e M^+	M^+–NSO	M^+–$(NSO)_2$	Ref.
$P(CH_3)_3$	62	154 to 157	1245 s 1230 vs	1061 m 1047 vs	567 w —	471	409	346	[10]
$P(C_2H_5)_3$	54	124 to 125	1243 vs 1228 vs	1057 s 1050 s	567 m 554 m	555	493	—	[10]
$P(C_3H_7\text{-}n)_3$	59	110 to 113	1249 vs 1221 m	1058 vs —	545 m —	639	576	513	[10]
$P(CH_3)_2C_6H_5$	65	107 to 110	1242 vs —	1055 vs 1020 m	555 w —	596	533	—	[10]
$PCH_3(C_6H_5)_2$	57	121 to 123	1242 vs —	1062 s —	569 w —	720	658	596	[10]
$P(C_6H_5)_3$ *)	62 [10], ~70 [8]	—	1243 vs [8, 10] 1228 s [8, 10]	1098 m [8] 1092 m [10] 1062 [8, 10]	547 m [10] —	—	—	—	[8, 10]
$P(C_6H_5)_3$ (*trans*)	40	—	1218	1099	590	—	—	—	[7]
$((C_6H_5)_2PCH_2)$	43	—	1239 vs —	1057 s 1012 m	562 w —	—	—	—	[10]

*) IR (Nujol): 14 absorptions between 1243 and 497 cm^{-1} [8].

References on p. 263

Table 56
^{1}H and ^{31}P NMR Data of *cis*-Pt(N=S=O)$_2$(PR$_3$)$_2$ Complexes and *trans*-Pt(N=S=O)$_2$(P(C$_6$H$_5$)$_3$)$_2$.

PR_3	^{1}H NMR (solvent?/TMS) δ (in ppm), J (in Hz)	^{31}P NMR (solvent?/external 85% H_3PO_4) δ (in ppm)	$^{1}J(Pt,P)$ (in Hz)	Ref.
$P(CH_3)_3$	1.17 (d), $^{2}J(P,H)=10.8$, $^{3}J(Pt,H)=34.1$	−32.6	3027	[10]
$P(C_2H_5)_3$	1.15 (d of t, 18H), $^{3}J(P,H)=17.4$, $^{2}J(H,H)=7.1$; 1.78 (m, 12 H)	1.1	3070	[10]
$P(C_3H_7\text{-}n)_3$	1.61 (m, 24H), 1.00 (t, 18H), $^{2}J(H,H)=7.0$	−7.8	3069	[10]
$P(C_4H_9\text{-}n)_3$	—	−7.0	3072	[10]
$P(CH_3)_2C_6H_5$	1.48 (d, 12H), $^{2}J(P,H)=10.9$, $^{3}J(Pt,H)=32.3$; 7.42 (m, 10H)	−22.1	3088	[10]
$PCH_3(C_6H_5)_2$	1.76 (d, 6H), $^{2}J(P,H)=10.2$, $^{3}J(Pt,H)=33.0$; 7.4 (m, 20H)	−8.1	3119	[10]
$P(C_6H_5)_3$	7.3 (m, 30H) [10]	8.8 [8, 10] 8.12 [7]	3190 [10] 3191 [8] 3257 [7]	[7, 8, 10]
$P(C_6H_5)_3$ (*trans*)*)	—	19.03	2854	[7]
$((C_6H_5)_2PCH_2)$	7.8 (m, 8H), 7.4 (m, 12H), 2.28 (m, 4H)	33.9	3112	[10]

*) ^{31}P NMR in C_6H_6.

Table 57
Crystal Data of *cis*-$Pt(N{=}S{=}O)_2(P(CH_3)_3)_2$ and *trans*-$Pt(N{=}S{=}O)_2(P(C_6H_5)_3)_2$.

	cis-$Pt(N{=}S{=}O)_2(P(CH_3)_3)_2$	*trans*-$Pt(N{=}S{=}O)_2(P(C_6H_5)_3)_2$
symmetry	monoclinic	monoclinic
space group	$P2_1/n{-}C^5_{2h}$ (No. 14)	$P2_1/n{-}C^5_{2h}$ (No. 14)
a in Å	9.963(2)	9.427(1)
b in Å	13.903(2)	20.123(3)
c in Å	11.406(2)	9.746(19)
β in °	109.85(2)	67.84(1)
V in Å^3	1486	1712
Z	4	2
D_x in g/cm^3	2.11	1.636
R	0.052	0.0317
R_w	0.055	0.0264
observed reflections	1387	1961
Ref.	[10]	[7]

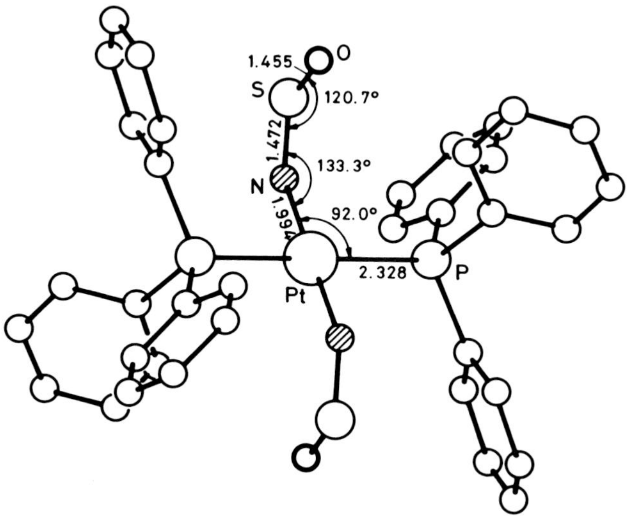

Fig. 22. Molecular structure of *trans*-$Pt(N{=}S{=}O)_2(P(C_6H_5)_3)_2$; bond lengths in Å, bond angles in ° [7].

References:

[1] Plenio, H.; Roesky, H. W.; Noltemeyer, M.; Sheldrick, G. M. (J. Chem. Soc. Chem. Commun. **1987** 1483/4).

[2] Herberhold, M.; Neumann, F.; Süss-Fink, G.; Thewalt, U. (Inorg. Chem. **26** [1987] 3612/5).

[3] Pandey, K. K.; Nehete, D. T.; Tewari, S. K.; Rewari, S.; Bhardwaj, R. (Polyhedron **7** [1988] 709/14).

[4] Jain, K. C.; Pandey, K. K.; Agarwala, U. C. (Indian J. Chem. A **20** [1981] 1022/3).

[5] Roesky, H. W.; Pandey, K. K.; Krebs, B.; Dartmann, M. (J. Chem. Soc. Dalton Trans. **1984** 2271/3).
[6] Pandey, K. K.; Agarwala, U. C. (Indian J. Chem. A **20** [1981] 906/7).
[7] Short, R.; Hursthouse, M. B; Purcell, T. G.; Woollins, J. D. (J. Chem. Soc. Chem. Commun. **1987** 407/8).
[8] Chivers, T.; Dhathathreyan, K. S.; Lensink, C.; Richardson, J. F. (Inorg. Chem. **27** [1988] 1570/4).
[9] Chivers, T.; Dhathathreyan, K. S. (unpublished results from [8]).
[10] Parkin, I. P.; Slawin, A. M. Z.; Williams, D. J.; Woollins, J. D. (Polyhedron **8** [1989] 835/9).

[11] Plenio, H.; Roesky, H. W.; Edelmann, F. T.; Noltemeyer, M. (J. Chem. Soc. Dalton Trans. **1989** 1815/8).
[12] Parkin, I. P.; Slawin, A. M. Z.; Williams, D. J.; Woollins, J. D. (J. Chem. Soc. Chem. Commun. **1989** 1060/1).
[13] Kelly, P. F.; Parkin, I. P.; Woollins, J. D. (Phosphorus Sulfur Silicon Relat. Elem. **41** [1989] 223/8).

2.7.3 Complexes of Bis(sulfinyl-amino)sulfane, $(O{=}S{=}N)_2S$

$[Zn((O{=}S{=}N)_2S)_2]^{2+}(AsF_6^-)_2 \cdot 2SO_2$

The complex compound was prepared in 95% yield by the reaction of $(O{=}S{=}N)_2S$ and $[Zn(SO_2)_2]^{2+}(AsF_6^-)_2$ (mole ratio 11:5) in liquid SO_2 at room temperature for 12 h in a pressure flask. After removal of SO_2 the bright yellow residue was repeatedly washed with dry CH_2Cl_2, and recrystallized from liquid SO_2.

The yellow crystals belong to the monoclinic system, space group $P2_1/n\text{–}C_{2h}^5$ (No. 14), with the lattice parameters a = 8.442(1), b = 13.106(2), c = 10.997(1) Å, β = 101.47(1)°; Z = 2, V = 1192.3(5) Å^3; D_x = 2.461 g/cm^3; R = 0.051, R_w = 0.059. The structure of the unit cell is shown in **Fig. 23**. The Zn atom is located at the inversion center. It is octahedrally coordinated by four O atoms of different $(O{=}S{=}N)_2S$ groups and by two F atoms of AsF_6^- groups. The $(O{=}S{=}N)_2S$ ligands are planar within 0.067 Å. They form chains between the Zn atoms leading to a two-dimensional network in the $\bar{1}01$ plane. The AsF_6^- group is disordered or must be assumed to rotate around the F(1)–As–F(2) axis; therefore in Fig. 23 the positions of four F atoms are represented by a wheel. For the SO_2 group only the O(3) atom is well-localized, while two positions are found for both S(4) and O(4). These atoms were described with a split-atom model. Both positions for the disordered atoms are included in Fig. 23; they are labeled S(4), S(4′), O(4), and O(4′). The crystal packing shows a number of intermolecular interactions (S···O and S···F), which are slightly shorter (up to 0.15 Å) than the sum of the van der Waals radii of the atoms. Bond distances and bond angles are given in Table 58. Positional parameters and nonbonded short contacts are given in the original paper.

$[Zn((O{=}S{=}N)_2S)_2]^{2+}(AsF_6^-)_2 \cdot 2SO_2$ is stable under an SO_2 atmosphere, but decomposes in moist air. The crystals lose SO_2 at room temperature with decomposition of the surface and formation of $[Zn((O{=}S{=}N)_2S)_2]^{2+}(AsF_6^-)_2$. The crystals decompose in the temperature range of 170 to 185°C.

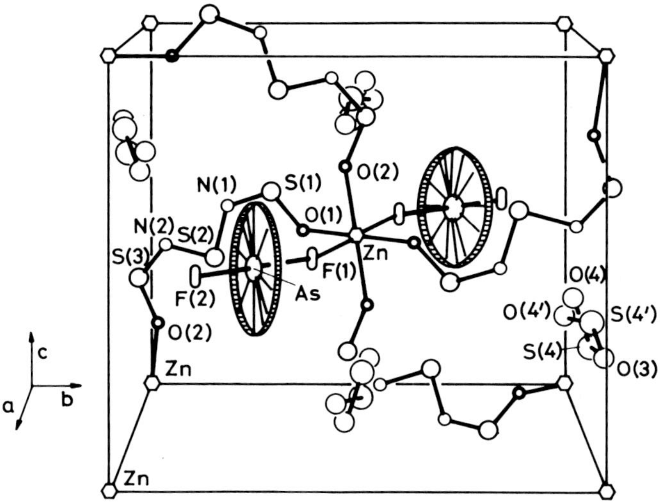

Fig. 23. Unit cell of $[Zn((O{=}S{=}N)_2S)_2]^{2+}(AsF_6^-)_2 \cdot 2SO_2$. The disordered or rotating part of AsF_6^- is represented by a wheel. For S(4) and O(4) the two possible positions are plotted.

Table 58

Bond Distances and Bond Angles in $[Zn((O{=}S{=}N)_2S)_2]^{2+}(AsF_6^-)_2 \cdot 2SO_2$.

bond distance	in Å	bond distance	in Å
Zn–O(1)	2.070(2)	S(3)–O(2)	1.468(2)
Zn–O(2)	2.030(2)	As–F(1)	1.756(2)
Zn–F(1)	2.082(2)	As–F(2)	1.687(2)
S(1)–O(1)	1.475(2)	S(4)–(3)	1.341(6)
S(1)–N(1)	1.514(3)	S(4)–O(4)	1.39(1)
S(2)–N(1)	1.641(3)	S(4')–O(3)	1.310(7)
S(2)–N(2)	1.647(3)	S(4')–O(4')	1.32(1)
S(3)–N(2)	1.513(3)		

bond angle	in °	bond angle	in °
O(1)–Zn–O(1)	180.0	Zn–O(1)–S(1)	129.1(1)
O(1)–Zn–O(2)	90.8(1)	Zn–O(2)–S(3)	145.6(2)
O(1)–Zn–O(2)	89.2(1)	O(1)–S(1)–N(1)	115.5(1)
O(1)–Zn–F(1)	91.6(1)	N(1)–S(2)–N(2)	97.0(1)
O(1)–Zn–F(1)	88.4(1)	N(2)–S(3)–O(2)	112.7(2)
O(2)–Zn–O(2)	180.0	S(1)–N(1)–S(2)	125.9(2)
O(2)–Zn–F(1)	91.4(1)	S(2)–N(2)–S(3)	123.3(2)
O(2)–Zn–F(1)	88.6(1)	Zn–F(1)–As	150.0(2)
F(1)–Zn–F(1)	180.0	F(1)–As–F(2)	179.2(1)
O(3)–S(4)–O(4)	116(1)	O(3)–S(4')–O(4')	135(1)

Reference:

Roesky, H. W.; Thomas, M.; Bats, J. W.; Fuess, H. (Inorg. Chem. **22** [1983] 2342/3).

$[Ag_4((O{=}S{=}N)_2S)_9]^{4+}(AsF_6^-)_4 \cdot SO_2$

The complex compound was obtained when $(O{=}S{=}N)_2S$ and $Ag^+AsF_6^-$ (mole ratio 4:1) react in liquid SO_2 at room temperature for 2 h in a pressure flask. During slow removal of SO_2 orange crystals form, m.p. 82 to 84°C. The preparation must be done with careful exclusion of moisture in an inert nitrogen atmosphere and in the absence of light.

The crystals are rhombohedral, space group R3c–C_{3v}^6 (No. 161), with a = b = 18.249(5), c = 35.95(2) Å, Z = 6, V = 10368 Å^3, D_x = 2.55 g/cm^3; R = 0.070 for 1653 observed reflections (F > 4σ (F)). There are two crystallographically independent silver atoms. One (Ag(2)) lies on a threefold axis, and is octahedrally coordinated by the terminal oxygen atoms of six $(O{=}S{=}N)_2S$ ligands, as shown in **Fig. 24**. The other (Ag(1)) occupies a general position and is unsymmetrically coordinated by the six nitrogen atoms of three $(O{=}S{=}N)_2S$ ligands, as shown in **Fig. 25**. There are also three relatively short Ag(1)···S interactions and a weak Ag(1)···O(12) interaction. Atomic distances and bond angles are given in Table 59; atomic coordinates are listed in the paper. The $(O{=}S{=}N)_2S$ ligands are all planar within 0.12 Å. If the weak Ag(1)···O(12) interaction is included each ligand is coordinated to silver atoms by two N and one of its O atoms. The S^{IV}–N bonds (mean 1.530 Å) are appreciably shorter than the S^{II}–N bonds (mean 1.649 Å), indicating fairly localized double and single bonds. Two of the three $(O{=}S{=}N)_2S$ ligands attached to Ag(1) are also bound via terminal oxygens to Ag(2), producing a polymeric cationic network. The AsF_6^- anions and SO_2 solvent molecules occupy holes in this network without themselves acting as ligands. (IR spectrum in the paper.)

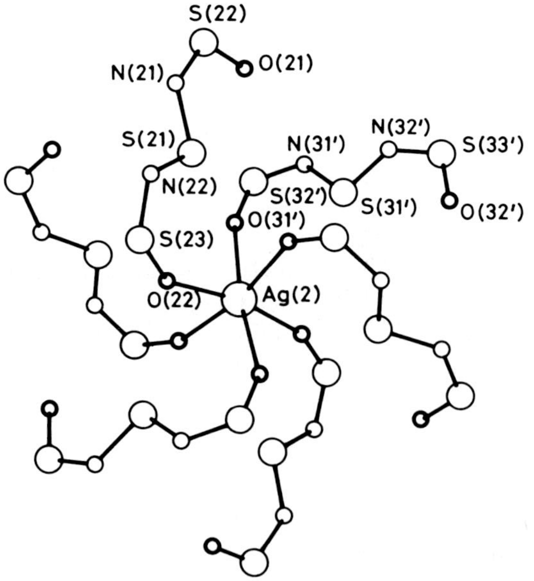

Fig. 24. The coordination of Ag(2) in $[Ag_4((O{=}S{=}N)_2S)_9]^{4+}(AsF_6^-)_4 \cdot SO_2$ viewed down the crystallographic threefold axis.

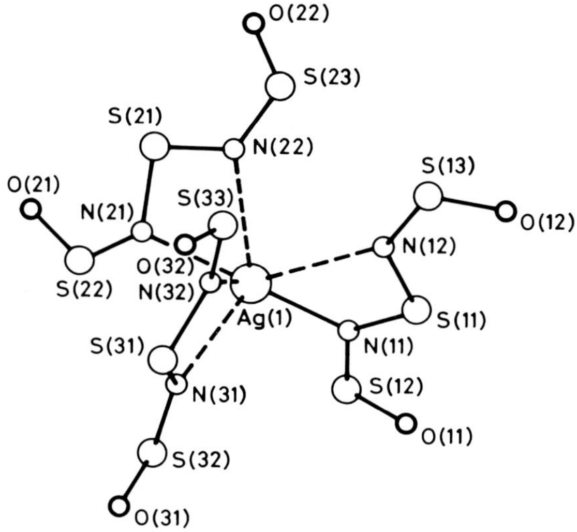

Fig. 25. The coordination of Ag(1) in $[Ag_4((O{=}S{=}N)_2S)_9]^{4+}(AsF_6^-)_4 \cdot SO_2$. An additional weak Ag···O interaction of 2.903 Å is omitted for clarity. Ag–N bonds given as full bonds are shorter than the other four Ag–N bonds.

Table 59

Atomic Distances and Bond Angles in $[Ag_4((O{=}S{=}N)_2S)_9]^{4+}(AsF_6^-)_4 \cdot SO_2$.

atomic distance	in Å	atomic distance	in Å
N(11)–S(11)	1.629(20)	N(11)–S(12)	1.562(15)
N(12)–S(11)	1.647(17)	N(12)–S(13)	1.510(16)
O(11)–S(12)	1.428(19)	O(12)–S(13)	1.474(22)
N(21)–S(21)	1.663(15)	N(21)–S(22)	1.523(20)
N(22)–S(21)	1.674(19)	N(22)–S(23)	1.508(18)
O(21)–S(22)	1.485(18)	O(22)–S(23)	1.423(15)
N(31)–S(31)	1.656(19)	N(31)–S(32)	1.522(15)
N(32)–S(31)	1.621(19)	N(32)–S(33)	1.554(16)
O(31)–S(32)	1.435(23)	O(32)–S(33)	1.437(28)
Ag(1)···S(11)	3.382(8)	Ag(1)···S(12)	3.363(7)
Ag(1)–N(11)	2.321(21)	Ag(1)–N(12)	2.948(18)
Ag(1)···S(21)	3.310(6)	Ag(1)–N(21)	2.441(19)
Ag(1)–N(22)	2.626(13)	Ag(1)–N(31)	2.616(20)
Ag(1)–N(32)	2.760(19)	Ag(1)···O(12)	2.903(18)
Ag(2)–O(22)	2.530(15)		

bond angle	in °	bond angle	in °
N(11)–Ag(1)–N(21)	154.5(7)	N(21)–Ag(1)–N(22)	58.7(6)
N(21)–Ag(1)–N(31)	84.4(7)	O(22)–Ag(2)–O(31^{II})	94.4(7)

Table 59 (continued)

bond angle	in °	bond angle	in °
O(22^{IV})–Ag(2)–O(31^{II})	168.6(6)	O(31^{II})–Ag(2)–O(31^{V})	83.7(6)
N(11)–S(11)–N(12)	97.7(10)	N(12)–S(13)–O(12)	118.2(10)
Ag(1)–N(11)–S(12)	118.7(12)	S(11)–N(12)–S(13)	122.8(14)
N(21)–S(22)–O(21)	116.1(9)	Ag(1)–N(21)–S(21)	106.0(10)
S(21)–N(21)–S(22)	121.0(11)	Ag(1)–N(22)–S(23)	135.1(11)
Ag(2)–O(22)–S(23)	148.1(12)	N(32)–S(33)–O(32)	117.7(12)
Ag(1)–N(31)–S(32)	128.3(11)	S(31)–N(32)–S(33)	121.4(14)
N(11)–Ag(1)–N(22)	120.4(7)	N(11)–Ag(1)–N(31)	109.7(7)
N(22)–Ag(1)–N(31)	125.9(7)	O(22^{III})–Ag(2)–O(31^{II})	107.3(7)
O(22)–Ag(2)–O(22^{III})	75.4(5)	N(11)–S(12)–O(11)	111.8(11)
Ag(1)–N(11)–S(11)	116.7(9)	S(11)–N(11)–S(12)	122.4(13)
N(21)–S(21)–N(22)	95.6(9)	N(22)–S(23)–O(22)	116.9(11)
Ag(1)–N(21)–S(22)	133.0(8)	Ag(1)–N(22)–S(21)	98.2(7)
S(21)–N(22)–S(23)	123.9(10)	N(31)–S(31)–N(32)	96.3(10)
Ag(1)–N(31)–S(31)	107.3(8)	S(31)–N(31)–S(32)	124.1(14)
N(31)–S(32)–O(31)	116.2(11)		

Symmetry operators: I $-\frac{1}{3}+x$, $\frac{1}{3}+x-y$, $-\frac{1}{6}+z$; II $\frac{5}{3}-y$, $\frac{4}{3}-x$, $-\frac{1}{6}+z$; III $2-y$, $1+x-y$, z; IV $1+y-x$, $2-x$, z; V $\frac{2}{3}+x$, $\frac{4}{3}+x-y$, $-\frac{1}{6}+z$.

Reference:

Roesky, H. W.; Thomas, M.; Jones, P. G.; Pinkert, W.; Sheldrick, G. W. (J. Chem. Soc. Dalton Trans. **1983** 1211/3).

2.7.4 Complexes of N-Sulfinyl-p-toluenesulfonamide, O=S=NSO$_2$C$_6$H$_4$CH$_3$-4

OsCl(NO)(P(C$_6$H$_5$)$_3$)$_2$(O=S=NSO$_2$C$_6$H$_4$CH$_3$-4)

The complex was prepared by the reaction of (η^2-C$_2$H$_4$)OsCl(NO)(P(C$_6$H$_5$)$_3$)$_2$ or OsCl(NO)(P(C$_6$H$_5$)$_3$)$_3$ with O=S=NSO$_2$C$_6$H$_4$CH$_3$-4 (mole ratio ~1:1) in toluene or THF. After 20 min stirring the reaction mixture was worked up and the yellow complex was isolated in 92% yield. The complexes OsCl(NO)(CO)(P(C$_6$H$_5$)$_3$)$_2$ and OsCl(NO)(P(C$_6$H$_5$)$_3$)$_2$(SO$_2$) also react with O=S=NSO$_2$C$_6$H$_4$CH$_3$-4 with substitution of CO and SO$_2$, respectively, to give the title compound. The melting point is 149°C with decomposition.

^{31}P {^{1}H} NMR (CDCl$_3$/external D$_3$PO$_4$) at 25°C: δ (in ppm) = 9.1; −15.5; ^{2}J(P,P′) = 350 Hz.

IR (Nujol): ν (in cm^{-1}) = 1768 ν(NO); 1088, 920 ν(SO/SN).

Based on IR data the N=S=O group is assumed to be coordinated via the N=S bond.

The complex is stable to atmospheric moisture. For reactions with the complex the authors found it convenient to prepare it in situ from the ethylene complex and O=S=NSO$_2$C$_6$H$_4$CH$_3$-4. The complex is then converted to the corresponding SO$_2$ complex OsCl(NO)(P(C$_6$H$_5$)$_3$)$_2$(SO$_2$)

by treatment with wet [$^2H_7$]-dimethylformamide at 60°C. Treatment of the title compound with H_2S and subsequently adding 1,8-diazabicyclo[5.4.0]undec-7-ene yielded $OsCl(NO)(P(C_6H_5)_3)_2(S_2)$ [1].

$RhCl(CO)(P(C_6H_5)_3)_2(O{=}S{=}NSO_2C_6H_4CH_3\text{-}4)$

The complex was formed by the reaction of $RhCl(CO)(P(C_6H_5)_3)_2$ with $O{=}S{=}NSO_2C_6H_4CH_3$-4. IR: ν (in cm^{-1}) = 1126, 1007 ν(N=S=O); 1316, 1306, 1293, 1161, and 1154 ν(SO_2); 2027 ν(CO). Based on the IR results, a pyramidally coordinated imino-oxo-λ^4-sulfane was suggested [1].

$IrCl(CO)(P(C_6H_5)_3)_2(O{=}S{=}NSO_2C_6H_4CH_3\text{-}4)$

The title compound was prepared by the reaction of $IrCl(CO)(P(C_6H_5)_3)_2$ with $O{=}S{=}NSO_2$-$C_6H_4CH_3$-4 (mole ratio ~1:1) in THF. The complex was isolated after 5 min stirring as bright orange-yellow microcrystalline solid in a yield >95%. The melting point is 126°C with decomposition.

^{31}P {^1H} NMR ($CDCl_3$/external D_3PO_4) at 25°C: δ = 7.0 ppm. IR (Nujol): ν (in cm^{-1}) = 2034 ν(CO); 1118, 1014, 1023 ν(SO, SN). Based on IR data a pyramidal coordination of the NSO group was suggested. The complex is very sensitive to hydrolysis. It is converted to the corresponding SO_2 complex $IrCl(CO)(P(C_6H_5)_3)_2(SO_2)$ by traces of moisture. The complex reacts with H_2S in THF to yield $IrCl(CO)(P(C_6H_5)_3)_2(S_2O)$ in 76% yield [1].

$Pt(P(C_6H_5)_3)_2(O{=}S{=}NSO_2C_6H_4CH_3\text{-}4)$

The compound was prepared by reaction of $(\eta^2\text{-}C_2H_4)Pt(P(C_6H_5)_3)_2$ with $O{=}S{=}NSO_2C_6H_4CH_3$-4 (mole ratio ~1:1.4) in C_6H_6 for 3 to 4 min. After adding $(C_2H_5)_2O$ a heavy white precipitate formed (yield 51%); m.p. 190 to 194°C. IR (Nujol): ν (in cm^{-1}) = 1525 s; 1350 s, 1308 m, 1150 vs, 1070 s, 880 m, br, 850 m [2].

References:

[1] Herberhold, M.; Hill, A. F. (J. Chem. Soc. Dalton Trans. **1988** 2027/31).
[2] Blake, D. M.; Reynolds, J. R. (J. Organometal. Chem. **113** [1976] 391/6).

2.7.5 Complexes of N-Sulfinyl-alkanamines, O=S=NR, R = alkyl

$Ni(P(C_6H_5)_3)_2(O{=}S{=}NR)$, R = n-$C_4H_9$, c-$C_6H_{11}$

The orange complexes were prepared in analogy to the corresponding O=S=NAr complexes (see p. 282) by treating of $(\eta^2\text{-}C_2H_4)Ni(P(C_6H_5)_3)_2$ with O=S=NR (R = n-C_4H_9 and c-C_6H_{11}) in THF. After ethylene evolution has ceased, the reaction mixture was stirred for 2 h.

The diamagnetic complexes are very soluble in THF and benzene and insoluble in ether. The O=S=NR ligand can be replaced by phosphite or 1,4-diazabutadiene, even at room temperature.

Reference:

Walther, D.; Pfützenreuter, C. (Z. Chem. [Leipzig] **17** [1977] 426/7).

2.7.6 Complexes of N-Sulfinyl-arenamines, O=S=NR, R = aryl

2.7.6.1 N-Sulfinyl-aniline Complexes of Ti

$TiCl_4(O{=}S{=}NC_6H_5)$

The addition compound was prepared by adding a solution of $O{=}S{=}NC_6H_5$ in pentane to a solution of $TiCl_4$ in pentane (mole ratio ~0.66:1). The yellow compound was isolated by filtration after 24 h.

IR (Nujol): ν (in cm^{-1}) = 1089s, 985s, ν(N=S=O); 395s, br, 335br, sh, ν(TiCl) [1].

$TiCl_4(O{=}S{=}NC_6H_5)_2$

The title compound was obtained by shaking a mixture of $TiCl_4$ and $O{=}S{=}NC_6H_5$ (mole ratio ~1:3) in pentane for 24 h. The yellow complex was isolated by filtration.

IR (Nujol): ν (in cm^{-1}) = 1081s, 994s, ν(N=S=O); 375br, 335br, sh, 305sh, ν(TiCl) [1].

$TiCl_3(OCH_3)(O{=}S{=}NC_6H_5)$

The compound was synthesized by the reaction of $TiCl_3OCH_3$ with $O{=}S{=}NC_6H_5$ (mole ratio ~1:1) in CH_2Cl_2 (24 h). The pale yellow addition compound was isolated by filtration.

^1H NMR (C_6H_6/TMS): δ = 3.87 ppm (OCH_3). IR (Nujol): ν (in cm^{-1}) = 1095s, 990s, ν(N=S=O); 385s, br, 355sh, ν(TiCl); 630m, ν(TiO). The color of the complex changed on standing at room temperature from yellow to green [1].

2.7.6.2 N-Sulfinyl-aniline Complex of Zr

$ZrCl_4(O{=}S{=}NC_6H_5)$

The title compound was prepared by shaking $ZrCl_4$ and $O{=}S{=}NC_6H_5$ (mole ratio 1:1.25 or 1:5) in CCl_4 for 24 h. The yellow-orange addition compound was isolated by filtration. When $O(C_2H_5)_2$ was used as solvent, the reaction led to the production of only $ZrCl_4 \cdot 2\,O(C_2H_5)_2$.

IR (Nujol): ν (in cm^{-1}) = 1095s, 975s ν(N=S=O); 370sh, 335sh, br, 300sh, ν(ZrCl) [1].

2.7.6.3 N-Sulfinyl-aniline Complex of V

$V(O)Cl_3(O{=}S{=}NC_6H_5)_2$

The addition compound was synthesized by adding a solution of $O{=}S{=}NC_6H_5$ in benzene to a solution of $VOCl_3$ in benzene (mole ratio 2:1); decomposition point 77°C (sealed tube).

The IR spectrum shows 26 bands in the range 3050 to 680 cm^{-1}.

The complex is very hygroscopic and decomposes slowly at room temperature. The compound decomposes on boiling in benzene with evolution of SO_2 gas, forming a black powder with composition $(VCl_3NC_6H_5)_x$ [2].

2.7.6.4 N-Sulfinyl-aniline Complex of Mn

$MnCl_2(O{=}S{=}NC_6H_5)_2$

A complex with this composition was reportedly obtained by adding a solution of $MnCl_2 \cdot 4H_2O$ in C_2H_5OH to a solution of $O{=}S{=}NC_6H_5$ in C_2H_5OH (mole ratio 1:4). After 1 h a pale pink complex was centrifugated, yield ~55%. The compound does not melt up to 300°C. The magnetic moment was measured to be 5.8 B.M. (Gouy's method) at 25 ± 3°C. Based on IR data, a coordination of the $O{=}S{=}NC_6H_5$ ligand via N was proposed [3].

2.7.6.5 N-Sulfinyl-arenamine Complexes of Fe

$FeCl_2(P(C_6H_5)_3)(O{=}S{=}NC_6H_5)$

The title compound was formed by adding a solution of $FeCl_2(P(C_6H_5)_3)_2$ in benzene to a solution of $O{=}S{=}NC_6H_5$ in benzene (mole ratio ~1:5). The reaction mixture was refluxed for ~1 h and subsequently concentrated. On addition of petroleum ether the greyish green complex precipitated, yield ~60%; m.p. 185 to 186°C. The magnetic moment is 5.18 B.M. (Gouy's method) at 25 ± 3°C. Based on IR data, a coordination of the $O{=}S{=}NC_6H_5$ ligand via N was suggested [3].

$Fe(CO)_2(PR_3)_2(O{=}S{=}NAr)$, Ar = C_6H_5, 4-FC_6H_4, 4-ClC_6H_4, 4-BrC_6H_4, 4-$CH_3C_6H_4$, 4-$CH_3OC_6H_4$, 4-$NO_2C_6H_4$; PR_3 = $P(C_6H_5)_3$, $P(C_6H_4Cl\text{-}4)_3$, $P(C_6H_4CH_3\text{-}4)_3$, $P(C_6H_4OCH_3\text{-}4)_3$, $P(C_6H_4CH_3\text{-}4)(C_6H_5)_2$, $P(C_6H_4OCH_3\text{-}4)(C_6H_5)_2$, $P(CH_2C_6H_5)_3$

The title complexes were prepared by the reactions of $Fe_2(CO)_9$ with $O{=}S{=}NAr$ and PR_3 (mole ratio 1:1:2) in THF at room temperature for 1 to 4 h. The yields, melting points, and IR data are compiled in Table 60, p. 272. The reactions of $Fe_2(CO)_9$ with $O{=}S{=}NC_6H_5$ in THF at room temperature yield $Fe(CO)_5$ and red oils which could be converted to $Fe(CO)_2(P(C_6H_5)_3)_2$-$(O{=}S{=}NC_6H_5)$ by adding $P(C_6H_5)_3$. The complex $Fe(CO)_2(P(C_6H_5)_3)_2(O{=}S{=}NC_6H_4NO_2\text{-}4)$ was also obtained from the reaction of Fe(benzylideneacetone)$(CO)_3$ with $O{=}S{=}NC_6H_4NO_2$-4 and $P(C_6H_5)_3$ (mole ratio 1:1:2) in THF, yield 73%. The compounds with Ar = C_6H_5 or 4-$NO_2C_6H_4$ and PR_3 = $P(C_6H_5)_3$ were also prepared by stirring equimolar amounts of $Fe(CO)_2(P(C_6H_5)_3)_3$ and $O{=}S{=}NAr$ for 2 h (yield ~75%, both). UV irradiation of $Fe(CO)_5$, $O{=}S{=}NC_6H_4NO_2$-4, and PR_3 (R = C_6H_5, C_6H_4Cl-4, and $C_6H_4CH_3$-4) in THF (mole ratio 1:1:2) for 3 h yielded the corresponding title complexes in ~30% yield.

Based on IR data, the authors assume a η^2-coordination of the $O{=}S{=}NAr$ ligand via N and S to the Fe atom.

The complexes are brown to red microcrystalline solids, which are insoluble in water and soluble to some extent in most organic solvents. They are reasonably stable in air. The $O{=}S{=}N$-$C_6N_4NO_2$-4 derivatives are sparingly soluble crimson solids that are uneffected by air even after 3 years. The $O{=}S{=}NC_6H_4OCH_3$-4 compounds are red-brown soluble solids that show evidence of decomposition by air after a few weeks.

References on p. 287

Table 60

Yields, Melting Points, and IR Spectra of $Fe(CO)_2(PR_3)_2(O{=}S{=}NAr)$ Complexes [4].

Ar	PR_3	yield in %	m.p. in °C	IR, ν in cm^{-1} ν(CO)[a]	ν(CO)[a]	ν(SO)[b]	ν(NS)[b]
C_6H_5	$P(C_6H_5)_3$	36	108 to 110	1918(10)	1980(7.3)	1063	950
$4\text{-}CH_3OC_6H_4$	$P(C_6H_5)_3$	30	124 to 126	1915(10)	1975(6.4)	1059	940
$4\text{-}CH_3C_6H_4$	$P(C_6H_5)_3$	32	92 to 94	1915(10)	1979(8.0)	1068	952
$4\text{-}FC_6H_4$	$P(C_6H_5)_3$	32	123 to 124	1917(10)	1981(7.1)	1062	951
$4\text{-}ClC_6H_4$	$P(C_6H_5)_3$	40	110 to 112	1920(10)	1982(6.7)	1066	953
$4\text{-}BrC_6H_4$	$P(C_6H_5)_3$	43	108 to 110	1921(10)	1983(6.7)	1062	950
$4\text{-}NO_2C_6H_4$	$P(C_6H_5)_3$	45	141 to 143	1931(10)	1992(6.5)	1063	938
$4\text{-}NO_2C_6H_4$	$P(C_6H_4OCH_3\text{-}4)(C_6H_5)_2$	49	120 to 123	1930(10)	1992(6.4)	1063	934
$4\text{-}NO_2C_6H_4$	$P(C_6H_4OCH_3\text{-}4)_3$	44	121 to 124	1927(10)	1992(6.2)	1059	935
$4\text{-}NO_2C_6H_4$	$P(C_6H_4CH_3\text{-}4)(C_6H_5)_2$	49	130 to 132	1931(10)	1993(6.6)	1062	937
$4\text{-}NO_2C_6H_4$	$P(C_6H_4CH_3\text{-}4)_3$	55	146 to 147	1931(10)	1993(6.9)	1066	937
$4\text{-}NO_2C_6H_4$	$P(C_6H_4Cl\text{-}4)_3$	46	168 to 170	1933(10)	1993(7.8)	1074	939
$4\text{-}NO_2C_6H_4$	$P(CH_2C_6H_5)_3$	39	dec. 115	1941(10)	1999(8.2)	—	930
$4\text{-}ClC_6H_4$	$P(C_6H_4CH_3\text{-}4)_3$	40	dec. 124	1915(10)	1979(6.8)	1062	941
$4\text{-}BrC_6H_4$	$P(C_6H_4CH_3\text{-}4)_3$	43	dec. 120	1917(10)	1980(6.6)	1063	938

[a] Measured in $CHCl_3$ solution. – [b] Measured in CsBr disks.

Reactions of $Fe(CO)_2(PR_3)_2(O{=}S{=}NAr)$ with a reactant L (L = trialkylphosphites, $P(OR')_3$ (R' = CH_3, C_2H_5, i-C_3H_7, or $CH_3C(CH_2)_3$), and alkyl isocyanides, R'NC (R' = CH_3, C_2H_5, c-C_6H_{11}, or $C_6H_5CH_2$)), displace $P(C_6H_5)_3$ and/or CO to form analogous complexes of the type $Fe(P(C_6H_5)_3)$-$(L)(CO)_2(O{=}S{=}NAr)$, $Fe(L)_2(CO)_2(O{=}S{=}NAr)$, or $Fe(P(C_6H_5)_3)_2(L)(CO)(O{=}S{=}NAr)$ (Ar = C_6H_5, 4-Cl-C_6H_4, 4-BrC_6H_4, 4-$CH_3C_6H_4$, 4-$NO_2C_6H_4$). A solution of $Fe(CO)_2(P(C_6H_5)_3)_2(O{=}S{=}NC_6H_5)$ in benzene reacts with $[4\text{-}FC_6H_4N_2]^+BF_4^-$ in acetone (mole ratio 1:1) for 30 min to give $[Fe(CO)_2(P(C_6H_5)_3)_2(NNC_6H_4F\text{-}4)]^+BF_4^-$. $Fe(CO)_2(P(C_6H_5)_3)_2(O{=}S{=}NC_6H_5)$ reacts with $C_2(CN)_4$ (mole ratio 1:1) in toluene to give $Fe(CO)_2(P(C_6H_5)_3)_2(C_2(CN)_4)$ in ~54% yield. The same complex yields $Fe(CO)_2(P(C_6H_5)_3)_2(O_2CCH_3)_2$ when it is treated with glacial acetic acid in benzene (~35% yield). $Fe(CO)_2(P(C_6H_5)_3)_2(O{=}S{=}NAr)$ (Ar = C_6H_5 or 4-BrC_6H_4) did not react with CH_3I, C_2H_5I, or CH_3SO_3F in benzene solution with reactant mole ratios up to 1:10, long reaction times (up to 24 h) or elevated temperatures (50°C for CH_3SO_3F). $Fe(CO)_2(P(C_6H_5)_3)_2$-$(O{=}S{=}NC_6H_4Cl\text{-}4)$ reacts with CO gas in benzene even in the dark to yield $Fe(CO)_4P(C_6H_5)_3$ as the major product and $Fe(CO)_3(P(C_6H_5)_3)_2$ as by-product. $Fe(CO)_2(P(C_6H_5)_3)_2(O{=}S{=}NC_6H_4NO_2\text{-}4)$ also converts to $Fe(CO)_3(P(C_6H_5)_3)_2$ when it is dissolved in CS_2 after 4 h (yield ~50%). The same complex did not react with $(C_6H_5)_2C_2$ at 60°C after 4 h [4].

$Fe(CO)_2(P(C_6H_5)_3)(P(OCH_2)_3CCH_3)(O{=}S{=}NAr)$, Ar = C_6H_5, 4-ClC_6H_4, 4-BrC_6H_4, 4-$CH_3OC_6H_4$, 4-$NO_2C_6H_4$

The complexes were prepared by reaction of equimolar amounts of $Fe(CO)_2(P(C_6H_5)_3)_2$-$(O{=}S{=}NAr)$ and $P(OCH_2)_3CCH_3$ in benzene. The reactions, which were completed after 5 min, are virtually quantitative. For melting points and IR data see Table 61 [4].

Table 61

Melting Points and IR Data of $Fe(CO)_2(P(C_6H_5)_3)(P(OCH_2)_3CCH_3)(O{=}S{=}NAr)$ [4].

Ar	m.p. in °C	IR, ν in cm^{-1}			
		ν(CO)[a]	ν(CO)[a]	ν(SO)[b]	ν(NS)[b]
C_6H_5	dec. 101	1950(10)	2006(8.1)	—	—
4-$CH_3C_6H_4$	dec. 72	1948(10)	2003(8.7)	—	—
4-ClC_6H_4	114 to 116	1952(10)	2008(8.8)	—	—
4-BrC_6H_4	dec. 128	1952(10)	2008(8.6)	1020	925
4-$NO_2C_6H_4$	134 to 135	1964(10)	2020(6.6)	1020	929

[a] Measured in $CHCl_3$. – [b] Measured in CsBr disks.

$Fe(CO)(P(C_6H_5)_3)_2(CNR)(O{=}S{=}NC_6H_4NO_2\text{-}4)$, R = CH_3, C_2H_5, c-C_6H_{11}, $CH_2C_6H_5$

The complexes were prepared by stirring a solution of $Fe(CO)_2(P(C_6H_5)_3)_2(O{=}S{=}NC_6H_4NO_2\text{-}4)$ and RNC (mole ratio 1:1) in benzene for 6 to 8 h. The yields are in the range 40 to 50% (after purification). Melting points and IR data are listed in Table 62, p. 274.

A solution of $Fe(CO)(P(C_6H_5)_3)_2(CNCH_2C_6H_5)(O{=}S{=}NC_6H_4NO_2\text{-}4)$ in benzene reacted with a solution of $[4\text{-}FC_6H_4N_2]^+BF_4^-$ in acetone (mole ratio 1:1) over a period of 30 min to produce $[Fe(CO)(P(C_6H_5)_3)_2(CNCH_2C_6H_5)(NNC_6H_4F\text{-}4)]^+BF_4^-$ (95% yield) [4].

$Fe(CO)_2(P(OR)_3)_2(O{=}S{=}NC_6H_4NO_2\text{-}4)$, $P(OR)_3$ = $P(OCH_3)_3$, $P(OC_2H_5)_3$, $P(OC_3H_7\text{-}i)_3$, $P(OCH_2)_3CCH_3$

The complexes were obtained by reacting $Fe(CO)_2(P(C_6H_5)_3)_2(O{=}S{=}NC_6H_4NO_2\text{-}4)$ and $P(OR)_3$ (mole ratio 1:2) in benzene for 30 min; yields >90%. IR data and melting points see Table 63, p. 274 [4].

Table 62
Melting Points and IR Data of $Fe(CO)(P(C_6H_5)_3)_2(CNR)(O{=}S{=}NC_6H_4NO_2\text{-}4)$ [4].

R	m.p. in °C	IR, ν in cm^{-1}			
		ν(CO)[a]	ν(CO)[a]	ν(SO)[b]	ν(SO)[b]
CH_3	62 to 64	1927(10)	2157(4.5)	1039	930
C_2H_5	dec. 126	1925(10)	2146(6.2)	1039	929
$c\text{-}C_6H_{11}$	101 to 103	1925(10)	2137(4.5)	1039	927
$CH_2C_6H_5$	dec. 132	1929(10)	2137(5.0)	1045	933

[a] Measured in $CHCl_3$. – [b] Measured in CsBr disks.

Table 63
Melting Points and IR Data of $Fe(CO)_2(P(OR)_3)_2(O{=}S{=}NC_6H_4NO_2\text{-}4)$ [4].

$P(OR)_3$	m.p. in °C	IR ($CHCl_3$), ν in cm^{-1}	
		ν(CO)	ν(CO)
$P(OCH_3)_3$	115 to 118	1965(10)	2022(9.4)
$P(OC_2H_5)_3$	95 to 97	1958(10)	2019(6.0)
$P(OC_3H_7\text{-}i)_3$	75 to 77	1955(10)	2017(7.0)
$P(OCH_2)_3CCH_3$	dec. 150	1972(10)	2023(7.0)

2.7.6.6 N-Sulfinyl-aniline Complexes of Ru

$RuCl_2(P(C_6H_5)_3)_2(O{=}S{=}NC_6H_5)_2$

The title compound was prepared by adding a solution of $O{=}S{=}NC_6H_5$ in $CHCl_3$ to a solution of $RuCl_2(P(C_6H_5)_3)_2$ in the same solvent (mole ratio ~6:1). The blue complex was filtered after 1 h stirring (~65% yield). The melting point of the diamagnetic complex is 190 to 192°C. The $O{=}S{=}NC_6H_5$ ligand is assumed to be coordinated via the N atom [3].

$RuCl_2(As(C_6H_5)_3)_2(O{=}S{=}NC_6H_5)_2$

The title complex was formed by adding a solution of $O{=}S{=}NC_6H_5$ in benzene to a solution of $RuCl_3(As(C_6H_5)_3)_2 \cdot CH_3OH$ in CH_2Cl_2 (mole ratio ~6:1). After stirring the mixture for 2 h and subsequently concentrating, a brown solid was filtered (yield ~60%). The melting point of the diamagnetic complex is 176 to 177°C. UV spectrum ($CHCl_3$): A weak band at 550 nm was assigned to a $^1T_{2g} \rightarrow {}^1A_{1g}$ transition [3].

$Ru(CO)(P(C_6H_5)_3)_2(L)(O{=}S{=}NR)$, L = CO, $CNC_4H_9\text{-}t$, $CNC_6H_3(CH_3)_2\text{-}2,6$; R = $C_6H_4CH_3\text{-}2$, $C_6H_4CH_3\text{-}4$

The orange complexes $Ru(CO)(P(C_6H_5)_3)_2(L)(O{=}S{=}NR)$ (L = CO, R = $C_6H_4CH_3\text{-}4$; L = $CNC_4H_9\text{-}t$, R = $C_6H_4CH_3\text{-}4$; L = $CNC_6H_3(CH_3)_2\text{-}2,6$, R = $C_6H_4CH_3\text{-}2$; L = $CNC_6H_3(CH_3)_2\text{-}2,6$, R = $C_6H_4CH_3\text{-}4$) were prepared by the reactions of $Ru(CO)(P(C_6H_5)_3)_3L$ (L = CO, $CNC_4H_9\text{-}t$, $CNC_6H_3(CH_3)_2\text{-}2,6$) with $O{=}S{=}NR$ (R = $C_6H_4CH_3\text{-}2$, $C_6H_4CH_3\text{-}4$) (mole ratio 1:2) in toluene (or THF) for 1 h. Exemplary preparative details are given for $Ru(CO)(CNC_6H_3(CH_3)_2\text{-}2,6)(P(C_6H_5)_3)_2(O{=}S{=}NC_6H_4CH_3\text{-}2)$, yield 91%; m.p. 148°C with decomposition. The yield of $Ru(CO)(CNC_4H_9\text{-}t)(P(C_6H_5)_3)_2(O{=}S{=}NC_6H_4CH_3\text{-}4)$ is 87%.

Table 64

$^{31}P\{^1H\}$, ^1H NMR, and IR Spectra of $Ru(CO)L(P(C_6H_5)_3)_2(O{=}S{=}NR)$ [11].

L	R	$^{31}P\{^1H\}$ NMR ($CDCl_3/D_3PO_4$–D_2O) δ in ppm	^1H NMR ($CDCl_3$/TMS) δ in ppm	IR (Nujol), ν in cm^{-1} ν(C≡N)	ν(CO)	ν(C–N)	ν(S=O)	ν(N=S)
CO	$C_6H_4CH_3$-4	35.6	2.16 (s, $C_6H_4C\underline{H}_3$)	—	2015 1941	1310 1283	1040	938
CNC_4H_9-t	$C_6H_4CH_3$-4	38.3	1.01 (s, $C(CH_3)_3$) 2.15 (s, $C_6H_4C\underline{H}_3$)	2155	1920	1310 1295	1029	928
$CNC_6H_3(CH_3)_2$-2,6	$C_6H_4CH_3$-2	37.5	1.96 (s, $C_6H_4C\underline{H}_3$) 2.05 (s, $C_6H_3(C\underline{H}_3)_2$)	2106	1927	1300 1285	1029	939
$CNC_6H_3(CH_3)_2$-2,6	$C_6H_4CH_3$-4	36.8	2.05 (s, $C_6H_3(C\underline{H}_3)_2$) 2.16 (s, $C_6H_4C\underline{H}_3$)	2107	1927	1306	1029	932

References on p. 287

Based on the spectroscopic data of the complexes it was concluded that the O=S=NR ligand is bound to ruthenium through both nitrogen and sulfur. The sulfur atom is supposed to be coordinated trans to the isonitrile ligand.

NMR and IR data of the complexes are compiled in Table 64, p. 275 [11].

2.7.6.7 N-Sulfinyl-aniline Complex of Co

$CoH(P(OC_6H_5)_3)_2(O{=}S{=}NC_6H_5)$

Adding a solution of $CoH(P(OC_6H_5)_3)_4$ in CH_3CN to a solution of $O{=}S{=}NC_6H_5$ in benzene (mole ratio ~1:6), concentrating this mixture after 1.5 h and adding ether led to the precipitation of a brown solid which was recognized to be the title compound based on elemental analysis (yield ~48%). The melting point of the diamagnetic complex is 158 to 160°C. Detailed spectroscopic data are not communicated [3].

2.7.6.8 N-Sulfinyl-arenamine Complexes of Rh

$RhCl(PR_3)_2(O{=}S{=}NAr)$, Ar = C_6H_5, 4-ClC_6H_4, 4-$CH_3C_6H_4$, 2,4,6-$(CH_3)_3C_6H_2$; PR_3 = $P(C_3H_7\text{-}i)_3$, $P(C_6H_{11}\text{-}c)_3$

The compounds were prepared by adding PR_3, and after 15 min O=S=NAr, to a solution of $(RhCl(C_8H_{14}\text{-}c)_2)_2$ in toluene (mole ratio 4:2:1). The complexes were isolated by removal of the solvent.

Each of the complexes exists as two isomers, one with the O=S=N ligand coordinated to the rhodium atom via the sulfur atom (isomer A) and one probably via the π-N=S bond (isomer B), both in the solid state and in solution. In the case of $O{=}S{=}NC_6H_2(CH_3)_3$-2,4,6 only one isomer was observed. Spectroscopic data and ratios of relative abundances of the complexes (at −3 and −56°C) are listed in Table 65.

A single structure analysis of $RhCl(P(C_3H_7\text{-}i)_3)_2(O{=}S{=}NC_6H_4CH_3\text{-}4)$ was performed. The complex crystallizes in the triclinic space group $P\bar{1}-C_i^1$ (No. 2) with a = 9.779(1), b = 9.875(1), c = 16.101(1) Å; α = 96.49(1)°, β = 97.18(1)°, γ = 99.29(1)°; Z = 2. R = 0.047 for 2498 reflections. Atomic coordinates are given in the paper [5]. The crystal structure shows that the O=S=N-$C_6H_4CH_3$-4 ligand is in the cis configuration and is coordinated to rhodium via sulfur. The Rh–N=S=O group is planar. The coordination around rhodium is approximately square-planar. The geometry of the $O{=}S{=}NC_6H_4CH_3$-4 ligand is not appreciably influenced by coordination to rhodium. The molecular structure with selected bond lengths and bond angles is shown in **Fig. 26**, p. 279.

The thermal properties of $RhCl(P(C_3H_7\text{-}i)_3)_2(O{=}S{=}NC_6H_4CH_3\text{-}4)$ and $RhCl(P(C_3H_7\text{-}i)_3)_2(O{=}S{=}N\text{-}C_6H_2(CH_3)_3\text{-}2,4,6)$ were studied under N_2 and CO atmosphere using thermogravimetry (TG) and differential thermal analysis (DTA). The complexes melt with decomposition. The results are tabulated in the paper. Under a CO atmosphere the formation of $RhCl(CO)(P(C_3H_7\text{-}i)_3)_2$ from $RhCl(P(C_3H_7\text{-}i)_3)_2(O{=}S{=}NC_6H_4CH_3\text{-}4)$ was observed. The same compound was also produced when CO gas was bubbled through a solution of the O=S=NAr complexes. The complex RhCl-$(P(C_3H_7\text{-}i)_3)_2(O{=}S{=}NC_6H_2(CH_3)_3\text{-}2,4,6)$ converted almost completely into $RhCl(CO)(P(C_3H_7\text{-}i)_3)_2$ when it was exposed to the air for 75 d [5].

Table 65

Spectroscopic Data and Relative Abundances of the Isomers A and B of $RhCl(PR_3)_2(O{=}S{=}NAr)$ [5].

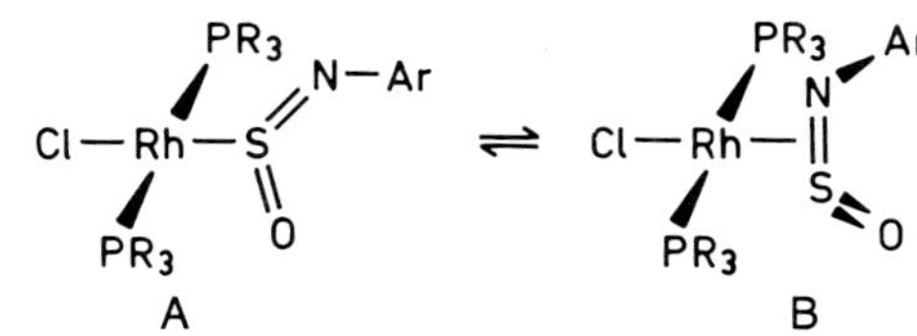

isomer	R	Ar	IR, ν (in cm^{-1}) (Nujol mull, KBr disks)	^{31}P NMR ($C_6D_5CD_3$/ext. 85% H_3PO_4) δ (in ppm) ($^1J(Rh-P)$ in Hz) t = −3°C[b]	^1H NMR[d] ($C_6D_5CD_3$/TMS) δ (in ppm)	relative abundances in %, in $C_6D_5CD_3$ t = −3°C	t = −56°C
A[a]	i-C_3H_7	C_6H_5	1287 m, ν(C_6H_5–N); 1269 s, ν(NS); 1108 s, ν(SO); 338 m, ν(RhCl); 306 m, ν(RhS)	40.2 (110)	[c]	33	20
B	i-C_3H_7	C_6H_5	1057 s, ν(SO); 959 s, ν(NS); 226 w, ν(RhCl)	32.7 (109)	—	67	80
A	i-C_3H_7	$C_6H_4CH_3$-4	1292 s, ν(C_6H_4–N); 1268 s, ν(NS); 1148 s, ν(SO); 338 m, ν(RhCl); 284 w, ν(RhS)	40.1 (112)	7.65 (d, C_6H_4, ortho H), 6.83 (d, C_6H_4, meta H), 1.98 (s, CH_3) (A and B)	51	31
B	i-C_3H_7	$C_6H_4CH_3$-4	—	32.6 (109)		49	69
A	i-C_3H_7	C_6H_4Cl-4	1286 s, ν(C_6H_4–N); 1264 s, ν(NS); 1150 s, ν(SO); 338 w, ν(RhCl); 284 w, ν(RhS)	40.3 (110)	7.20 (d, C_6H_4, ortho H), 6.60 (d, C_6H_4, meta H) (A and B)	21	16
B	i-C_3H_7	C_6H_4Cl-4	—	32.8 (108)		79	84
A	i-C_3H_7	$C_6H_2(CH_3)_3$-2,4,6	1306 m, ν(C_6H_2–N); 1248 s, ν(NS); 1142 s, ν(SO); 336 m, ν(RhCl); 283 m, ν(RhS)	41.7 (113) (t = 37°C)	6.67 (s, C_6H_4, meta H), 2.07 (s, CH_3, para), 2.32 (s, CH_3, ortho)	100	100

References on p. 287

Table 65 (continued)

isomer	R	Ar	IR, ν (in cm^{-1}) (Nujol mull, KBr disks)	^{31}P NMR ($C_6D_5CD_3$/ext. 85% H_3PO_4) δ (in ppm) ($^1J(Rh–P)$ in Hz) t = −3°C[b]	^1H NMR[d] ($C_6D_5CD_3$/TMS) δ (in ppm)	relative abundances in %, in $C_6D_5CD_3$ t = −3°C	t = −56°C
A	c-C_6H_{11}	$C_6H_4CH_3$-4	1292sh, ν(C_6H_4–N); 1266s, ν(NS); 1124s, ν(SO); 304m, ν(RhCl); 270w, ν(RhS)	31.0 (111)	7.68 (d, C_6H_4, ortho H), 6.85 (d, C_6H_4, meta H), 1.97 (s, CH_3) (for A and B)	71	—
B	c-C_6H_{11}	$C_6H_4CH_3$-4	—	22.8 (110)		29	—

[a] IR data are also given in the paper for O=S=$^{15}NC_6H_5$ species. – [b] With exception of Ar = $C_6H_3(CH_3)_3$-2,4,6, t = 37°C. – [c] ^1H NMR for O=S=$^{15}NC_6H_5$: δ = 7.64 ppm (d, C_6H_5, ortho H); $^3J(H,^{15}N)$ = 7 Hz; ^{15}N NMR (in $CDCl_3$/ext. HN=CHNH$_2$) at 0°C: δ = −1 ppm (d); $^1J(^{103}Rh,^{15}N)$ = 15.5 Hz. – [d] Although two isomers are present as shown by ^{31}P NMR, only one type of aryl group is observed in the ^1H NMR which shows that the two isomers are already in fast exchange.

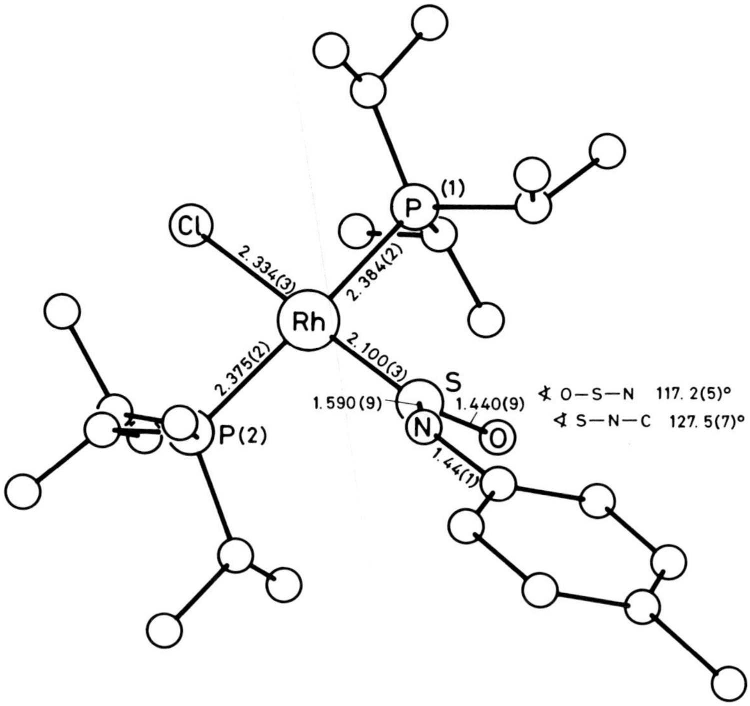

Fig. 26. Molecular structure of *trans*-$RhCl(P(C_3H_7\text{-}i)_3)_2(O{=}S{=}NC_6H_4CH_3\text{-}4)$; bond lengths in Å, bond angles in ° [5].

$RhCl(P(C_6H_5)_3)(O{=}S{=}NC_6H_5)_2$

The title compound was obtained by adding a solution of $RhCl(P(C_6H_5)_3)_3$ in CH_2Cl_2 to a solution of $O{=}S{=}NC_6H_5$ in benzene (mole ratio ~1:6). The mixture was refluxed for 2 h. After concentration and addition of petroleum ether, the light brown complex precipitated; yield ~60%. The melting point of the diamagnetic complex is 210 to 212°C. Detailed spectroscopic data are not reported [3].

$RhCl(CO)(P(C_6H_5)_3)(O{=}S{=}NC_6H_5)$

The complex was formed by adding a solution of $O{=}S{=}NC_6H_5$ in benzene to a stirred solution of $RhCl(CO)(P(C_6H_5)_3)_2$ in benzene (mole ratio ~6:1). After 1 h stirring, concentrating, and addition of petroleum ether a yellowish brown solid was isolated from the mixture which was recognized to be the title compound by elemental analysis; yield ~68%. The melting point of the diamagnetic complex is 238 to 240°C. Detailed spectroscopic data are not reported [3].

2.7.6.9 N-Sulfinyl-arenamine Complexes of Ir

$IrCl(PR_3)_2(O{=}S{=}NAr)$, Ar = 4-$CH_3C_6H_4$, 2,4,6-$(CH_3)_3C_6H_2$; R = i-$C_3H_7$, c-$C_6H_{11}$

The complexes were prepared in analogy to the corresponding rhodium complexes, see p. 276 [5].

In agreement with those compounds for Ar = 2,4,6-$(CH_3)_3C_6H_2$ and R = i-C_3H_7 only one isomer was detected in the ^{31}P NMR spectrum. However, three isomers were observed for Ar = 4-$CH_3C_6H_4$; R = i-C_3H_7 and c-C_6H_{11}. Two of them were identified as complexes with η^1-S (isomer A) and η^2-N=S (isomer B) bonded O=S=NAr ligand. The origin of the third isomer (C) could not be explained. IR and NMR data, as well as relative abundances in $C_6D_5CD_3$ solution are presented in Table 66.

The thermal properties of the complexes were studied by TG and DTA under N_2 and CO atmosphere. The data are given in the paper. The complexes melt with decomposition. Under CO atmosphere the decomposition product of $IrCl(P(C_3H_7\text{-}i)_3)_2(O{=}S{=}NC_6H_4CH_3\text{-}4)$ probably form $IrCl(CO)(P(C_3H_7\text{-}i)_3)_2$ identified by analysis, ^{31}P NMR and IR spectroscopy. The same complex was also produced when CO gas was bubbled through solutions of the corresponding O=S=NAr complexes. Heating $IrCl(P(C_3H_7\text{-}i)_3)_2(O{=}S{=}NC_6H_4CH_3\text{-}4)$ in the solid state under N_2 to 150°C led to the formation of a brown powder with the suggested structure based on IR and ^{31}P NMR data [5]:

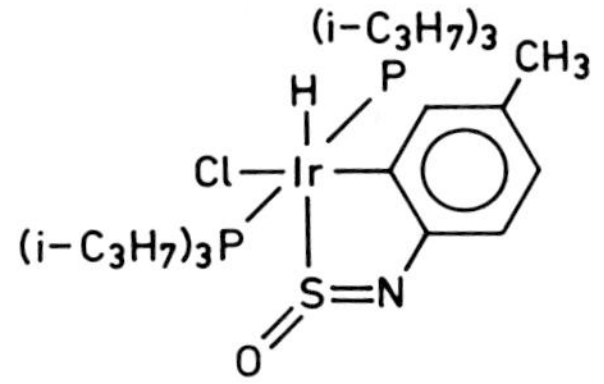

$IrCl(P(C_6H_5)_3)_2(O{=}S{=}NC_6H_4CH_3\text{-}4)$

The title compound was formed by the reaction of $IrCl(P(C_6H_5)_3)_2N_2$ with $O{=}S{=}NC_6H_4CH_3\text{-}4$ in toluene. However, the complex was too unstable to be isolated [5].

$Ir(H)Cl(P(CH(CH_3)CH_2)(C_3H_7\text{-}i)_2)(P(C_3H_7\text{-}i)_3)(O{=}S{=}NC_6H_2(CH_3)_3\text{-}2,4,6)$

The complex was formed during the preparation of $IrCl(P(C_3H_7\text{-}i)_3)_2(O{=}S{=}NC_6H_2(CH_3)_3\text{-}2,4,6)$ (with S-coordinated N=S=O ligand). Two isomers (relative abundances 77 : 23 at 37°C) with the proposed structures were detected by ^{31}P NMR spectroscopy:

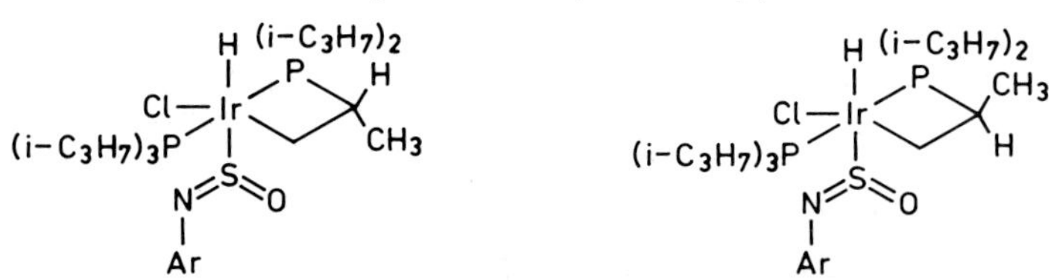

Ar = $C_6H_2(CH_3)_3$-2,4,6

No details of preparation were communicated in the paper.

^{31}P {^1H} NMR ($C_6D_5CD_3$/external 85% H_3PO_4) at 37°C: δ (in ppm) = 11.8, −34.5 (metallated phosphine) and 9.5, −31.6 (metallated phosphine), respectively; ^2J(P–Ir–P) = 318 and 324 Hz, respectively; ^1H NMR ($C_6D_5CD_3$/TMS): δ (in ppm) = 6.65 (s, C_6H_2, meta H); 2.07 (s, CH_3, para); 2.28 (s, CH_3 ortho); ca. −18.1 (m, IrH). IR (Nujol): ν (in cm^{-1}) = 2236 m, ν(IrH); 1306 m, ν(C_6H_2–N); 1248 s, ν(NS); 1144 s, ν(SO); 251 m, ν(IrCl); 307 w, ν(IrS) [5].

Table 66

Spectroscopic Data and Relative Abundances of the Isomers A, B, and C of $IrCl(PR_3)_2(O{=}S{=}NAr)$ [5].
(For isomers A and B, see the corresponding Rh complexes in Table 65, pp. 277/8. The origin of isomer C could not be explained.)

isomer	R	Ar	IR, ν (in cm^{-1}) (Nujol mull, KBr disks)	^{31}P NMR ($C_6D_5CD_3$/ext. 85% H_3PO_4) δ (in ppm) t = 37°C	^1H NMR ($C_6D_5CD_3$/TMS) δ (in ppm)	relative abundances in %, in $C_6D_5CD_3$ t = 37°C	t = −56°C
A	$i\text{-}C_3H_7$	$C_6H_4CH_3\text{-}4$	1290 s, $\nu(C_6H_4{-}N)$; 1264 s, ν(NS); 1148 s, ν(SO); 328 w, ν(IrCl); 312 w, ν(IrS)	29.9	7.67 (d, C_6H_4, ortho H), 6.93 (d, C_6H_4, meta H), 1.95 (s, CH_3, para) (for A, B, C)	74	68
B	$i\text{-}C_3H_7$	$C_6H_4CH_3\text{-}4$	—	15.9		12	17
C	$i\text{-}C_3H_7$	$C_6H_4CH_3\text{-}4$	—	14.8		14	15
A	$i\text{-}C_3H_7$	$C_6H_2(CH_3)_3\text{-}2,4,6$	1304 m, $\nu(C_6H_2{-}N)$; 1242 s, ν(NS); 1142 s, ν(SO); 328 w, ν(IrCl); 308 w, ν(IrS)	30.3	6.65 (s, C_6H_2, meta H), 2.07 (s, CH_3, para), 2.28 (s, CH_3, ortho)	100	—
A	$c\text{-}C_6H_{11}$	$C_6H_4CH_3\text{-}4$	—	19.6	7.63 (d, C_6H_4, ortho H), 6.78 (d, C_6H_4, meta H), 1.98 (s, CH_3, para) (for A, B, C)	84	—
B	$c\text{-}C_6H_{11}$	$C_6H_4CH_3\text{-}4$	—	7.0		11	—
C	$c\text{-}C_6H_{11}$	$C_6H_4CH_3\text{-}4$	—	1.6		5	—

References on p. 287

$IrX(CO)(P(C_6H_5)_3)_2(O{=}S{=}NC_6H_4NO_2\text{-}4)$, X = Cl, Br

The complexes were prepared quantitatively by stirring a mixture of *trans*-$[IrX(CO)(P(C_6H_5)_3)_2]$ (X = Cl, Br) and $O{=}S{=}NC_6H_4NO_2$-4 (mole ratio ~1:2.5) in benzene in a dry box for 1 h. Then the mixture was filtered (open to air), yielding bright yellow and orange products, respectively. The complexes were washed with n-hexane; the washing with benzene or ether resulted in extraction of the N-sulfinylaniline ligand from the complex. The complex with X = Cl was also formed by the reaction of $IrCl(CO)(P(C_6H_5)_3)_2O_2$ with $O{=}S{=}NC_6H_4NO_2$-4. However, unreacted IrCl(CO)-$(P(C_6H_5)_3)_2O_2$ was still present under the reaction conditions used. The melting points are 140 to 145°C for X = Cl and 175°C for X = Br. IR (Nujol) for X = Cl: ν (in cm^{-1}) = 2000 vs, ν(CO); 270 w, ν(IrCl); unassigned: 1585 s; 1290 vs, br; 1178 m; 1110 m, sh; 1080 sh; 940 w; 850 w. The IR absorptions of the Br complex are close to those of the Cl complex [6].

2.7.6.10 N-Sulfinyl-arenamine Complexes of Ni

$Ni(P(C_6H_5)_3)_2(O{=}S{=}NR)$, R = C_6H_5, 4-$CH_3C_6H_4$, 4-$NO_2C_6H_4$, 4-$O{=}S{=}NC_6H_4$, 4-(4′-$O{=}S{=}NC_6H_4$)C_6H_4

The complexes were prepared by adding a solution of O=S=NR in THF to a solution of $(\eta^2\text{-}C_2H_4)Ni(P(C_6H_5)_3)_2$ in THF at 10°C (mole ratio ~1:1). After ethylene evolution ceased, the reaction mixture was stirred for another 2 h. The reaction is said to succeed also with other Ni complexes such as L'_2NiL'' (L′ = PR_3 or 2,2′-bipyridyl, L″ = COD, $C_6H_5COC_6H_5$ etc.). To achieve a better yield, argon was bubbled through the mixture to remove the resulting ethene. The title compounds precipitated on addition of ether and they were isolated after several hours by filtration. Properties and spectra of the complexes are listed in the following table.

R	properties and spectra
C_6H_5	yield 80%; red crystals; IR (Nujol) in the range 1300 to 1000 cm^{-1}: ν (in cm^{-1}) = 1301, 1189, 1120, ν(NS); 1095, 1065, 1045 ν(SO); 1025 UV: λ_{max} = 463 nm
4-$CH_3C_6H_4$	blood red crystals; IR (Nujol): ν (in cm^{-1}) = 1308, 1191, 1123 ν(NS); 1098, 1067, 1041 ν(SO); 1017 UV: λ_{max} = 463 nm
4-$NO_2C_6H_4$	brown-red crystals UV: λ_{max} = 431 nm
4-$O{=}S{=}NC_6H_4$	brown-red crystals UV: λ_{max} = 410 nm
4-(4′-$O{=}S{=}NC_6H_4$)C_6H_4	deep red crystals UV: λ_{max} = 368 nm

The complexes are diamagnetic. They are soluble in THF and insoluble in ether. They quickly exchange their ligands against phosphites or 1,4-diazabutadienes at room temperature [7].

$Ni(bipy)(O{=}S{=}NC_6H_5)_x$, bipy = 2,2′-bipyridyl, x = 1.5 to 2

The complex was obtained by adding $O{=}S{=}NC_6H_5$ to Ni(bipy)(COD), COD = 1,5-cyclooctadiene. An exact stoichiometric complex could not be isolated. The red-brown compound is difficultly soluble. It is paramagnetic (μ_{eff} = 2.75 B.M.). $O{=}S{=}NC_6H_5$ can be thermally liberated below 100°C and $C_6H_5N{=}S{=}NC_6H_5$ is formed at ~100°C. On heating the substance above 100°C with excess $O{=}S{=}NC_6H_5$ SO_2 is eliminated and $C_6H_5N{=}S{=}NC_6H_5$ forms. This reaction is very slow. The $O{=}S{=}NC_6H_5$ ligand may be partly connected to a dimeric anion [8].

$Ni(P(OC_6H_5)_3)_2(O{=}S{=}NC_6H_5)_2$

A solution of $Ni(P(OC_6H_5)_3)_4$ in $CHCl_3$ was refluxed with a solution of $O{=}S{=}NC_6H_5$ in benzene (mole ratio ~1:6) for 2 h. After cooling, a brown solid was centrifugated which was recognized as the title compound (yield ~38%). The diamagnetic compound melts in the range of 169 to 179°C. No detailed spectroscopic data were given [3].

2.7.6.11 N-Sulfinyl-arenamine Complexes of Pd

$Pd(P(C_6H_5)_3)_2(O{=}S{=}NAr)$, Ar = C_6H_5, $C_6H_4CH_3$-4

The complexes were prepared by adding $Pd(P(C_6H_5)_3)_4$ to a nitrogen-degassed $(C_2H_5)_2O$ solution of O=S=NAr (mole ratio ~1:1). They were isolated by filtration. $Pd(P(C_6H_5)_3)_2(O{=}S{=}NC_6H_5)$ is a yellow solid, m.p. 106°C. $Pd(P(C_6H_5)_3)_2(O{=}S{=}NC_6H_4CH_3\text{-}4)$ is a yellow solid, m.p. 150°C; ^{1}H NMR spectrum ($CDCl_3$/TMS): $\delta = 2.16$ ppm (CH_3). The reaction with dioxygen in benzene solution led to the formation of $Pd(P(C_6H_5)_3)_2SO_4$ and 4-$CH_3C_6H_4N{=}NC_6H_4CH_3$-4 [9].

$Pd(P(C_6H_5)_3)_2(O{=}S{=}NC_6H_5)_2$

The title compound was reported to be formed by the reaction of $Pd(P(C_6H_5)_3)_4$ with $O{=}S{=}NC_6H_5$ in benzene (mole ratio 1:6). The red-orange mixture was concentrated and ether was added. An orange-colored solid precipitated, which was recognized as the title compound based on its elemental analysis, yield ~55%. The melting point of the diamagnetic compound is 138 to 140°C. No detailed spectroscopic data were published [3].

2.7.6.12 N-Sulfinyl-arenamine Complexes of Pt

$Pt(P(C_6H_5)_3)_2(O{=}S{=}NAr)$, Ar = C_6H_5, 2-$CH_3C_6H_4$, 4-$CH_3C_6H_4$, 4-$CH_3OC_6H_4$, 4-$NO_2C_6H_4$, 3,5-$(CH_3)_2$-C_6H_3, 2,4,6-$(CH_3)_3C_6H_2$

The complexes were obtained by two different methods: The reaction of $(\eta^2\text{-}C_2H_4)Pt$-$(P(C_6H_5)_3)_2$ with O=S=NAr (Ar = C_6H_5 [6, 10], 2-$CH_3C_6H_4$ [10], 4-$CH_3C_6H_4$, 4-$NO_2C_6H_4$ [6, 10], 3,5-$(CH_3)_2C_6H_3$, 2,4,6-$(CH_3)_3C_6H_2$ [10]) in toluene or benzene (Ar = C_6H_5 yield: 80% [10]), or by the reaction of $Pt(P(C_6H_5)_3)_3$ with O=S=NAr (Ar = C_6H_5 [6, 9], 4-$CH_3C_6H_4$, and 4-$CH_3OC_6H_4$ [9]). Different experimental conditions have been used by applying the latter method. $Pt(P(C_6H_5)_3)_2$-$(O{=}S{=}NC_6H_5)$ was obtained by condensing $O{=}S{=}NC_6H_5$ and toluene onto $Pt(P(C_6H_5)_3)_3$ at 77 K and subsequently warming the mixture to ice temperature [6]. Other workers carried out the reaction (Ar = C_6H_5, 4-$CH_3C_6H_4$, and 4-$CH_3OC_6H_4$) in carefully deoxygenated benzene solution (at room temperature ?) [9].

The molecular geometry of $Pt(P(C_6H_5)_3)_2(O{=}S{=}NC_6H_2(CH_3)_3\text{-}2,4,6)$ was determined by an X-ray crystal structure analysis. The complex crystallizes in the triclinic space group $P\bar{1}$–C_i^1 (No. 2) with a = 9.830(2), b = 10.761(3), c = 20.184(5) Å; $\alpha = 87.93(4)°$, $\beta = 88.98(4)°$, $\gamma = 68.47(3)°$; Z = 2, R = 0.66 for 3422 reflections. The atomic coordinates are given in the paper. The coordination around the platinum is approximately square-planar. The P_2Pt plane is almost perpendicular to

References on p. 287

the NSO plane, showing that the NSO ligand is bonded via the N=S bond to the Pt atom. The aryl group is located cis to S=O. The molecular structure is shown in **Fig. 27**. Selected bond distances and bond angles are given in Table 67 [10].

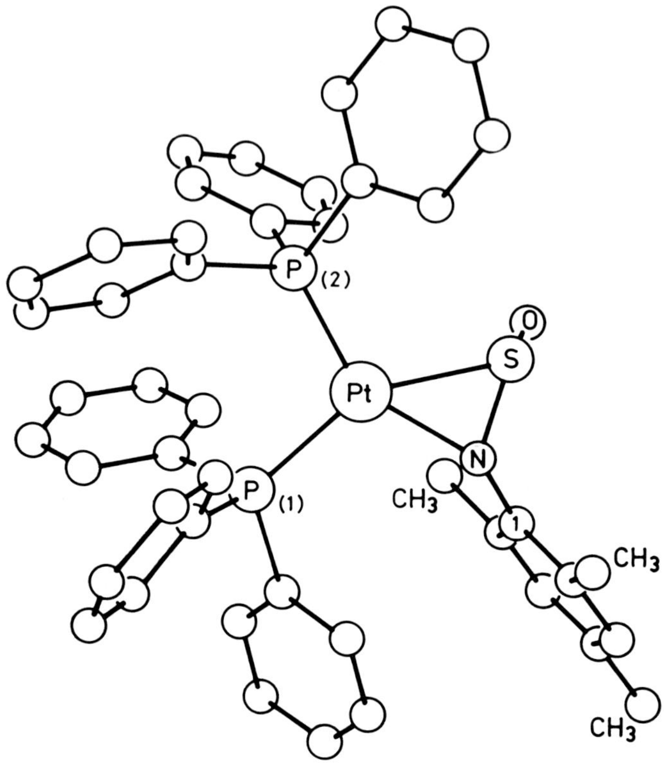

Fig. 27. Molecular structure of $Pt(P(C_6H_5)_3)_2(O{=}S{=}NC_6H_2(CH_3)_3\text{-}2,4,6)$ [10].

Table 67
Bond Distances and Bond Angles of $Pt(P(C_6H_5)_3)_2(O{=}S{=}NC_6H_2(CH_3)_3\text{-}2,4,6)$ [10].

bond distance	in Å	bond angle	in °
O–S	1.455(12)	O–S–N	116.4(7)
S–N	1.629(12)	S–N–C(1)	126.4
N–C(1)	1.44	N–Pt–S	43.1(3)
Pt–P(1)	2.263(4)	Pt–N–S	77.8(5)
Pt–P(2)	2.296(4)	Pt–S–N	59.1(5)
Pt–N	2.056(13)	Pt–S–O	114.3(7)
Pt–S	2.332(4)		

Melting points and spectroscopic properties of the complexes are compiled in Table 68.

Table 68

Properties and Spectra of $Pt(P(C_6H_5)_3)_2(O{=}S{=}NAr)$.

complex (yield in %)	appearance m.p. in °C	spectra: NMR, δ in ppm, J in Hz IR (Nujol mull or KBr disk), ν in cm^{-1} UV (CH_2Cl_2), λ in nm
$Pt(P(C_6H_5)_3)_2(O{=}S{=}NC_6H_5)$ [10], $Pt(P(C_6H_5)_3)_2(O{=}S{=}NC_6H_5)\cdot\frac{1}{2}C_6H_6$ (85 and 98) [6], $Pt(P(C_6H_5)_3)_2(O{=}S{=}NC_6H_5)\cdot C_6H_6$ [9]	yellow solid 115 [6], 130 [9]	IR: 1256 m, ν(C_6H_5–N); 1051 s, ν(SO); 926 m, ν(NS) [10]; 1255 s, 1055 vs, 930 w [6]
$Pt(P(C_6H_5)_3)_2(O{=}S{=}^{15}NC_6H_5)$ [10]	—	^{195}Pt NMR ($CDCl_3$/*cis*-$PtCl_2(S(CH_3)_2)_2$), at −30°C: δ = −660, J(Pt, N) = 136 ± 10 ^{31}P NMR ($CDCl_3$/external 85% H_3PO_4), at −30°C: δ = 20.0, ^{1}J(Pt, P) = 3407, ^{2}J(N–Pt–P) = 2; δ = 16.3, ^{1}J(Pt, P) = 4587, ^{2}J(N–Pt–P) = 25, ^{2}J(P–Pt–P) = 9 IR: 1248 m, ν(C_6H_5–N); 1050 s, ν(SO); 916 m, ν(NS) [10]
$Pt(P(C_6H_5)_3)_2(O{=}S{=}NC_6H_4CH_3\text{-}2)$ [10]	—	^{1}H NMR ($CDCl_3$/TMS): δ = 2.28 (CH_3) ^{31}P NMR ($CDCl_3$/external 85% H_3PO_4), at −30°C: δ = 18.8, ^{1}J(Pt, P) = 3498; δ = 15.5, ^{1}J(Pt, P) = 4531, ^{2}J(P–Pt–P) = 14 IR: 1288 m, ν(C_6H_5–N); 1058 s, ν(SO); 912 w, ν(NS) [10]
$Pt(P(C_6H_5)_3)_2(O{=}S{=}NC_6H_4CH_3\text{-}4)$ [9, 10], $Pt(P(C_6H_5)_3)_2(O{=}S{=}NC_6H_4CH_3\text{-}4)\cdot C_6H_6$ (76) [6]	yellow solid 107 [9], 170 [6]	^{1}H NMR ($CDCl_3$/TMS): δ = 6.78 (H, ortho), 6.48 (H, meta), 2.05 (CH_3) [10]; δ = 2.13 (CH_3) [9] ^{31}P NMR ($CDCl_3$/external 85% H_3PO_4) at −30°C: δ = 20.0, ^{1}J(Pt, P) = 3483; δ = 16.6, ^{1}J(Pt, P) = 4580, ^{2}J(P–Pt–P) = 11 [10] IR: 1500 s, 1240 m, br, 1048 vs, 920 w, 820 w [6]; 1250 m, ν(C_6H_5–N); 1056 s, ν(SO); 922 m, ν(SN) [10] UV: 333 (π → π*), ε = 9000 $L\cdot mol^{-1}\cdot cm^{-1}$ [10]
$Pt(P(C_6H_5)_3)_2(O{=}S{=}NC_6H_4NO_2\text{-}4)$ (100) [6]	orange solid 170 [6]	IR: 1575 vs, 1290 vs, 1280 vs, 1175 m, 1108 s, 1050 vs, 918 s, 830 w [6]
$Pt(P(C_6H_5)_3)_2(O{=}S{=}NC_6H_4OCH_3\text{-}4)$ [9]	110 [9]	^{1}H NMR ($CDCl_3$/TMS): δ = 3.75 (OCH_3) [9]

References on p. 287

Table 68 (continued)

complex (yield in %)	appearance m.p. in °C	spectra: NMR, δ in ppm, J in Hz IR (Nujol mull or KBr disk), ν in cm^{-1} UV (CH_2Cl_2), λ in nm
$Pt(P(C_6H_5)_3)_2(O{=}S{=}NC_6H_3(CH_3)_2$-3,5) [10]	—	^1H NMR ($CDCl_3$/TMS): δ = 6.45 (H, ortho), 6.23 (H, para), 1.83 (CH_3) ^{31}P NMR ($CDCl_3$/external 85% H_3PO_4) at −30°C: δ = 19.7, $^1J(Pt,P)$ = 3388; δ = 16.0, $^1J(Pt,P)$ = 4560, $^2J(P\text{–}Pt\text{–}P)$ = 9 IR: 1314 s, ν(C_6H_3–N); 1040 s, ν(SO); 968 m, ν(NS) UV: 320 ($\pi \rightarrow \pi^*$) [9]
$Pt(P(C_6H_5)_3)_2(O{=}S{=}NC_6H_2(CH_3)_3$-2,4,6) [10]	—	^1H NMR ($CDCl_3$/TMS): δ = 6.47 (H, meta), 2.13 (CH_3, para), 2.22 (CH_3, ortho) ^{31}P NMR ($CDCl_3$/external 85% H_3PO_4) at −30°C: δ = 16.6, $^1J(Pt,P)$ = 3560; δ = 15.1, $^1J(Pt,P)$ = 4545, $^2J(P\text{–}Pt\text{–}P)$ = 13 IR: 1220 m, ν(C=$C_{arom.}$); 1050 s, ν(SO); 906 w, ν(NS) UV: 377 ($\pi \rightarrow \pi^*$), ε = 2350 $L \cdot mol^{-1} \cdot cm^{-1}$ [10]

Solutions of the complexes $Pt(P(C_6H_5)_3)_2(O{=}S{=}NAr)$ in $CDCl_3$, aged for a few days, formed *cis*-$Pt(P(C_6H_5)_3)_2Cl_2$ and $Pt(P(C_6H_5)_3)_2SO_4$ [10]. When $Pt(P(C_6H_5)_3)_2(O{=}S{=}NAr)$ (Ar = C_6H_5, 4-CH_3-C_6H_4, 2-$CH_3C_6H_4$, 3,5-$(CH_3)_2C_6H_3$, and 2,4,6-$(CH_3)_3C_6H_2$) were dissolved in incompletely dried and deoxygenated solvents, such as C_6H_5Cl, the complex $Pt(P(C_6H_5)_3)_2SO_4 \cdot C_6H_5Cl$ formed [10]. The same reaction occurred when dioxygen was bubbled through a benzene solution of $Pt(P(C_6H_5)_3)_2(O{=}S{=}NC_6H_4CH_3\text{-}4)$ to yield $Pt(P(C_6H_5)_3)_2SO_4$ as white precipitate. 4-$CH_3C_6H_4N{=}N$-$C_6H_4CH_3$-4 was recognized in the mother liquor. When the same complex was treated with mineral acids, HX (X = BF_4, ClO_4) precipitation of $[Pt_2(OH)_2(P(C_6H_5)_3)_4]X_2$ in $(C_2H_5)_2O$ and evolution of SO_2 was observed. 4-$CH_3C_6H_4NH_2 \cdot HX$ was recognized in the mother liquor. The reaction with HCl under analogous conditions yielded *cis*-$Pt(P(C_6H_5)_3)_2Cl_2$ besides SO_2 and 4-$CH_3C_6H_4NH_2 \cdot HCl$. When $Pt(P(C_6H_5)_3)_2(O{=}S{=}NC_6H_4CH_3\text{-}4)$ was treated with CO (in hexane–cyclohexane (10:1)), $P(C_6H_5)_3$ (in nitrogen-degassed C_2H_5OH), fumaronitrile (in benzene–$(C_2H_5)_2O$), and maleic anhydride (in nitrogen-degassed C_2H_5OH), respectively, the N-sulfinylaniline ligand was displaced, and $Pt(P(C_6H_5)_3)_2L_n$ complexes were formed (L = CO, $P(C_6H_5)_3$, n = 2; L = maleic anhydride or fumaronitrile, n = 1) [9].

$Pt(P(C_6H_5)_3)_2(O{=}S{=}NC_6H_5)_2$

The title complex was obtained by reaction of $Pt(P(C_6H_5)_3)_4$ with $O{=}S{=}NC_6H_5$ in benzene (mole ratio ~1:6). After stirring for 1 h and concentration of the mixture, ether was added to precipitate the pale yellow solid which was recognized to be the title compound (~60% yield). The melting point is 172 to 173°C. The complex is diamagnetic. Detailed spectroscopic data were not communicated [3].

2.7.6.13 N-Sulfinyl-aniline Complex of Cu

$CuCl(P(C_6H_5)_3)_2(O{=}S{=}NC_6H_5)$

The compound was prepared by the reaction of $CuCl(P(C_6H_5)_3)_3$ dissolved in CH_2Cl_2 with a solution of $O{=}S{=}NC_6H_5$ in the same solvent (mole ratio ~1:6.5). After stirring and concentration a pale yellow solid precipitated when petroleum ether was added. The substance was identified to be the title compound (yield ~43%). The melting point of the diamagnetic complex is 220 to 223°C. No detailed spectroscopic data were given in the paper [3].

2.7.6.14 N-Sulfinyl-aniline Complex of Ag

$[Ag(P(C_6H_5)_3)(O{=}S{=}NC_6H_5)]^+NO_3^-$

The complex was synthesized by refluxing a mixture of $[Ag(P(C_6H_5)_3)_4]^+NO_3^-$ and $O{=}S{=}N$-C_6H_5 in CH_3CN (mole ratio ~1:5) for 1 h. On cooling, a gray-colored solid precipitated which was recognized as the title compound (yield ~38%). The melting point of the diamagnetic complex is >300°C. No detailed spectroscopic data were given [3].

References:

[1] Wilkins, J. D. (J. Inorg. Nucl. Chem. **38** [1976] 673/5).
[2] Slawisch, A. (Naturwissenschaften **56** [1969] 396; Z. Anorg. Allgem. Chem. **374** [1970] 291/6).
[3] Arulsamy, K. S.; Ashok, R. F. N.; Agarwala, U. C. (Indian J. Chem. A **23** [1984] 21/5).
[4] Ashton, H. C.; Manning, A. R. (Inorg. Chem. **22** [1983] 1440/5).

[5] Meij, R.; Stufkens, D. J.; Vrieze, K.; Van Gerresheim, W.; Stam, C. H. (J. Organometal. Chem. **164** [1979] 353/70).
[6] Blake, D. M.; Reynolds, J. R. (J. Organometal. Chem. **113** [1976] 391/6).
[7] Walther, D.; Pfützenreuter, C. (Z. Chem. [Leipzig] **17** [1977] 426/7).
[8] Walther, D.; Dinjus, E.; Wolf, H. (Z. Chem. [Leipzig] **19** [1979] 381/2).
[9] La Monica, G.; Pizzotti, M.; Cenini, S. (Gazz. Chim. Ital. **108** [1978] 611/3).
[10] Meij, R.; Stufkens, D. J.; Vrieze, K.; Roosendaal, E.; Schenk, H. (J. Organometal. Chem. **155** [1978] 323/36).

[11] Herberhold, M.; Hill, A. F. (J. Organometal. Chem. **368** [1989] 111/7).

2.8 *tert*-Butylimino-hydroxy-λ^4-sulfanyl Radical, $HO{-}\dot{S}{=}NC_4H_9$-t, and *tert*-Butylamino-sulfinyl Radical, $O{=}\dot{S}{-}NHC_4H_9$-t

The radical $HO\dot{S}{=}NC_4H_9$-t was deduced from ESR spectroscopic observations on photolysis of a mixture containing $O{=}S{=}NC_4H_9$-t, t-$C_4H_9OOC_4H_9$-t, and isopropyl alcohol. It arises by a direct hydrogen transfer from $(CH_3)_2\dot{C}OH$ (obtained from t-$C_4H_9\dot{O}$ + $(CH_3)_2CHOH \rightarrow (CH_3)_2\dot{C}OH$ + t-C_4H_9OH) according to $(CH_3)_2\dot{C}OH + O{=}S{=}NC_4H_9\text{-t} \rightarrow (CH_3)_2CO + HO\dot{S}{=}NC_4H_9$-t, or by an addition elimination sequence according to $(CH_3)_2\dot{C}OH + O{=}S{=}NC_4H_9\text{-t} \rightarrow (CH_3)_2C(OH)S(O)\dot{N}C_4H_9\text{-t} \rightarrow (CH_3)_2CO + HO\dot{S}{=}NC_4H_9$-t. It is difficult to eliminate with certainty the tautomeric aminosulfinyl radical $O{=}\dot{S}NHC_4H_9$-t. Formation of the deuterated species was not given. ESR data in cyclopropane at 200 K are given below:

radical	g factor	hyperfine splitting in G	
		a(N)	a(H)*)
$HO\dot{S}{=}NC_4H_9$-t	2.0071	1.4	9.5 (1)
$DO\dot{S}{=}NC_4H_9$-t	2.0072	1.4	1.4 (1 D)

*) The number of nuclei coupling is given in parentheses.

Reference:

Baban, J. A.; Roberts, B. P. (J. Chem. Soc. Perkin Trans. II **1978** 678/83).

2.9 Dialkylamino-sulfinyl Radicals, $O{=}\dot{S}{-}NR^1R^2$, $R^1 = R^2 = CH_3$, C_2H_5; R^1, $R^2 = -(CH_2)_4-$

The radicals were generated by photolysis of the corresponding diaminosulfoxides $(R^1R^2N)_2SO$ in cyclopropane (sometimes with CH_2Cl_2 to increase solubility) or by irradiation of mixtures of di-*tert*-butyl peroxide or bis(trimethylsilyl)peroxide and the corresponding $(R^1R^2N)_2SO$ compound. The radical with $R^1{=}R^2 = CH_3$ was also detected during photolysis of $(CH_3)_2NS(O)Cl$ in cyclopropane–CH_2Cl_2 solvent. The species (for ESR data, see Table 69) are closely related to the alkoxysulfinyl radicals $O{=}\dot{S}{-}OR$; g values are intermediate between those found for $O{=}\dot{S}{-}R$ and $O{=}\dot{S}{-}OR$ radicals.

$O{=}\dot{S}{-}N(C_2H_5)_2$ presumably dimerises to give the S–S- or S–O-coupled product. Spectroscopically, the decay of the radical in cyclopropane at 163 K was deduced to be of second order with a rate constant of 1.1×10^9 $L \cdot mol^{-1} \cdot s^{-1}$.

Table 69

ESR[a] Parameters for Dialkylamino-sulfinyl Radicals at ~170 K.

radical	solvent	g factor	hyperfine splitting in G a(N)	a(H)[c]
$O{=}\dot{S}{-}N(CH_3)_2$[b]	cyclopropane–dichloromethane (1:1)	2.0060	6.8	3.8 (6)
$O{=}\dot{S}{-}N(C_2H_5)_2$	cyclopropane	2.0060	6.1	2.3 (4)
$O{=}\dot{S}{-}N(CH_2)_4$	cyclopropane–ethylene oxide (6:1)	2.0061	6.3	3.7 (4)

[a] The spectra of the radicals were most intense at low temperatures, and the decrease in amplitude at higher temperatures was not accounted for by an increase in linewidth. – [b] A figure of the ESR spectrum at 160 K is given in the paper. – [c] The number of nuclei coupling is given in parentheses.

Reference:

Baban, J. A.; Roberts, B. P. (J. Chem. Soc. Perkin Trans. II **1978** 678/83).

2.10 Salts of the N-Sulfinyl-aminium Ions (1+), $[O{=}S{\doteq}NR^1R^2]^+Y^-$,

R^1 = alkyl, R^2 = alkyl, perfluoroalkyl, phenyl; R^1, R^2 = $-(CH_2)_n-$; Y = BF_4, AsF_6, SbF_6, $SbCl_6$

The formation methods for these salts are shown in the scheme below (LA = Lewis acid: BF_3, AsF_5, SbF_5, and $SbCl_5$; X = F, Cl).

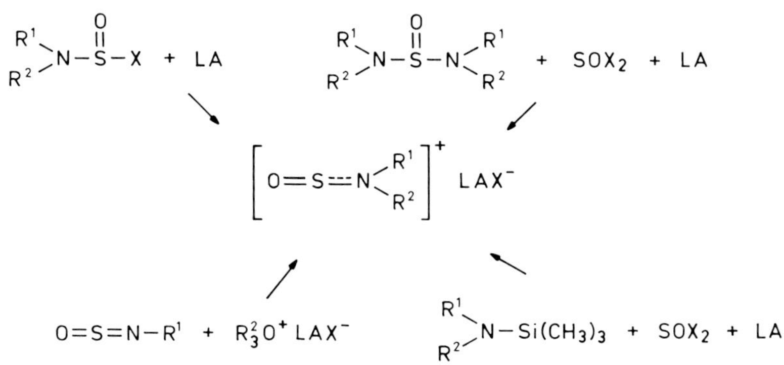

For structural considerations, see $[OSN(CH_3)_2]^+BF_4^-$ (p. 290). The salts are colorless, very moisture-sensitive crystalline products, which are stable and can be stored under N_2 at room temperature for more than a year without decomposition (exceptions: $[OSN(C_2H_5)_2]^+BF_4^-$ and $[OSN(C_6H_5)CH_3]^+SbCl_6^-$). In aprotic solvents as CH_3CN or CH_3NO_2 the salts are slightly soluble, in CH_2Cl_2, the solubility is weak. The chemical reactivity is similar to that of N-sulfinyl-4-methylbenzenesulfonamide, most reactions of the salts proceeding smoothly at room temperature. Thus, azasulfonium salts may be obtained with sulfoxides, cycloadducts with dienes, and iminium salts with aldehydes or thiocarbonyl compounds.

2.10.1 Salts of the N-Sulfinyl-N,N-dimethyl-aminium Ion (1+), $[O{=}S{\cdots}N(CH_3)_2]^+Y^-$, $Y = BF_4$, AsF_6, SbF_6, $SbCl_6$

$\mathbf{[OSN(CH_3)_2]^+BF_4^-}$ is quantitatively formed by condensing $(CH_3)_2NS(O)F$ with equimolar amounts of BF_3 in liquid SO_2 at $-196°C$ and warming to $-80°C$ and then slowly to room temperature [1]. $[OSN(CH_3)_2]^+BF_4^-$ was also obtained in 97% [2] to 99% yield [3] by methylation of $O{=}S{=}NCH_3$ with $(CH_3)_3O^+BF_4^-$ for 10 h at 15°C without a solvent [2, 3].

Solid, m.p. 72 to 75°C [2], 75°C [1, 3] (from CH_3CN–CH_2Cl_2–$(C_2H_5)_2O$) [3]. IR (Nujol ?): 14 absorption bands between 3040 and 515 cm^{-1}, one band region 1140 to 1000 cm^{-1}, unassigned, recorded [1]. ^1H NMR (SO_2/TMS): $\delta = 3.84$ ppm (CH_3) [1]; (CD_3CN, at room temperature): $\delta = 3.85$ ppm (s, CH_3), and two signals with distance 0.06 ppm at $-55°C$ [3]; (CD_3NO_2/TMS): $\delta = 4.02$ ppm (s, CH_3) [2]; (CH_3CN/TMS): $\delta = 4.00$ ppm, measured during reaction pursuit [4]. ^{19}F NMR ($SO_2/CFCl_3$): no signal observed [1].

The N,N-dimethylsulfoxoimonium ion structure $O{=}S{=}\overset{\oplus}{N}(CH_3)_2$ was concluded from preparative considerations [2] and by IR data [5]. But from comparing studies (with the isosteric system $SO_2/CH_3OSO^+/CH_3SOF$) it was shown that a participation of the sulfur cation $O{=}\overset{\oplus}{S}{-}N(CH_3)_2$ has to be discussed [1].

From temperature dependence of the ^1H NMR spectrum, it was concluded that the rotation around the S–N bond is hindered. From the coalescence temperature $T_c \approx -22°C$, the free activation enthalpy for the rotation around the S–N bond, or inversion at N, was calculated to be 56.9 ± 0.8 kJ/mol, indicating an at least partial double-bond character of the S–N bond [3].

$[OSN(CH_3)_2]^+BF_4^-$ is an extremely moisture-sensitive compound puffing up at once with evolution of SO_2 when exposed to moist air [3]. As an electron-poor compound, it is suitable for cycloadditions; for instance, $[OSN(CH_3)_2]^+BF_4^-$ reacts with dienes $CH_2{=}C(R^1)C(R^2){=}CR^3$ (~10% excess) (R^1, $R^2 = CH_3$, $R^3 = H$, or $R^1 = CH_3$, R^2, $R^3 = H$: in CH_3CN, ice-cooling and 1 h at room temperature; yield 95 or 91%, respectively; R^1, $R^2 = H$, $R^3 = CH_3$: in CH_2Cl_2 and 1 h at room temperature; yield 92%; R^1, R^2, $R^3 = H$: in CH_3CN and 1 h at room temperature; yield 13%) with 1,4-addition to form the corresponding tetrafluoroborates of 3,6-dihydro-2H-1,2-thiazinium 1-oxides [3], see also [2]:

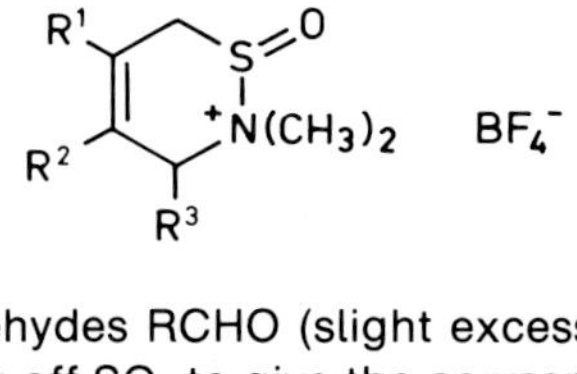

$[OSN(CH_3)_2]^+BF_4^-$ reacts with aldehydes RCHO (slight excess, mole ratio 1:1.1 to 1.2, R see Table 70) in dry CH_3CN by splitting off SO_2 to give the corresponding aldimonium tetrafluoroborates: $RC(H){=}O + [OSN(CH_3)_2]^+BF_4^- \rightarrow [RC(H){=}N(CH_3)_2]^+BF_4^- + SO_2$ [6, 7]. With thione compounds $[OSN(CH_3)_2]^+BF_4^-$ reacts analogously with a formal S_2O splitting off: $C_6H_5C(R){=}S$ + $[OSN(CH_3)_2]^+BF_4^- \rightarrow [C_6H_5C(R){=}N(CH_3)_2]^+BF_4^- +$ "S_2O" [7].

For details (nature of the substituents R, reaction conditions, and yield) see Table 70.

When adding a 2M solution of $[OSN(CH_3)_2]^+BF_4^-$ in CH_3CN to a 0.1 to 1M solution of the ketimine $R^1R^2C{=}C{=}NC_6H_5$ ($R^1 = R^2 = CH_3$, C_6H_5, 4-ClC_6H_4, 4-$CH_3OC_6H_4$; $R^1 = CH_3$, $R^2 = SC_6H_5$) in CH_2Cl_2 at $-20°C$, stirring the mixture for 0.5 h and allowing to come to room temperature the corresponding tetrafluoroborates of 3-dimethylimonio-1,2-thiazetidine 1-oxides I (see p. 292) were formed in 52 to 88% yield [8].

Table 70

Reactions of $[OSN(CH_3)_2]^+BF_4^-$ with Aldehydes RCHO and Thione Compounds, e.g., $C_6H_5C(S)R$.

reaction partners	reaction conditions	products	yield (in %)	Ref.
$[OSN(CH_3)_2]^+BF_4^- + RC(H){=}O$				
R				
C_2H_5	5 min at 20°C	$C_2H_5CH{=}N(CH_3)_2^+BF_4^-$	72	[6]
$i\text{-}C_3H_7$	5 min at 20°C	$i\text{-}C_3H_7CH{=}N(CH_3)_2^+BF_4^-$	85	[6]
$CH_3CH{=}CH$	1 h at 20°C	$CH_3CH{=}CHCH{=}N(CH_3)_2^+BF_4^-$	100 (crude)	[6]
$CCl_3CH{=}CH$	3 h at 20°C	$CCl_3CH{=}CHCH{=}N(CH_3)_2^+BF_4^-$	64	[6]
C_6H_5	5 min at reflux	$C_6H_5CH{=}N(CH_3)_2^+BF_4^-$	97	[7]
$C_6H_5C(CH_3)H$	10 min at 20°C	$C_6H_5C(CH_3)HCH{=}N(CH_3)_2^+BF_4^-$	72	[6]
$C_6H_5CH{=}CH$	2 h at 20°C	$C_6H_5CH{=}CHCH{=}N(CH_3)_2^+BF_4^-$	75	[6]
$2,4,6\text{-}(CH_3O)_3C_6H_2$	2 h at room temperature	$2,4,6\text{-}(CH_3O)_3C_6H_2CH{=}N(CH_3)_2^+BF_4^-$	99	[7]
$4\text{-}(CH_3)_2NC_6H_4$	1 h at room temperature	$4\text{-}(CH_3)_2NC_6H_4CH{=}N(CH_3)_2^+BF_4^-$	86	[7]
$[OSN(CH_3)_2]^+BF_4^- + C_6H_5C(R){=}S$				
R				
$N(CH_3)_2$	in CH_3CN, 20 d at room temperature	$(CH_3)_2NC(C_6H_5){=}N(CH_3)_2^+BF_4^-$	87	[7]
SCH_3	in $CD_3CN\text{–}CH_2Cl_2$	$CH_3SC(C_6H_5){=}N(CH_3)_2^+BF_4^-$	80	[7]
C_6H_5		no analytically pure product		[7]
$[OSN(CH_3)_2]^+BF_4^-$ + 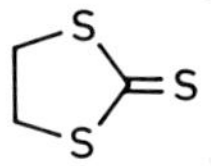	in CH_3CN, 24 h at room temperature	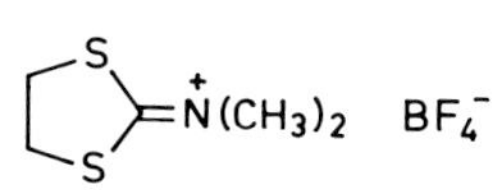	61	[7]
$[OSN(CH_3)_2]^+BF_4^-$ + 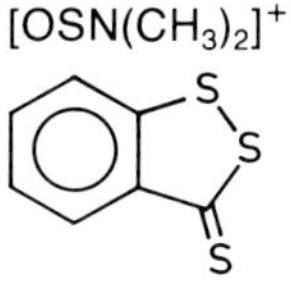	in $CH_3CN\text{–}CHCl_3$, 2 d at room temperature	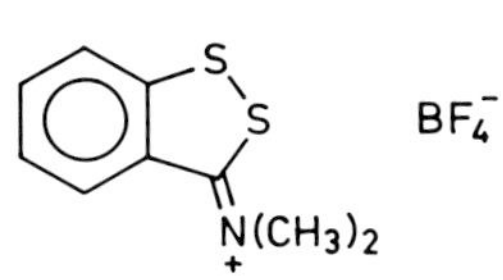	87	[7]
$[OSN(CH_3)_2]^+BF_4^-$ + 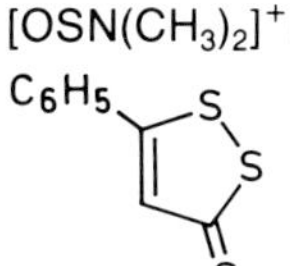	in $CH_3CN\text{–}CH_2Cl_2$, 3 d at room temperature	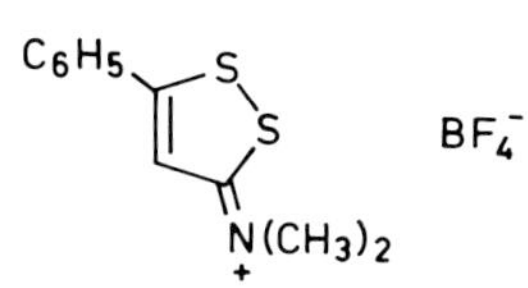	72	[7]

References on p. 297

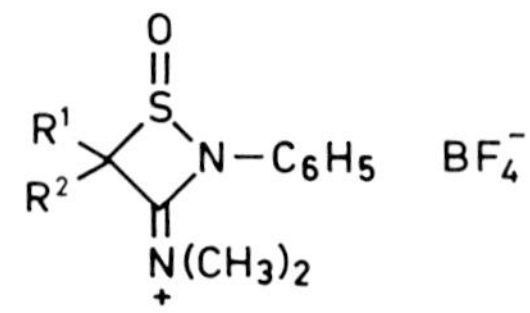

I

$R^1 = R^2 = CH_3$, C_6H_5, 4-ClC_6H_4, 4-$CH_3OC_6H_4$; $R^1 = CH_3$, $R^2 = SC_6H_5$

$[OSN(CH_3)_2]^+BF_4^-$ reacts with the aliphatic carbodiimides i-$C_3H_7N{=}C{=}NC_3H_7$-i or c-$C_6H_{11}N{=}C{=}N$-C_6H_{11}-c (equimolar amounts, addition at 20°C and 1 h furthermore stirring) in anhydrous CH_3CN to form the 3-dimethylamino-1,2,4-thiadiazetium 1-oxide tetrafluoroborate II; yield: 46 or 52%, respectively [9].

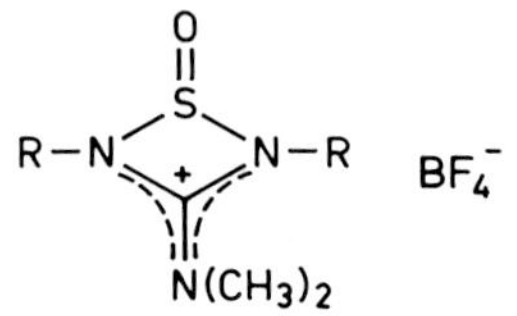

II

R = i-C_3H_7 or c-C_6H_{11}

The compound reacts in CH_3NO_2 with equimolar amounts of diarylcarbodiimides, 4-RC_6H_4-$N{=}C{=}NC_6H_4R$-4 (R = H: 16 h at 40°C; R = OCH_3: 72 h at room temperature) to form 3-dimethylimonio-3,4-dihydro-2H-1,2,4-benzothiadiazine 1-oxide tetrafluoroborate derivatives III; yield 54 or 74%, respectively.

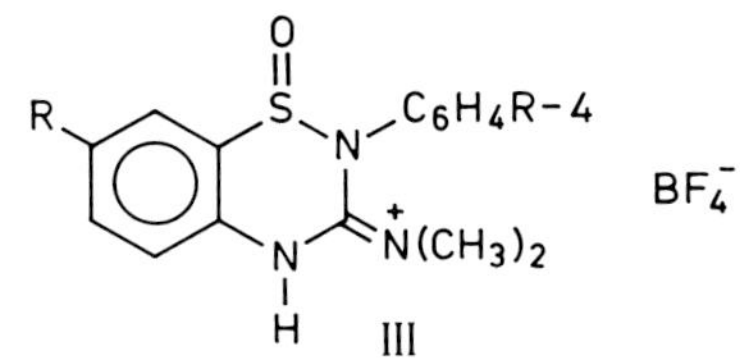

III

R = H or OCH_3

With the bromo derivative (R = Br) no reaction was observed even at 60 to 80°C for 16 h. When N,N'-bis(4-methoxyphenyl)carbodiimide was reacted with equimolar amounts of $[OSN(CH_3)_2]^+$-BF_4^- in CH_3NO_2 at −40°C for 4 h and then a saturated aqueous solution of K_2CO_3 was added and the mixture stirred for 24 h at 20°C SO_2 and $(4\text{-}CH_3OC_6H_4NH)_2C{=}N(CH_3)_2^+BF_4^-$ (30%) was obtained [10].

When adding a solution of the sulfinyl compounds R^1R^2SO (5% excess; $R^1 = R^2 = CH_3$, C_2H_5; R^1, $R^2 = -[CH_2]_4-$; $R^1 = CH_3$, $R^2 = 4\text{-}CH_3OC_6H_4$, or $N(CH_3)_2$) in CH_3CN to a solution of $[OSN(CH_3)_2]^+BF_4^-$ in CH_3CN at −25°C, and allowing the mixture to react for 1 h at −25°C and another 2 to 12 h at room temperature the corresponding aminosulfonium salts $[(CH_3)_2NSR^1R^2]^+BF_4^-$ are formed with evolution of SO_2 [4], see also [11]: $[OSN(CH_3)_2]^+$-$BF_4^- + OSR^1R^2 \rightarrow [(CH_3)_2NSR^1R^2]^+BF_4^- + SO_2$, $R^1 = R^2 = CH_3$ (46%), C_2H_5 (57%); R^1, $R^2 = -[CH_2]_4-$ (54%); $R^1 = CH_3$, $R^2 = 4\text{-}CH_3OC_6H_4$ (52%); $R^1 = CH_3$, $R^2 = N(CH_3)_2$ (67%).

$[OSN(CH_3)_2]^+BF_4^-$ reacts with pairs of stereoisomeric cyclic sulfoxides to give one product in each case. The reactions are summarized in Table 71.

$[OSN(CH_3)_2]^+AsF_6^-$ was quantitatively formed by condensing $(CH_3)_2NS(O)F$ with a 10% excess of AsF_5 in liquid SO_2 at −196°C and warming to −80°C and then slowly to room temperature. Solid, m.p. 125°C. IR (Nujol?): 12 unassigned absorption bands between 3040 and 512 cm^{-1} recorded. ^1H NMR (SO_2/TMS): δ = 3.83 ppm (s). ^{19}F NMR ($SO_2/CFCl_3$): δ = −64.2 ppm (1:1:1:1 q, As–F), $^1J(As,F)$ = 935 Hz [1].

Table 71
Reactions of $[OSN(CH_3)_2]^+BF_4^-$ with Cyclic Sulfoxides [12], see also [13].

cyclic sulfoxide	reaction conditions, yield	product
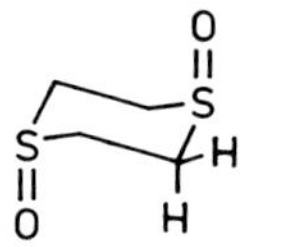	2:1 mole ratio, in CH_3CN, −30°C and 24 h at room temperature; 18% yield	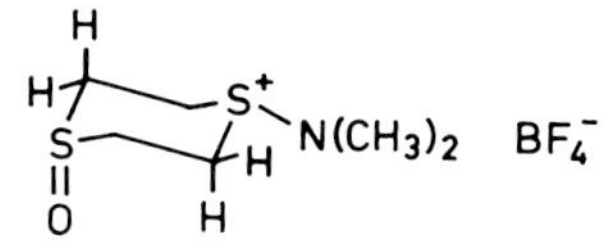
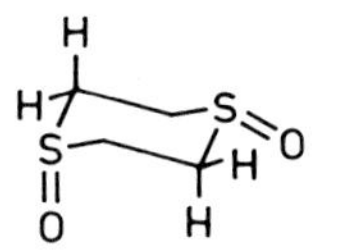	1:1 mole ratio, in CH_3CN, −40°C and 4 d at 20°C; 15% yield	
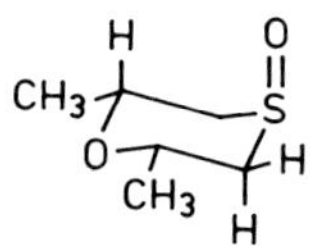 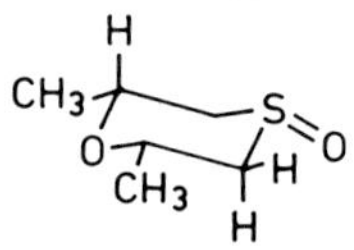	1:1 mole ratio, in CH_3CN, 3 h at −30°C and 3 h at 20°C; yield 29 or 27%, respectively	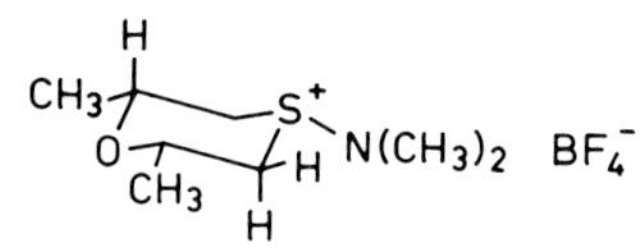
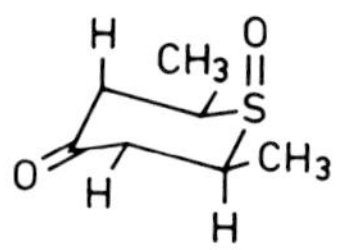 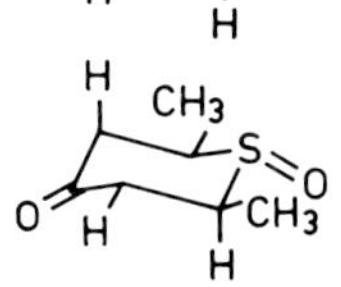	1:1 mole ratio in CH_3CN–CH_2Cl_2, 3 h at −30°C and 16 h at 20°C; quantitative yield	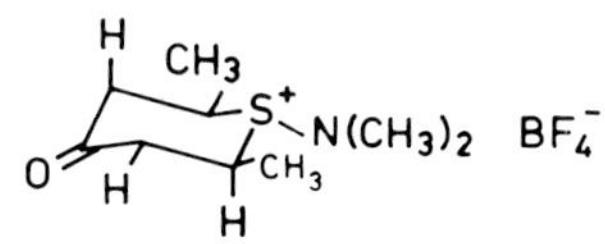

$[OSN(CH_3)_2]^+SbF_6^-$ was quantitatively formed by condensing $(CH_3)_2NS(O)F$ with equimolar amounts of SbF_5 in liquid SO_2 at −196°C and warming to −80°C and then slowly to room temperature. Solid, m.p. 157°C. IR (Nujol?): 12 absorption bands (unassigned) between 3040 and 512 cm^{-1} recorded. ^1H NMR (SO_2/TMS): δ = 3.98 ppm (s). ^{19}F NMR ($SO_2/CFCl_3$): no signal observed [1].

$[OSN(CH_3)_2]^+SbCl_6^-$ was formed by adding a slight excess of $SbCl_5$ dissolved in CH_2Cl_2 to a solution of $(CH_3)_2NS(O)Cl$ in CH_2Cl_2 at room temperature; yield 94%. In a 91% yield $[OSN(CH_3)_2]^+SbCl_6^-$ was obtained when equimolar amounts of $(CH_3)_2NS(O)N(CH_3)_2$ and $SOCl_2$ in CH_2Cl_2 were brought together, stirred for 1 h and then added to a double molar amount of $SbCl_5$ in CH_2Cl_2. Reacting $(CH_3)_2NS(O)N(CH_3)_2$ in CH_2Cl_2 with $SbCl_5$ (mole ratio 1:2) in CH_2Cl_2 gave the hexachloroantimonate in an 68% yield [5].

$[OSN(CH_3)_2]^+SbCl_6^-$ is an extremely moisture-sensitive, colorless fine-crystalline solid. It melts at 80°C with brown coloring and decomposition. IR (Nujol): ν (in cm^{-1}) = 3038 m, 2950 m, ν(CH); 1470 sh, 1435 m, 1425, 1408 m, δ(CH); 1460 s to vs, ν_{as}(O=S=N); 1290 ms, ϱ, γ(CH); 1243 vs, ν_s(O=S=N); 1175, ϱ, γ(CH); 1035 m, $\nu_{as}(NC_2)$; 945 s, $\nu_s(NC_2)$; 515 m, δ(O=S=N); 335 vs, $\nu(SbCl_6)$; Raman: ν (in cm^{-1}) (rel. int.) = 335 (100), 292 (24), 73 (32), $\nu(SbCl_6)$ [5].

References on p. 297

The compound reacts with 1,3-dienes $CH_2{=}C(R^1)C(R^2){=}CHR^3$ (~10% excess) ($R^1 = R^2 = CH_3$, $R^3 = H$: in CH_3CN, ice-cooling and 1 h at room temperature, yield 46%; $R^1 = CH_3$, $R^2 = R^3 = H$: in CH_3CN at −25°C and 1 h at −10°C, yield 44%; $R^1{=}R^2{=}H$, $R^3 = CH_3$: in CH_3CN at −40°C for 1 h, yield 44.5%; $R^1 = R^2 = R^3 = H$: in CH_3CN at −10°C for 0.5 h, yield 51%) with 1,4-addition to form the corresponding hexachloroantimonates of 3,6-dihydro-2H-1,2-thiazinium 1-oxides [3]:

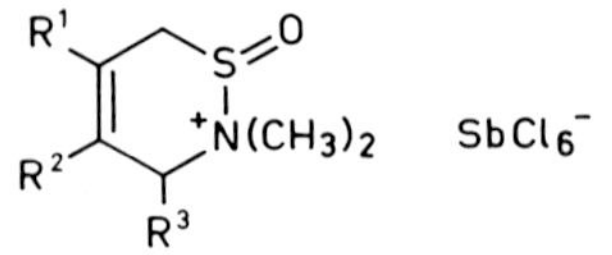

Passing CH_2N_2 through a solution of $[OSN(CH_3)_2]^+SbCl_6^-$ in CH_3CN cooled to −30°C gives, 1,1-dimethylazetidinium hexachloroantimonate in a 19% yield [3]:

Adding a solution of the sulfinyl compounds R^1R^2SO (5% excess, $R^1 = R^2 = CH_3$, $N(CH_3)_2$, $R^1 = CH_3$, $R^2 = N(CH_3)_2$) in CH_3CN to a solution of $[OSN(CH_3)_2]^+SbCl_6^-$ in CH_3CN at −25°C, and allowing the mixture to react for 1 h at −25°C and for another 2 to 12 h at room temperature gives the corresponding aminosulfonium hexachloroantimonates according to $[OSN(CH_3)_2]^+$-$SbCl_6^- + R^1R^2SO \rightarrow [(CH_3)_2NSR^1R^2]^+SbCl_6^- + SO_2$; $R^1 = R^2 = CH_3$ (43% yield); $N(CH_3)_2$ (20% yield); $R^1 = CH_3$, $R^2 = N(CH_3)_2$ (69% yield) [4], see also [11]. It was suggested that the salt undergoes a back scission according to $[OSN(CH_3)_2]^+SbCl_6^- \rightleftharpoons (CH_3)_2NS(O)Cl + SbCl_5$ (see formation on p. 293) [11].

2.10.2 Salts of the N-Sulfinyl-N,N-diethyl-aminium Ion (1+), $[O{=}S{\doteq}N(C_2H_5)_2]^+Y^-$, $Y = BF_4$, $SbCl_6$

$[OSN(C_2H_5)_2]^+BF_4^-$ was formed by adding excess (mole ratio 10:7.7) $O{=}S{=}NC_2H_5$ to $(C_2H_5)_3O^+BF_4^-$ at 150°C and allowing the mixture to stand for 10 h at this temperature. The compound was obtained as an oil; yield 90%. ^1H NMR (CD_3CN/TMS): δ (in ppm) = 1.50 (t, 6H, CH_3), 4.30 (q, 4H, CH_2), $^3J(CH_3,CH_2) = 6.5$ Hz. The compound reacts with optically active sulfoxides, $R(CH_3)SO$ (R = n-C_4H_9, i-C_4H_9, 4-$CH_3C_6H_4$) to give the corresponding optically active aminosulfonium salts, $[(C_2H_5)_2NS(CH_3)R]^+BF_4^-$; the mechanism of the reaction is discussed [14].

$[OSN(C_2H_5)_2]^+SbCl_6^-$ was prepared by adding a solution of $(C_2H_5)_2NH$ (excess) in CH_2Cl_2 to a solution of $SOCl_2$ in CH_2Cl_2 at −78°C under N_2. The hydrochloride was filtered off at room temperature, and a solution of $SbCl_5$ (equimolar to $SOCl_2$) in CH_2Cl_2 was dropped into the filtrate cooled to −30°C. Then the mixture was reacted for 15 min at room temperature to give the salt in a 75% yield (method I) [15]. $[OSN(C_2H_5)_2]^+SbCl_6^-$ was also formed by adding a solution of $SOCl_2$ in CH_2Cl_2 to an equimolar amount of the sulfinyldiamide $(C_2H_5)_2NS(O)N(C_2H_5)_2$ (method II) or to an equimolar amount of $(C_2H_5)_2NSi(CH_3)_3$ (method III) in CH_2Cl_2 at −78°C. After allowing the mixture to come to room temperature it was stirred for 1 h and cooled again to −78°C. Then a solution of $SbCl_5$ (in excess) in CH_2Cl_2 was dropped slowly into the reaction mixture, which was finally stirred for another 1 h at room temperature. Yield 85% (method III) [16].

Colorless crystals, m.p. 148°C [15] or 147°C with decomposition [16]. ^1H NMR (CD_3CN/TMS): δ (in ppm) = 1.52 [15], 1.586 [16] (dt, 6H, CH_3); 4.31 [15], 4.366 [16] (q, 4H, CH_2). ^{13}C NMR (CD_3CN/TMS): δ (in ppm) = 12.17, 12.82 (C2, C2′), 48.53 (C1) [15]. IR (KBr): 12 absorption bands (unassigned) between 2993 and 335 cm^{-1} recorded. MS (100 eV): m/e (rel. int. in %) = 120 (5) $[OSN(C_2H_5)_2]^+$, 104 (3), 72 (100) [16].

Reacting excess 2,3-dimethyl-1,3-butadiene with a solution of $[OSN(C_2H_5)_2]^+$ $SbCl_6^-$ for 2 h at 20°C gives under cycloaddition I (44%) [16].

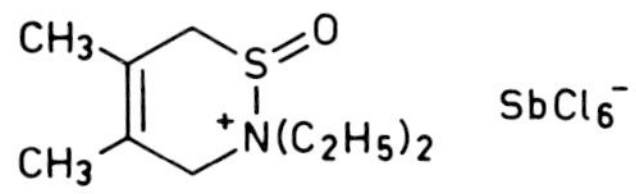

I

With $(CH_3)_2SO$ (5% excess) in CH_3CN the compound reacts analogously to $[OSN(CH_3)_2]^+SbCl_6^-$ to give $[(C_2H_5)_2NS(CH_3)_2]^+SbCl_6^-$ (63% yield) with evolution of SO_2 [4].

2.10.3 N-Sufinyl-N,N-diisopropyl-aminium Ion (1+) Hexachloroantimonate, $[O{=}S{\doteq}N(C_3H_7\text{-}i)_2]^+SbCl_6^-$

$[OSN(C_3H_7\text{-}i)_2]^+SbCl_6^-$ was prepared analogously to $[OSN(C_2H_5)_2]^+SbCl_6^-$ (see p. 294); yield: 71% by method I [15], 79% by method II, 84% by method III [16].

Colorless crystals, m.p. 142°C [15] or 142.7°C [16] with decomposition. ^1H NMR (CD_3CN/TMS): δ (in ppm) = 1.66 [15, 16] (d, 12H, CH_3), 4.80 [15], 4.838 [16] (m, 2H, CH). ^{13}C NMR (CD_3CN/TMS): δ (in ppm) = 19.88, 20.95 (C2, C2′), 59.69, 61.91 (C1, C1′) [15]. IR (KBr): 12 absorption bands (unassigned) between 3001 and 335 cm^{-1} recorded. MS (100 eV): m/e (rel. int. in %) = 148 (10) $[OSN(C_3H_7)_2]^+$, 132 (2), 100 (7), 74 (100) [16].

With $(CH_3)_2SO$ (5% excess) in CH_3CN $[OSN(C_3H_7\text{-}i)_2]^+SbCl_6^-$ reacts analogously to $[OSN(CH_3)_2]^+SbCl_6^-$ (see p. 294) to give $[(i\text{-}C_3H_7)_2NS(CH_3)_2]^+SbCl_6^-$ (42% yield) with evolution of SO_2 [4].

2.10.4 N-Sulfinyl-pyrrolidinium Ion (1+) Hexachloroantimonate, $[O{=}S{\doteq}N(CH_2)_4]^+SbCl_6^-$

$[OSN(CH_2)_4]^+SbCl_6^-$ was prepared analogously to $[OSN(C_2H_5)_2]^+SbCl_6^-$ (see p. 294); yield: 44% by method II, 71% by method III.

Colorless crystals, m.p. 135.1°C with decomposition. ^1H NMR (CD_3CN/TMS): δ (in ppm) = 2.27 (m, broad, 4H, CH_2), 4.457 (m, broad, 4H, NCH_2). IR (KBr): 11 absorption bands (unassigned) between 2983 and 335 cm^{-1} recorded. MS (100 eV): m/e (rel. int. in %) = 118 (6) $[OSN(CH_2)_4]^+$, 70 (2), 45 (100) [16].

2.10.5 N-Sulfinyl-piperidinium Ion (1+) Hexachloroantimonate, $[O{=}S{\doteq}N(CH_2)_5]^+SbCl_6^-$

The title compound was formed analogously to $[OSN(C_2H_5)_2]^+SbCl_6^-$ (see p. 294); yield: 71% by method I [15], 63% by method II, 88% by method III [16].

Colorless crystals, m.p. 138°C [15], 138.1°C [16] with decomposition. ^1H NMR (CD_3CN/TMS): δ (in ppm) = 1.88 [15], 1.972 [16] (m, 6H, CH_2), 4.35 [15], 4.399 [16] (m, 4H, NCH_2). ^{13}C NMR (CD_3CN/TMS): δ (in ppm) = 22.30 (C3), 26.32 (C2), 51.68, 51.84 (C1, C1′) [15]. IR (KBr): 13 unas-

References on p. 297

signed absorption bands between 2979 and 335 cm^{-1} recorded [16]. MS (100 eV): m/e (rel. int. in %) = 132 (24) $[OSN(CH_2)_5]^+$, 116 (2), 84 (100) [16].

The compound reacts with a slight excess of $(CH_3)_2SO$ analogously to $[OSN(CH_3)_2]^+SbCl_6^-$ (see p. 294) with evolution of SO_2 to give $[(CH_2)_5NS(CH_3)_2]^+SbCl_6^-$ in a 66% yield (in CH_3CN solution at −25°C) [4] to 74% yield (in CH_2Cl_2 at room temperature) [16].

2.10.6 N-Sulfinyl-morpholinium Ion (1+) Hexachloroantimonate, $[O{=}S{\doteq}N(CH_2)_4O]^+SbCl_6^-$

The title compound was prepared analogously to $[OSN(C_2H_5)_2]^+SbCl_6^-$ (see p. 294); yield: 66% by method I [15], 65% by method II, 83% by method III [16].

Colorless crystals, m.p. 139°C [15] or 137.4°C [16] with decomposition. ^1H NMR (CD_3CN/TMS): δ (in ppm) = 3.96 [15], 4.007 [16] (m, 4H, OCH_2); 4.33 [15], 4.372 [16] (m, 4H, NCH_2). ^{13}C NMR (CD_3CN/TMS): δ (in ppm) = 50.75, 50.82 (C1, C1′), 66.13, 66.32 (C2, C2′) [15]. IR (KBr): 11 unassigned absorption bands between 2940 and 335 cm^{-1} recorded. MS (100 eV): m/e (rel. int. in %) = 134 (2) $[OSN(CH_2)_4O]^+$, 118 (2), 86 (48) 57 (100) [16].

With $(CH_3)_2SO$ (5% excess) in CH_3CN the compound reacts with evolution of SO_2 to give $[O(CH_2)_4NS(CH_3)_2]^+SbCl_6^-$ in an 78% yield [4].

2.10.7 N-Sulfinyl-N,N-dibenzyl-aminium Ion (1+) Hexachloroantimonate, $[O{=}S{\doteq}N(CH_2C_6H_5)_2]^+SbCl_6^-$

The title compound was obtained analogously to $[OSN(C_2H_5)_2]^+SbCl_6^-$ (see p. 294); yield: 81% by method II.

Colorless crystals, m.p. >100°C (carbonization). ^1H NMR (CD_3CN/TMS): δ (in ppm) = 5.246 (s, 4H, CH_2), 7.480 (s, 10 arom. H). IR (KBr): 15 unassigned absorption bands between 3019 and 335 cm^{-1} recorded. MS (100 eV): m/e (rel. int. in %) = 244 (4) $[OSN(CH_2C_6H_5)_2]^+$, 228 (40), 196 (23), 91 (34), 62 (100) [16].

2.10.8 N-Sulfinyl-N-methyl-N-ethyl-aminium Ion (1+) Tetrafluoroborate, $[O{=}S{\doteq}N(CH_3)C_2H_5]^+BF_4^-$

The title compound was synthesized at 15°C for 10 h without a solvent from the reactions of $O{=}S{=}NCH_3$ with $(C_2H_5)_3O^+BF_4^-$ or $O{=}S{=}NC_2H_5$ with $(CH_3)_3O^+BF_4^-$, alternatively [2, 3].

At room temperature the compound was obtained as an oil [2], at −20°C it is crystalline, m.p. >0°C. ^1H NMR (CD_3CN/TMS): δ (in ppm) = 1.52 (t, 3H), 3.82 (s, 3H), 4.35 (q, 2H), $^3J(CH_3,CH_2)$ = 7 Hz [3], see also [2].

2.10.9 N-Sulfinyl-N-methyl-N-benzyl-aminium Ion (1+) Hexachloroantimonate, $[O{=}S{\doteq}N(CH_3)CH_2C_6H_5]^+SbCl_6^-$

The title compound was prepared analogously to $[OSN(C_2H_5)_2]^+SbCl_6^-$ (see p. 294), from the reaction of $CH_3(C_6H_5CH_2)NS(O)N(CH_2C_6H_5)CH_3$ with $SOCl_2$ and $SbCl_5$, method II (yield 69%) or from the reaction of $CH_3(C_6H_5CH_2)NSi(CH_3)_3$ with $SOCl_2$ and $SbCl_5$, method III (yield 91%).

Colorless crystals, m.p. >110°C (carbonization). ^1H NMR (CD_3CN/TMS): δ (in ppm) = 3.690 (s, 3H, CH_3), 5.398 (s, 2H, CH_2), 7.540 (s, 5 arom. H). IR (KBr): 13 unassigned absorption bands between 3100 and 335 cm^{-1} recorded. MS (100 eV): m/e (rel. int. in %) = 168 (34) $[OSN(CH_2C_6H_5)CH_3]^+$, 120 (15), 105 (2), 91 (100) [16].

2.10.10 Salts of N-Sulfinyl-N-methyl-N-perfluoroalkyl-aminium Ions (1+), $[O{=}S{\doteq}N(CH_3)R_f]^+Y^-$, $R_f = CF_3$, C_2F_5; Y = AsF_6, SbF_6

$[OSN(CH_3)CF_3]^+AsF_6^-$ was prepared by condensing $CF_3(CH_3)NS(O)F$ with AsF_5 (mole ratio 4:7) and a big excess of SO_2 in a glass bomb and stirring the mixture for 2 h at room temperature. After removal of the volatile components at 10^{-3} Torr the compound was quantitatively obtained.

Colorless solid, m.p. 74°C. ^1H NMR (SO_2/TMS): δ = 4.1 ppm (s). ^{19}F NMR ($SO_2/CFCl_3$): δ = −61.79 ppm (s). IR (Kel-F/Nujol): 23 unassigned bands between 3070 and 400 cm^{-1} [17].

$[OSN(CH_3)CF_3]^+SbF_6^-$ was quantitatively formed by stirring $CF_3(CH_3)NS(O)F$ with SbF_5 (mole ratio 4:3) in a big excess of SO_2 for 0.5 h at room temperature and removing the volatile components in vacuum. It was also obtained from $O{=}S{=}NCF_3$ by methylation with $CH_3OSO^+SbF_6^-$.

Pale yellow solid, m.p. 53°C. ^1H NMR (SO_2/TMS): δ = 4.1 ppm (s). ^{19}F NMR ($SO_2/CFCl_3$): δ = −60.11 ppm (s).

The compound reacts with Nujol [17].

$[OSN(CH_3)C_2F_5]^+SbF_6^-$ was formed by condensing SbF_5 in a dynamic vacuum into a glass bomb and adding at −40°C about equimolar amounts of CH_3F in SO_2. The mixture was then cooled to −80°C and slowly treated with equimolar amounts of $O{=}S{=}NC_2F_5$. After 1 h at room temperature SO_2 and excess reaction partners were removed at 10^{-3} Torr to get the product in a quantitative yield.

Solid, m.p. 57°C. ^1H NMR (SO_2/TMS): δ = 4.0 ppm (s). ^{19}F NMR ($SO_2/CFCl_3$): δ (in ppm) = −80.80 (s, CF_3), −96.06 (s, CF_2). IR (Kel-F): 27 unassigned bands between 3070 and 305 cm^{-1}.

The compound reacts with Nujol [17].

2.10.11 N-Sulfinyl-N-methyl-N-phenyl-aminium Ion (1+) Hexachloroantimonate, $[O{=}S{\doteq}N(CH_3)C_6H_5]^+SbCl_6^-$

The title compound was prepared analogously to $[OSN(C_2H_5)_2]^+SbCl_6^-$ (see p. 294), from the reaction of $CH_3(C_6H_5)NS(O)N(C_6H_5)CH_3$ with $SOCl_2$ and $SbCl_5$, method II (yield 51%) or from the reaction of $CH_3(C_6H_5)NSi(CH_3)_3$ with $SOCl_2$ and $SbCl_5$, method III (yield 81%).

Colorless crystals; the melting point could not be determined because of its extremely sensitivity to moisture. ^1H NMR (CD_3CN/TMS): δ (in ppm) = 4.15 (s, 3H, CH_3), 7.68 (m, 5 arom. H) [16].

References:

[1] Mews, R.; Henle, H. (J. Fluorine Chem. **14** [1979] 495/510).
[2] Kresze, G.; Rössert, M. (Angew. Chem. **90** [1978] 61; Angew. Chem. Intern. Ed. Engl. **17** [1978] 63).
[3] Kresze, G.; Rössert, M. (Liebigs Ann. Chem. **1981** 58/64).

[4] Schwöbel, A.; Pérez, M. A.; Rössert, M.; Kresze, G. (Liebigs Ann. Chem. **1982** 723/8).
[5] Warthmann, W.; Schmidt, A. (Z. Anorg. Allgem. Chem. **418** [1975] 61/4).
[6] Schwöbel, A.; Kresze, G. (Synthesis **1984** 944/5).
[7] Pérez, M. A.; Rössert, M.; Kresze, G. (Liebigs Ann. Chem. **1981** 65/9).
[8] Lux, R.; Kresze, G. (Liebigs Ann. Chem. **1987** 527/9).
[9] Schwöbel, A.; Kresze, G. (Liebigs Ann. Chem. **1984** 900/3).
[10] Kresze, G.; Schwöbel, A.; Hatjiissaak, A.; Ackermann, K.; Minami, T. (Liebigs Ann. Chem. **1984** 904/10).

[11] Kresze, G.; Rössert, M. (Angew. Chem. **90** [1978] 61/2; Angew. Chem. Intern. Ed. Engl. **17** [1978] 64).
[12] Schwöbel, A.; Kresze, G.; Pérez, M. A. (Liebigs Ann. Chem. **1985** 72/7).
[13] Schwöbel, A.; Kresze, G.; Pérez, M. A. (Tetrahedron Letters **1982** 1243/6).
[14] Drabowicz, J.; Bujnicki, B.; Mikolajczyk, M. (J. Org. Chem. **46** [1981] 2788/90).
[15] Pérez, M. A.; Kresze, G. (Synthesis **1981** 707/8).
[16] Neidlein, R.; Walser, P. (Chem. Ber. **115** [1982] 2428/36).
[17] Henle, H.; Mews, R. (Chem. Ber. **115** [1982] 2935/42).

2.11 Thiosulfinyl Imides, Sulfinothioyl Imides, N-Thiosulfinyl-amines and Derivatives

The following reviews which deal with N-thiosulfinylamines have been published: Inagaki, Y.; Okazaki, R. (Sulfur Rept. **2** [1982] 137/68), Kutney, G. W.; Turnbull, K. (Chem. Rev. **82** [1982] 333/57), Bleisch, S.; Mayer, R. (Houben-Weyl, Methoden Org. Chem. 4th Ed. E **11** [1985] 579/84), Chivers, T. (Sulfur Rept. **7** [1986] 89/107).

2.11.1 Thiosulfinyl Imide, Thiothionyl Imide, S=S=NH

The parent compound, S=S=NH has not been synthesized until now, but an ab initio MO study of the Z and E isomers of S=S=NH (see **Fig. 28**) was carried out.

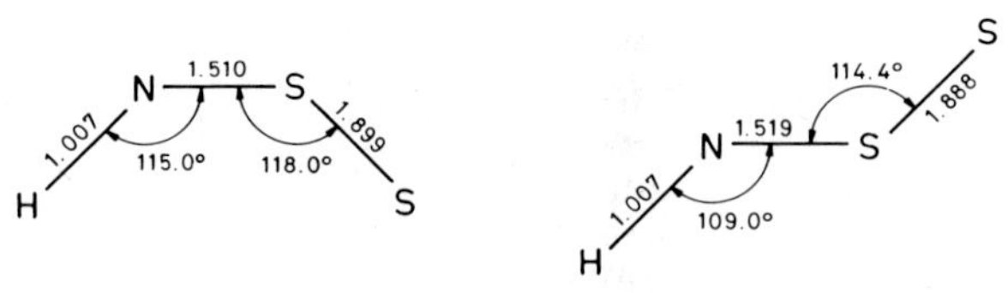

Z E

Fig. 28. Z and E isomers of S=S=NH. Calculated bond lengths in Å and bond angles in °.

The theoretical prediction that the Z form is energetically more stable than the E form (3.8 kcal/mol) is consistent with X-ray structural analysis, performed for S=S=N–S–N=P$(C_6H_5)_3$ (see pp. 312/3), the 1-(thiosulfinylaminothio)-piperidine derivative I (see pp. 314, 317) and S=S=N-$C_6H_2CH_3$-6-$(C_4H_9$-t$)_2$-2,4 (see pp. 322/3), which all have a Z configuration.

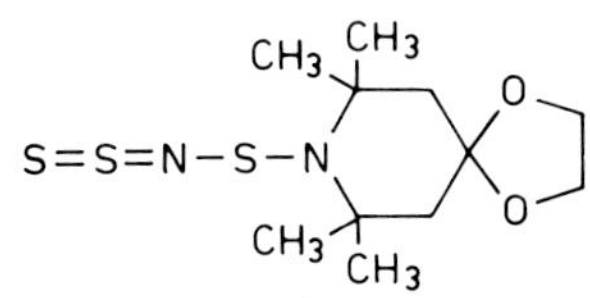

I

The geometrical parameters (see Fig. 28), optimized at the Hartree-Fock (HF)/4-31G* level are in good agreement with those obtained by X-ray analysis.

Reference:

Nakamura, S.; Takahashi, M.; Okazaki, R.; Morokuma, K. (J. Am. Chem. Soc. **109** [1987] 4142/8).

2.11.2 Thiosulfinyl Mercaptoimide, Thiosulfinyl-amino-sulfane, S=S=NSH

The parent compound, S=S=NSH has not been synthesized, but MO-SCF MNDO calculations for this model compound have been carried out. Calculated atomic charges and bond orders for planar S=S=NSH are given in **Fig. 29**. Energies and eigenvector coefficients on the π orbitals for S=S=NSH are compared with those of O=S=NSH, see the paper.

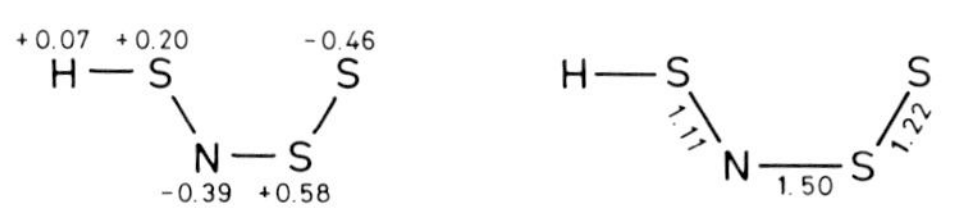

a b

Fig. 29. Atomic charges (a) and bond orders (b) for planar S=S=NSH.

Reference:

Chivers, T.; Oakley, R. T.; Pieters, R.; Richardson, J. F. (Can. J. Chem. **63** [1985] 1063/7).

2.11.3 Trisulfur Nitride Ion (1−), $[S\dot{=}S\dot{=}N\dot{=}S]^-$

2.11.3.1 Preparation of Salts of S_3N^-

$((C_6H_5)_3P{=})_2N^+S_3N^-$

The salt was prepared in ~57% yield by the rapid reaction of $((C_6H_5)_3P)_2N^+S_4N^-$ with $(C_6H_5)_3P$ in the mole ratio ~1:1.4 in CH_3CN at room temperature. Slow addition of $(C_2H_5)_2O$ precipitated a bright orange microcrystalline solid [1, 2, 3].

$((C_6H_5)_3P{=})_2N^+S_3{}^{15}N^-$

Isotopically enriched $((C_6H_5)_3P)_2N^+S_3{}^*N^-$ (where $^*N = 30\%\ ^{15}N$) was obtained from $((C_6H_5)_3P)_2N^+S_4{}^*N^-$ and $(C_6H_5)_3P$ in a manner similar to the unlabeled compound [1, 2].

$(C_6H_5)_4As^+S_3N^-$

The compound was prepared from $(C_6H_5)_4As^+S_4N^-$ and $(C_6H_5)_3P$ in the mole ratio 1:1.4 in CH_3CN at room temperature [1, 2].

$(C_6H_5)_4As^+S_3{}^{15}N^-$

Isotopically enriched $(C_6H_5)_4As^+S_3{}^*N^-$ (where $^*N = 30\%$ ^{15}N) was prepared from $(C_6H_5)_4As^+$-$S_4{}^*N^-$ and $(C_6H_5)_3P$ by a similar procedure as described for the unlabeled compound [1, 2].

2.11.3.2 Formation in Solutions

S_3N^- in very low concentration was detected by its Raman lines at 576, 695, and 898 cm^{-1} under resonance Raman conditions (excitation at 647.1 nm) in rather concentrated (e.g., 0.18 M) solutions of elemental sulfur in liquid NH_3 at temperatures below 230 K [8]. Formation of S_3N^- (along with S_7NH, S_4NH, and polysulfides H_2S_x) in solutions of sulfur in liquid ammonia at −35°C was established from UV-visible spectra. The light-induced changes of these spectra and those of $((C_6H_5)_3P)_2N^+S_4N^-$ in liquid ammonia indicate that, in a solution of sulfur in liquid NH_3, $6 \pm 2\%$ of sulfur is in the form of S_3N^- at −35°C [9]. This high content [9] is in contrast to results obtained from Raman spectroscopy [8]. S_3N^- in solutions of $((C_6H_5)_3P)_2N^+S_4N^-$ in liquid NH_3 forms by nucleophilic cleavage of one S–S bond in S_4N^-. This was detected by Raman spectroscopy [4]. The formation of S_3N^- in CH_3CN solution by electrochemical reduction of sulfur-nitrogen compounds is given in Table 72.

Table 72
Formation of S_3N^- by Electrochemical Reduction of Sulfur-Nitrogen Compounds in CH_3CN – 0.1 M $(C_2H_5)_4NClO_4$ Solutions at Room Temperature.
Potentials with reference to an Ag|0.1 M $AgClO_4$–CH_3CN Electrode.

starting solution	electrode	potential in V	number of electrons n	Ref.
0.11 mM $((C_6H_5)_3P{=})_2N^+S_4N^-$	Pt-gauze	−2.0	1.5	[5]
0.08 mM $((C_6H_5)_3P{=})_2N^+S_4N^-$	Hg-pool	−1.8	1.3 to 1.5	[5]
S_7NH	Hg-pool	−1.45	1.0	[5]
0.07 mM S_7NH	Pt-gauze	−1.75	4.0	[5]
1,4-$S_6(NH_2)_2$	Hg-pool	−1.5	1.3	[5]
0.13 mM 1,4-$S_6(NH_2)_2$	Pt-gauze	−1.8	3	[5]
0.30 mM $S_4N_4O_2$*)	Hg-pool	−1.7	2.8	[6]

*)
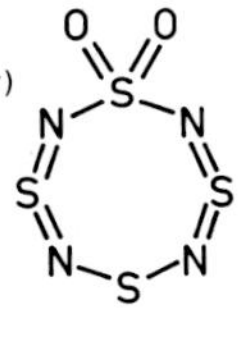

2.11.3.3 Molecular Structure and Electronic Structure

An unbranched arrangement of atoms $SNSS^-$ is supposed from the vibrational spectra of the ^{15}N-labeled anion $S_3{}^*N^-$ (where $^*N = 30\%$ ^{15}N) without distinguishing between cis, trans, or helical conformations [1]. Attempts to obtain a single-crystal analysis of S_3N^- were thwarted by disorder problems. At best $(C_6H_5)_4As^+S_3N^-$ has yielded to analysis only in terms of a model in which the S_3N^- anion exists in a nearly planar cis conformation, exhibiting fourfold disorder. Refinement of the model did not yield accurate structural parameters [7].

Ab initio Hartree-Fock-Slater SCF calculations on the basis of assumed bond lengths and bond angles similar to those for S_4N^-, viz., d(S–N) = 1.60 Å, d(S–S) = 1.90 Å, ∢(SSN) = 110°, ∢(SNS) = 120°, show that the statistical energy of the cis conformation is lower than the trans conformation by ca. 10 to 20 kcal/mol. The calculations also show that the ground-state electronic structure of S_3N^- is a 6π-electron system. The S–S linkage is primarily a σ bond whereas there is a strong π contribution to the SNS region. The charge distribution calculated for the cis conformer shows that the terminal sulfurs both carry a significant negative charge, S(1) (N-bonded) −0.28, S(3) −0.41, whereas the internal sulfur is essential neutral. The nitrogen center is also negative (−0.32) [1].

2.11.3.4 Spectra

^{14}N NMR. $((C_6H_5)_3P{=})_2N^+S_3N^-$ (CH_3CN/external CH_3NO_2): $\delta = +235$ ppm (S_3N^-) [10].

^{15}N NMR. The chemical shift of S_3N^- in a 4×10^{-2} M solution of $((C_6H_5)_3P{=})_2N^+S_3{}^*N^-$ (where $^*N = 30\%$ ^{15}N) in CH_3CN containing ca. 10^{-1} M $Cr(acac)_3$ referred to external liquid NH_3 at 25°C is $\delta = +617.6$ ppm (s) [1] or $\delta = +237.4$ ppm referred to external neat CH_3NO_2.

The vibrational spectra of the S_3N^- ion given in Table 73 were assigned on the basis of an acyclic planar structure with C_s symmetry and on comparison with the vibrational spectra of S_4N^-. However, the data do not permit a distinction between the alternative structures of C_s and C_1 symmetry.

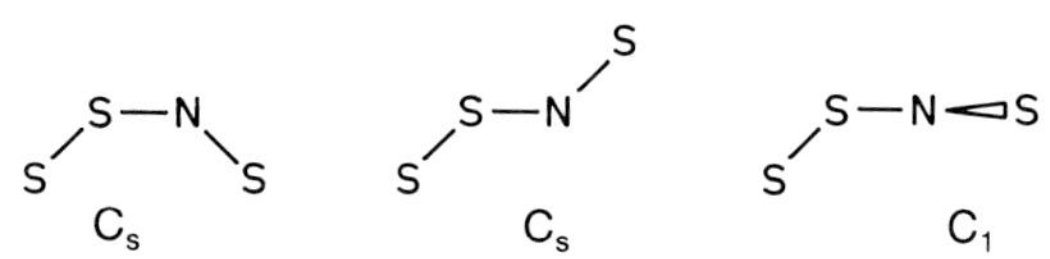

Table 73

Vibrational Spectra of the S_3N^- Ion in $((C_6H_5)_3P{=})_2N^+S_3N^-$ and $((C_6H_5)_3P{=})_2N^+S_3{}^*N^-$ (where $^*N = 30\%$ ^{15}N) [1, 2].

ν in cm^{-1} IR Nujol mull	Raman KBr	assignment
893 vs*⁾	894 m*⁾	ν_{as}(SNS)
666 s*⁾	686 s*⁾	ν_s(SNS)
564 s	574 vs	ν(S–S)

*⁾ Band shows satellite bands resulting from the presence of ^{15}N.

The resonance Raman (RR) spectrum of $((C_6H_5)_3P{=})_2N^+S_3N^-$ obtained with 488 nm excitation exhibits overtone progressions for all three fundamental stretching vibrations of S_3N^-, ν(S–S), ν_s(SNS), and ν_{as}(SNS), and several series of combination bands. For the S–S stretching vibration at 570 cm^{-1}, the progression can readily be observed up to the fourth overtone [1].

The UV-visible spectrum of S_3N^- in CH_2Cl_2 solution shows a smooth absorption band at $\lambda_{max} = 465$ nm [2] which can be assigned to a $\pi^* \rightarrow \pi^*$ transition [1]. A solution of $((C_6H_5)_3P{=})_2N^+$-S_3N^- in liquid NH_3 shows an absorption at $\lambda_{max} = 462$ nm and a shoulder at 290 nm which are assigned to the anion [9].

2.11.3.5 Electrochemical Behavior

The following polarographic half-wave potentials of 0.28 mM $((C_6H_5)_3P{=})_2N^+S_3N^-$ in CH_3CN $E_{1/2}$ vs. Ag|0.1M $AgClO_4$–CH_3CN have been measured in the presence of different supporting electrolytes: 0.05 M $(CH_3)_4NClO_4$, $E_{1/2} = -1.84$ V (i = 3.1 μA), and a small wave at $E_{1/2} = -2.20$ V (i = 0.9 μA); 0.1 M $(C_2H_5)_4NClO_4$, $E_{1/2} = -2.07$ V (i = 2.4 μA); 0.1 M $(n\text{-}C_4H_9)_4NClO_4$, $E_{1/2} = -2.09$ V (i = 1.0 μA).

A 0.28 M solution of $((C_6H_5)_3P{=})_2N^+S_3N^-$ in CH_3CN −0.1 M $(C_2H_5)_4NClO_4$ exhibited a greenish yellow color and gave two cathodic waves prior to the reduction of the $((C_6H_5)_3P{=})_2N^+$ cation on the dropping Hg electrode. The smaller first wave ($E_{1/2} = -1.63$ V, i = 0.3 μA) is due to the presence of a small amount of the S_4N^- ion, which is formed spontaneously from S_3N^- in solution. The second wave ($E_{1/2} = -2.065$ V, 2.55 μA) is due to the reduction of the S_3N^- ion. The addition of $(C_6H_5)_3P$ (1 mM in CH_3CN) to the solution converted all the S_4N^- to S_3N^-, as determined by the polarogram and the visible spectrum of the resulting yellow solution. The remaining polarographic wave ($E_{1/2} = -2.07$ V, i = 2.4 μA) was found to be diffusion-controlled.

Controlled potential electrolysis of $((C_6H_5)_3P{=})_2N^+S_3N^-$ on a Hg-pool electrode at −2.2 V gave coulometric n values of 1.4 to 1.8, and the UV-visible spectrum of the electrolyzed solution showed a single band at 375 nm, tentatively assigned to S_2N^-. It was proposed that the electrochemical reduction of S_3N^- results in the scission of an S–S bond to give S_2N^- and S_x^{2-}.

On the rotating Pt electrode, a 0.28 mM solution of $((C_6H_5)_3P{=})_2N^+S_3N^-$ in CH_3CN gave a single cathodic wave at −2.10 V with 0.1 M $(C_2H_5)_4NClO_4$ as supporting electrolyte. Irreversible and poorly behaved waves were observed at −2.04 and −2.4 V for $(CH_3)_4NClO_4$ and $(n\text{-}C_4H_9)_4$-$NClO_4$, respectively [5].

2.11.3.6 Chemical Reactions

On heating $((C_6H_5)_3P{=})_2N^+S_3N^-$ in the solid state, the sample darkens at 80 to 100°C, and becomes royal blue at ca. 150°C owing to the formation of S_4N^- [1]. This conversion of S_3N^- into S_4N^- was also observed during the preparation of KBr disks when the pressure exceeded 4×10^7 Pa [1, 2]. $((C_6H_5)_3P{=})_2N^+S_3N^-$ in CH_3CN solution reacts even at −50°C to form S_4N^- [1]. This reaction occurs also in solutions of $((C_6H_5)_3P{=})_2N^+S_3N^-$ and $(C_6H_5)_4As^+S_3N^-$ in CH_2Cl_2 at room temperature [2]. In solutions of $((C_6H_5)_3P{=})_2N^+S_3N^-$ in CH_3CN, S_3N^- reacts rapidly with S_8 to produce S_4N^- [5]. The photoreaction $S_3N^- + S + h\nu \rightarrow S_4N^-$ is revealed by increasing absorption at 578 nm of S_4N^- in either an 8.7×10^{-5} M $((C_6H_5)_3P{=})_2N^+S_3N^-$ solution in liquid NH_3 or a 1.21×10^{-3} M elemental sulfur solution in liquid NH_3 at −35°C using filtered light either around 290 or 460 nm. S_4N^- is formed in higher relative concentrations in the elemental sulfur –NH_3 system than in the $((C_6H_5)_3P{=})_2N^+S_3N^-$ solution in NH_3. A reservoir of sulfur in the form of polysulfides is present in the elemental sulfur–NH_3 system to react with the photoexcited S_3N^- ion [9].

$((C_6H_5)_3P{=})_2N^+S_3N^-$ and $(C_6H_5)_4As^+S_3N^-$ in CH_3CN solution react slowly with $(C_6H_5)_3P$ to give an SN product which was not identified [1, 2].

The reaction of ~1.47 mmol of $((C_6H_5)_3P{=})_2N^+S_3N^-$ with 2.52 mmol of $NiBr_2$ in CH_3CN at reflux for 2 h gives $Ni(S_3N)_2$ in 31% yield, along with $Ni(S_2N_2H)(S_3N)$ and $Ni(S_2N_2H)_2$, both in low yields [1, 2]. In a similar reaction of 1.70 mmol of $((C_6H_5)_3P{=})_2N^+S_3N^-$ with 2.29 mmol $CoBr_2$ in refluxing CH_3CN for 1 h $Co(S_3N)_2$, $Co(S_2N_2H)(S_3N)$, and $Co(S_2N_2H)_2$ are produced in low yields [1]. Treatment of ~2.1 mmol of $((C_6H_5)_3P{=})_2N^+S_3N^-$ with 2.5 mmol CuCl in CH_3CN for 2.5 h at ambient temperature gives $((C_6H_5)_3P{=})_2N^+$ $[CuCl(S_3N)]^-$ in 23% yield. With an excess of $((C_6H_5)_3P{=})_2N^+S_3N^-$ the complex is obtained in 46% yield. Reaction of $((C_6H_5)_3P{=})_2N^+S_3N^-$ with $AgNO_3$ in the mole ratio 1:1 in CH_3CN gives $((C_6H_5)_3P{=})_2N^+[Ag(S_3N)_2]^-$ in 17% yield. For a 2:1 mole ratio of S_3N^-:$AgNO_3$ the yield dropped to 12% and oxidation of S_3N^- to S_4N^- was observed. When an excess of $AgNO_3$ was used, the major products were S_4N_4 and Ag_2S [3]. The reaction of S_3N^- with Pd and Pt halides does not lead to the formation of $M(S_3N)_2$ complexes with M = Pd or Pt [1].

References:

[1] Bojes, J.; Chivers, T.; Laidlaw, W.G.; Trsic, M. (J. Am. Chem. Soc. **104** [1982] 4837/41).
[2] Bojes, J.; Chivers, T. (J. Chem. Soc. Chem. Commun. **1980** 1023/4).
[3] Bojes, J.; Chivers, T.; Codding, P. W. (J. Chem. Soc. Chem. Commun. **1981** 1171/3).
[4] Chivers, T.; Lau, C. (Inorg. Chem. **21** [1982] 453/5).
[5] Chivers, T.; Hojo, M. (Inorg. Chem. **23** [1984] 2738/42).
[6] Chivers, T.; Hojo, M. (Inorg. Chem. **23** [1984] 4088/93).
[7] Cordes, A. W.; Swepston, P. N. (private communication in [1]).
[8] Dubois, P.; Lelieur, J. P.; Lepoutre, G. (Inorg. Chem. **26** [1987] 1897/902).
[9] Prestel, H.; Schindewolf, U. (Z. Anorg. Allgem. Chem. **551** [1987] 21/32).
[10] Chivers, T.; Schmidt, K. J.; McIntyre, D. D.; Vogel, H. J. (Can. J. Chem. **67** [1989] 1788/94).

2.11.4 Tetrasulfur Nitride Ion (1−), $[S \overset{..}{-} S \overset{..}{-} N \overset{..}{-} S \overset{..}{-} S]^-$

2.11.4.1 Preparation of Salts of S_4N^-

2.11.4.1.1 $(n\text{-}C_4H_9)_4N^+S_4N^-$

The salt was prepared according to the equation $S_7NH + (n\text{-}C_4H_9)_4N^+OH^- \rightarrow (n\text{-}C_4H_9)_4N^+\text{-}S_4N^- + 3/8\,S_8 + H_2O$ by addition of $(n\text{-}C_4H_9)_4N^+OH^-$ in $CH_3OH-(C_2H_5)_2O$ to a solution of S_7NH in $(C_2H_5)_2O$ (mole ratio 1.25 : 1), cooled to −78°C. The greenish yellow precipitate changed to a deep purple solid when left for 3 d. Most of the sulfur can be removed from the solid by Soxhlet extraction with $n\text{-}C_5H_{12}$ or $n\text{-}C_6H_{14}$. Further purification was achieved by dissolving the dark blue solid in THF, filtering, and allowing the THF evaporate to give an oil. Addition of $(C_2H_5)_2O$ precipitated a fine blue-black powder. The yield is 71% [1, 2, 3]. The powder melts at 49°C [1, 2].

2.11.4.1.2 $((C_6H_5)_3P{=})_2N^+S_4N^-$

Preparation by Thermolysis of $((C_6H_5)_3P{=})_2N^+S_4N_5^-$

A slurry of $((C_6H_5)_3P{=})_2N^+S_4N_5^-$ in CH_3CN was heated to reflux to produce a deep blue solution after 6 h. Heating was continued for another 16 h and then the mixture was cooled to 0°C, whereupon a few crystals of S_8 appeared. The filtrate was concentrated and cooled to −10°C to give crystals of $((C_6H_5)_3P{=})_2N^+S_4N^-$ in 50% yield. The reaction was shown, with the aid of UV-visible spectra, to proceed according to the following sequence: $S_4N_5^- \rightarrow S_3N_3^- +$ "SN_2"; $S_3N_3^- \rightarrow S_4N^- + N_2 + 1/8\,S_8$ [4, 5].

Preparation by Thermolysis of $((C_6H_5)_3P{=})_2N^+S_3N_3^-$

A solution of $((C_6H_5)_3P{=})_2N^+S_3N_3^-$ in CH_3CN was heated to reflux for 7 h. The deep blue solution was cooled to −10°C to crystallize $((C_6H_5)_3P{=})_2N^+S_4N^-$ formed in 35% yield [4, 5].

Preparation by Reaction of $((C_6H_5)_3P{=})_2N^+N_3^-$ with S_4N_4

A slurry of equimolar amounts of $((C_6H_5)_3P{=})_2N^+N_3^-$ and S_4N_4 in CH_3CN was stirred for 1.5 h to give a very dark blue-green solution, which was heated at reflux for a further 2 h. The royal-blue solution was reduced and then cooled to −20°C to precipitate $((C_6H_5)_3P{=})_2N^+S_4N^-$. The yield was 77% [6].

Formation from $((C_6H_5)_3P{=})_2N^+S_3N^-$

The formation of S_4N^- from $((C_6H_5)_3P{=})_2N^+S_3N^-$ in the solid state during the preparation of KBr disks when the pressure exceeded 4×10^7 Pa was observed by Raman spectroscopy [7].

Formation by Thermolysis of $((C_6H_5)_3P{=})_2N^+S_3N_3O^-$

Thermolysis of $((C_6H_5)_3P{=})_2N^+S_3N_3O^-$ gave $((C_6H_5)_3P{=})_2N^+S_4N^-$ as the only identifiable product [29].

The deep blue crystals of $((C_6H_5)_3P{=})_2N^+S_4N^-$ have a copper-like metallic sheen and melt at 155 to 158°C. They are stable in air for several hours [4, 5].

2.11.4.1.3 $((C_6H_5)_3P{=})_2N^+S_4{}^{15}N^-$

Isotopically enriched $((C_6H_5)_3P{=})_2N^+S_4{}^*N^-$ (where $^*N = 30\%$ ^{15}N) was prepared by thermolysis of $((C_6H_5)_3P{=})_2N^+S_4{}^*N_5^-$ in CH_3CN, using the procedure described for $((C_6H_5)_3P{=})_2N^+S_4N^-$; see p. 303 [4].

2.11.4.1.4 $(C_6H_5)_4As^+S_4N^-$

The dark blue crystals of the salt were obtained from thermolysis of $(C_6H_5)_4As^+S_4N_5^-$ in CH_3CN according to the procedure described for $((C_6H_5)_3P)_2N^+S_4N^-$, see p. 303 [8].

2.11.4.1.5 $((CH_3)_2N)_3S^+S_4N^-$

The blue salt was produced by stirring a mixture of 2.4 mmol $((CH_3)_2N)_3S^+[O{=}S{=}N]^-$ and 0.59 mmol S_8 in 10 mL CH_3CN at 23°C for 24 h. The solid was recrystallized from CH_3CN–$(C_2H_5)_2O$ [28].

2.11.4.2 Formation in Solutions

The formation of S_4N^- in "solutions" of elemental sulfur in liquid NH_3 of varying concentrations of 9.4×10^{-3}, 1.3×10^{-1}, and 7.7×10^{-1} M at ambient temperatures is shown by Raman spectroscopic studies. The following scheme is proposed for the S_4N^- formation: $H_2N^- + S_8 \rightarrow H_2NS_8^-$, $H_2NS_8^- \rightarrow S_7N^- + H_2S$, $S_7N^- \rightarrow S_4N^- + {}^3\!/_8 S_8$. The blue color of "solutions" of sulfur in NH_3 is due to the presence of both S_4N^- and S_3^-. The relative amount of these ions is concentration dependent [9]. The results are confirmed by other Raman and absorption studies on 1.3×10^{-3} and 1.5×10^{-2} M solutions of sulfur in liquid NH_3 at 25°C [10]. It was concluded from an absorption study that the equilibrium in which S_4N^- is involved does not depend significantly upon the temperature of the solution from 200 K to room temperature [27]. It was estimated from the absorption intensity at 578 nm that $4 \pm 1\%$ of the dissolved sulfur in the sulfur–NH_3 system at −35°C is in the form S_4N^- [24]. When gaseous NH_3 was bubbled into a slurry of S_8 in $((CH_3)_2N)_3PO$ a blue solution characteristic of the S_4N^- ion was obtained [11].

S_4N^- forms by the photoreaction $S_3N^- + S + h\nu \rightarrow S_4N^-$ on illumination with filtered light either around 290 or 460 nm of the $((C_6H_5)_3P)_2N^+S_3N^-$–NH_3 system or the elemental sulfur–NH_3 system at −35°C. S_4N^- formation is established by increasing absorption at 578 nm. In the sulfur–NH_3 system, S_4N^- forms in higher relative concentrations than in $((C_6H_5)_3P)_2N^+S_3N^-$

solutions because a reservoir of sulfur in the form of polysulfides is present to react with the photoexcited S_3N^- ion [24].

The formation of S_4N^- was observed in $(NH_4)_2S_{10}-NH_3$ solutions at 200 K by its absorption band at 580 nm [25]. S_4N^- is produced in 1M solutions of Li_2S_5, Li_2S_6, and $Li_2S_{6.5}$ in liquid NH_3 at room temperature by disproportionation of S_6^{2-} and S_5^{2-}. The following reaction with NH_3 proceeds in several steps. It was concluded from absorption measurements that a 1.5×10^{-2}M $Li_2S_{6.5}$ solution in liquid NH_3 contains less than 10^{-5}M of S_4N^- [26].

The S_4N^- ion is produced as an intermediate by reaction of S_8 with excess LiN_3 in $((CH_3)_2N)_3PO$ [11] or NaN_3 in DMF [12] at room temperature in the synthesis of cyclic sulfur imides. In a similar way S_4N^- forms when S_8 is treated with $(CH_3)_3SiN_3$ in $((CH_3)_2N)_3PO$ at 65°C [11].

An inky-blue reaction mixture containing the S_4N^- ion is obtained by passing NH_3 gas through a solution of S_2Cl_2 in $((CH_3)_2N)_3PO$ [11] or DMF [13].

The presence of S_4N^- is inferred in blue solutions formed from mixtures of liquid NH_3 and liquid SO_2 with low $NH_3:SO_2$ ratio (<7) which react after 1.5 to 2 years at low temperatures in sealed tubes [14].

S_4N^- and $S_3N_3^-$ form by electrolysis of 0.11 mM S_4N_2 in CH_3CN – 0.1M $(C_2H_5)_4NClO_4$ on a Pt-gauze electrode at −1.4 V (towards Ag|0.1 $AgClO_4-CH_3CN$) according to the equation $9S_4N_2+10e^-\rightarrow6S_4N^-+4S_3N_3^-$.

S_4N^-, S_3N^-, and HgS are produced by electrochemical reduction of S_7NH on a Hg-pool electrode at −1.45 V (towards Ag|0.1 $AgClO_4-CH_3CN$) in CH_3CN containing 0.1M $(C_2H_5)_4NClO_4$. S_4N^- forms according to $S_7NH+e^-\rightarrow\frac{1}{2}H_2+S_7N^-$, $S_7N^-\rightarrow S_4N^-+\frac{3}{8}S_8$. Coulometric reduction of 0.07 mM S_7NH in CH_3CN – 0.1M $(C_2H_5)_4NClO_4$ at −1.75 V on a Pt-gauze electrode gives mainly S_4N^- at the number of electrons n = 2.

The formation of S_4N^- by electrochemical reduction of 1,4-$S_6(NH)_2$ in CH_3CN in the presence of $R_4N^+ClO_4^-$, where R = CH_3, C_2H_5, n-C_4H_9, on a Hg-pool electrode at −1.5 V and on a Pt-gauze electrode at −1.8 V can be explained by the following reduction processes: 1,4-$S_6(NH)_2+e^-\rightarrow\frac{1}{2}H_2+$1,4-$S_6N_2H^-$; 1,4-$S_6N_2H^-\rightarrow S_4N^-+$"$S_2NH$"; 1,4-$S_6(NH)_2+2e^-\rightarrow H_2+S_6N_2^{2-}$; $S_6N_2^{2-}\rightarrow S_4N^-+S_2N^-$ [15].

It is well established that S_4N^- forms by rearrangement of the S_7N^- ion on warming solutions which are obtained by deprotonation of S_7NH with bases at low temperatures [11], e.g., with LiOH in THF at 0°C [16], KOH in THF at −62°C [17], and n-C_4H_9Li in THF at −60°C [18] or in $(C_2H_5)_2O$ at −78°C [19]. The equilibrium $S_7N^-\rightleftharpoons S_4N^-+\frac{3}{8}S_8$ prevails in these solutions [11].

S_4N^- was detected by absorption spectroscopy in blue solutions of S_7NH [2, 3] and 1,4-$S_6(NH)_2$ [2] in $((CH_3)_2N)_3PO$. During one week, one equivalent of S_4N^- is produced per 1,4-$S_6(NH)_2$ [2]. From spectrometric and conductometric results on solutions of S_7NH in DMF, the following equilibria are postulated: $S_7NH\rightleftharpoons H^++S_7N^-$, $S_7N^-\rightleftharpoons S_4N^-+\frac{3}{8}S_8$ with an equilibrium constant $K=0.13\pm0.06$ for the reaction $S_7N^-\rightleftharpoons S_4N^-+\frac{3}{8}S_8$ [2].

The formation of S_4N^- by electrolysis of 0.30 mM S_3N_2O (I) in CH_3CN – 0.1M $(C_2H_5)_4NClO_4$ on a Pt-gauze electrode was detected by its absorption spectrum. The concentration of S_4N^- increases with the number of electrons $n\leq1.3$ [20].

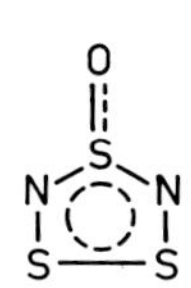

I

S_4N^- forms from $((C_6H_5)_3P{=})_2N^+S_3N^-$ in CH_3CN solution even at −50°C [6]. The formation of S_4N^- from $((C_6H_5)_3P{=})_2N^+S_3N^-$ in CH_2Cl_2 solutions at ambient temperature [7] and of $((C_6H_5)_3P{=})_2N^+S_3N_3^-$ in refluxing CH_3CN [4] is observed by absorption spectroscopy. S_4N^- is produced by reaction of S_3N^- with S_8 in CH_3CN [15].

A small amount of S_4N^- formed on addition of S_4N_4 to an electrolyzed solution of $(O{=}S{=}N)_2S$ in CH_3CN [20].

In the solid state, the formation of S_4N^- is observed on heating $((C_6H_5)_3P{=})_2N^+S_3N^-$ at ca. 150°C [6]. This formation of S_4N^- from $((C_6H_5)_3P{=})_2N^+S_3N^-$ and $(C_6H_5)_4As^+S_3N^-$ was also detected during the preparation of KBr disks when the pressure exceeded 4×10^7 Pa [6, 7].

2.11.4.3 Crystal and Molecular Structures

The existence of different geometries for the acyclic S_4N^- anion, which was predicted by MO calculations [4], is established by studies of the crystal structures of $(C_6H_5)_4As^+S_4N^-$ [8] and $((C_6H_5)_3P{=})_2N^+S_4N^-$ [4, 8].

A change in the crystal environment could lead to different geometries of the S_4N^- anion with different "long-short" S–N bond frequencies. The S_4N^- anion is nearly planar and possesses a "sickle shaped" cis, trans conformation.

2.11.4.3.1 $((C_6H_5)_3P{=})_2N^+S_4N^-$

The crystal and molecular structure of $((C_6H_5)_3P{=})_2N^+S_4N^-$ have been determined by X-ray crystallography. The compound crystallizes in the space group $P\bar{1}$–C_i^1 (No. 2), triclinic, a = 11.220(2), b = 16.860(3), c = 9.862(2) Å, α = 99.17(1)°; β = 98.04(1)°, γ = 71.56(1)°, V = 1739.2 Å^3; Z = 2, D_x = 1.300 g/cm^3. An ordered anion model was used first in the refinement of the structure, and from 2987 observed reflections the final residuals were R = 0.064 and R_w = 0.083 [4, 5]. Atomic positional and thermal parameters as well as bond lengths and bond angles are given in the paper [4]. In a re-refinement using a disordered anion model (i.e. two orientations of the anions), the final residues were R = 0.052 and R_w = 0.063. The fractional atomic coordinates are compiled in the paper [8]. Refinement of the sulfur population parameters indicated an occupancy factor ratio of 0.9 : 0.1 for the two orientations of the S_4N^- anion. Selected bond lengths and bond angles of the S_4N^- anion with orientation A and orientation B are given in Table 74. The molecular structure of the disordered anion model is shown in **Fig. 30** [8]. The S_4N^- anions are essentially planar cis, trans chains with N as the central atom [4].

The inclusion of small amounts of anion disorder in the re-refinement resulted in several significant changes of structural parameters for S_4N^-. The biggest change is the diminished difference in the SN bond lengths: the 1.667(5) and 1.521(5) Å values obtained from the ordered refinement [4] have become 1.625(5) and 1.572(6) Å in the disordered model [8]. The short S–S bond length of 1.879(3) Å [4] has changed to a more typical value of 1.914(3) Å in the disordered model [8]. The dihedral angle between the planes of the two disordered anions is 148(1)° [8].

Table 74

Selected Bond Lengths (in Å) and Bond Angles (in °) of S_4N^- in $((C_6H_5)_3P{=})_2N^+$-S_4N^- for the Disordered Anion Model [8].
For the enumeration of the atoms see Fig. 30.

orientation A (mult. = 0.9)		orientation B (mult. = 0.1)	
S(1)–S(2)	1.930(3)	S(6)–S(7)	1.97(4)
S(2)–N(3)	1.572(6)	S(7)–N(8)	1.57(2)
N(3)–S(4)	1.625(5)	N(8)–S(9)	1.59(2)
S(4)–S(5)	1.914(3)	S(9)–S(10)	1.95(2)
S(1)–N(3)	2.885(6)	S(6)–N(8)	2.99(4)
S(2)–S(4)	2.777(3)	S(7)–S(9)	2.67(3)
S(2)–S(5)	3.141(3)	S(7)–S(10)	3.21(3)
S(1)–S(2)–N(3)	110.5(3)	S(6)–S(7)–N(8)	115(2)
S(2)–N(3)–S(4)	120.5(4)	S(7)–N(8)–S(9)	115(1)
N(3)–S(4)–S(5)	111.0(3)	N(8)–S(9)–S(10)	119(1)

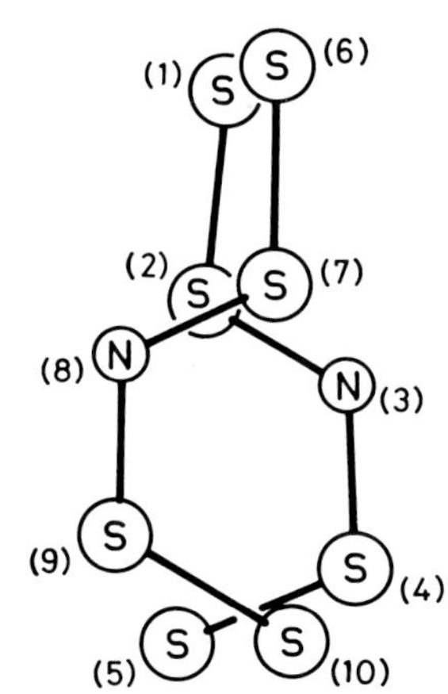

Fig. 30. Molecular structure of S_4N^- in $((C_6H_5)_3P{=})_2N^+S_4N^-$ for the disordered anion model [8].

2.11.4.3.2 $(C_6H_5)_4As^+S_4N^-$

The crystal and molecular structure of $(C_6H_5)_4As^+S_4N^-$ have been determined by X-ray diffraction analysis. The crystals are monoclinic, space group $P2_1/n$–C^5_{2h} (No. 14) with a = 10.560(2), b = 16.273(8), c = 13.994(4) Å, β = 91.50(2)°, V = 2404 Å^3; Z = 4, D_x = 1.45 g/cm^3. The refinement of the structure using a twofold disordered model for the S_4N^- anion and 2171 reflections gave the final residuals R = 0.040 and R_w = 0.056. The occupancy factor ratio for the two observations of the anion (0.6 : 0.4) is based on the relative heights of the nitrogen Fourier peaks. The fractional atomic coordinates are given in the paper. Selected bond lengths and bond angles of the S_4N^- anion are compiled in Table 75, p. 308. The molecular structure of the S_4N^- anion is shown in **Fig. 31**, p. 308. Both orientations A and B of the anion are nearly planar and possess a "sickle-shaped" cis, trans conformation. The largest deviation from the best plane of orientation A is 0.04 Å for atom S(1) and from the best plane of orientation B is 0.09 Å for atom S(6). The S(1)···S(6) separation is only 0.28(2) Å. The dihedral angle between the planes of the two disordered anions is 147(1)° [8].

Table 75

Selected Bond Lengths (in Å) and Bond Angles (in °) of S_4N^- in $(C_6H_5)_4As^+S_4N^-$ for the Disordered Anion Model [8].
For the enumeration of the atoms see Fig. 31.

orientation A (mult. = 0.6)		orientation B (mult. = 0.4)	
S(1)–S(2)	1.96(1)	S(6)–S(7)	1.89(2)
S(2)–N(3)	1.62(1)	S(7)–N(8)	1.63(2)
N(3)–S(4)	1.55(1)	N(8)–S(9)	1.57(2)
S(4)–S(5)	1.928(6)	S(9)–S(10)	1.913(9)
S(1)–N(3)	2.97(2)	S(6)–N(8)	2.91(2)
S(2)–S(4)	2.753(6)	S(7)–S(9)	2.77(1)
S(2)–S(5)	3.181(6)	S(7)–S(10)	3.17(1)
S(1)–S(2)–N(3)	111.6(5)	S(6)–S(7)–N(8)	111.4(9)
S(2)–N(3)–S(4)	120.6(8)	S(7)–N(8)–S(9)	119.5(9)
N(3)–S(4)–S(5)	114.0(5)	N(8)–S(9)–S(10)	114.1(7)

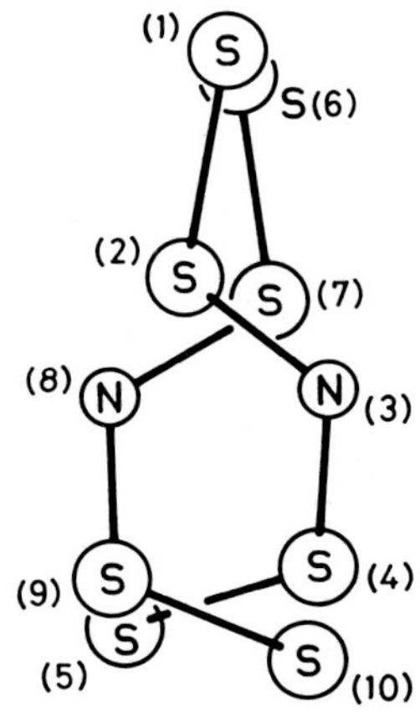

Fig. 31. Molecular structure of S_4N^- in $(C_6H_5)_4As^+S_4N^-$ for the disordered anion model [8].

2.11.4.3.3 Calculations

Geometrical parameters for cis, trans isomers, the cis, cis isomer and the trans, trans isomer of S_4N^- were calculated using the MNDO model, and were optimized with respect to the energy. The potential energy was derived as a function of angles, in plane and out-of-plane movement starting from the cis, cis conformation. The energy difference between the minima is relatively small (20 to 30 kJ/mol). It was found that the cis, trans isomer (sickle-like structure) is more stable by 20 to 30 kJ/mol than the trans, trans isomer and the cis, cis isomer [21].

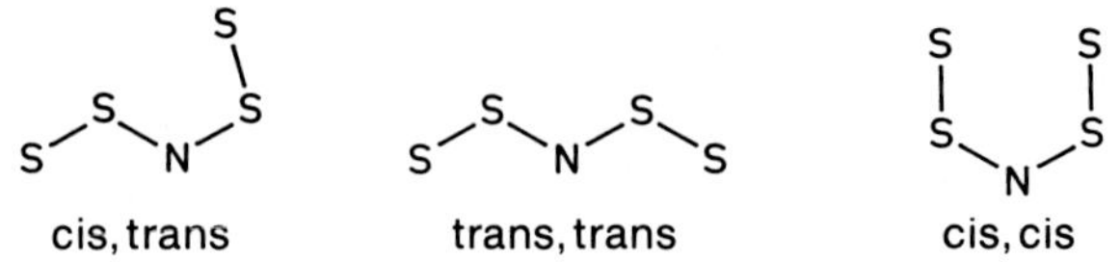

Similar results have been obtained from ab initio Hartree-Fock-Slater SCF calculations. The calculated statistical energies of the experimental cis, trans, the symmetrical cis, trans, and the symmetrical trans, trans conformation are within 10 kcal/mol of one another [4].

The stability of the cis, trans configuration is tracked back to stabilization of the HOMO and to a Coulomb attraction [21].

2.11.4.4 Electronic Structure

Ab initio Hartree-Fock-Slater SCF calculations have been carried out in order to obtain an MO description of the ground-state electronic structure. The calculated statistical energy of the experimentally found cis, trans conformation is not substantially lower than that of the symmetrical cis, trans or trans, trans conformations of the anion. The ground-state electronic structure of S_4N^- is an 8 π electron system [4]. Four of the five π levels are occupied and thus the π bond order is reduced because one π^* orbital is occupied as well [21]. S–N linkages can be described approximately as net $\frac{2}{3}$ σ plus $\frac{1}{2}$ π bonds, while the S–S bonds are approximately $\frac{5}{4}$ σ plus $\frac{1}{4}$ π. The terminal sulfur atoms each carry two lone pairs of electrons and bear large negative charges, −0.42 and −0.44, as does the nitrogen, −0.35; the internal sulfur atoms have one lone pair of electrons and have small positive charges, +0.01 and +0.20 [4].

2.11.4.5 Spectra

^{14}N NMR. $((C_6H_5)_3P{=})_2N^+S_4N^-$ (CH_3CN/external CH_3NO_2): $\delta = +106$ ppm (S_4N^-) [28].

^{15}N NMR. The chemical shift of S_4N^- of $((C_6H_5)_3P{=})_2N^+S_4{}^*N^-$ (where $^*N = 30\%$ ^{15}N) in CH_3CN referred to external liquid NH_3 at 25°C $\delta = +244.2$ ppm (s) [6] or $\delta = -136.0$ ppm referred to external neat CH_3NO_2, is incorrectly given [28]. The value $\delta = +212.8$ ppm in CH_3Cl referred to external liquid NH_3 quoted in Ref. [22] is in error [6].

The vibrational spectra of the S_4N^- ion given in Table 76 were assigned on the basis of the planar structure with cis, trans conformation obtained by X-ray diffraction analysis [4]. Tentative vibrational assignments for the S_4N^- ion made on the basis of a branched-chain structure [2] are incorrect [4].

Table 76

Vibrational Spectra of the S_4N^- Ion in $((C_6H_5)_3P{=})_2N^+S_4N^-$ and $((C_6H_5)_3P{=})_2N^+S_4{}^*N^-$ (where $^*N = 30\%$ ^{15}N) [4].

IR ν in cm^{-1}		Raman ν in cm^{-1}	assignment
Nujol mull	CH_3CN solution	solid	
893 w*)	892 w*)	892 w*)	ν_{as}(SNS)
711 m*)	709 m*)	710 vs*)	ν_s(SNS)
594 s	592 ms	592 s	ν(S(3)–S(4))
567 vs	562 vs	570 s	ν(S(1)–S(2))
416 w	415 w	—	—
—	—	227 vw	—

*) Band shows satellite bands resulting from the presence of ^{15}N in the spectra of $((C_6H_5)_3P{=})_2N^+S_4{}^*N^-$.

Almost identical Raman and IR frequencies for S_4N^- of $((C_6H_5)_2N)_3S^+S_4N^-$ as for S_4N^- of $((C_6H_5)_3P{=})_2N^+S_4N^-$ were measured [28].

Unassigned IR frequencies are given for $(n\text{-}C_4H_9)_4N^+S_4N^-$ [1, 3] and $(C_6H_5)_4As^+S_4N^-$ [8].

UV-visible spectrum of $((C_6H_5)_3P{=})_2N^+S_4N^-$ in CH_2Cl_2: $\lambda_{max} = 582$ nm, $\varepsilon \approx 1.6 \times 10^4$ $L \cdot mol^{-1} \cdot cm^{-1}$ [4, 5]. The strong absorption was tentatively assigned to a π^*(HOMO) $\rightarrow \pi^*$(LUMO) transition [4].

Solutions of $(n\text{-}C_4H_9)_4N^+S_4N^-$ in either $((CH_3)_2N)_3PO$ or THF do not obey Beer's law. At concentrations below ca. 1×10^{-5} M, the molar extinction of the band at $\lambda_{max} = 595$ nm starts to decrease, and a new peak appears at 470 nm, but at concentrations close to 1×10^{-4} M there is no sign of this second peak and the molar extinction $\varepsilon = (1.90 \pm 0.03) \times 10^4$ $L \cdot mol^{-1} \cdot cm^{-1}$ [2].

2.11.4.6 Solubility

$(n\text{-}C_4H_9)_4N^+S_4N^-$ is very soluble in THF but almost insoluble in $(C_2H_5)_2O$ [1].

2.11.4.7 Electrochemical Behavior

Polarographic data for $((C_6H_5)_3P{=})_2N^+S_4N^-$ in CH_3CN at 23°C containing various supporting electrolytes are given in Table 77.

Table 77

Polarographic Half-Wave Potentials of $((C_6H_5)_3P{=})_2N^+S_4N^-$ (0.40 mM in CH_3CN), $E_{1/2}$ vs. Ag|0.1 M $AgClO_4$ – CH_3CN, as a Function of the Supporting Electrolyte [15].

supporting electrolyte	anodic $E_{1/2}$ in V (i in μA)	cathodic $E_{1/2}$ in V (i in μA)	$E_{1/2}$ in V (i in μA)
0.1 M $LiClO_4$	−0.83 (0.7)	−1.30 (5.9)	−1.89 (2.1)
0.1 M $NaClO_4$	−0.74 (0.8)	−1.24 (2.0) −1.47 (2.3)	−1.62 (3.8)
0.05 M $(CH_3)_4NClO_4$	−0.71 (1.3)	−1.70 (7.7)	−2.17 (1.3)
0.1 M $(C_2H_5)_4NClO_4$	−0.64 (1.7)	−1.575 (3.1)	−2.03 (3.5)
0.1 M $(n\text{-}C_4H_9)_4NClO_4$	−0.64 (1.5)	−1.565 (2.3)	−2.46 (3.3)

Voltammograms for $((C_6H_5)_3P{=})_2N^+S_4N^-$ in CH_3CN in the presence of 0.1 M $(n\text{-}C_4H_9)_4NClO_4$ and 0.1 M $(C_2H_5)_4NClO_4$ at 23°C on the dropping Hg electrode and the rotating Pt electrode were measured. In the case of 0.1 M $(C_2H_5)_4NClO_4$ two cathodic waves were observed at −1.58 and −2.03 V for the dropping Hg electrode. The $E_{1/2}$ value for the first wave shifted positively when $c(S_4N^-) \leq 0.2$ mM. The first wave was shown to be diffusion-controlled, but the second wave was neither reversible nor diffusion-controlled. With 0.1 M $(n\text{-}C_4H_9)_4NClO_4$ the first wave for $((C_6H_5)_3P{=})_2N^+S_4N^-$ also occurred at −1.57 V, but the second wave was shifted to more negative potentials, −2.46 V, so that it overlapped the reduction wave for the $((C_6H_5)_3P{=})_2N^+$ cation. With 0.05 M $(CH_3)_4NClO_4$, the height of the first wave at −1.70 V was considerably enhanced and it was accompanied by a polarographic maximum. A complicated sequence of polarographic waves, suggesting decomposition, was observed with $LiClO_4$, or $NaClO_4$ as supporting electrolytes. On the rotating Pt electrode, $((C_6H_5)_3P{=})_2N^+S_4N^-$ gave a single wave at ca. −1.8 V with

$(C_2H_5)_4NClO_4$ or $(n\text{-}C_4H_9)_4NClO_4$ and at −1.74 V with $(CH_3)_4NClO_4$. Smaller waves were observed for $LiClO_4$ and $NaClO_4$.

Controlled potential coulometry of 0.11 mM $((C_6H_5)_3P{=})_2N^+S_4N^-$ in CH_3CN – 0.1 M $(C_2H_5)_4$-$NClO_4$ was performed on a Pt-gauze electrode at −2.0 V. After the uptake of one electron, the blue solution became green. At the number of electrons n = 1.5, the UV-visible spectrum showed strong absorption of S_3N^- (λ_{max} = 465 nm) and a medium intensity band at 375 nm. When the electrolysis of S_4N^- was carried out at −2.3 V, the only absorption band in the UV-visible spectrum of the electrolyzed solution was at 375 nm, tentatively assigned to the S_2N^- ion, and the coulometric n value was 3.

The electrolysis of 0.08 mM $((C_6H_5)_3P{=})_2N^+S_4N^-$ on a Hg-pool electrode at −1.8 V in CH_3CN – 0.1 M $(C_2H_5)_4NClO_4$ gave n values of 1.3 to 1.5 and produced S_3N^- and the 375 nm species, probably S_2N^-. When the electrolysis of $((C_6H_5)_3P{=})_2N^+S_4N^-$ was carried out at −2.2 V on a Hg-pool, the coulometric n value was 3. During the course of the electrolysis the color of the solution changed from blue (n = 0) through pale green (n = 1) and yellow (n = 1.5) to colorless (n = 3), and the UV-visible spectrum of the final solution exhibited a single peak at 375 nm [15].

2.11.4.8 Chemical Reactions

In solutions, S_4N^- is very sensitive to traces of O_2 and moisture, decomposing to as-yet-uncharacterized oxyanions of sulfur. S_4N^- in CH_2Cl_2 decomposes slowly at room temperature to give $S_3N_3^-$ and S_8 [4]. S_4N^- produced from S_8 and liquid NH_3 is photosensitive in this solution. It decomposes by irradiation with white light [10]. On illumination with light matching the absorption band at 578 nm, S_4N^- which is formed in liquid NH_3 solutions of elemental sulfur and in NH_3 solution of $((C_6H_5)_3P{=})_2N^+S_4N^-$, reacts to give S_3N^- at −35°C [24]. Under resonance Raman conditions (excitation at 647.1 nm) S_4N^- in the elemental sulfur–NH_3 system reacts at temperatures below 230 K to give S_3N^- in very low yields [27].

Equilibria of the type $S_7N^- \rightleftharpoons S_4N^- + {}^3/_8 S_8 \rightleftharpoons {}^1/_2 S_6N_2^{2-} + {}^1/_2 S_8$ in solutions of S_7N^- in $((CH_3)_2$-$N)_3PO$, DMF [11], THF [16, 18], and $(C_2H_5)_2O$ [19] are established from preparative work [11]. S_4N^- generated in "solutions" of S_8 in liquid NH_3 is transformed to S_6^{2-} by reduction with alkali metals, e.g. Li [10].

An 8×10^{-3} M solution of $(n\text{-}C_4H_9)_4N^+S_4N^-$ in THF is hydrolyzed with 5% aqueous HCl to give S_8, S_7NH, and $S_6(NH)_2$ isomers [2].

$((C_6H_5)_3P{=})_2N^+S_4N^-$ in liquid NH_3 reacts to give S_3N^- by nucleophilic cleavage of one S–S bond [9].

S_4N^- present in "solutions" of S_8 in liquid NH_3 reacts with added CH_3NH_2 to give unknown products [10].

$((C_6H_5)_3P{=})_2N^+S_4N^-$ [6, 7, 23] and $(C_6H_5)_4As^+S_4N^-$ [6, 7] react with $P(C_6H_5)_3$ in the mole ratio of ~1:1.4 in CH_3CN at 23°C to give the appropriate salts of S_3N^- and $(C_6H_5)_3PS$ [6, 7, 23].

The reaction of S_4N^- with $(CH_3)_3SiCl$ to give the intermediate product $S{=}S^{\oplus}{-}N(R){-}S{-}S^{\ominus}$ is proposed in the mechanism for the formation of $1,4\text{-}S_6N_2(Si(CH_3)_3)_2$. The S_4N^- solution is obtained by reaction of S_7NH with $n\text{-}C_4H_9Li$ in THF at −60°C [18].

References:

[1] Bojes, J.; Chivers, T.; Drummond, I. (Inorg. Syn. **18** [1978] 205/6).
[2] Chivers, T.; Drummond, I. (Inorg. Chem. **13** [1974] 1222/5).
[3] Chivers, T.; Drummond, I. (J. Chem. Soc. Chem. Commun. **1973** 734/5).

[4] Chivers, T.; Laidlaw, W. G.; Oakley, R. T.; Trsic, M. (J. Am. Chem. Soc. **102** [1980] 5773/81).
[5] Chivers, T.; Oakley, R. T. (J. Chem. Soc. Chem. Commun. **1979** 752/3).
[6] Bojes, J.; Chivers, T.; Laidlaw, W. G.; Trsic, M. (J. Am. Chem. Soc. **104** [1982] 4837/41).
[7] Bojes, J.; Chivers, T. (J. Chem. Soc. Chem. Commun. **1980** 1023/4).
[8] Burford, N.; Chivers, T.; Cordes, A. W.; Oakley, R. T.; Pennington, W. T.; Swepston, P. N. (Inorg. Chem. **20** [1981] 4430/2).
[9] Chivers, T.; Lau, C. (Inorg. Chem. **21** [1982] 453/5).
[10] Bernard, L.; Lelieur, J. P.; Lepoutre, G. (Nouv. J. Chim. **9** [1985] 199/203).

[11] Bojes, J.; Chivers, T. (J. Chem. Soc. Dalton Trans. **1975** 1715/7).
[12] Bojes, J.; Chivers, T. (Inorg. Nucl. Chem. Letters **10** [1974] 735/7).
[13] Heal, H. G.; Kane, J. (Inorg. Syn. **11** [1968] 184/94).
[14] Meyer, B.; Mulliken, B.; Weeks, H. (Phosphorus Sulfur **8** [1980] 281/90, 289).
[15] Chivers, T.; Hojo, M. (Inorg. Chem. **23** [1984] 2738/42).
[16] Olsen, B. A.; Olsen, F. P. (Inorg. Chem. **8** [1969] 1736/41).
[17] Mendelsohn, M. H.; Jolly, W. L. (J. Inorg. Nucl. Chem. **35** [1973] 95/9).
[18] Bruce, R. B.; Stephan, D. W.; McGlinchey, M. J. (Inorg. Chim. Acta **53** [1981] L19/L20).
[19] Woollins, J. D. (Polyhedron **3** [1984] 1365/7).
[20] Chivers, T.; Hojo, M. (Inorg. Chem. **23** [1984] 4088/93).

[21] Gleiter, R.; Bartetzko, R. (Z. Naturforsch. **36b** [1981] 492/7).
[22] Chivers, T.; Oakley, R. T.; Scherer, O. J.; Wolmershäuser, G. (Inorg. Chem. **20** [1981] 914/7).
[23] Bojes, J.; Chivers, T.; Codding, P. W. (J. Chem. Soc. Chem. Commun. **1981** 1171/3).
[24] Prestel, H.; Schindewolf, U. (Z. Anorg. Allgem. Chem. **551** [1987] 21/32).
[25] Dubois, P.; Lelieur, J. P.; Lepoutre, G. (Inorg. Chem. **27** [1988] 1883/90, 1885).
[26] Dubois, P.; Lelieur, J. P.; Lepoutre, G. (Inorg. Chem. **27** [1988] 73/80).
[27] Dubois, P.; Lelieur, J. P.; Lepoutre, G. (Inorg. Chem. **26** [1987] 1897/902).
[28] Chivers, T.; Schmidt, K. J.; McIntyre, D. D.; Vogel, H. J. (Can. J. Chem. **67** [1989] 1788/94).
[29] Shevalier, M. (Diss. Univ. Calgary 1981 from [28]).

2.11.5 Thiosulfinyl-amino-sulfane Derivatives, N-Thiosulfinyl-sulfenamides

2.11.5.1 Thiosulfinyl-amino-(triphenyl-λ^5-phosphanyliden)-amino-sulfane, $S{=}S{=}NSN{=}P(C_6H_5)_3$

The title compound formed in 36% yield by heating a slurry of $S_3N_3{-}N{=}P(C_6H_5)_3$ (see "Sulfur-Nitrogen Compounds" Part 2, 1985, pp. 82/7) in CH_3CN at reflux for 4 h. Subsequent workup of the reaction mixture afforded dark red crystals of $S{=}S{=}NSN{=}P(C_6H_5)_3$; m.p. 116 to 118°C [1, 2].

An X-ray diffraction study at 22±1°C shows the crystals to be monoclinic, space group $P2_1/c{-}C_{2h}^5$ (No. 14) with a = 12.573(1), b = 11.170(1), c = 14.816(2) Å and β = 115.96(1)°, V = 1872.2 Å^3; Z = 4, D_x = 1.37 g/cm^3. R = 0.040, R_w = 0.062 for 1747 reflections [1, 2].

The crystal structure consists of discrete molecules; the molecular structure is shown in **Fig. 32**. The atomic coordinates are listed in the paper [1]. The molecule consists of an open-chain PNSNSS sequence, the N and S atoms all being coplanar to within 0.04 Å. The phosphorus is rotated away from this plane, producing an N(3)–S(4)–N(5)–P(6) torsion angle of 136.4(3)°. The S(1)–S(2) bond length of 1.908 Å is remarkably short, indicating a bond order greater than one. The S(2)–N(3) and S(4)–N(3) bond lengths are almost equal within experimental error (1.592(4) and 1.587(4) Å). Selected bond lengths and bond angles are summarized in Table 78 [1, 2].

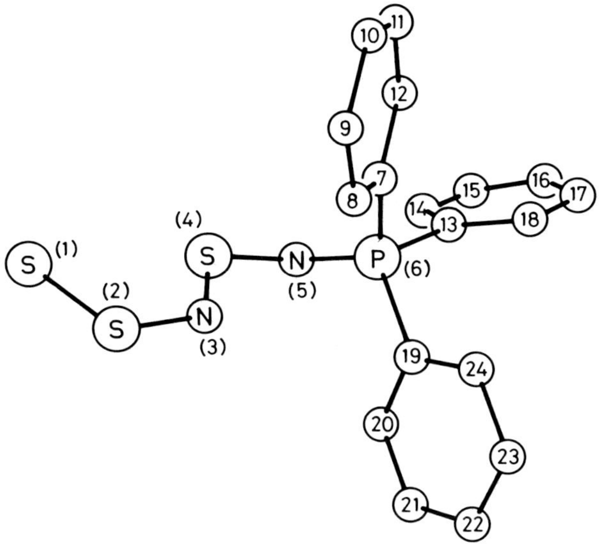

Fig. 32. Molecular structure of S=S=NSN=P(C_6H_5)$_3$ [1].

Table 78
Selected Bond Lengths and Bond Angles of S=S=NSN=P(C_6H_5)$_3$ [1].

bond length	in Å	bond angle	in °
S(1)–S(2)	1.908(2)	S(1)–S(2)–N(3)	111.4(2)
S(2)–N(3)	1.592(4)	S(2)–N(3)–S(4)	120.9(2)
N(3)–S(4)	1.587(4)	N(3)–S(4)–N(5)	107.7(2)
S(4)–N(5)	1.607(4)	S(4)–N(5)–P(6)	124.0(2)
N(5)–P(6)	1.598(3)	N(5)–P(6)–C(7)	114.3(2)
P(6)–C(7)	1.799(4)	N(5)–P(6)–C(13)	105.2(2)
P(6)–C(13)	1.800(5)	N(5)–P(6)–C(19)	113.2(2)
P(6)–C(19)	1.807(4)	C(7)–P(6)–C(13)	108.6(2)
S(1)···S(4)	3.131(2)	C(7)–P(6)–C(19)	108.1(2)

^{15}N NMR spectrum of 30% ^{15}N-enriched S=S=NSN=P(C_6H_5)$_3$ ($CHCl_3$/NH_3(l)) at 25°C: δ (in ppm) = 472.0 (1N, s); 86.9 (1N, d), ^1J(P, N) = 58.0 Hz [3]; referred to CH_3NO_2: δ (in ppm) = +91.8, −293.3.

^{31}P NMR ($CDCl_3$/external H_3PO_4): δ = 23.1 ppm (s).

UV-visible (CH_2Cl_2): λ_{max} (in nm) (ε in L·mol^{-1}·cm^{-1}) = 491 (14000) and 314 (2300).

IR (Nujol): 27 absorptions between 1600 and 250 cm^{-1} (1075 ν(P=N)). Raman (powder): Three strong bands characteristic of the SN=S=S group at 589 cm^{-1} ν(S=S) and 722 and 921 cm^{-1} ν(S=N) are recorded [1].

2.11.5.2 1-(Thiosulfinylaminothio)-piperidines, N-Thiosulfinyl-1-piperidine-sulfenamides, S=S=NSR, R=1-piperidinyl, subst. 1-piperidinyl

The compounds belonging to this section are summarized in Table 79.

Preparation. The following methods of preparation were used for the 1-(thiosulfinylamino-thio)-piperidines listed in the following table.

Method I: The compounds No. 1 to 6 were obtained by treating the corresponding piperidines with an equimolar amount of S_2Cl_2 at 10 to 15°C in $CH_3C_6H_5$–c-C_6H_{12} for 2 h or in $CHCl_3$ for 20 min. The reaction mixture was poured into ice and 35% aqueous NH_3 [6], see also [7].

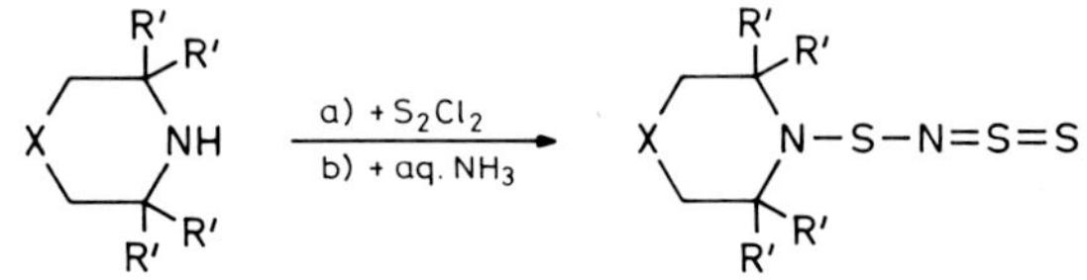

Method II: The reaction of bis(1-piperidinyl) disulfide (for No. 1) and bis(2,2,6,6-tetramethyl-1-piperidinyl) disulfide (for No. 3) or bis(2,2,6,6-tetramethyl-1-piperidinyl) trisulfide (for No. 3) with equimolar amounts of S_2Cl_2 followed by treatment with 10% aqueous NH_3 gave the compounds No. 1 and 3 [6].

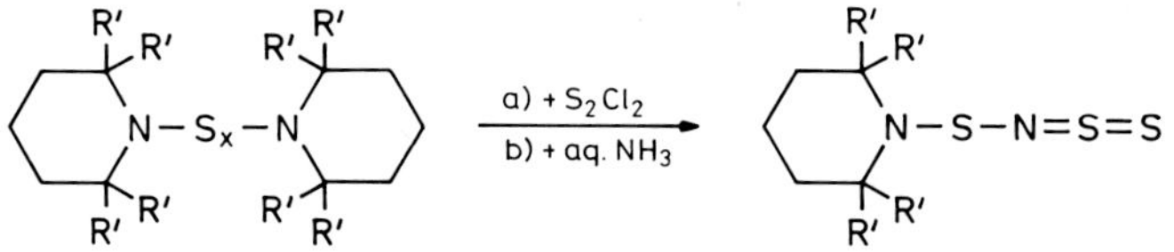

R' = H, x = 2
R' = CH_3, x = 2, 3

Method III: Compound No. 1 was also obtained by reaction of the corresponding 1-piperidinesulfenyl chloride with equimolar amounts of S_2Cl_2 followed by treating with 10% aqueous NH_3 [6].

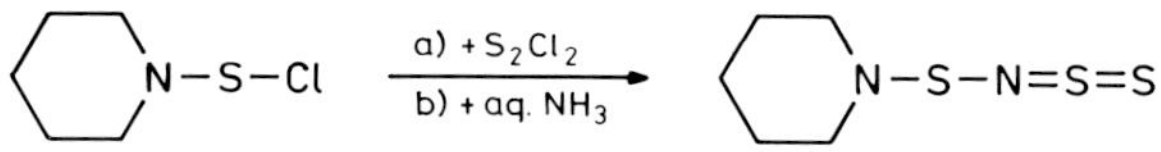

Crystal Structure. The structure of 7,7,9,9-tetramethyl-8-thiosulfinylaminothio-1,4-dioxa-8-azaspiro[4,5]decane (No. 5 in Table 79) was determined by X-ray crystallographic analysis. The crystals of the compound are monoclinic, space group $P2_1/n$–C^5_{2h} (No. 14) with a = 15.310(2), b = 9.477(1), c = 10.860(2) Å and β = 104.58(2)°; Z = 4, D_x = 1.32 g/cm³. R = 0.067.

The structure of the compound is shown in **Fig. 33**, p. 317. S(1), N(2), S(2), and S(3) are fairly planar with a maximum deviation of 0.001 Å from the least-squares plane. N(1) deviates 0.084 Å from this plane. The S(2)–S(3) bond length of 1.912 Å is significantly shorter than the usual S–S single bond distance of 2.0 to 2.1 Å indicating considerable double-bond character. Selected bond lengths and bond angles are listed in Table 80, p. 317. Final atomic coordinates are listed in the paper [8]. The structures of the remaining compounds (No. 1 to 4 and No. 6 in Table 79) were analogously assigned on the basis of ^{1}H NMR, UV, and mass spectra and/or elemental analysis [6].

Table 79

Methods of Preparation, Yields, Properties, and Spectra of 1-(Thiosulfinylaminothio)-piperidines [6].

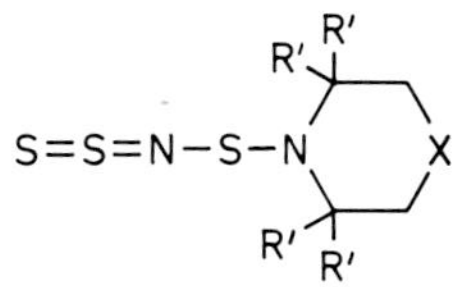

No.	X	R′	meth. of prep. (yield in %)	appearance, m.p. in °C	R^f-SiO_2 (solvent)	spectra
1	CH_2	H	I (~0.4), II (2.8), III (1.2)	—	0.25 (hexane)	UV (C_2H_5OH): $\lambda_{max}=410$ nm MS: m/e=194 (M^+)
2	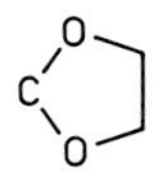	H	I (~0.4)	—	—	UV (C_2H_5OH): $\lambda_{max}=410$ nm MS: m/e=252 (M^+)
3	CH_2	CH_3	I (8.6), II (35.2 and 32.7)	red oil	0.75 (C_6H_6–c-C_6H_{12})	^1H NMR ($CDCl_3$/TMS): δ (in ppm)=1.40 (s, 12H), 1.5 to 1.8 (m, 6H) UV (C_2H_5OH): $\lambda_{max}=413$ nm MS: m/e=250 (M^+)
4	C=O	CH_3	I (14)	red prisms, 116.5 to 118	0.13 (C_6H_6)	^1H NMR (CCl_4/TMS): δ (in ppm)=1.47 (s, 12H), 2.60 (s, 4H) UV: λ_{max} (in nm) (log ε)=409 (4.11) in hexane; 411 (4.11) in C_2H_5OH MS: m/e=264 (M^+)
5	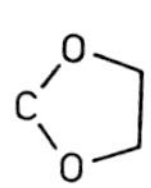	CH_3	I (7)	red prisms, 98 to 99	0.50 (C_6H_6)	^1H NMR ($CDCl_3$/TMS): δ (in ppm)=1.45 (s, 12H) 1.92 (s, 4H), 3.97 (s, 4H) UV: λ_{max} (in nm) (log ε)=412 (4.14) in c-C_6H_{12}; 414 (4.14) in C_2H_5OH

Table 79 (continued)

No.	X	R′	meth. of prep. (yield in %)	appearance, m.p. in °C	R^f-SiO_2 (solvent)	spectra
6	$CHO(O)CC_6H_5$	CH_3	I (6)	red prisms, 99 to 100	0.62 (C_6H_6)	^1H NMR ($CDCl_3$/TMS): δ (in ppm) = 1.46 (s, 6H), 1.54 (s, 6H), 1.75 (br, d, 2H), 2.27 (dd, 2H), 5.58 (tt, 1H); $^2J(H,H)$ = 12 and 4.5 Hz; 7.3 to 8.3 (m, 5H) UV: λ_{max} (in nm) (log ε) = 411 (4.15) in c-C_6H_{12}; 413 (4.15) in C_2H_5OH MS: m/e = 370 (M^+)

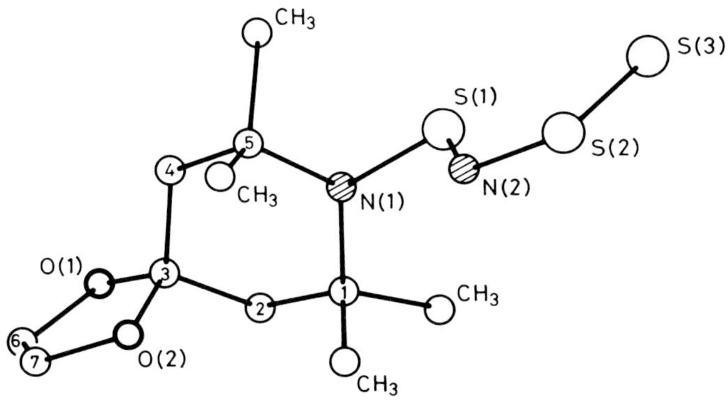

Fig. 33. Molecular structure of 7,7,9,9-tetramethyl-8-thiosulfinyl-aminothio-1,4-dioxa-8-azaspiro[4,5]decane (No. 5 in Table 79) [8].

Table 80

Selected Bond Lenghts and Bond Angles of 7,7,9,9-Tetramethyl-8-thiosulfinyl-aminothio-1,4-dioxa-8-azaspiro[4,5]decane (No. 5 in Table 79) [8].

bond length	in Å	bond angle	in °
S(1)–N(1)	1.634(5)	N(1)–S(1)–N(2)	105.7(1)
S(1)–N(2)	1.657(6)	C(3)–O(2)–C(7)	108.3(3)
S(2)–S(3)	1.912(3)	C(3)–O(1)–C(6)	108.8(3)
S(2)–N(2)	1.569(7)	C(1)–N(1)–C(5)	120.1(4)
O(1)–C(3)	1.421(8)	S(3)–S(2)–N(2)	114.9(1)
O(1)–C(6)	1.373(10)	S(1)–N(1)–C(5)	118.8(3)
O(2)–C(3)	1.433(7)	N(1)–C(1)–C(2)	107.0(4)
O(2)–C(7)	1.373(11)	N(1)–C(5)–C(4)	108.5(5)
N(1)–C(1)	1.500(7)	S(1)–N(1)–C(1)	117.6(4)
N(1)–C(5)	1.505(8)	S(1)–N(2)–S(2)	119.2(4)

Reactions. The compounds No. 1 to 6 in Table 79 are thermally stable. Thus, refluxing of 5 to 10×10^{-5} M solutions in $CHCl_3$ for 3 h showed no detectable change in the visible absorption at 410 nm. Whereas, a decrease in the 410 nm band of 5 to 7 nm was observed when 5 to 10×10^{-5} M solutions of the compounds in c-C_6H_{12} or C_2H_5OH were allowed to stand under diffused light for 1 h [6]. The compounds show fungicidal activity [7].

2.11.5.3 N-Thiosulfinyl-trifluoromethanesulfenamide, S=S=NSCF$_3$

The thermally unstable title compound was prepared in 20% yield by treating a solution of CF_3SNH_2 and quinoline (or acridine) in CS_2 at 20°C with a CS_2 solution of S_2Cl_2 (mole ratio 1:2:1) for 0.5 h. After filtration, the compound was isolated as a red liquid by fractionated condensation in the −50°C trap in 98% purity.

^{19}F NMR (solvent ?/$CFCl_3$): $\delta = -48.2$ ppm (s, CF_3); the unusual shift to low field is explained by the lower shielding of the F atoms caused by the neighboring N=S=S system.

^{13}C NMR (solvent ?/TMS): δ = 129.1 ppm (q, CF_3), ^1J (C, F) = 312 ± 2 Hz.

IR (liquid): ν (in cm^{-1}) = 1175 s, 1125 s, 715 s, 640 m, 505 m.

MS (70 eV): m/e (rel. int. in %) = 179 (74) M^+, 110 (79) S_3N^+, 101 (2) CF_3S^+, 82 (4) SCF_2^+, 78 (34) S_2N^+, 69 (36) CF_3^+, 64 (46) S_2^+, 46 (100) SN^+.

Thermolysis of S=S=NSCF$_3$ at 60°C for 4 h afforded S_8 and $CF_3SN{=}S{=}NSCF_3$. The chlorination with Cl_2 in CCl_4 at −30°C for 30 min yielded S_2Cl_2 and unstable $CF_3CN{=}SCl_2$, which decomposed to give $S_3N_3Cl_3$ and CF_3SCl. Warming a mixture of S=S=NSCF$_3$ and HgF_2 (mole ratio ~1:3.75) from −196 to −30°C in CCl_4 for 30 min produced $CF_3SN{=}SF_2$ and HgS. If S=S=N-SCF$_3$ was reacted with cyclopentadiene in $CHCl_3$ with ice-cooling for 45 min the isomeric adducts I and II were formed [4].

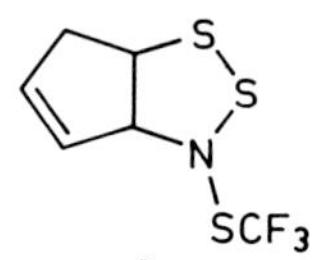

I

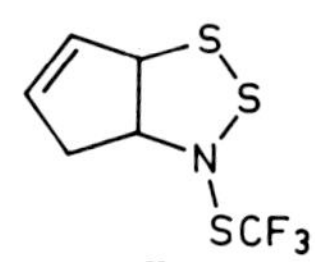

II

2.11.5.4 N-Thiosulfinyl-pentachloroethanesulfenamide, S=S=NSC$_2$Cl$_5$

The title compound was obtained in 96% yield (crude product) by adding a solution of S_2Cl_2 in CS_2 to a mixture of $C_2Cl_5SNH_2$ and quinoline dissolved in CS_2 (mole ratio 1:1:2). A precipitate, formed during the reaction, was removed by filtration and after evaporation of the filtrate the compound was remained as a red oil.

^{13}C NMR ($CDCl_3$/TMS, $Cr(CH_3C(O)CHC(O)CH_3)_3$): δ (in ppm) = 103.46, 103.68, 103.93, 104.46, 105.28 (unassigned peaks).

IR (neat): ν (in cm^{-1}) = 1520, 785, 760, 705, 670, and 630.

UV (c-C_6H_{12}): λ_{max} (in nm) = 400 and 281.

MS (70 eV): m/e = 309 (M^+), 199 ($C_2Cl_5^+$), 164 ($C_2Cl_4^+$), 145 ($ClNS_3^+$), 117 (CCl_3^+), 110 (NS_3^+) [5].

2.11.5.5 N, N′-Bis(thiosulfinyl)-bicyclo[2.2.1]hept-5-ene-2,3-disulfenamide, 2,3-(S=S=NS)$_2$C$_7$H$_8$

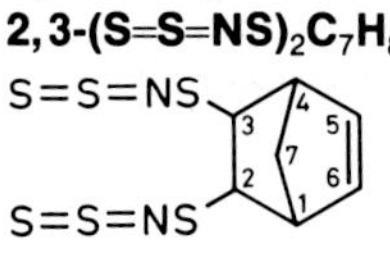

The reaction of equimolar solutions of S_4N_2 and 2,5-norbornadiene in CH_2Cl_2 under conditions of high dilution yielded the 1:1 adduct $S_4N_2 \cdot C_7H_8$ in 15% yield and small quantities of 2,3-(S=S=NS)$_2$C$_7$H$_8$. When a tenfold excess of norbornadiene was added rapidly to a solution of S_4N_2 in CH_2Cl_2, the yield of $S_4N_2 \cdot C_7H_8$ was reduced (less than 1%), and 2,3-(S=S=NS)$_2$C$_7$H$_8$ was formed in <1% yield. The two compounds were separated by gel-permeation chromatography.

2,3-(S=S=NS)$_2$C$_7$H$_8$ crystallizes as orange rectangular blocks. An X-ray diffraction study shows that the crystals are monoclinic, space group $P2_1/c-C_{2h}^5$ (No. 14) with a = 9.216(1), b = 14.318(2), c = 10.229(1) Å and β = 104.57(1)°; Z = 4; V = 1306.3(5) Å^3, D_x = 1.59 g/cm^3; R = 0.037,

$R_w = 0.053$ for 1627 observed reflections. The molecular structure is shown in **Fig. 34**. The molecule consists of a norbornenyl group with two S_3N residues bound in an exo-fashion at the 2,3-positions. Both S_3N units are planar (to within 0.006 and 0.007 Å) and their mean planes intersect with a dihedral angle of 82.4°. The bond lengths and bond angles within the two S_3N groups are remarkably similar. Selected bond lengths and bond angles are given in Table 81. Positional parameters are listed in the paper [9].

M.p. 72.5 to 73.5°C.

^{1}H NMR ($CDCl_3$/TMS): δ (in ppm) = 1.78, 1.98 (H7,H7′; AB, ^2J(H7,H7′) = 10.1 Hz); 3.35 (H2,H3; t, ^3J = 1.8 Hz); 4.23 (H1,H4; d, ^4J = 2.1 Hz); 6.43 (H5,H6; t, ^3J = 1.8 Hz).

UV (CH_2Cl_2): λ_{max} (in nm) (log ε) = 433 (4.2), 408 (4.3), 291 (3.7). The absorptions at 433 and 408 nm, due to $\pi^* \rightarrow \pi^*$ transitions, appear as a pair of overlapping bands. This effect is assumed to be a result of an excitonic coupling of the two S_3N chromophors. The weak energy band at 291 nm is probably a $\sigma \rightarrow \sigma^*$ transition [9].

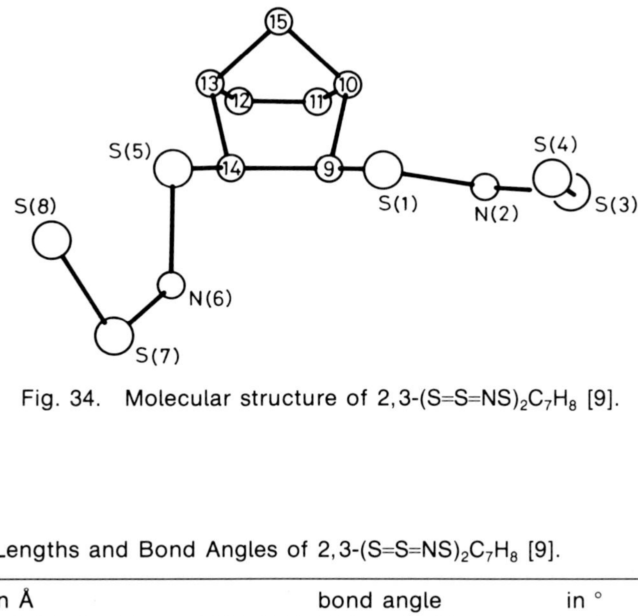

Fig. 34. Molecular structure of 2,3-(S=S=NS)$_2C_7H_8$ [9].

Table 81
Selected Bond Lengths and Bond Angles of 2,3-(S=S=NS)$_2C_7H_8$ [9].

bond length	in Å	bond angle	in °
S(1)–N(2)	1.639(3)	N(2)–S(1)–C(9)	96.4(1)
S(1)–C(9)	1.819(3)	S(4)–S(3)–N(2)	114.8(1)
S(3)–S(4)	1.910(1)	N(6)–S(5)–C(14)	98.1(1)
S(3)–N(2)	1.573(3)	S(8)–S(7)–N(6)	114.9(1)
S(5)–N(6)	1.644(3)	S(1)–N(2)–S(3)	119.1(2)
S(5)–C(14)	1.821(3)	S(5)–N(6)–S(7)	118.9(2)
S(7)–S(8)	1.896(1)	S(1)–C(9)–C(10)	112.1(2)
S(7)–N(6)	1.571(3)	S(1)–C(9)–C(14)	112.4(2)
		S(5)–C(14)–C(9)	116.6(2)
		S(5)–C(14)–C(13)	108.9(2)
		C(10)–C(9)–C(14)	102.4(2)

References:

[1] Chivers, T.; Cordes, A. W.; Oakley, R. T.; Swepston, P. N. (Inorg. Chem. **20** [1981] 2376/80).
[2] Chivers, T.; Oakley, R. T.; Cordes, A. W.; Swepston, P. N. (J. Chem. Soc. Chem. Commun. **1980** 35/6).
[3] Chivers, T.; Oakley, R. T.; Scherer, O. J.; Wolmershäuser, G. (Inorg. Chem. **20** [1981] 914/7).
[4] Haas, A.; Walz, R. (Chem. Ber. **118** [1985] 3248/57).
[5] Boureghda, A.; Abdel-Megeed, M. F.; Ghattas, A.-B. A. G.; Jensen, B.; Senning, A. (Sulfur Letters **5** [1987] 159/63).
[6] Morimura, S.; Horiuch, H.; Tamura, C.; Yoshioka, T. (Bull. Chem. Soc. Japan **53** [1980] 1666/9).
[7] Soma, N.; Morimura, S.; Yoshioka, T.; Nishimura, T. (Japan Kokai Tokkyo Koho 53-124273 [1978] 1/4; C.A. **90** [1979] No. 121433).
[8] Tamura, C.; Aiba, K.; Sato, S.; Hata, T.; Morimura, S.; Yoshioka, T. (Acta Cryst. B **33** [1977] 3918/20).
[9] Cordes, A. W.; Koenig, H.; Noble, M. C.; Oakley, R. T. (Inorg. Chem. **22** [1983] 3375/8).

2.11.6 N-Thiosulfinyl-organylamines

2.11.6.1 N-Thiosulfinyl-alkanamines, S=S=NR, R = t-C_4H_9, $C(CH_3)_2CN$

The compounds were prepared in nearly quantitative yields by adding $S(Si(CH_3)_3)_2$ to stoichiometric amounts of RN=SCl_2 (R = t-C_4H_9 or $C(CH_3)_2CN$) at −30°C. After the mixture was allowed to stand at 20°C for 24 h $(CH_3)_3SiCl$ was evaporated, leaving viscous yellow liquids, which decomposed during vacuum distillation.

The IR spectrum of S=S=NC$(CH_3)_2$CN exhibits 22 absorptions between 540 and 2990 cm^{-1}.

The thermolysis of S=S=NC_4H_9-t at 100°C for 3 h leads to elimination of sulfur and conversion to t-C_4H_9N=S=NC_4H_9-t. Treatment of solutions of S=S=NR (R = t-C_4H_9, $C(CH_3)_2CN$) in CCl_4 with Cl_2 for 10 min yields RN=SCl_2. The reaction of S=S=NC_4H_9-t with Br_2 in CCl_4 at 20°C for 24 h affords Br_2S=NC_4H_9-t and S_2Br_2.

Reference:

Shermolovich, Yu. G.; Vasil'ev, V. V.; Markovskii, L. N. (Zh. Org. Khim. **13** [1977] 727/30; J. Org. Chem. [USSR] **13** [1977] 664/7).

2.11.6.2 N-Thiosulfinyl-arenamines

2.11.6.2.1 N-Thiosulfinyl-2,4,6-tribromo-aniline, S=S=N$C_6H_2Br_3$-2,4,6

S=S=N$C_6H_2Br_3$-2,4,6 was obtained quantitatively by heating a mixture of equimolar amounts of 2,4,6-$Br_3C_6H_2$N=SCl_2 and $S(Si(CH_3)_3)_2$ at 70°C for 20 min. After removal of $(CH_3)_3SiCl$ in vacuum, the compound was isolated as brown prisms; m.p. 85 to 86°C.

It decomposes on heating at 100°C for 3 h to give sulfur and 2,4,6-$Br_3C_6H_2$N=S=NC_6H_2-Br_3-2,4,6. Passing Cl_2 into a solution of S=S=N$C_6H_2Br_3$-2,4,6 in $C_2H_4Cl_2$ for 10 min yields 2,4,6-$Br_3C_6H_2$N=SCl_2 and SCl_2.

Reference:

Shermolovich, Yu. G.; Vasil'ev, V. V.; Markovskii, L. N. (Zh. Org. Khim. **13** [1977] 727/30; J. Org. Chem. [USSR] **13** [1977] 664/7).

2.11.6.2.2 N′-Thiosulfinyl-N, N-dimethyl-1,4-benzenamine, $S{=}S{=}NC_6H_4N(CH_3)_2$-4

The compound was formed by reaction of 4-$(CH_3)_2NC_6H_4NH_2$ with S_2Cl_2 or bis(phthalimido) disulfide (I) [1]. Refluxing of equimolar amounts 4-$(CH_3)_2NC_6H_4NH_2$ and $C_2H_5OSSOC_2H_5$ in C_6H_6 for 6 h afforded the compound in low yield (~3%) [2]. It was prepared in 15% yield by adding P_2S_5 to a solution of 4-$(CH_3)_2NC_6H_4N{=}O$ in C_6H_6–CH_2Cl_2 under nitrogen. Subsequently, dry pyridine was added over 13 min and then the reaction mixture was stirred at 24 to 32°C for ~35 min. The compound was isolated by extraction with light petroleum [1]. It was obtained in 50% yield by treating a solution of 4-$(CH_3)_2NC_6H_4N{=}O$ in C_6H_6–CH_2Cl_2 with an equimolar amount of 2,4-bis(4-methoxyphenyl)-1,3,2,4-dithiadiphosphetane 2,4-disulfide (II) at 0°C for 1.5 h. Subsequently the compound was isolated by extraction with light petroleum [3]. It formed also in low yield (1.3%) by reaction of $O{=}S{=}NC_6H_4N(CH_3)_2$-4 in C_6H_6 with P_2S_5 at room temperature for 26 h and extraction with boiling petroleum [1].

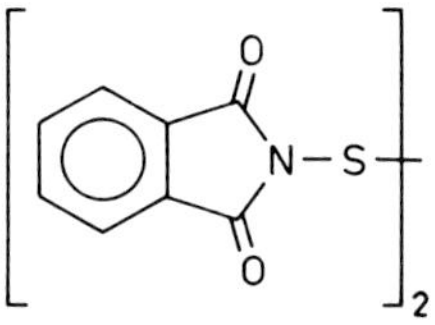

I

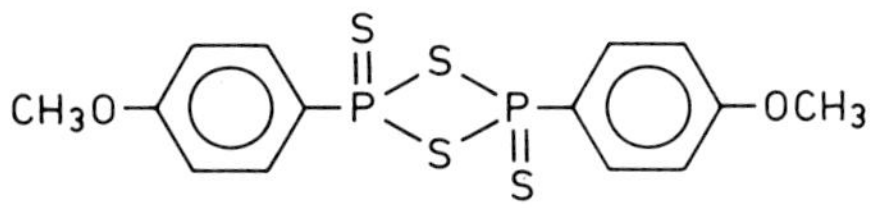

II

The deep violet crystals, which appear to be air-sensitive [4], melt at 110 [3], 112 to 113 [2], 110 to 115 [1], and 113 to 115°C with slow decomposition [1].

^1H NMR ($CDCl_3$/TMS): δ (in ppm) = 3.1 (s, $N(CH_3)_2$), 6.65 (d, 2 arom. H), 8.6 (d, 2 arom. H) [1].

IR (KBr and Nujol): ν (in cm^{-1}) = 1605, 1535, 1315, 1290, 1180, 830, and 680 [1, 2].

UV (c-C_6H_{12}): λ_{max} (in nm) (ε in $L\cdot mol^{-1}\cdot cm^{-1}$) = 255 (7400), 284 (7100), 348 (3300), and 510 (38400); (CH_3OH): 238sh (4600), 262sh (5900), 285 (7200), 350 (1700), and 538 (39000).

MS: m/e = 198 M^+ [1].

Pyrolysis of $S{=}S{=}NC_6H_4N(CH_3)_2$-4 at 200°C under N_2 for 10 min gives $(4\text{-}(CH_3)_2NC_6H_4N{=})_2$ and sulfur. It is hydrolyzed with aqueous 2N HCl in ether and reduced with H_2S in CH_2Cl_2–ether or with $NaBH_4$ in THF, followed by treatment with $C_6H_5C(O)Cl$ in the presence of $(C_2H_5)_3N$ or pyridine to give 4-$(CH_3)_2NC_6H_4NHC(O)C_6H_5$. The compound reacts with cyclopentadiene in CH_2Cl_2 at 0°C and then at 4°C for 74 h or with norbornadiene in CH_2Cl_2 at room temperature in the dark for 50 h to give the 1,3-cycloaddition products IIIa or IIIb and IV, respectively [1].

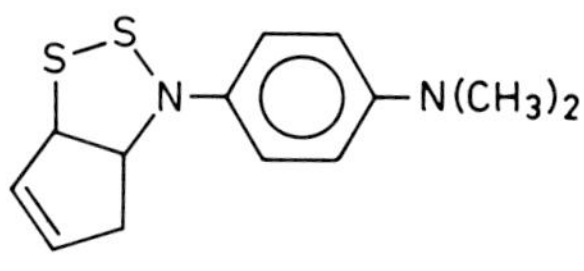

IIIa

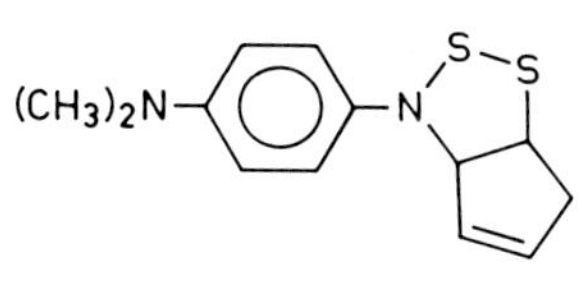

IIIb

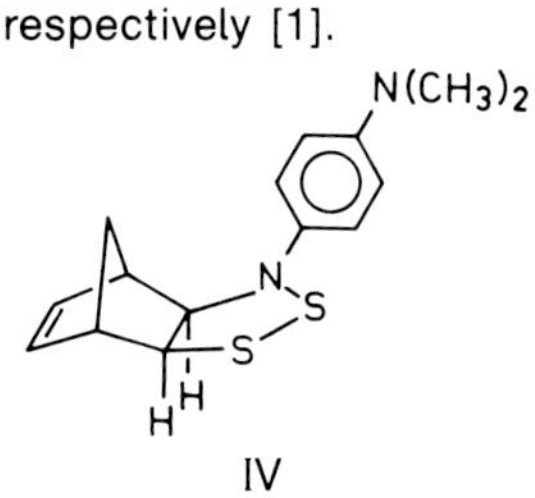

IV

References:

[1] Barton, D. H. R.; Robson, M. J. (J. Chem. Soc. Perkin Trans. I **1974** 1245/7).
[2] Kagami, H.; Motoki, S. (J. Org. Chem. **42** [1977] 4139/41).
[3] Jørgensen, K. A.; Shabana, R.; Scheibye, S.; Lawesson, S.-O. (Bull. Soc. Chim. Belges **89** [1980] 247/53).
[4] Inagaki, Y.; Okazaki, R.; Inamoto, N. (Bull. Chem. Soc. Japan **52** [1979] 1998/2001).

2.11.6.2.3 N-Thiosulfinyl-2,4,6-trialkyl-benzenamines

2.11.6.2.3.1 N-Thiosulfinyl-2,4,6-trimethyl-benzenamine, $S{=}S{=}NC_6H_2(CH_3)_3$-2,4,6

Treatment of a mixture of 2,4,6-$(CH_3)_3C_6H_2NH_2$ and $(C_2H_5)_3N$ in ether with an ethereal solution of S_2Cl_2 (mole ratio ~1:2:1) at −78°C, followed by stirring at −78°C for 40 min yielded the compound in 2% yield after working up of the reaction mixture at −2°C.

The reddish purple oil is extremely unstable. ^1H NMR (CCl_4/TMS): δ (in ppm) = 2.01 (s, 2 CH_3), 2.31 (s, CH_3), 6.90 (s, arom. H); UV (pentane): λ_{max} (in nm) (ε in $L \cdot mol^{-1} \cdot cm^{-1}$) = 336 (5700), 462 (1900), 540sh (870) [1, 2].

2.11.6.2.3.2 N-Thiosulfinyl-2,4-di-*tert*-butyl-6-methyl-benzenamine, $S{=}S{=}NC_6H_2CH_3$-6-$(C_4H_9$-t$)_2$-2,4

The compound was prepared in 80.1% yield by adding a solution of S_2Cl_2 in ether to a solution of 2,4-(t-$C_4H_9)_2$-6-$CH_3C_6H_2NH_2$ and $(C_2H_5)_3N$ (mole ratio ~1:1:2) in ether at 0°C and subsequently stirring for 1 h. The reaction mixture was poured into water, the solvent of the organic layer removed and the residue separated by chromatography (silica gel, hexane) [1, 2]. It formed in 16.9% yield, along with (2,4-(t-$C_4H_9)_2$-6-$CH_3C_6H_2N)_2S$ (50.1%) and 2,4-(t-$C_4H_9)_2$-6-$CH_3C_6H_2NH_2$ (26.3%) by reaction of SCl_2 with a mixture of 2,4-(t-$C_4H_9)_2$-6-$CH_3C_6H_2NH_2$ and $(C_2H_5)_3N$ (mole ratio ~1:1:2) in ice-cold ether and stirring of the reaction mixture for 1.5 h [1].

The deep purple crystals, stable at ambient temperature melt at 67 to 68°C [1, 2].

An X-ray diffraction study shows that the crystals are monoclinic, space group $P2_1/a-C^5_{2h}$ (No. 14), with a = 19.050(6), b = 9.614(2), c = 10.084(3) Å and β = 118.25(3)°; Z = 4, V = 1629.9 Å^3, D_x = 1.149 g/cm^3. R = 0.057, R_w = 0.044 for 2049 observed reflections. The structure of the compound is shown in **Fig. 35**. The compound has a Z configuration. The S(1)–S(2) and N–S(1) double-bond distances are 1.898(2) and 1.548(3) Å, respectively. The terminal S atom is cis to C across the N=S bond with a torsion angle of 4.7(4)°. The dihedral angle between the planes of the benzene ring and the N-thiosulfinyl group is 68(3)°. Selected bond lengths and bond angles are given in Table 82. Atomic parameters are listed in the paper [3].

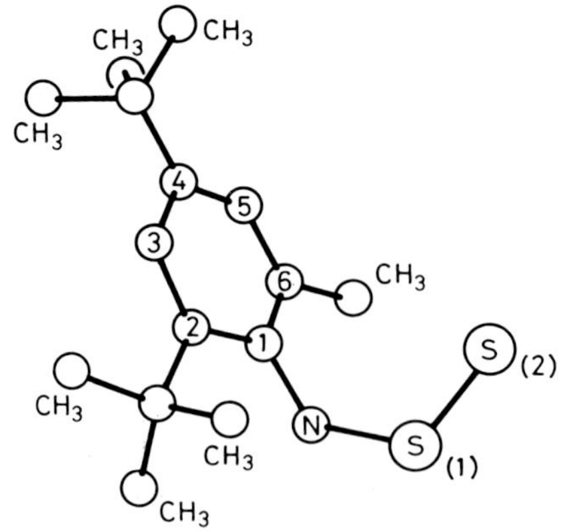

Fig. 35. Molecular structure of $S{=}S{=}NC_6H_2CH_3$-6-$(C_4H_9$-t$)_2$-2,4 [3].

Table 82
Selected Bond Lengths and Bond Angles of $S{=}S{=}NC_6H_2CH_3$-6-$(C_4H_9$-t$)_2$-2,4 [3].

bond length	in Å	bond angle	in °
S(1)–S(2)	1.898(2)	S(2)–S(1)–N	119.7(1)
S(1)–N	1.548(3)	S(1)–N–C(1)	126.4(3)
N–C(1)	1.420(5)	N–C(1)–C(2)	119.5(3)
		N–C(1)–C(6)	118.6(3)

The fact that the IR spectrum in solution is identical with that in a KBr disk and also that the ^1H NMR spectrum shows only one species over a temperature range of −102 to 70°C suggests that the compound has a Z configuration in solution also [4].

Electric dipole moment: 1.51 D (determined by measuring the dielectric constants in dilute benzene solutions at 25°C). From estimated bond moments of S=S (2.60 D) and N=S (0.55 D) the charge separation in the N=S=S group is about half that of the N=S=O group [4].

^1H NMR (CCl_4/TMS): δ (in ppm) = 1.33 (s, 2 t-C_4H_9), 2.02 (s, CH_3), 7.22 (q, arom. H), J = 2 Hz [1, 2].

IR (KBr): 14 absorption bands between 2950 and 620 cm^{-1}.

UV (hexane): λ_{max} (in nm) (ε in $L \cdot mol^{-1} \cdot cm^{-1}$) = 270sh (1470), 310 (1940), 343 (5590), 476 (2190), 536 (1190) [1, 2]; (CCl_4): 345 (5830), 480 (2230), 548sh (1280); (C_2H_5OH): 343 (6000), 478 (2170), 544sh (1260) [1].

MS: m/e (rel. int. in %) = 281 (0.5) M^+, 266 (20), 224 (100), 217 (18), 209 (18), 204 (20) [1]; 247 (26) $M^+ - H_2S$, 232 (100), 204 (95) [2].

Heating of $S{=}S{=}NC_6H_2CH_3$-6-$(C_4H_9$-t$)_2$-2,4 in refluxing C_6H_6 for 21 h [5, 6] or in toluene at 70°C for 28 h [5] afforded 2,1-benzisothiazole I, the corresponding aniline II and sulfur:

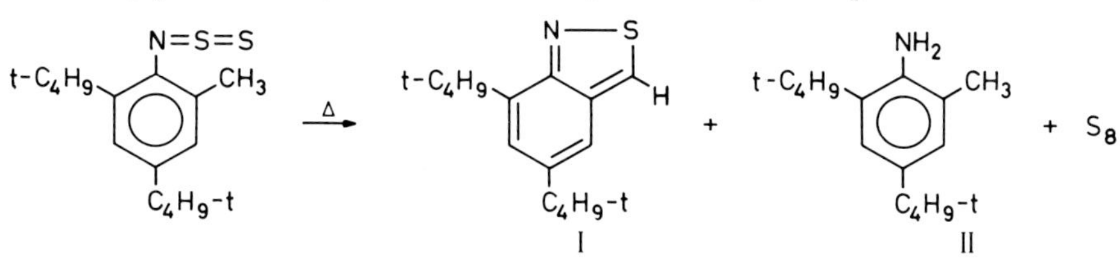

I ~32 or 40% II ~26 or 46%

Irradiation of the compound with pyrex filtered light from a medium pressure mercury lamp below 25°C afforded the sulfur diimide III, the corresponding aniline II, and sulfur [5, 6]:

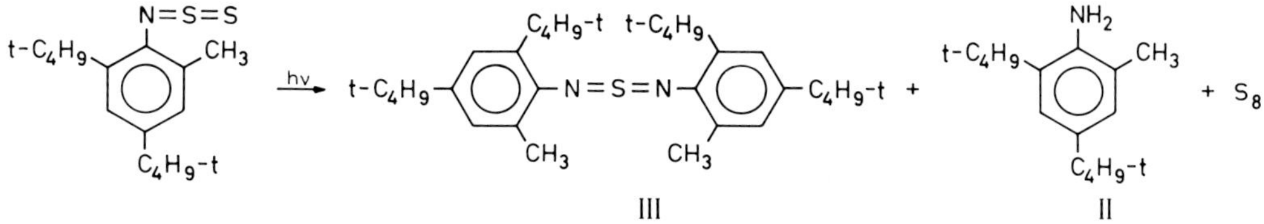

III 34.7 to 70.9% II 17.1 to 28.5%

The yields of the reaction products were dependent on the solvent (pentane, benzene or 2-propanol), concentration of the starting compound (0.7 to 16.7 g/L), and irradiation time

(4 to 35.3 h). The yield of product III increased with increasing concentration and irradiation time. When the compound was irradiated in an EPA matrix at 77 K, an intermediate was observed by a change in the intermittently monitored electronic spectra. It was assumed to be a dithionitro compound IV or a dithiaziridine V [5].

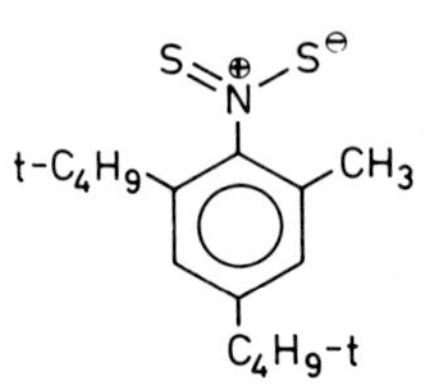

IV

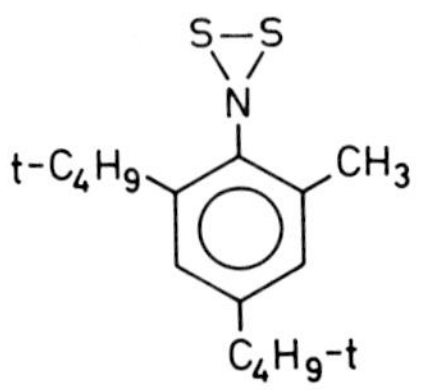

V

Reaction of the compound with an equimolar amount of Br_2 in CCl_4 at 0°C for 1.5 h afforded the sulfur diimide III (64.4%) and sulfur (17.2%). No reaction was observed with $(CH_3)_3O^+BF_4^-$ and $(CH_3)_3SiCl$ at room temperature [7]. The compounds II and III were obtained in 7 and 59% yields, respectively, by the reaction with *tert*-butyl radicals (generated by photolysis of 2,2'-dimethyl-2,2'-azopropane) [13]. The reactions with alkylamines, pyrrolidine, and thiourea, affording the aniline II and other products, which could not be identified are summarized in Table 83 [8].

Table 83

Reaction Conditions and Yields of 2,4-(t-C_4H_9)$_2$-6-$CH_3C_6H_2NH_2$ (II) in the Reactions of Alkylamines, Pyrrolidines, and Thiourea (A) with S=S=$NC_6H_2CH_3$-6-(C_4H_9-t)$_2$-2,4 (B) [8].

reactant A	reaction time	solvent	mole ratio A:B	yield of II in %
n-$C_4H_9NH_2$	88 h	C_2H_5OH	1.08	41.1
t-$C_4H_9NH_2$	22 h	$(C_2H_5)_2O$	1.39	30.6
$C_{10}H_{15}NH_2$-1 ($C_{10}H_{15}$ = 1-adamantanyl)	11 d	CH_2Cl_2	1.01	81.1
$C_6H_5NH_2$	7 d	$(C_2H_5)_2O$	0.94	no reaction
$(CH_2)_4NH$	2.5 h	CH_2Cl_2	2.23	63.7
$(CH_2)_4NC_5H_7$ (C_5H_7 = 1-cyclopenten-1-yl)	immediately	CH_2Cl_2	1.02	57.7
$SC(NH_2)_2$	1 d	C_2H_5OH	1.0	87.7

Passing H_2S into a benzene solution of the compound at ~10°C for 3.5 h yielded the corresponding aniline II in 50.8%. Treatment of the compound with an equimolar amount of $(C_6H_5)_3P$ in ether at −78°C and subsequently stirring at this temperature for 4 h and then at −5°C for 13.5 h gave the corresponding sulfur diimide III (21%) and the phosphine imide VI (66%) (yields are based on consumed educt) [8]:

III VI

It reacted with $3\text{-ClC}_6H_4CO_3H$ in CH_2Cl_2 at 0°C followed by stirring at 0°C for 1.5 h and then at room temperature for further 1.5 h to give $O{=}S{=}NC_6H_2CH_3\text{-}6\text{-}(C_4H_9\text{-t})_2\text{-}2,4$ (30.8%) and the aniline II (41.3%) [7]. Heating the compound with cyclopentadiene gave the cycloadduct VIIa or VIIb in low yield. No adducts were produced with $C_6H_5C(O)C{\equiv}CC(O)C_6H_5$ or $CH_3O(O)CC{\equiv}C\text{-}C(O)OCH_3$. Heating the title compound in 2,3-dimethyl-1,3-butadiene afforded I and II (15 and 77%, respectively) and no adduct [13].

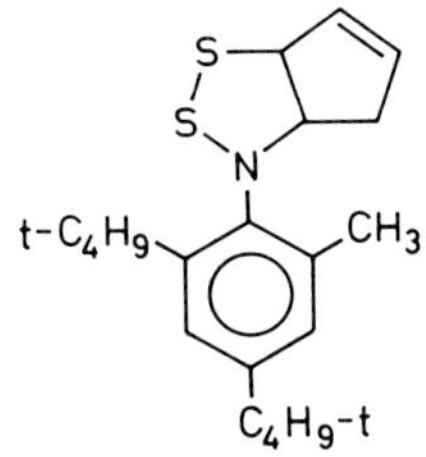
VIIa

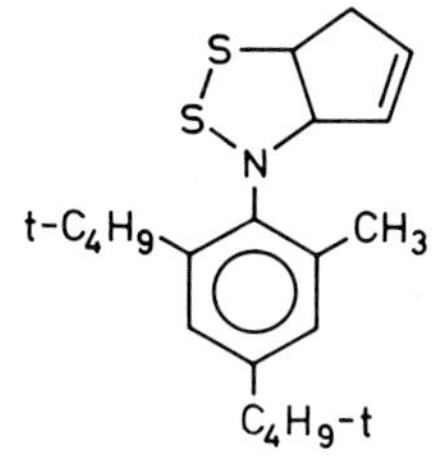
VIIb

It reacted with *trans*-cyclooctene, *trans,cis*-1,5-cyclooctadiene (in $CHCl_3$ at 20°C for 4 h), norbornene (at 50°C for 3 d), and $O(CH_2)_4NC_5H_7$ (C_5H_7 = 1-cyclopenten-1-yl) (in ether at 20°C for 3.5 h) to give the 1:1 adducts VIII (50%), IX (83%), X (73%), and XI (80%), respectively [9]. The reactions with equimolar amounts of $LiC_4H_9\text{-n}$ or CH_3MgI in ethereal solution at 0°C afforded the aniline II (54.1 or 65.2%) and the sulfur diimide III (23.6 or 8.5%, respectively). If the compound was reacted with excess $i\text{-}C_3H_7MgBr$ in ether at 0°C for 1 h only the aniline II (59.6%) was formed [8].

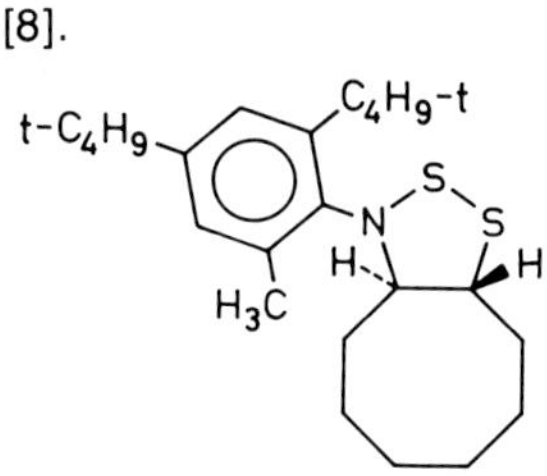
VIII

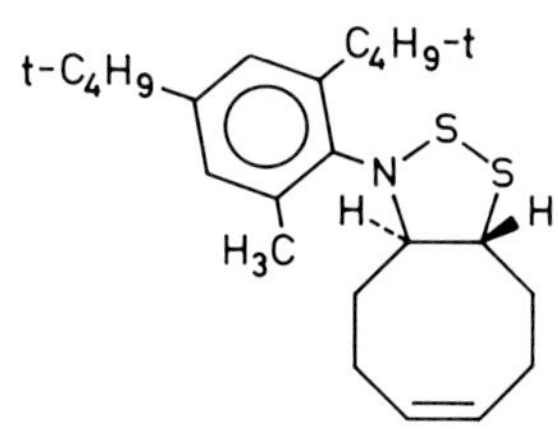
IX

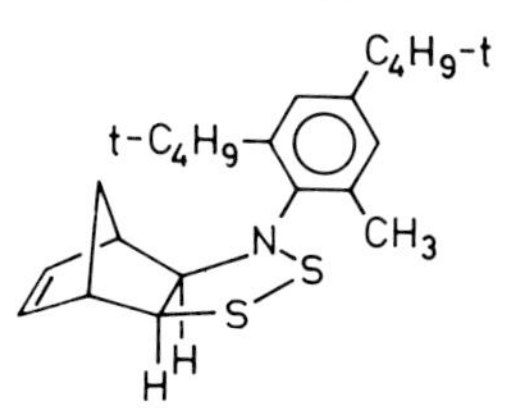
X

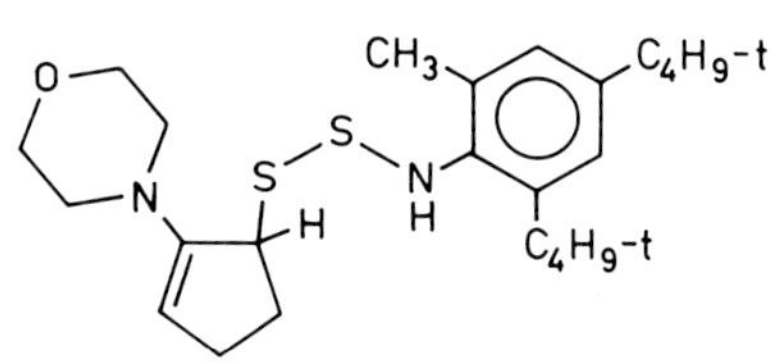
XI

2.11.6.2.3.3 N-Thiosulfinyl-2,4-di-*tert*-butyl-6-isopropyl-benzenamine, $S{=}S{=}NC_6H_2C_3H_7\text{-i-}6\text{-}(C_4H_9\text{-t})_2\text{-}2,4$

Adding of an ethereal solution of S_2Cl_2 to a mixture of $2,4\text{-}(t\text{-}C_4H_9)_2\text{-}6\text{-i-}C_3H_7C_6H_2NH_2$ and $(C_2H_5)_3N$ in ether (mole ratio ~1:1:2) at 0°C, stirring of the reaction mixture for 45 min and then usual work up gave $S{=}S{=}NC_6H_2C_3H_7\text{-i-}6\text{-}(C_4H_9\text{-t})_2\text{-}2,4$ as a purple oil (14.4% yield).

^1H NMR (CCl_4/TMS): δ (in ppm) = 1.12 (d, $2CH_3$), ^3J = 6.5 Hz; 1.34 (s, $3CH_3$); 1.36 (s, $3CH_3$); 2.58 (sept, CH), ^3J = 6.5 Hz; 7.27 (q, 2 arom. H), J = 2 Hz.

IR (neat): ν (in cm^{-1}) = 1595, 1360, 1128, 1000.

UV (pentane): λ_{max} (in nm) (ε in $L \cdot mol^{-1} \cdot cm^{-1}$) = 272sh (2060), 305sh (2330), 343 (6960), 483 (1810), 542sh (1130).

It decomposed slowly at room temperature [1].

2.11.6.2.3.4 N-Thiosulfinyl-2,4,6-tri-*tert*-butyl-benzenamine, $S{=}S{=}NC_6H_2(C_4H_9\text{-}t)_3\text{-}2,4,6$

The title compound was only detected in solution as part of an equilibrium mixture; it could not be isolated in the solid state.

The reaction of 2,4,6-(t-$C_4H_9)_3C_6H_2NH_2$ with equimolar S_2Cl_2 in the presence of a twofold quantity of $(C_2H_5)_3N$ in ether at 0°C for 1 h, followed by dry column chromatography (silica gel, n-hexane) afforded a yellow crystalline compound with the molecular formula $C_{18}H_{29}NS_2$ in 70% yield [1, 2]. It was also formed in 23.7% yield, along with the corresponding sulfur diimide (12.6%) and N-sulfinylaniline (11.9%), by treatment of an ice-cold ethereal solution of 2,4,6-(t-$C_4H_9)_3C_6H_2NH_2$ and $(C_2H_5)_3N$ with SCl_2 in ether (mole ratio ~1:2:1) [1].

In solution, this compound is reddish and was found to be a mixture of two isomers, the major one XIIa and the minor one XIIb [1]. This fact was confirmed by others using NMR spectroscopy [14]. The mole ratio of the two species varies reversible with temperature. In the solid state the structure of XIIa was confirmed by X-ray analysis [1].

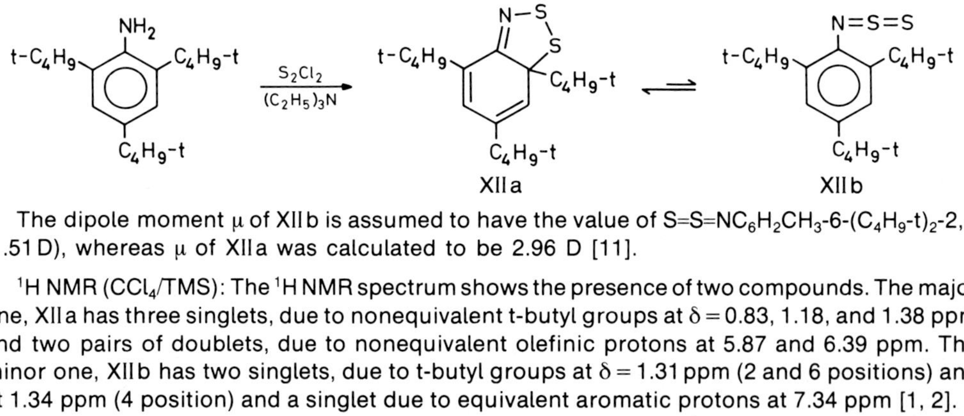

XIIa XIIb

The dipole moment μ of XIIb is assumed to have the value of $S{=}S{=}NC_6H_2CH_3\text{-}6\text{-}(C_4H_9\text{-}t)_2\text{-}2,4$ (1.51 D), whereas μ of XIIa was calculated to be 2.96 D [11].

^1H NMR (CCl_4/TMS): The ^1H NMR spectrum shows the presence of two compounds. The major one, XIIa has three singlets, due to nonequivalent t-butyl groups at δ = 0.83, 1.18, and 1.38 ppm and two pairs of doublets, due to nonequivalent olefinic protons at 5.87 and 6.39 ppm. The minor one, XIIb has two singlets, due to t-butyl groups at δ = 1.31 ppm (2 and 6 positions) and at 1.34 ppm (4 position) and a singlet due to equivalent aromatic protons at 7.34 ppm [1, 2].

Equilibrium constants, K = [XIIa]/[XIIb] were calculated based on the NMR signal density of the two olefinic protons of XIIa and the aromatic protons of XIIb. The following table shows that in CCl_4 solution the relative amount of the minor compound XIIb increases with increasing temperature. The free energy of isomerization, ΔG correlates with the temperatures linearly [10]. Measurements in CD_2Cl_2 at 0 and 35°C (concentration 0.48 mol/L) give K values of 33.7 and 14.2, respectively [1, 2] and in C_6D_6 at 25°C, K = 11.3 [11]. Temperature dependence of K and ΔG in CCl_4 between 11 and 63°C [10]:

t in °C	11	23	30	33	38	43	48	52	58	63
$K = \frac{[XIIa]}{[XIIb]}$	19.5	14.1	12.0	10.7	9.9	9.2	8.2	6.6	5.7	5.1
−ΔG in kcal/mol	1.68	1.56	1.50	1.44	1.42	1.39	1.34	1.22	1.14	1.09

Based on these data, the enthalpy (ΔH) and the entropy (ΔS) of isomerization were estimated to be −4.9 $kcal \cdot mol^{-1}$ and −11.3 $cal \cdot mol^{-1} \cdot K^{-1}$, respectively [10].

Equilibrium constants, determined at 43°C in various solvents demonstrate, that polar solvents greatly enhance the preference of the cyclic isomer XIIa [10]:

solvent	$n-C_6H_{14}$	$c-C_6H_{12}$	CCl_4	$CH_3C(O)OC_2H_5$	CD_2Cl_2	$CD_3C(O)CD_3$	CH_3CN
$K=\frac{[XIIa]}{[XIIb]}$	6.3	8.6	9.2	14.5	12.2	17.7	25

UV (hexane) for XIIa and XIIb: λ_{max} (in nm) (ε in $L \cdot mol^{-1} \cdot cm^{-1}$) = 268 (2610), 292 (2580), 340 (2050), 410 (3480), and 535 weak. The very weak absorption at 535 nm in fairly concentrated solutions (ca. 2.5×10^{-2} M) is considered to be characteristic of the N-thiosulfinylanilines [1, 2].

MS for XIIa and XIIb: m/e (rel. int. in %) = 323 (trace) M^+, 291 (44), 266 (72), 251 (49), 244 (100), 236 (49) [1].

Thermolysis of XIIb equilibrated with XIIa in refluxing C_6H_6 for 9.7 h or in boiling $C_2H_5OH-C_6H_6$ in the presence of catalytic amounts of p-toluenesulfonic acid gave XIII (20.6 or 27.5%), the corresponding aniline XIV (34.2 or 39.6%) and sulfur [5].

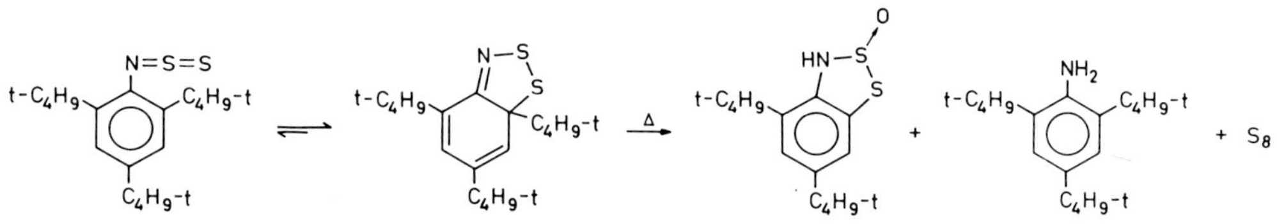

XIIb XIIa XIII XIV

Thermolysis in degassed C_6H_6 at 110°C for 5 min afforded a persistent ESR signal (a_N = 8.2 G, g = 2.008) which was assigned to the radical XVa or XVb [5], see also [15]. Thermolysis in refluxing C_6H_6 for 5 h under a stream of oxygen gave a complex reaction mixture, from which the compounds XIV (11.5%), XVI (12.2%), and XVII (9.5%) could be isolated.

Photolysis ot the equilibrium mixture of the compound in pentane with pyrex filtered light from a medium pressure mercury lamp for 32.8 h at 20°C yielded the aniline XIV (10.6%) and the corresponding sulfur diimide XVIII (7.5%) [5]. The reaction of the equilibrium mixture (XIIa and XIIb) with $(C_6H_5)_3P$ in ether (mole ratio ~1:1) at −78°C for 1.5 h, at 0°C for 2 h and then at room temperature for 3 h afforded XIV (38.6%), XVIII (34.4%), and XX (11%). No sulfur diimide XVIII was obtained when this reaction was performed under a stream of oxygen in ether at −78°C for 5 h and at 0°C for 2 h (XIV: 35.4%, XX: 38.6%). (The yields are based on consumed XII.) The reaction with excess $P(OCH_3)_3$ in ether at room temperature for 3 d produced XXI in 56% yield [8].

If the tautomeric mixture XIIa und XIIb was oxidized with $3-CH_3C_6H_4CO_3H$ in CH_2Cl_2 at temperatures between −78 and 39°C, the reaction products XIX (44.8 to 20.6%), XIII (43.9 to 62.2%), and XX (1.3 to 6.2%) were obtained. The latter product was attributable to the oxidation of XIIb [7, 12].

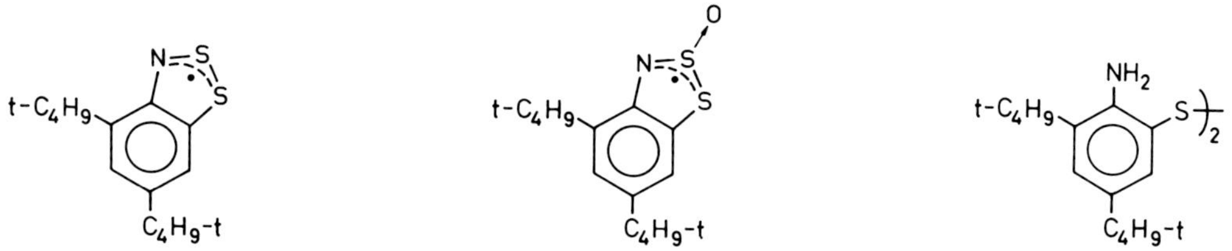

XVa XVb XVI

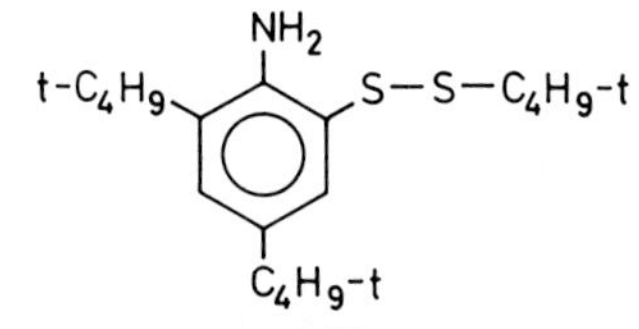

XVII

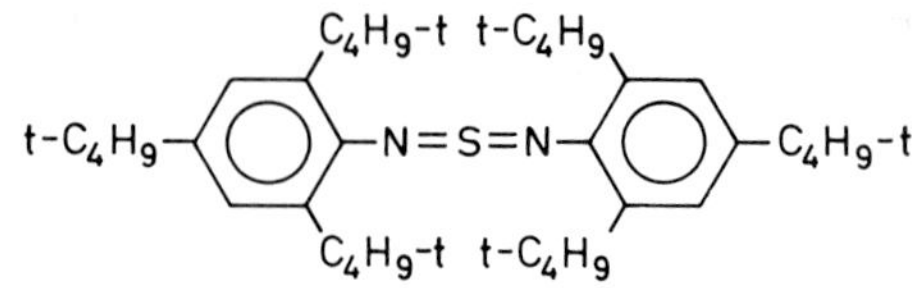

XVIII

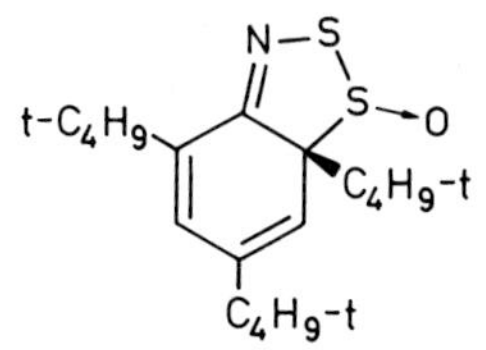

XIX

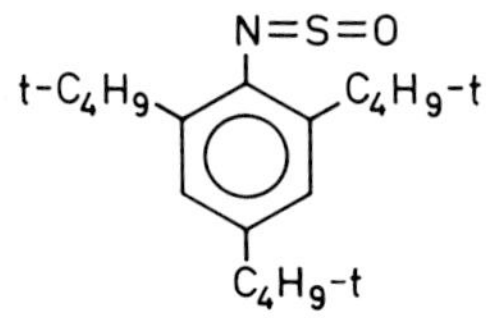

XX

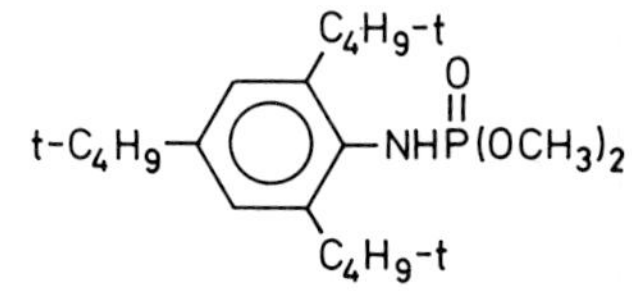

XXI

References:

[1] Inagaki, Y.; Okazaki, R.; Inamoto, N. (Bull. Chem. Soc. Japan **52** [1979] 1998/2001).
[2] Inagaki, Y.; Okazaki, R.; Inamoto, N. (Tetrahedron Letters **1975** 4575/8).
[3] Iwasaki, F. (Acta Cryst. B **35** [1979] 2099/103).
[4] Inagaki, Y.; Okazaki, R.; Inamoto, N.; Shimozawa, T. (Chem. Letters **1978** 1217/8).
[5] Inagaki, Y.; Okazaki, R.; Inamoto, N. (Bull. Chem. Soc. Japan **52** [1979] 2002/7).
[6] Inagaki, Y.; Okazaki, R.; Inamoto, N. (Tetrahedron Letters **1977** 293/4).
[7] Inagaki, Y.; Okazaki, R.; Inamoto, N. (Bull. Chem. Soc. Japan **52** [1979] 3615/8).
[8] Inagaki, Y.; Hosogai, T.; Okazaki, R.; Inamoto, N. (Bull. Chem. Soc. Japan **53** [1980] 205/9).
[9] Huisgen, R.; Peng, X. (Tetrahedron Letters **27** [1986] 6063/6).
[10] Inagaki, Y.; Okazaki, R.; Inamoto, N.; Yamada, K.; Kawazura, H. (Bull. Chem. Soc. Japan **52** [1979] 2008/9).

[11] Inagaki, Y.; Okazaki, R.; Inamoto, N. (Heterocycl. **9** [1978] 1613/6).
[12] Inagaki, Y.; Okazaki, R.; Inamoto, N. (Chem. Letters **1978** 1095/8).
[13] Inagaki, Y.; Okazaki, R. (Sulfur Rept. **2** [1982] 137/68, 157, 160).
[14] Mayer, R.; Bleisch, S.; Domschke, G. (Z. Chem. [Leipzig] **18** [1978] 323/9, 328).
[15] Mayer, R.; Domschke, G.; Bleisch, S.; Fabian, J.; Barth, A.; Staško, A. (Collection Czech. Chem. Commun. **49** [1984] 648/703, 688).

Physical Constants and Conversion Factors

Avogadro constant N_A (or L) $= 6.02214 \times 10^{23}$ mol^{-1}

Faraday constant $F = 9.64853 \times 10^{4}$ C/mol

molar gas constant $R = 8.31451$ J·mol^{-1}·K^{-1}

molar volume (ideal gas) $V_m = 2.24141 \times 10^{1}$ L/mol
(273.15 K, 101325 Pa)

Planck constant $h = 6.62608 \times 10^{-34}$ J·s

elementary charge $e = 1.60218 \times 10^{-19}$ C

electron mass $m_e = 9.10939 \times 10^{-31}$ kg

proton mass $m_p = 1.67262 \times 10^{-27}$ kg

1 kg = 2.205 pounds

1 m = 3.937×10^{1} inches = 3.281 feet

1 m^3 = 2.642×10^{2} gallons (U.S.)

1 m^3 = 2.200×10^{2} gallons (Imperial)

Force	N	dyn	kp
1 N	1	10^{5}	1.019716×10^{-1}
1 dyn	10^{-5}	1	1.019716×10^{-6}
1 kp	9.80665	9.80665×10^{5}	1

Pressure	Pa	bar	kp/m^2	at	atm	Torr	lb/in^2
1 Pa = 1 N/m^2	1	10^{-5}	1.019716×10^{-1}	1.019716×10^{-5}	9.86923×10^{-6}	7.50062×10^{-3}	1.450378×10^{-4}
1 bar = 10^{6} dyn/cm^2	10^{5}	1	1.019716×10^{4}	1.019716	9.86923×10^{-1}	7.50062×10^{2}	1.450378×10^{1}
1 kp/m^2 = 1 mm H_2O	9.80665	9.80665×10^{-5}	1	10^{-4}	9.67841×10^{-5}	7.35559×10^{-2}	1.422335×10^{-3}
1 at (technical)	9.80665×10^{4}	9.80665×10^{-1}	10^{4}	1	9.67841×10^{-1}	7.35559×10^{2}	1.422335×10^{1}
1 atm = 760 Torr	1.01325×10^{5}	1.01325	1.033227×10^{4}	1.033227	1	7.60×10^{2}	1.469595×10^{1}
1 Torr = 1 mm Hg	1.333224×10^{2}	1.333224×10^{-3}	1.359510×10^{1}	1.359510×10^{-3}	1.315789×10^{-3}	1	1.933678×10^{-2}
1 lb/in^2 = 1 psi	6.89476×10^{3}	6.89476×10^{-2}	7.03069×10^{2}	7.03069×10^{-2}	6.80460×10^{-2}	5.17149×10^{1}	1

Work, Energy, Heat	J	kW·h	kcal	Btu	eV
1 J = 1 W·s = 1 N·m = 10^7 erg	1	2.778×10^{-7}	2.39006×10^{-4}	9.4781×10^{-4}	6.242×10^{18}
1 kW·h	3.6×10^6	1	8.604×10^2	3.41214×10^3	2.247×10^{25}
1 kcal	4.1840×10^3	1.1622×10^{-3}	1	3.96566	2.6117×10^{22}
1 Btu (British thermal unit)	1.05506×10^3	2.93071×10^{-4}	2.5164×10^{-1}	1	6.5858×10^{21}
1 eV	1.602×10^{-7}	4.450×10^{-14}	3.8289×10^{-11}	1.51840×10^{-10}	1

1 $cm^{-1} = 1.239842 \times 10^{-4}$ eV   1 Hz $= 4.135669 \times 10^{-15}$ eV

1 hartree = 27.2114 eV   1 eV $\triangleq$ 23.0578 kcal/mol

Power	kW	hp	kp·m·s^{-1}	kcal/s
1 kW = 10^3 J/s	1	1.35962	1.01972×10^2	2.39006×10^{-1}
1 hp (horsepower, metric)	7.3550×10^{-1}	1	7.5×10^1	1.7579×10^{-1}
1 kp·m·s^{-1}	9.80665×10^{-3}	1.333×10^{-2}	1	2.34384×10^{-3}
1 kcal/s	4.1840	5.6886	4.26650×10^2	1

References:

International Union of Pure and Applied Chemistry, Manual of Symbols and Terminology for Physicochemical Quantities and Units, Pergamon, London 1979; Pure Appl. Chem. **51** [1979] 1/41.

The International System of Units (SI), National Bureau of Standards Spec. Publ. 330 [1972].

Landolt-Börnstein, 6th Ed., Vol. II, Pt. 1, 1971, pp. 1/14.

ISO Standards Handbook 2, Units of Measurement, 2nd Ed., Geneva 1982.

Cohen, E. R., Taylor, B. N., Codata Bulletin No. 63, Pergamon, Oxford 1986.